EQUILIBRIUM STATISTICAL MECHANICS

EQUILIBRIUM STATISTICAL MECHANICS

Gene F. Mazenko
University of Chicago

A WILEY-INTERSCIENCE PUBLICATION
JOHN WILEY & SONS, INC.
New York • Chichester • Weinheim • Brisbane • Singapore • Toronto

This book is printed on acid-free paper. ∞

Copyright © 2000 by John Wiley & Sons, Inc. All rights reserved.

Published simultaneously in Canada.

No part of this publication may be reproduced, stored in a retrieval system or transmitted in any form or by any means, electronic, mechanical, photocopying, recording, scanning or otherwise, except as permitted under Sections 107 or 108 of the 1976 United States Copyright Act, without either the prior written permission of the Publisher, or authorization through payment of the appropriate per-copy fee to the Copyright Clearance Center, 222 Rosewood Drive, Danvers, MA 01923, (978) 750-8400, fax (978) 750-4744. Requests to the Publisher for permission should be addressed to the Permissions Department, John Wiley & Sons, Inc., 605 Third Avenue, New York, NY 10158-0012, (212) 850-6011, fax (212) 850-6008, E-Mail: PERMREQ@WILEY.COM.

For ordering and customer service, call 1-800-CALL-WILEY.

Library of Congress Cataloging-in-Publication Data:

Mazenko, G. (Gene)
    Equilibrium statistical mechanics / Gene F. Mazenko.
      p.   cm.
    Includes index.
    ISBN 0-471-32839-1 (v. 1 : cloth)
    1. Statistical mechanics.   I. Title
    QC174.8.M39   2000   00-025718
    530.13—dc21

Printed in the United States of America.

10 9 8 7 6 5 4 3 2 1

CONTENTS

Preface   xiii

1 General Principles of Statistical Mechanics   1

    1.1 More Is Different!   1
    1.2 Macroscopic Systems   2
    1.3 Measurements   3
        1.3.1 Ideal Statistical Systems   3
        1.3.2 System of Identical Point Particles   6
    1.4 Variables in Statistical Mechanics   10
    1.5 Macrovariables   13
        1.5.1 Two Points of View   13
        1.5.2 Boltzmann Paradigm   14
        1.5.3 Langevin Paradigm   17
    1.6 System Preparation   18
    1.7 Random Variables and Information   20
        1.7.1 Introduction   20
        1.7.2 Missing Information: Equal Probabilities   21
        1.7.3 Missing Information: Variable Probabilities   24
    1.8 Maximum Entropy and the Second Law of Thermodynamics   28
    1.9 Microcanonical Ensemble   29
        1.9.1 Introduction   29
        1.9.2 Ideal Gas   33
    1.10 Open Ensembles   36
        1.10.1 Local Conservation Laws   36
        1.10.2 Open Boundaries   40
        1.10.3 Introduction of Temperature and Chemical Potential   40
        1.10.4 Grand Canonical Ensemble   45
        1.10.5 Ideal Gas in the Grand Canonical Ensemble   51
    1.11 Fluctuations   53
    1.12 Canonical Ensemble   54
    1.13 Symmetry and Equilibrium Ensembles   58
    1.14 Mechanical Forces: Solids and Liquids   69
        1.14.1 Shear Forces   69
        1.14.2 Distortions   71

vi   CONTENTS

		1.14.3	External Forces	79
		1.14.4	Fluids	82
		1.14.5	Solids	84
		1.14.6	Adiabatic Processes	85
		1.14.7	Pressure, Thermodynamic Derivatives, and Equations of State	89
	1.15	Nosé Dynamics		90
	1.16	Ensembles and Independent Variables		93
References				94
Problems				102

2 Principles of Thermodynamics — 117

	2.1	Introduction		117
	2.2	General Postulates of Thermodynamics		118
	2.3	First-Derivative Information		120
		2.3.1	Conjugate Forces	120
		2.3.2	Energy Representation	121
		2.3.3	Euler Equation	122
	2.4	Legendre Transforms		124
		2.4.1	Introduction	124
		2.4.2	Application in Classical Mechanics	126
		2.4.3	Multiple Legendre Transformations	127
		2.4.4	Examples	128
	2.5	Second-Derivative Information		131
		2.5.1	Physical Observables	131
		2.5.2	Maxwell Relations	132
		2.5.3	Jacobians and Transformations	133
		2.5.4	Example: Expressing $C_V = C_V(C_p, \alpha_T, \kappa_T)$	134
	2.6	Fluctuations and Stability		135
		2.6.1	Introduction	135
		2.6.2	Single-Component Subsystem	142
		2.6.3	Phase Equilibrium	144
		2.6.4	Mixtures	151
		2.6.5	Chemical Reactions	154
	2.7	Landau Theory of Phase Transitions		156
		2.7.1	Order Parameters	156
		2.7.2	Free Energy	158
		2.7.3	Critical Points	160
		2.7.4	Effective Free Energy	160
		2.7.5	Zero Conjugate Field	163
		2.7.6	Nonzero Conjugate Field	165
		2.7.7	Phase Separation and a Conservation Law	166

	2.8	Phase Separation in Binary Alloys	169
	2.9	van der Waals Equation of State	177
		2.9.1 Motivation	177
		2.9.2 Phase Diagram	179
		2.9.3 Maxwell Construction	181
		2.9.4 Critical Point	184
	References		189
	Problems		191

3 Quantum Statistical Mechanics — 199

	3.1	General Principles	199
	3.2	Equilibrium Probability Operator	202
	3.3	Statistical Mechanics in Second-Quantized Language	205
	3.4	Quantum Hamiltonians	208
		3.4.1 Electron Gas	209
		3.4.2 Helium	209
		3.4.3 Electromagnetic Radiation	210
		3.4.4 Quadratic Hamiltonians	210
	3.5	Partition Function for a Quadratic Hamiltonian	211
	3.6	Single-Particle Density of States	214
	3.7	The Classical Limit	218
	3.8	Equation of State for Ideal Quantum Systems	219
	3.9	Chemical Potential for Massive Nonrelativistic Particles	220
	3.10	Blackbody Radiation: Photon Gas	222
	3.11	Properties of Ideal Fermi Systems at Low Temperatures	226
		3.11.1 Introduction	226
		3.11.2 $T = 0$	226
		3.11.3 $T > 0$	229
	3.12	Photon–Electron–Positron Gas	232
	3.13	Bose–Einstein Condensation	236
		3.13.1 Uniform Systems	236
		3.13.2 Bose–Einstein Condensation in a Harmonic Trap	244
	References		248
	Problems		250

4 Statistical Mechanics of Fluids — 259

	4.1	Ideal Gases with Internal Degrees of Freedom	259
		4.1.1 Internal Partition Function	259
		4.1.2 Equipartition Theorem	261
		4.1.3 Quantum Mechanical Rotation and Vibration	262
	4.2	Averages over Quantities Depending only on Momentum	268
	4.3	Mayer Cluster Expansion	270

	4.3.1	Partition Function for Classical Interacting Fluids	270
	4.3.2	Linked Cluster Expansions	274
4.4	Equation of State		285
4.5	Low-Density Limit		288
	4.5.1	First-Density Correction	288
	4.5.2	Temperature Dependence of $b_2(T)$	289
	4.5.3	van der Waals Equation of State	292
	4.5.4	Range of Validity of the Cluster Exapansion	294
4.6	Structural Information in Fluids		294
	4.6.1	X-ray Scattering	294
	4.6.2	Structure Factor and Radial Distribution Function	296
	4.6.3	Low-Density Limit	299
	4.6.4	Compressibility Limit	303
4.7	Beyond Perturbation Theory		304
	4.7.1	Introduction	304
	4.7.2	Extrapolation of Theories	305
	4.7.3	Numerical Work	307
	4.7.4	Principle of Corresponding States	308
4.8	Charged Systems		313
	4.8.1	Introduction	313
	4.8.2	One-Component Plasma	314
	4.8.3	Spatial Correlations	316
	4.8.4	Debye–Hückel Theory	316
4.9	Coarse Graining and Effective Hamiltonians		320
	4.9.1	Introduction	320
	4.9.2	Formal Development	321
	4.9.3	van der Waals Interaction	324
	4.9.4	Vibrating-Rotor Model for Diatomic Molecules	328
4.10	Composite Particles: Dissociation and Ionization		332
	4.10.1	Introduction	332
	4.10.2	Ionization of Atomic Hydrogen	332
References			338
Problems			340

5 Equilibrium Properties of Dielectric and Magnetic Materials — 347

5.1	Introduction		347
5.2	Perturbation Theory		348
5.3	Nonpolar–Nonmagnetic Materials		354
	5.3.1	Introduction	354
	5.3.2	Molecular Polarizabilities	355
	5.3.3	Diamagnets	357

5.4	Permanent Dipole Moments		359
	5.4.1 Coarse Graining		359
	5.4.2 Atomic Systems		359
5.5	Statistical Mechanical Development for Dipoles		363
5.6	Free Moments		366
	5.6.1 Separation of Translational and Magnetic Degrees of Freedom		366
	5.6.2 Classical Rigid Moments		367
	5.6.3 Quantum Mechanical Theory		370
	5.6.4 Fluctuating Dipole Model		373
	5.6.5 General Phenomenology of Dilute Paramagnets and Dielectrics		374
	5.6.6 Magnetic Structure Factor		375
5.7	Free Charged Particles in a Static Magnetic Field		376
	5.7.1 Classical Case		376
	5.7.2 Quantum Mechanical Case		377
	5.7.3 Density of States		379
	5.7.4 Dilute Limit		382
	5.7.5 Calculation for Arbitrary Density		384
5.8	Interacting Electric Dipole Moments		389
	5.8.1 Dipole–Dipole Interaction		389
	5.8.2 Mean-Field Theory		391
5.9	Magnetic Lattice Models		395
	5.9.1 Heisenberg Model		395
	5.9.2 Mean-Field Theory Solution		397
	5.9.3 Antiferromagnetism		406
	5.9.4 Magnetic Structure		410
	5.9.5 Magnetic Models		414
5.10	Solutions for the Ising Model		415
	5.10.1 Introduction		415
	5.10.2 One-Dimensional Case		416
	2.10.3 Ising Model in Higher Dimensions		421
5.11	High-Temperature Expansions		425
5.12	Monte Carlo Simulations		431
	5.12.1 Basic Technique		431
	5.12.2 Detailed Balance		433
	5.12.3 Computation of Averages		436
5.13	Transition from Classical to Modern Theory of Critical Phenomena		438
References			440
Problems			443

CONTENTS

6 Statistical Mechanics of Solids — 451

    6.1 Introduction — 451
        6.1.1 What Is a Solid? — 451
        6.1.2 Building Up a Theory of Solids — 453
    6.2 Lattice Structures — 454
    6.3 Ground-State Lattice Structure — 458
    6.4 Lattice Vibrations — 466
        6.4.1 Harmonic Hamiltonian — 466
        6.4.2 Normal Modes of Vibration — 468
        6.4.3 Example: Cubic Crystal with Nearest-Neighbor Interactions — 471
        6.4.4 Long-Wavelength Limit — 472
        6.4.5 Thermal Excitations: Classical Theory — 474
        6.4.6 Spatial Correlations: Classical Treatment — 476
    6.5 Quantum Mechanical Theory — 480
        6.5.1 Quantization — 480
        6.5.2 Phonons — 483
        6.5.3 Spatial Structure — 484
        6.5.4 The Debye Theory — 487
    6.6 Elastic Theory of Solids — 495
        6.6.1 Single-Component Systems — 495
        6.6.2 General Treatment of Elastic Theory — 500
        6.6.3 Bulk Modulus for Cubic Systems — 505
        6.6.4 Isotropic Limit — 507
        6.6.5 Compliance Coefficients — 509
        6.6.6 Distortion by an Elastic Solid — 511
    References — 516
    Problems — 518

Appendices — 523

A Introduction to Probability Theory — 523
    Permutations and Combinations — 523
    Probability Distributions — 524
    Examples of Distributions — 527

B Transformation Theory, Conservation Laws, and Invariance Principles — 530
    Phase-Space Transformations — 530
    Time Translations — 530
    Canonical Transformation — 531
    Symmetry Principles and Constants of the Motion — 533
    Distortions — 537
    Evolution Equations and Averages — 540

C	Motion in Phase Space	542
	Deterministic Mechanics	542
	Types of Flow in Phase Space	543
	Ergodic Theorems	546
	Physical Systems	551
	KAM Theorem	555
	System of Hard Spheres	557
D	Inequality Used to Maximize Entropy	557
E	Volume and Surface Area of a d-Dimensional Sphere of Unit Radius	558
F	Local Conservation Laws	560
G	Method of Lagrange Multipliers	565
H	Stationary-Phase Treatment of the Partition Function in the Microcanonical Ensemble	566
I	Properties of Jacobians	570
J	Abstract Vector Spaces	571
K	Second Quantization	574
	Spin and Statistics	574
	Creation and Annihilation Operators	574
	Field Operators	579
L	Evaluation of Bose Integrals	581
M	Fermi Integrals at Low Temperatures	582
N	Thermal Perturbation Theory	586
	Introduction	586
	Effective Hamiltonians	591
O	Statistical Mechanics of Systems with a Classical Quadratic Hamiltonian	594
	References	600
Index		**605**

PREFACE

It is the intention of this graduate-level text to present statistical mechanics from the modern condensed matter physics point of view. This approach emphasizes symmetry principles, conservation laws, and the consequences of broken symmetry. Pioneered by Landau and emphasized by Anderson, the notions of symmetry and broken symmetry are now understood as crucial to a fundamental understanding of statistical physics. The existence of a closed set of macrovariables, along with the second law of thermodynamics, forms the fundamental basis for statistical mechanics and thermodynamics. The general identification of the macrovariables for complex systems requires an understanding of conservation laws and the consequences of a broken continuous symmetry. The focus here is on the role of broken translational symmetry in treating solids. This involves an appreciation of the fundamental difference in the ability of solids to sustain certain applied mechanical forces compared to fluids.

Another motivation for writing this book is to highlight the approach of coarse graining in statistical mechanics. Coarse graining is a shorthand term for averaging over a set of microscopic degrees of freedom to obtain a self-consistent description of the same system at longer length scales. This approach becomes essential as physicists treat more complex systems on the mesoscopic and macroscopic spatial scales. Although there has been recent focus on coarse graining due to the brilliant success of the renormalization group approach to critical phenomena, this approach has a long, rich, and less specialized history. Indeed, hydrodynamics and elastic theory are classic examples of coarse graining and were developed as theories almost two hundred years ago. More recently, we have come to recognize thermodynamics as a coarse-grained version of statistical mechanics. In Chapter 4 the ideas of coarse graining are discussed in some generality, and then a number of examples are discussed.

The first three chapters, with some selection, make up a conventional one-quarter or one-semester course in statistical mechanics for a general collection of physics and chemistry students. There is more than enough material in Chapters 4 to 6 to fill out the rest of a year-long course. The level of discussion in Chapters 4 to 6 is in somewhat more depth than in the earlier chapters, but there is an effort to keep the mathematical presentation in the book at a level where experimental students are comfortable. More sophisticated material is treated in appendices.

Thermodynamics is treated at two levels. In Chapter 1 it is shown how thermodynamics follows smoothly from statistical mechanics in the case of fluids. Then, in Chapter 2, following the presentation in the classic text by Callen, the general formal structure of thermodynamics is presented. This approach will not appeal to all since it appears to be highly abstract and removed from any specific application. Such is the burden of a universal theory. In my own opinion, a less structured approach to thermodynamics has a tendency to circle around and hide the more general structure.

It is important, as emphasized in Chapters 4 through 6, that students understand the interplay between theory and experiment. This requires some discussion of historical development and I try to give some feeling for the enduring quality of a good, simple idea, such as the Debye approximation in treating solids. By giving some historical background on various topics, I hope to expose students to how science plays out over time. When starting a project there should be a constructive mixture of appreciation for what has been accomplished previously and the need for a fresh point of view.

In presenting statistical mechanics one always has a problem deciding what to do with ergodic theory. My own experience has been that physics students enjoy being introduced to this material when mixed in with some ideas from dynamical systems. On the other hand, if one treats this material in a one-quarter or one-semester course, one has to cut other important topics. Since the basics of statistical mechanics can be developed without a substantial reference to ergodic theory, I have chosen to put this material into an appendix. Several other more mathematical topics have also been put into appendices.

In Chapters 1 and 6 I have included material associated with the very different response properties of fluids and solids to applied mechanical forces. The emphasis is on the new aspects of the problem associated with broken translational symmetry and the need to change thermodynamics from the form describing the fluid phase to the description for solids, which requires including elastic degrees of freedom. In my experience this very general connection between broken continuous symmetry and need to modify thermodynamics is not widely appreciated.

The plan is to follow this introductory volume, with three additional volumes:

2. Fluctuations, Phase Transitions, and Defects
3. Nonequilibrium Statistical Mechanics
4. Field-Theoretic Methods in Equilibrium and Nonequilibrium Statistical Mechanics

Since some topics will be treated in detail in subsequent volumes, they are deemphasized in this beginning volume. For example, equilibrium spatial structure is not emphasized in Volume 1 since it will be treated in detail in Volume 2. Similarly, detailed discussions of nonequilibrium behavior and field theoretical ideas are delayed until later volumes. Certain very important material is not treated in this series. For example, the currently very active fields of random or quenched systems and strongly coupled electronic systems are mentioned only in passing.

I thank Professor Oriol Valls for detailed comments on an earlier version of this book. My wife, Judy, has given me, in her own special way, substantial support during this project. I dedicate this work to her.

EQUILIBRIUM STATISTICAL MECHANICS

1 General Principles of Statistical Mechanics

1.1 MORE IS DIFFERENT!

Matter collected in bulk can show mysterious and wonderful properties [1]. Indeed, the same material, the same collection of a large number of atoms or molecules, can show fundamentally different properties as one changes the temperature or applied external forces. The most familiar example is the comparison between a liquid and a solid. Solids are rigid, whereas liquids are not. This is why we fill swimming pools with liquids, not solids. Liquids, although buoyant, can receive divers. Solids can stand alone, whereas liquids require a container. There are many other important examples of physical systems that display phases with qualitatively different physical behavior. Ferromagnets are familiar from the bottle opener attached to the side of your refrigerator. The opener has a net magnetization which enables it to be attached to a metal surface. If one raises the temperature high enough, the opener will lose its net magnetization and become paramagnetic. There are other materials which, although *normal* at higher temperatures, become *superfluids* at low temperature with new exotic flow properties. For example, superconductors can carry electric current over long distances without dissipation.

What controls whether a particular physical system is in a particular phase? From the discussion above we see that temperature enters consistently into the discussion. Temperature is a very interesting concept that is not easily tied to the traditional Newtonian mechanical picture of the universe. If one looks at the most microscopic level, one has only a vague idea of the temperature of the entire sample of material. Yet the idea of temperature is something for which we have a rather precise intuitive feeling. Our bodies have built-in thermometers.

Statistical mechanics and thermodynamics in general, and the concept of temperature in particular, are not founded on a foreign set of principles beyond traditional physics. The main ideas used in treating large systems are the conservation laws for energy and momentum. In particular, thermodynamics was developed in the eighteenth and nineteenth centuries during a long odyssey [2] to understand the connections between work, temperature, and the establishment of the general principal of conservation of energy beyond conservation of mechanical energy. The importance of conservation laws in the nineteenth century evolved naturally into the related concepts of invariance principles and symmetry considerations, which have emerged as among the most dominant themes in twentieth-century physics. As we discuss in more detail below, the change of a thermodynamic phase for a system can in many (not all) cases be associated with a change in symmetry. Thus a uniform fluid freezes to form a reproducible crystalline lattice with reduced translational symmetry. The idea of symmetry breaking and the associated changes in the phase structure of a physical system has been very important in both statistical and high-energy physics.

It is amazing that the macroscopic properties of simple fluids in thermal equilibrium can be specified by giving only the temperature and pressure of the fluid. For a simple fluid system at a given temperature and pressure, one can reproducibly obtain the same values for the heat capacity, the thermal expansion coefficient, and the isothermal compressibility. Why is this? The situation changes when one lowers the temperature and the fluid solidifies. In a solid it is not sufficient simply to specify the temperature and pressure in order to understand the response of the system to external forces. The difference is associated with rigidity. Fluids flow when subjected to weak forces, whereas solids, which are rigid, do not. If one takes a pencil and pushes it into a bowl of fluid, not much happens. The pencil simply displaces the fluid whose level moves up to accommodate the pencil. The pressure is not changed. Clearly, if one pushes a pencil against a solid, the response [3] is much different. The pencil will not, for modest applied force, enter the solid. Instead, one can push the solid with the pencil. If one applies the force at an angle to the face of the rectangular block (apply a shear), you can rotate it. The macroscopic state of a solid is specified by giving not only the pressure but also the vector nature of the applied forces. We quantify this idea later.

Before we can study these variations between phases, we must develop methods for dealing with the properties within a phase. For most of this chapter we focus on the general principles of statistical mechanics as they apply to simple systems such as noble gases. The basic ideas we develop center around the treatment of a large number of degrees of freedom in the context of satisfying the conservation laws. New ingredients enter the analysis as we treat more complex physical systems such as ferromagnets.

1.2 MACROSCOPIC SYSTEMS

The world of our senses consists of *macroscopic* objects built up from huge numbers of atoms or molecules. These objects can be extremely complex, ranging from simple gases to biological organisms. It is appealing to believe that we can understand the properties of these systems if we understand the structure of the constituent individual atoms and molecules and the interactions among them. For microscopic systems consisting of a few atoms we can calculate their properties straightforwardly. We prepare the system by specifying the initial conditions and then compute the subsequent behavior by solving the associated equations of motion (Newton's law for the classical case). Using modern high-speed computers, we can carry out such calculations [4] for a relatively large number of particles (10^4) for a simple classical gas.

Suppose that we could extend this procedure to macroscopic systems. We would then generate a vast amount of data detailing the phase-space coordinates (positions and momenta) of all the particles in the system as a function of time. As we discuss in detail below, the precise nature of these data, however, is very sensitive to the initial conditions, and from an experimental point of view, the precise specification of initial conditions seems hopeless. The idea that we can prepare a macroscopic system so that we know (in a classical system) the phase-space coordinates of all the particles at a given initial time is absurd. Thus our numerical experiment is useless unless we can determine the set of initial conditions appropriate to experimental observations. This appears to be a daunting task.

We might be ready to give up on macroscopic systems as hopeless if it were not for the obvious: We see regularity all about us. Macroscopic systems are *not* totally chaotic. We do make measurements on large systems, and these measurements appear to have not only regularity but essentially definite values. Consider the obvious fact that water freezes at 1 atmosphere of pressure at zero degrees Celsius. This result is clearly independent of any detailed specification of the phase-space coordinates of all the water molecules in a macroscopic sample. There are, of course, many other examples of reproducible macroscopic phenomena that are discussed in subsequent chapters: Condensation of gases into liquids, alignment of cigar-shaped molecules at lower temperatures in liquid crystals, formation of a lattice structure during solidification, development of extended objects such as bubbles in certain magnetic systems, and superfluid flow at low temperatures in superfluids and superconductors. These phenomena can be connected to the microscopic forces governing the basic constituents (atoms or molecules) in such systems. This connection is through the laws of statistical mechanics [5] — the field that involves the study of reproducible measurable properties of macroscopic systems.

The key words in the last sentence are *reproducible* and *measurable*. Generally in science, if an experiment carried out in one laboratory cannot be reproduced in laboratories elsewhere in the world, it is useless. When we try to measure the properties of macroscopic systems, the problem of initial conditions mentioned above severely limits the number of reproducible experiments. We return to this very important constraint later. We first must work to quantify the act of measurement on a system that has many degrees of freedom.

1.3 MEASUREMENTS

1.3.1 Ideal Statistical Systems

The concept of measurement is strongly coupled to probability theory. This connection comes through our requirement of reproducibility and the associated necessity of multiple determinations of the value of an observable. Once we have multiple observations, the idea of an average and an associated probability distribution of observed values is natural. Some basic aspects of probability theory [6] are discussed in Appendix A. We begin our discussion here with a treatment of an idealized statistical system. This will position us to treat more realistic physical situations.

Let us consider a system that consists of the physical observable A. Then a measurement on this system results in the determination of the value of A. We assume that A can be observed to have n possible values $(A_1, A_2, \ldots, A_n)$. We assume that each distinct value of A corresponds to a different *configuration* of the system. Thus this system can be in n possible configurations. A simple example is the roll of a die. In that case $A = (1, 2, 3, 4, 5, 6)$ and the different configurations correspond to the six possible values of the up face of the die. We cannot predict with certainty the result of any one observation (roll of the die). Although it is not possible to predict the outcome of a single experiment, it is possible to make precise statements about the results of a large number of similar experiments. These precise statements are statistical in nature. Instead of focusing attention on a single system A (say, one die), consider a large number, N, of similar but independent systems (N dice). Such an assembly is called an *ensemble* [7]. Ideally, N is supposed to be arbitrarily large. Each system is assumed to satisfy the same conditions known to be satisfied by the original system A. We imagine that each system was prepared in the same way and subjected to the same experiment as in the original system. If the system being studied is time independent (the value of a die after a roll), we could repeat an experiment N times in succession using the same system (being careful to prepare the system in the same way before each sampling). Suppose that we make N total measurements. We will then obtain the result A_i some N_i times and the set of results $(N_1, N_2, \ldots, N_n)$. All of the N_i's must add up to N:

$$\sum_{i=1}^{n} N_i = N. \qquad (1)$$

The *average* value of A is defined as

$$\langle A \rangle = \frac{1}{N} \sum_{i=1}^{n} A_i N_i. \qquad (2)$$

The fraction $P_i = N_i/N$ approaches [8] the probability of the occurrence of the outcome i in the limit as $N \to \infty$. To the extent that N is made very large, a repetition of the same experiment on the ensemble is expected to lead with increasing reproducibility to the same ratio N_i/N. It is worth emphasizing that this determination of the P_i can be obtained to arbitrary precision. Any imprecision is involved with the experimental error in determining A_i. Using the definition of the probabilities P_i for obtaining the configuration i given above, the average of the variable A can be written

$$\langle A \rangle = \sum_{i=1}^{n} A_i P_i. \qquad (3)$$

Dividing Eq. (1) by N and taking the large-N limit, we have the normalization condition on the probabilities,

$$\sum_{i=1}^{n} P_i = 1, \qquad (4)$$

which means that each sampling must give one of the n configurations.

A reproducible measurement corresponds to carrying out a sufficiently large number of samples N such that N_i/N has approached P_i to an accuracy greater than the intrinsic uncertainty in the experiment. Under these circumstances, if we carry out a new set of observations, the same distribution P_i will be reproduced, to within the same experimental uncertainty. Suppose that we measure the variable A a total of N times and obtain an average $\langle A \rangle_N$ and then measure A in a second series of experiments M times to obtain the average $\langle A \rangle_M$. *Reproducibility* means that

$$\lim_{N \to \infty} \langle A \rangle_N = \lim_{M \to \infty} \langle A \rangle_M. \tag{5}$$

It is worth pointing out that this statistical development includes deterministic situations as a special case. Suppose we prepare a system such that A takes the value A_j and our system is time independent. Then, trivially, if we perform a measurement of A a total of N times, we obtain $A = A_j$ each time, and the average is given by

$$\langle A \rangle = \lim_{N \to \infty} \langle A \rangle_N = A_j \tag{6}$$

and the probability distribution is such that the system is found only in the state j.

In addition to the average of a stochastic variable A, we are interested in the *fluctuation* in the measured values of A defined by

$$\langle (\delta A)^2 \rangle = \sum_{i=1}^{n} (A_i - \langle A \rangle)^2 P_i \tag{7}$$

where we have introduced the notation

$$\delta A = A - \langle A \rangle. \tag{8}$$

The fluctuation in A gives a measure of the deviations of the observed values of A from its average value. If there is some preferred state, say j and $P_i \approx \delta_{i,j}$, then $\langle A \rangle = A_j$ and $\langle (\delta A)^2 \rangle = (A_j - A_j)^2 = 0$. Closely related to the fluctuation in the variable A is the *dispersion* in A, defined by

$$\Delta_A = \frac{\sqrt{\langle (\delta A)^2 \rangle}}{\langle A \rangle}. \tag{9}$$

If we have a second stochastic variable B in the problem, we can define the *correlation function* for A and B:

$$C_{AB} = \langle \delta A \delta B \rangle. \tag{10}$$

We see then that the fluctuation in A is just the *autocorrelation function* C_{AA}.

Example: Coin Flipping. The classic and simplest example of a measurement is coin flipping. Let us characterize the flipping event by assigning $+1$ to a head and -1 to a tail. Then the observed variable is $A = (+1, -1)$. If we flip a coin N times, we will obtain N_1 heads and N_2 tails, where

$$N = \sum_{i=1}^{2} N_i = N_1 + N_2. \tag{11}$$

The probability of obtaining a head is for N very large estimated to be $P_1 = N_1/N$, while for a tail, $P_2 = N_2/N$. The average of A is

$$\langle A \rangle = \sum_{i=1}^{2} A_i P_i = P_1 - P_2. \tag{12}$$

If it is equally probable to obtain a head as a tail—the flip is random—then $\langle A \rangle = 0$:

$$P_1 = P_2 \equiv P. \tag{13}$$

Using the normalization condition

$$1 = \sum_{i=1}^{2} P_i = P_1 + P_2, \tag{14}$$

gives $P = \frac{1}{2}$. We can test the equal probability hypothesis by flipping a coin a very large number of times and checking whether $\langle A \rangle$ is going to zero as the number of flips increases. To gain some feeling for this, you can explore the applet as stated in Ref 9, which allows one to see how the average is approached and gain a feeling for the level of fluctuations.

1.3.2 System of Identical Point Particles

The first step in extending our analysis of an ideal statistical system to real physical systems is to identify the appropriate variables characterizing a particular physical systems. We must identify those coordinates that label the configurations available to the system. Let us for the moment restrict [10] our analysis to a set of N classical identical particles forming an isolated simple fluid. In this case a particular configuration corresponds to the set of phase-space coordinates $(\mathbf{r}_1, \mathbf{r}_2, \ldots, \mathbf{r}_N; \mathbf{p}_1, \mathbf{p}_2, \ldots, \mathbf{p}_N)$, where $\mathbf{r}_1$ and $\mathbf{p}_1$ are, for example, the position and momentum of particle 1 in this configuration. When the $\mathbf{r}$'s and $\mathbf{p}$'s change (i.e., when the particles move) we have a new configuration (see Fig. 1.1). The stochastic variables in this case are the $6N$ phase-space coordinates and any functions of them. Simple examples of *stochastic variables* of physical interest are the kinetic energy

$$K = \sum_{i=1}^{N} \frac{\mathbf{p}_i^2}{2m} \tag{15}$$

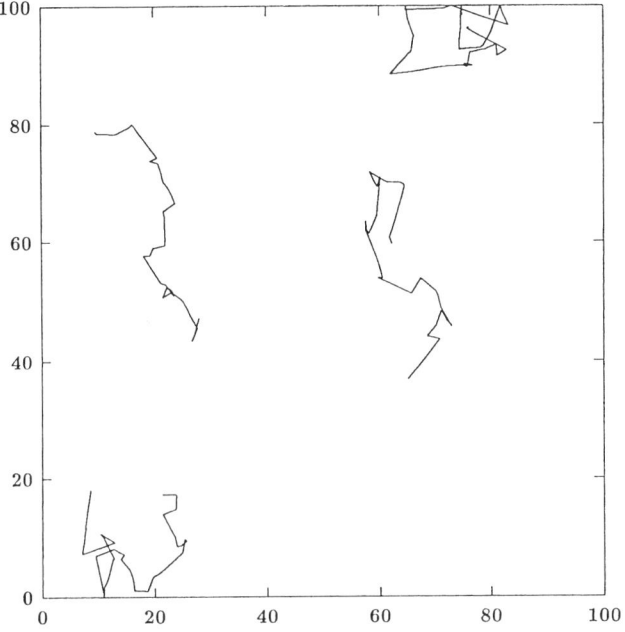

FIGURE 1.1 Trajectories of four randomly selected particles out of a system of 1000 interacting particles moving in two spatial dimensions.

where m is the mass of each particle; the total momentum of the system,

$$\mathbf{P} = \sum_{i=1}^{N} \mathbf{p}_i, \tag{16}$$

the total angular momentum,

$$\mathbf{L} = \sum_{i=1}^{N} \mathbf{r}_i \times \mathbf{p}_i, \tag{17}$$

and the Hamiltonian H, which for a given set of phase-space coordinates, gives the energy. A more complicated stochastic variable which depends only on the positions of the particles is the *particle density*, which is a sum of Dirac δ-functions:

$$n(\mathbf{x}) = \sum_{i=1}^{N} \delta(\mathbf{x} - \mathbf{r}_i). \tag{18}$$

We discuss this quantity in much more detail below.

8 GENERAL PRINCIPLES OF STATISTICAL MECHANICS

It will be convenient and will save some writing if we introduce the shorthand notation

$$q_N = (\mathbf{r}_1, \mathbf{r}_2, \ldots, \mathbf{r}_N; \mathbf{p}_1, \mathbf{p}_1, \ldots, \mathbf{p}_N)$$

and

$$dq_N \equiv d^3r_1 d^3r_2 \cdots d^3r_N d^3p_1 d^3p_2 \cdots d^3p_N$$

and treat q_N as a 6N-dimensional vector. It will also be useful at times to make plots in this 6N-dimensional space, where a single point denotes a particular state or configuration of the system. This phase space is conventionally [11] known as Γ-space.

Since we are dealing with continuous stochastic variables (see Appendix A) it is necessary to introduce the appropriate joint probability distribution $P(\mathbf{r}_1, \mathbf{r}_2, \ldots, \mathbf{r}_N; \mathbf{p}_1, \mathbf{p}_2, \ldots, \mathbf{p}_N) \equiv P[q_N]$. Then $P[q_N]dq_N$ is the probability of finding the fluid system in a configuration in the region between q_N and $q_N + dq_N$. The average value of a stochastic variable A, such as the kinetic energy, is given by multiplying A by the probability for being in configuration q_N and then *summing* over all possible configurations:

$$\langle A \rangle = \int dq_N \, P[q_N] A(q_N). \tag{19}$$

We require that the probability distribution be properly normalized:

$$\int dq_N \, P[q_N] = 1. \tag{20}$$

If we prepare a system in exactly [12] the same fashion M times and observe the configuration of the system, then in the limit as $M \to \infty$, we will construct $P[q_N]dq_N$ just as in the case of the ideal statistical system. The deterministic limit corresponds to knowing that we have prepared the system in the configuration q_N^I, and

$$P[q_N] = \delta(q_N - q_N^I), \tag{21}$$

where

$$\delta(q_N - q_N^I) = \delta(\mathbf{r}_1 - \mathbf{r}_1^I) \cdots \delta(\mathbf{r}_N - \mathbf{r}_N^I) \delta(\mathbf{p}_1 - \mathbf{p}_1^I) \cdots \delta(\mathbf{p}_N - \mathbf{p}_N^I) \tag{22}$$

is a 6N-dimensional Dirac δ-function.

As we have described this situation, it corresponds to specifying our system at a fixed time t_0 (let us set $t_0 = 0$) by specifying the phase-space coordinates

$$q_N(0) = (\mathbf{r}_1(0), \mathbf{r}_2(0), \ldots, \mathbf{r}_N(0); \mathbf{p}_1(0), \mathbf{p}_2(0), \ldots, \mathbf{p}_N(0)) \tag{23}$$

The state of the system is specified at a later time $t > 0$ by the phase-space coordinates

$$q_N(t) = (\mathbf{r}_1(t), \mathbf{r}_2(t), \ldots, \mathbf{r}_N(t); \mathbf{p}_1(t), \mathbf{p}_2(t), \ldots, \mathbf{p}_N(t)) \quad (24)$$

A key result of Newtonian mechanics [13] is that if we know the phase-space coordinates $q_N(0)$ at a time, then by solving Newton's equations we can determine the phase-space coordinates at any later (or earlier) time. A slightly more formal way of saying this is that $q_N(t)$ can be expressed in terms of the initial conditions and time:

$$q_N(t) = q_N(q_N(0), t). \quad (25)$$

This statement is easy to make explicit in the case of noninteracting particles where

$$\mathbf{r}_i(t) = \mathbf{r}_i(0) + \frac{\mathbf{p}_i(0)t}{m} \quad (26)$$

$$\mathbf{p}_i(t) = \mathbf{p}_i(0). \quad (27)$$

More generally, Eq. (25) tells us that any variable of interest (observable) $A(q_N(t))$ can, in principle, be expressed in the form

$$A(q_N(t)) = A(q_N(0), t) \quad (28)$$

and the $q_N(0)$ can be thought of as the stochastic variables in the problem. Once $q_N(0)$ is fixed, $q_N(t)$ follows from the equations of motion. Any uncertainty we have is tied to determination of the initial conditions. Consequently, the average observed value of $A(q_N(t))$ is given by averaging it over the probability distribution $P[q_N(0)]$ governing the original sample preparation at time $t_0 = 0$:

$$\langle A(t) \rangle = \int dq_N(0) A(q_N(0), t) P[q_N(0)]. \quad (29)$$

This is a fundamental result and focuses us on the need to be able to deal with sets of initial conditions that correspond to physical situations.

In the deterministic limit where

$$P[q_N(0)] = \delta(q_N(0) - q_N^I), \quad (30)$$

we have

$$\langle A(t) \rangle = \int dq_N(0) A(q_N(0), t) \delta(q_N(0) - q_N^I) \quad (31)$$

$$= A(q_N^I, t). \quad (32)$$

which means that at time t we will measure *precisely* the value of the variable $A(q_N)$ that has evolved from the initial value $A(q_N^I)$. In our general development we acknowledge that typically we do not know $P_N[q(0)]$ precisely, nor are we in a position to prepare the system in a pure, completely characterized configuration.

1.4 VARIABLES IN STATISTICAL MECHANICS

The set of phase-space coordinates for a collection of identical point particles is conceptually the simplest set of variables characterizing the state of a large system. As we proceed to more complex situations, the set of variables specifying the system becomes more involved. Here we give a brief overview of some of the other sets of physical configurations of interest. The simplest complication on our simplest system is that the point particles are not identical. For example, we can have a mixture of two different types of particle with masses m_1 and m_2. Clearly, one can move on from *binary mixtures* to mixtures with three components, *tertiary mixtures*, and so on. A simple generalization of this situation is to assume that the particles carry charge. For a system that can achieve equilibrium, one must have overall charge neutrality. Typically, one has charge bound in molecules or dissociated into a plasma. The cases of particles with different masses and charges require one to group the associated phase-space coordinates into sets of point particles with the same mass and charge. If a particle carries a magnetic moment, it has internal degrees of freedom. Although such magnetic moments are intrinsically quantum mechanical, we can model them (see Fig. 1.2) with a classical vector $\mathbf{m}_i$ associated with particle i. Typically, we take this vector to have a fixed length

$$\mathbf{m}_i^2 = m_0^2. \tag{33}$$

but be free to rotate. There are some circumstances where these moments are only free to rotate in an *easy plane* (called *XY models*) and others where the moments can

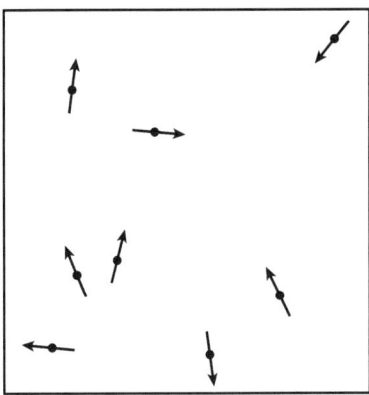

FIGURE 1.2 Particles carrying a magnetic moment indicated by an arrow.

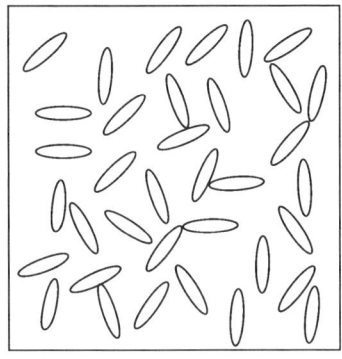

FIGURE 1.3 Cylindrical molecules in an isotropic high-temperature phase. At lower temperatures such systems can form liquid crystal phases.

only flip about an easy axis. This is called an *Ising* (or *uniaxial*) *magnet*. In these cases the system is specified by giving the set of variables $\{\mathbf{r}_i, \mathbf{p}_i, \mathbf{m}_i\}$ for each particle. Such systems are discussed in some detail in Chapter 5.

Systems made up of molecules offer a wider range of statistical variables. Molecules in their simplest form are just two-point particles tied together, and one can separate out the center-of-mass motion of the two particles and the internal relative motion (rotation and vibration). If the particles carry charge, they can produce an electric dipole moment.

More complicated molecules force one to treat extended objects. Cylindrical rodlike molecules (Fig. 1.3) are interesting because they can form liquid-crystal phases [14]. At first glance this may appear to be very similar to the magnetic case discussed above. The symmetry of the set of cigar-shaped molecules is different from the magnet. In a magnet the magnetic moment is a vector with a sense of direction. North differs from south. For a cigar there is no *sense* to the direction, no head on the arrow along the direction of the rod. This may seem like a small point but it has important physical ramifications. For these reasons, the specification of the configuration for a system of rodlike molecules requires some care.

Polymers [15] are very long sequences of chemically connected atoms or molecules. The specification of such systems is clearly very complex, even for a single polymer chain. As systems become more complex, there are circumstances where systems can be described approximately by rather simple models. A number of important examples arise for systems in their solid phase [16]. At sufficiently low temperatures the atoms at each lattice site in the solid move very little. For simple solids one has only the harmonic vibrations of the atoms about their equilibrium sites. In the case of solid mixtures it is not the jiggling at each site that is typically of interest, but the hopping and exchange of atoms leading to mixing or separation. As a simple model consider a solid with lattice positions occupied by two different types of atoms: A or B. These atoms can slowly *hop* around on the lattice (trading places), as shown in Fig. 1.4. At very high temperatures the A's and B's will mix, but as we lower the temperature the interaction between atoms becomes important. Consider first the case where A's *prefer* to be near B's (see Fig. 1.5*b*). In these cases,

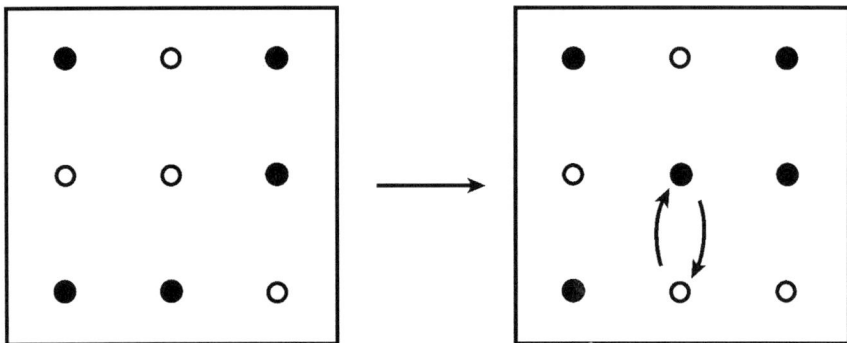

FIGURE 1.4 Atoms hopping and exchanging sites in a binary solid mixture.

as for example in β-brass, a mixture of copper and zinc (CuZn), as one lowers the temperature below $T_c = 742$ K with a 50–50 concentration one develops a sublattice ordering. A's want to be next to B's. One can also find the case where it is energetically more favorable for A's to be near A's and B's near B's than for A's to be near B's. Then, at sufficiently low temperatures, the system will tend to phase separate [17] or *precipitate* into A-rich and B-rich regions (see Fig. 1.5a).

This physical situation can be modeled very simply. We associate with each lattice site, labeled by **R**, a variable $n(\mathbf{R})$. This variable can take on only two values: $n(\mathbf{R}) = n_A$ corresponds to the site occupied by an A atom, and $n(\mathbf{R}) = n_B$ indicates that the site is occupied by a B atom. This model system with N lattice sites is then specified by the discrete set of variables $\{n(\mathbf{R}_1), n(\mathbf{R}_2), \ldots, n(\mathbf{R}_N)\}$ and has a total of 2^N degrees of freedom.

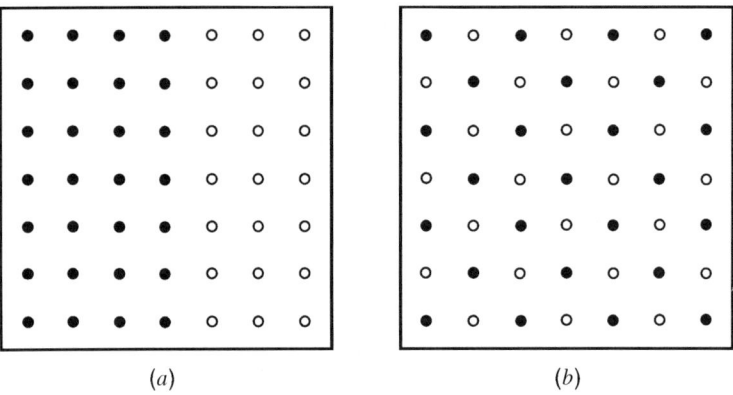

FIGURE 1.5 Ordered low-temperature phases for a binary solid mixture of two types of atoms. Atoms of type A are represented by solid circles; atoms of type B are represented by open circles. (*a*) Phase separation-A's want to be next to A's and B's; (*b*) sublattice ordering-A's want to be next to B's.

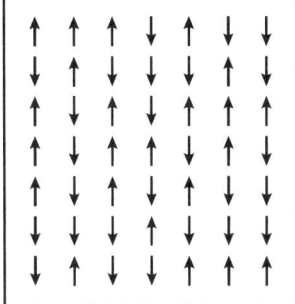

FIGURE 1.6 Ising (up–down) magnetic moments set on a lattice.

For solid systems where the atom or molecule at each site carries a magnetic moment, we can ignore the jiggling and look only at the magnetic degrees of freedom represented by the magnetic moment $\mathbf{m}(\mathbf{R})$ [18] located at each site $\mathbf{R}$. In the simplest case of an Ising or uniaxial magnet, the magnetic moment can take on only two values,

$$m_z(\mathbf{R}) = \pm m_0 \tag{34}$$

as shown in Fig. 1.6. Notice the similarities with the binary mixture case. In each case one has a variable with just two values at each lattice site.

As we move on to more complex systems such as those forming superfluids at low temperatures or to develop a quantum mechanical treatment, we must take into account the uncertainty principle in characterizing the system. This is discussed in some detail in Chapter 3, where we introduce the use of second-quantized field operators. These developments lead eventually to the use of field-theory descriptions for a wide range of problems in statistical physics.

1.5 MACROVARIABLES

1.5.1 Two Points of View

One of the most important ideas in statistical mechanics is that out of the huge number of dynamical variables characterizing a macroscopic system, only a few are important in understanding the reproducible properties of the system. We have already discussed that the macroscopic properties of a simple fluid system can be specified by giving the pressure and temperature. Why is this? This question can be addressed from two apparently very different points of view. The first approach, which we will refer to as the *Boltzmann paradigm*, evolves out of the nineteenth-century notion that the fundamentals of statistical mechanics could be derived, without further assumptions, from classical mechanics. This leads one to a detailed study of the dynamics of isolated systems. This mature and challenging

14 GENERAL PRINCIPLES OF STATISTICAL MECHANICS

field, called *ergodic theory*, is surveyed in Appendix C. Below we summarize some of the main ideas from ergodic theory that are important to the basis of statistical mechanics.

While ergodic theory played a central role in the early years in the development of statistical mechanics, over the past 50 years, a second approach, which we refer to as the *Langevin paradigm*, has grown in favor. In this approach one thinks of a statistical mechanical system of interest as being embedded in a complex environment. Thus, as for example in the classic experiments of Perron [19] on Brownian motion, one can study the average properties of large particles random walking through a bath of smaller particles. In this approach, which emphasizes *open* systems, one finds reproducible results because of balances of motion and flows associated with local conservation laws. Although these two approaches seem incompatible, we shall see, after journeys along very different paths, that they lead to the same basic results.

1.5.2 Boltzmann Paradigm

As mentioned above, ergodic theory evolved out of the belief that we can establish statistical mechanics from classical mechanics. The first step in this approach is to classify the possible types of flow of systems in phase space. Given an initial configuration or set of configurations in Γ-space, how does the system evolve in time? It turns out that such flows can be divided into three basic types: nonergodic, ergodic, and mixing. The procedure is to consider the motion of a typical small volume in phase space as a function of time (see Fig. C.1). For *nonergodic* systems the specified volume moves through Γ-space *without distortion* (see Fig. C.1a) and is restricted to a particular portion of Γ-space. Typically, the motion will be quasiperiodic and, as time evolves, the motion is restricted to a finite fraction of Γ-space. In this case there will be a part of phase space that will not be visited by the specified volume. *Ergodic* flows correspond to volumes that are only slightly distorted (see Fig. C.1b) as the system moves with time, but the distortion is significant enough that the volume does not return to its initial position in Γ-space. As time evolves, the volume element sweeps out essentially the entire phase space. For *mixing* systems the volume element is violently distorted (see Fig. C.1c) as it evolves in time. Individual points that start out close together in phase space have widely diverging trajectories. The mixing property is associated with systems in which the distance between initially neighboring points diverges exponentially in time. Eventually, in a mixing system, a volume will evolve into extremely fine filaments that spread uniformly over the entire Γ-space. Thus in a mixing system, in a coarse-grained sense, all portions of phase space are visited as the system evolves.

Physically, the nonergodic motion is distinct from the ergodic and mixing motions. In the nonergodic case things are periodic and there is no separation of nearby points with time. The system just keeps repeating itself. However, in the ergodic and mixing cases the system appears to become more distorted and chaotic as time evolves. Thus the position of the system in Γ-space at long times for these cases will be essentially randomly distributed.

If a system is ergodic, then, according to the *ergodic theorem*, time and phase-space averages are equal:

$$\int dq\, P[q] A(q) = \lim_{T \to \infty} \frac{1}{T - t_0} \int_{t_0}^{T} dt\, A(q(t)). \tag{35}$$

The ergodic theorem is important, when it holds, because it tells us that experimentally the process of time averaging is equivalent to ensemble averaging. Clearly, the ensemble averaging requires much more work since we must prepare the system many times. In the time average we prepare it once and let it evolve in time.

The notion of ergodic motion was introduced by Boltzmann [20] in 1871. He originally proposed that if left to itself, every system will sooner or later, pass through every point in Γ-space consistent with conservation of energy. This *ergodic hypothesis* does *not* hold in general. One can, however, prove a more restrictive *quasiergodic theorem*. If the probability distribution is stationary and constant in phase space (an invariant ensemble) and if the phase space of the system *cannot* be partitioned into invariant regions, the time average equals the phase-space average. The condition that the probability distribution is *invariant* is very restrictive. We return to this point later. The condition that phase space cannot be separated into invariant regions requires that no trajectory can be confined to a finite portion of phase space but, rather, must wander through the entire Γ-space. How can this partitioning come about? To address this question requires that we discuss constants of the motion for a system.

For a given dynamical system, located at time $t = 0$ at the phase-space point $q(0)$, we can determine the position in phase space $q(t) = q(t, q(0))$ at some later time t by running the equations of motion forward in time. On the other hand, we know that the inverse transformation exists since our isolated physical system is time-reversal invariant and we can write

$$q(0) = q_0(t, q(t)). \tag{36}$$

This gives us $6N$ equations identifying constants along any trajectory of the dynamical system. These $6N$ equations depend on t parametrically. We can, however, solve one of these equations for t as $t = t(q(t))$ and then substitute for t in the other $6N - 1$ equations to obtain $6N - 1$ constants of the motion. By means of this very formal procedure, we can, in principle, reduce the problem of deterministic classical mechanics to the determination of the constants of the motion. This analysis is worked out explicitly in Appendix C for the case of noninteracting particles. Although this program seems attractive, the situation is extremely complicated even in very simple situations. In particular, all of these constants of the motion should *not* be treated on the same footing.

As a very important example, we note that a system with energy conservation can certainly *not* be ergodic in Γ-space. The energy is an isolating constant that confines every trajectory to a hypersurface labeled by the energy. If there are other isolating constants the process of reducing the region of phase space that is sampled is continued. One of the main problems in ergodic theory is to determine the number of isolating constants in a given system. It is important to understand the necessity of dividing the constants of the motion into those that are isolating and those that are nonisolating. Isolating constants divide Γ-space into invariant regions, and a system with such isolating constants is not ergodic. Nonisolating constants do not partition Γ-space. In general, the rigorous identification of the isolating constants for a specific system is extremely difficult.

For mixing systems we understand, after identifying a small number of isolating constants, that the mixing behavior holds in the reduced phase space corresponding to a set of particular values for these isolating constants. Thus one can have mixing in the region of phase space corresponding to a particular value of the energy, for example. The key point for the mixing system is that essentially all initial configurations, if left alone, will approach, in a coarse-grained sense that allows one to perform averages, a time-independent or *invariant ensemble*. This is not generally true for ergodic systems. It is important to remember that mixing gives a *coarse-grained approach* to a stationary probability distribution. Therefore, if we average over small volumes of Γ-space, coarse grain, the distribution looks stationary.

It is at this stage that we must remember that in a typical physical situation for a large system, averages for an isolated system do become independent of time. Thus we expect under such circumstances that one will have, for example, the pressure for a gas that is constant over the sample. Such situations exist for mixing systems. Therefore, the long-time average of an observable in a mixing system is equal to the invariant ensemble average where the ensemble is governed by the constant probability distribution.

In ergodic theory one considers isolated systems and rather painstakingly sorts through all the constants of the motion to identify those that are isolating, and destroy mixing, and those that are not isolating. An important conclusion is that isolating constants of the motion can be divided into those that are *sensitive* to boundary conditions and small changes in the Hamiltonian and those that are not sensitive. A key realization is that these sensitive isolating constants do not survive a weak interaction between the system of interest and the environment. Only the robust constants of the motion survive the weak perturbations that we must assume exist in physical systems.

Another complication in judging whether a system is mixing or not is that the answer may change as the number of particles increases. There is a prejudice that most isolating constants are destroyed as N increases. We are then left with those isolating constants, such as the energy, which survive the large-N limit. The exceptional macrovariable, such as the variable I for the Toda Hamiltonian [21], does not correspond to a global symmetry in the problem. For an isolated system such constants of motion depend on very careful tuning of a Hamiltonian. Any fuzziness in the form of the interactions would destroy this constant of motion.

Let us summarize the ideas, relevant for large systems, which evolved from our discussion of flows in Γ-space. If we have $6N$ degrees of freedom, we must be careful to separate the associated $6N - 1$ constants of the motion into isolating and nonisolating. That is, we can classify the various constants of the motion into those that correspond to smooth, regular hypersurfaces in phase space (isolating constants) and those that correspond to highly erratic hypersurfaces in phase space (nonisolating constants). To simplify matters a bit, let us refer to the isolating constants of the motion as *macrovariables* and the nonisolating constants of the motion as *microvariables*. We assume that our large system is ergodic and mixing on the space spanned by the microvariables.

The most detailed preparation of the system would involve knowledge of all the phase-space coordinates of the system or, alternatively, knowledge of the $6N - 1$ constants of the motion. We must realize that knowledge of the macrovariables is much easier to obtain than knowledge of the microvariables. The energy of a system can be defined without reference to the detailed initial conditions and we can obtain rough estimates in terms of the average kinetic and potential energies. In contrast, the construction of the microvariables is extremely sensitive to initial and boundary conditions. If we change the initial conditions only slightly, the detailed structure of the microvariables will be changed radically. Thus if we have a small spread of uncertainty in the initial configuration of the system, we have a huge uncertainty about the structure of the microvariables. Small changes in the initial or boundary conditions will not result in large changes in the macrovariables (like the energy) for a large system.

1.5.3 Langevin Paradigm

From a more modern point of view, macrovariables can be identified from an appeal to a general symmetry principle. The idea, which is developed in some formal detail in Appendix B, is rather simple. A key component of this analysis is the identification of invariance or symmetry principles that can be used to characterize transformations in phase space. For this purpose it is useful to think of time evolution $[q(o) \to q(t)]$ as a sequence of time translations. One can then show that there are various other transformations that can be identified with physical symmetry operations in Γ-space (translations and rotations in space, etc.). We can associate with various transformations (translations in space and time and rotations in space) a variable $G(q(t))$ which is said to *generate* the transformation. If the system is invariant under the transformation (e.g., translationally invariant in space), the associated generator is a conserved variable! Thus we can associate each invariance with a global *conservation law*. Conversely, each global constant of the motion implies an *invariance principle* [22].

We can then go through the various conserved quantities and identify the associated invariance principle: The conserved total momentum generates spatial translations. The conserved total angular momentum, for systems interacting with radial forces, generates rotations. Energy is conserved in isolated systems and the Hamiltonian generates time translations. Thus if the system is time translationally invariant, the energy is conserved, and vice versa, if the energy is conserved, the system is time-translationally invariant. The symmetry principle discussed here goes beyond energy, momentum, and angular momentum for more complex systems. A variety of new symmetries are important in understanding the

18   GENERAL PRINCIPLES OF STATISTICAL MECHANICS

macroscopic behavior of systems such as superfluids. The importance of these and other symmetries and invariances are central to the development of statistical mechanics. A more detailed discussion of symmetry and its role in statistical mechanics is given in Section 1.10.

Clearly, an invariance principle does *not* depend on the details of the Hamiltonian or the number of particles. Thus, once the number of particles is large, the existence and nature of a conserved macrovariable will not change if we double the number of particles. Later, when we discuss open systems more carefully, we will see that a conservation law has local consequences which makes the special nature of the conserved quantity robust against perturbations at the boundaries and changes in the size of the system.

In the Langevin paradigm, systems are interacting among themselves, with the boundaries of the container, and with *thermal baths*. Thus, unlike the case of isolated small systems, there is no reason to believe that there are any isolating constants of the motion except those that follow from symmetry principles and exist independent of any details of the system. The questions of equilibration and mixing barely enter from this point of view since one typically assumes contact with a thermal bath that essentially forces the system of interest to equilibrate.

To summarize, from both ergodic theory and general symmetry principles, the macrostate of a large system is characterized by a small set of macrovariables. These macrovariables are just the set of globally conserved variables. If we impose external macroscopic constraints on a system, this will lead to additional macrovariables describing the system. For example, for a fluid confined to a specific volume, the size of that volume must be included in the set of macrovariables. The identification of additional macrovariables arising from other constraints and broken symmetry will emerge as we proceed.

1.6   SYSTEM PREPARATION

The basic upshot of the discussion in Section 1.5 is that the dynamical variables in a large system can be divided into two sets. The first small set of variables, the macrovariables, include all of the globally conserved variables, variables associated with macroscopic constraints and, as we discuss below, variables associated with broken symmetries. For simple fluids the symmetries of translational invariance in space and time and rotational invariance in space lead to the globally conserved quantities energy, total linear momentum, and total angular momentum, which should be treated as macrovariables. We must be careful to include other conserved variables such as particle number and variables such as the volume that arises due to the macroscopic constraint of confining the system to a box. All of the rest of the variables are lumped together to form a second set, the microvariables.

The macrovariables are robust and do not depend on details of the system. Small changes in the initial or boundary conditions will not result in large changes in the macrovariables for a large system. In contrast to, for example the energy, the construction of the microvariables is extremely sensitive to initial and boundary

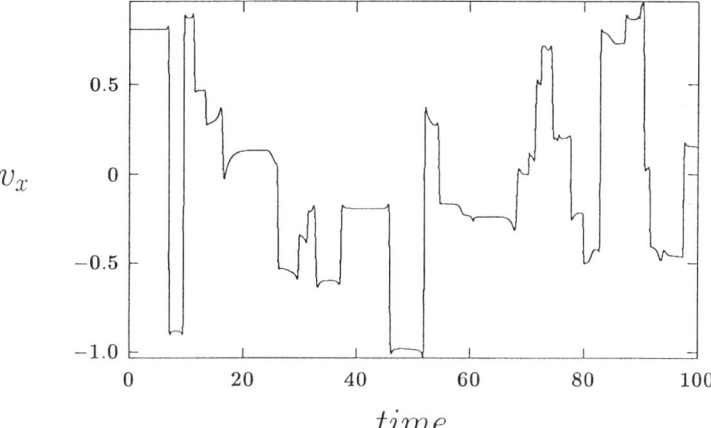

FIGURE 1.7 x-component of the velocity as a function of time for one of the particles shown in Figure 1.1.

conditions. If we change the initial conditions only slightly, the structure of the microvariables will be radically changed.

It is important to realize that in general we cannot expect to make a measurement of an arbitrarily observable in an arbitrary state and obtain reproducible results. Let us discuss the types of variables it makes sense to measure. Only measurements of *generic variables* [23] for large systems are generally reproducible. A generic variable is independent of any particular labeling of identical particles. This principle can be understood by considering an example. Suppose that we measure as a function of time the velocity of a particular particle we label as particle 1 (see Fig. 1.7). Clearly, $\mathbf{v}_1(t)$ will depend very sensitively on certain microvariables. In particular, $\mathbf{v}_1(t)$ depends on $\mathbf{v}_1(0)$, $\mathbf{r}_1(0)$ and the phase-space coordinates of all those particles with which particle 1 interacts. For a large system it is clearly not possible to repeat this experiment in such a way as to obtain the same velocity trajectory for a selected particle. It should also be clear that a velocity trajectory for a particular particle is not very interesting and clearly not reproducible. A related example of a generic variable is the kinetic energy:

$$K = \sum_{i=1}^{N} \tfrac{1}{2} m \mathbf{v}_i^2. \tag{37}$$

This is clearly a generic variable since it is invariant under exchange of any particle labels. In some circumstances the average (time or ensemble) [24] can be related to the average of a *tagged variable* that is not symmetric under exchange of particle labels. If we take the average of the kinetic energy

$$\langle K \rangle = \sum_{i=1}^{N} \tfrac{1}{2} m \langle \mathbf{v}_i^2 \rangle, \tag{38}$$

20 GENERAL PRINCIPLES OF STATISTICAL MECHANICS

we expect that all of the particles have the same average speed and

$$\langle K \rangle = \frac{N}{2} m \langle \mathbf{v}_1^2 \rangle, \tag{39}$$

where particle 1 is a typical particle.

Summarizing: We should only expect to obtain reproducible results from measurements of generic variables or variables that can be related to averages of generic variables.

Reproducibility clearly also depends on one's ability to reproduce the state of the system. Specification of the macrovariables is a necessary but not sufficient condition for reproducibility. Reproducibility means, of course, that if we prepare a system in exactly the same way, fixing certain experimental parameters as in a previous measurement, we measure the same value for a quantity. If we make a measurement in an arbitrary state of the system, it is possible that some knowledge of the microvariables may be necessary if we are to specify the state in which the measurement was made. The point is that in constructing the initial state, there will be a need to control some aspect of the microvariables if we are to reproduce the experiment. Since we usually do not have experimental methods for controlling the values of the microvariables, experiments requiring control over the initial correlations among the microvariables are very difficult to reproduce. There is, however, an initial state that we can reproduce and maintain while we perform our experiments. Experimentally, we first fix the values of the macrovariables, which are time independent, and then wait. Any correlations that exist among the microvariables at the time we isolate the system will eventually decay and the microvariables will become random variables. This special state, where the microvariables are random, is the *equilibrium* state [25]. Once the system is in the equilibrium state it will stay there and we can carry out our experiments. It should be clear that we can reproduce this state in a new experiment by adjusting the macrovariables to the same value as in the first experiment and then waiting [26].

Summarizing the basic points of this section: The dynamical variables in a large system can be separated into macrovariables and microvariables. The equilibrium state corresponds to the situation where we know the values of macrovariables but the microvariables are random variables. Measurements of generic variables for systems in thermal equilibrium state are reproducible.

1.7 RANDOM VARIABLES AND INFORMATION

1.7.1 Introduction

We can now move on to quantify our analysis. Our goal is to calculate the average value of some observable for a system in thermal equilibrium:

$$\langle A \rangle = \sum_i P_i A_i, \tag{40}$$

where i labels the configurations, A_i is the value of the observable in the ith configuration, P_i the equilibrium probability that the system is in the ith configuration, and we sum over all configurations. Clearly, we must determine the probability distribution P_i corresponding to a particular equilibrium macrostate. It may not yet be obvious, but we already have available the basic physics for determining P_i. This is contained in the statement: The equilibrium macrostate corresponds to the situation where we know the values of the macrovariables, but the microvariables are random variables. To quantify this statement we must quantify the notion of random variables.

Randomness is tied in with the notion of *information* [27] and its quantification. The information we have about a system is a measure of the degree of randomness of the system. Our statement that *in equilibrium the microvariables are random variables* is equivalent to the statement that we have a minimum amount of information about the microvariables.

1.7.2 Missing Information: Equal Probabilities

Information is the entity that makes the difference between knowing and not knowing, between being faced with a number of possibilities and knowing the one that actually prevails. To define it quantitatively, we consider first the simple case of choosing among n possibilities. As an example, we may consider an object hidden in one of n similar boxes; we do not know which one. We shall assume that the n possibilities are a complete set. The object must be in one of the boxes. We also assume that the n possibilities are mutually exclusive (i.e., the object cannot be inside more than one box). We shall also assume the n possibilities to be equally probable: No knowledge that we possess tends to make us prefer one possibility to another.

Our inability to decide among the n possibilities reflects a certain lack of information. The amount of *missing information*, which we denote by S (for historical reasons), must be a function of the number of possibilities:

$$S = S(n). \tag{41}$$

The larger n, the more information missing, that is,

$$S(n) > S(m) \quad \text{if} \quad n > m. \tag{42}$$

If there is only one possibility, $n = 1$, the object must be in that box and there is no missing information, that is,

$$S(1) = 0. \tag{43}$$

A key point in our development of the quantification of information concerns the combining of information among subsystems. Suppose that we break a system into independent subsystems. Thus a system with $2n$ objects can be broken up into two systems with n objects in each subsystem. How is the missing information for the entire system, $S(2n)$, related to the missing information associated with the object being in one of the two sets $S(2)$, and the missing information associated with each set, $S(n)$? We assume that in decisions of the same quality we receive the same amount of information. Suppose that we have four objects which can be divided up into two sets. We can determine the amount of information in a two-step process where we first choose between the sets and then choose between the objects in the chosen set. We can decide this by assuming that we obtain the same amount of information in each step $S(2)$ and

$$S(2 \times 2) = S(4) = S(2) + S(2). \tag{44}$$

More generally, we assume that

$$S(mn) = S(m) + S(n). \tag{45}$$

This rule for combining independent subsystems is consistent with the condition Eq. (43) since $S(1 \times n) = S(1) + S(n) = S(n)$. If, for example, we choose the rule $S(mn) = S(m)S(n)$, we would obtain nonsense since $S(1 \times n) = S(1)S(n) = 0$. Thus we have the three fundamental properties satisfied by missing information:

$$S(n) > S(m) \quad \text{if} \quad n > m \quad \text{(a)}$$
$$S(1) = 0 \quad \text{(b)}$$
$$S(mn) = S(m) + S(n) \quad \text{(c)}.$$

It is useful to extend our definition of S from the positive integers, n, to all positive rational arguments, $x \geq 1$, and define $S(x)$. We assume that $S(x)$ satisfies the same requirements generalized to continuous variables x and y:

$$S(x) > S(y) \quad \text{if} \quad x > y \quad \text{(a)}$$
$$S(1) = 0 \quad \text{(b)}$$
$$S(xy) = S(x) + S(y) \quad \text{(c)}$$

and (d) $S(x)$ is a continuous function of x for $x \geq 1$. These properties *uniquely* determine the functional form of S. To see this, consider a special value of x,

$$\ln x = \frac{m}{n}. \tag{46}$$

where m and n are positive integers. Exponentiating Eq. (46), we obtain $x = e^{m/n}$ or $x^n = e^m$ and we can write

$$S(x^n) \equiv S(e^m). \tag{47}$$

Using definition (c), the left-hand side of Eq. (47), can be written as

$$S(x^n) = S(x) + S(x^{n-1}), \qquad (48)$$

or, after repeating this process n times, $S(x^n) = nS(x)$. Similarly, we have that $S(e^m) = mS(e)$, so Eq. (47) reduces to

$$nS(x) = mS(e), \qquad (49)$$

or

$$\begin{aligned} S(x) &= \frac{m}{n} S(e) = S(e) \ln x \\ &\equiv k \ln x, \end{aligned} \qquad (50)$$

where $k = S(e)$ is a positive number determining the units of information.

Equation (50), giving the missing information, has only been established for the values of x given by $x = e^{m/n}$, but these special values are dense in the positive real numbers. They suffice to fix the value of $S(x)$ everywhere if $S(x)$ is continuous. Since we assume that $S(x)$ is continuous [condition (d)], we have, for all x,

$$S(x) = k \ln x. \qquad (51)$$

If we go back to our original problem of n choices, we see that the missing information is given by

$$S(n) = k \ln n. \qquad (52)$$

Thus far we have been considering the information missing before we examine a situation. How can we relate this to the information we gain by knowing the box containing the object? Initially, before we look at one of the boxes, we have no information, $I = 0$, and there is $S = k \ln n$ missing information. If we then observe which box contains the object, we clearly *obtain* an amount of information

$$I = k \ln n. \qquad (53)$$

As an example, let us consider the amount of information that can be conveyed by a page of symbols made by an English typewriter. Counting letters, upper and lower case, and numerals, there are about 75 different symbols. We suppose that our page may contain N such symbols. With 75 possibilities for each symbol, the total number of different pages that the typewriter may produce is 75^N. The information missing before we look at a page is

$$S = k \ln 75^N = Nk \ln 75. \qquad (54)$$

The information per space is $S/N = k \ln 75$. This number, by itself, does not mean too much. It takes on significance when we think about other assumptions we can make about the occurrence of letters on a page. In computing S we assumed that the various 75 letters occur with equal a priori probabilities in the language. This is, of course, not true. It is known, for example, that the letter E may occur with the probability 10.5%, whereas K appears with a probability of 0.3%. An important science in the communication and computer field is *information theory*. It deals with the problem of the most efficient modes of communication—in other words, how can we put the most information in the N spaces on a piece of paper?

1.7.3 Missing Information: Variable Probabilities

It is clear that we need to generalize our definition of information to include those situations where certain events occur with varying probabilities. It may appear that we need to start over in defining information, but we will find that this is not necessary.

Let us return to the problem of choice among n different configurations or possibilities. We label these configurations from 1 to n. We assume that the probability of finding the system in configuration i is P_i. These probabilities are nonnegative numbers

$$P_i \geq 0 \qquad i = 1, \ldots, n \tag{55}$$

which add up to 1:

$$\sum_{i=1}^{n} P_i = 1. \tag{56}$$

If we perform one experiment, sample the system once, we will find one of the n configurations, but we can make no definite statement about which configuration we find. If, however, we have an ensemble of N such systems, which we label from 1 to N, and N is very large, we *know* that we will find to good accuracy

$$N_i = P_i N \tag{57}$$

members of the ensemble in the ith configuration. Thus we *know* the numbers $(N_1, N_2, \ldots, N_i, \ldots, N_n)$ which add up to N:

$$\sum_{i=1}^{n} N_i = \sum_{i=1}^{N} P_i N = N. \tag{58}$$

What we do not know is the sequence of occurrence of configurations in the various ensemble members. For example, if we know that in flipping a coin six times we will obtain 3 heads and 3 tails, what we don't know is the sequence: In Table 1.1 we give the possible sequences in which heads and tails might occur.

TABLE 1.1 Possible Sequences in Flipping a Coin Six Times

H	H	H	T	T	T	T	H	H	H	T	T
H	H	T	H	T	T	T	H	H	T	H	T
H	H	T	T	H	T	T	H	H	T	T	H
H	H	T	T	T	H	T	H	T	T	H	H
H	T	H	H	T	T	T	H	T	H	T	H
H	T	H	T	H	T	T	H	T	H	H	T
H	T	H	T	T	H	T	T	T	H	H	H
H	T	T	T	H	H	T	T	H	H	T	H
H	T	T	H	T	H	T	T	H	H	T	H
H	T	T	H	H	T	T	T	H	H	H	T

We can obtain the total number of possible sequences by answering the more general question: If N objects are broken up into n groups (N_1 in subgroup 1, N_2 in subgroup 2, etc.) such that $\sum_{i=1}^{n} N_i = N$ and each member of the subgroup is equivalent, what are the possible distinct orderings? This is the problem mentioned in Appendix A of finding the number of permutations of a set of N objects that contain N_1 identical elements of one kind, N_2 identical elements of a second kind, up to n types of objects with the N_i adding up to N. If the N objects are *all* distinct ($N_i = 1$ for all i), there are $N!$ orderings. There are, for example, N_1 equivalent members of subgroup 1. If we interchange any of the results giving configuration 1 in a sequence, it will not change the sequence. For example, in the sequence

$$T\,H\,T\,H\,T\,H$$

if we exchange the fourth and sixth entries, we obtain

$$T\,H\,T\,H\,T\,H$$

and there is no distinguishable change in the sequence. There are $N_1!$ orderings of the N_1 members of a subgroup in a sequence that are equivalent. Thus we must divide $N!$ by $N_1!$ to obtain the distinct orderings—similarly for $N_2 \cdots N_n$. The total number of distinct orderings is

$$M \equiv \frac{N!}{\prod_{i=1}^{n}(N_i!)}. \tag{59}$$

We can immediately apply this to our example, where $N = 6$, $N_1 = 3$, and $N_2 = 3$. In this case

$$M = \frac{6!}{3!\,3!} = 20. \tag{60}$$

You can easily check that this gives the number of sequences we constructed explicitly in Table 1.1.

GENERAL PRINCIPLES OF STATISTICAL MECHANICS

We are now in a position to reevaluate our original problem: We have an ensemble of N systems that we label from 1 to N. Each system can be found in one of n possible configurations, which we label from 1 to n. We *know* that P_i is the probability of obtaining the configuration i for any one member of the ensemble. When we sample each member of the ensemble we will find, to a good approximation, $N_i = P_i N$ members of the ensemble in the ith configuration. What we do not know is which of the M sequences will correspond to the observed sequence of distribution of the n subgroups $(N_1, N_2, \ldots, N_n)$ into the N ensemble members. These sequences are *equally* probable. Consequently, our problem reduces to the choice between M equally probable results and is equivalent, from a statistical point of view, to our earlier problem of choosing among n boxes. We can therefore use our previous expression for the missing information to obtain

$$S_N = k \ln M = k \ln \frac{N!}{\prod_{i=1}^{n}(N_i!)}. \tag{61}$$

This is the missing information associated with the entire ensemble. The missing information per system is given, in the appropriate large N limit, as

$$S = \lim_{N \to \infty} \frac{1}{N} S_N = \lim_{N \to \infty} \frac{k}{N} \ln \frac{N!}{\prod_{i=1}^{n}(N_i)!}. \tag{62}$$

Fortunately, we can carry out this limit exactly. First we write

$$S = \lim_{N \to \infty} \frac{k}{N} \left[\ln N! - \sum_{i=1}^{n} \ln N_i! \right]. \tag{63}$$

You are asked in Problem 1.17 to derive *Stirling's formula*, given by

$$\ln N! = N \ln N - N + \frac{1}{2} \ln(2\pi N) + \mathcal{O}\left(\frac{1}{N}\right). \tag{64}$$

Using Stirling's formula and $N_i = P_i N$ in our expression for S, we obtain

$$S = \lim_{N \to \infty} \frac{k}{N} \left\{ N \ln N - N - \sum_{i=1}^{n}(NP_i)\ln(NP_i) - \sum_{i=1}^{n}(NP_i) \right\}$$

$$= \lim_{N \to \infty} \frac{k}{N} \left[N \ln N - N - \sum_{i=1}^{n}(NP_i \ln N + NP_i \ln P_i - NP_i) \right]$$

$$= \lim_{N \to \infty} \frac{k}{N} \left(N \ln N - N - N \ln N + N - N \sum_{i=1}^{n} P_i \ln P_i \right)$$

Then, using

$$\lim_{N\to\infty} \frac{\ln N}{N} = 0, \qquad (65)$$

we obtain our general expression for the missing information:

$$S = -k \sum_{i=1}^{n} P_i \ln P_i. \qquad (66)$$

Note that when all of the configurations are equally probable,

$$P_i = \frac{1}{n} \qquad (67)$$

for all i and we have

$$S(n) = -k \sum_{i=1}^{n} \frac{1}{n} \ln n^{-1} = kn \frac{1}{n} \ln n = k \ln n \qquad (68)$$

which agrees with Eq. (52). We also see that when we know definitely that the system is in configuration j, so that $P_j = 1$ and $P_i = 0$ if $i \neq j$, then $S = 0$. Thus if we know that the system is in the jth configuration, there is no missing information. There is nothing random about the situation.

To illustrate further the nature of S, let us focus for a moment on a system with only two configurations ($n = 2$). In this case because of the normalization ($P_1 + P_2 = 1$), there is only one independent probability. Clearly, S is a minimum ($= 0$) when $P_i = 1$, $P_j = 0$, and $i \neq j$. Let us investigate the maximum values of S (when we have the most information missing). The extremum values of

$$S = -k[P_1 \ln P_1 + (1 - P_1)\ln(1 - P_1)] \qquad (69)$$

as a function of P_1 are given by

$$\frac{\partial S}{\partial P_1} = -k[\ln P_1 + 1 - \ln(1 - P_1) - 1]$$

$$= -k \ln \frac{P_1}{1 - P_1} = 0. \qquad (70)$$

The solution of this equation satisfies

$$\frac{P_1}{1 - P_1} = 1, \qquad (71)$$

which reduces to the case of equal probabilities

$$P_1 = P_2 = \tfrac{1}{2} \tag{72}$$

indicating that the choices in the experiment are completely random. We can check that this is a maximum by computing the second derivative. Since

$$\frac{\partial^2 S}{\partial P_1^2} = -k\left(\frac{1}{P_1} + \frac{1}{1-P_1}\right) < 0, \tag{73}$$

we indeed have a maximum:

$$S_{\max} = -k(\tfrac{1}{2}\ln\tfrac{1}{2} + \tfrac{1}{2}\ln\tfrac{1}{2}) = k\ln 2. \tag{74}$$

The missing information is bounded in this case by

$$k\ln 2 \geq S \geq 0. \tag{75}$$

Thus we see that S is a measure of the degree of randomness of a system. In Problem 1.19 you are asked to show that for general n, S is maximized by

$$P_i = \frac{1}{n} \tag{76}$$

for all i. Thus the arguments made above generalize to arbitrary n.

As a very simple example, consider the case

$$P_1 = \tfrac{1}{3}, \qquad P_2 = \tfrac{2}{3}. \tag{77}$$

This system is neither completely random nor completely deterministic. In this case

$$\begin{aligned}
S &= k\left(\tfrac{1}{3}\ln\tfrac{1}{3} + \tfrac{2}{3}\ln\tfrac{2}{3}\right) \\
&= k\left(\ln 3 - \tfrac{2}{3}\ln 2\right) \\
&= k(0.6365) < k\ln 2 = k(0.6931)
\end{aligned} \tag{78}$$

which gives us a quantitative measure of the degree of randomness.

1.8 MAXIMUM ENTROPY AND THE SECOND LAW OF THERMODYNAMICS

We stop here to state prominently the most important physics associated with equilibrium statistical mechanics. We first note, within a physics context, that the missing information of a system is called the *entropy* [28]. It is a law of physics, the second law of thermodynamics [29], that an arbitrarily large, macroscopically isolated, stable system will eventually approach a state of maximum entropy. This state of maximum entropy is the equilibrium state. The implications of the second law in thermodynamics are discussed in Chapter 2. Here we are concerned with its central implications for statistical mechanics.

There are a number of conditions attached to the second law that require discussion. The second law holds in the context of arbitrarily large systems. In this limit any random fluctuations in the entropy are down compared to the value of the entropy by a factor that increases with the size of the system. We discuss such fluctuations later in this chapter. Our system must be macroscopically isolated. This means that we are not driving or pumping the system and the macrovariables are held constant in time. A major consequence of the second law is the acknowledgment that we cannot microscopically isolate physical systems. Small boundary and external perturbations have the consequence of destroying all pristine microscopic constraints (e.g., Poincaré recurrence cycles) on large systems.

The condition that a system be stable is simply that it be capable of sustaining an equilibrium state. Unstable systems, such as a collection of electrons, clearly will not equilibrate because of the lack of charge neutrality. In this context, gravitating systems must be treated with particular care.

Finally, consider the statement that an appropriate system will *eventually* equilibrate. There are two points to make. First, once the system is isolated macroscopically, it will not matter if the system is run forward or backward in time; in each case the entropy will increase or stay constant. Any arrow of time arguments have to do with sample preparation. The preparation process is causal. The time t_0 when we have finished preparing our system will be later than the time t_{-1} when we started building the container. Experimentally, we have no choice but to run the system forward in time. However, in principle, the second law allows one to run the system backward in time. In this case we must replace the physical situation earlier with the same conditions of isolation used when we run the system forward. Under these conditions we will obtain an increasing entropy as time decreases.

The second point to make about the eventual approach of systems to equilibrium is that the associated time scales are system dependent and can range from extremely fast for fluids at high temperatures to extremely slow for atoms hopping around in solids at low temperatures. Human time scales (minutes) sit within this span of time scales. Thus to understand whether a system has fully equilibrated, one must understand the basic equilibration times for that system.

1.9 MICROCANONICAL ENSEMBLE

1.9.1 Introduction

Now that we have an operational method for measuring the degree of randomness or entropy of a system and a statement of the second law of thermodynamics, we can quantify our discussion of the equilibrium state. Let us first discuss the case of an isolated system where the equilibrium state corresponds to the situation where we know the values of the macrovariables—the energy E, total momentum $\mathbf{P}$, total angular momentum $\mathbf{L}$, total number of particles N, and volume V—but we have a minimum amount of information about the microvariables in the system. These conditions specify the *microcanonical ensemble* [30] (we discuss other types of ensembles later). To simplify matters somewhat, we assume that the system is not moving or rotating, so $\mathbf{P} = \mathbf{L} = 0$. It is not difficult to come back and treat the more general case. Thus we seek to find the maximum entropy S consistent with a constant E, N, and V. We specialize our analysis to the case of a classical gas for definiteness. The general development will also hold for quantum systems, as discussed in Chapter 3.

We consider a system of N particles with phase-space coordinates; $(q_N) \equiv (\mathbf{r}_1, \mathbf{r}_2, \ldots, \mathbf{r}_N, \mathbf{p}_1, \ldots, \mathbf{p}_N)$. The probability distribution will in general be a function of q_N. We require that the Hamiltonian has a constant value E:

$$H = E. \tag{79}$$

The total number of distinguishable phase-space configurations for a system with energy E and N particles in volume V is proportional to

$$Z_M(E, N, V) \equiv \int_V \frac{dq_N}{h^{3N} N!} E_T \delta(E - H). \tag{80}$$

The δ-function in Eq. (80) constrains the phase-space configurations to a constant energy E. We measure the energy in units of

$$E_T = Nmc^2 \tag{81}$$

where mc^2 is the rest energy of each particle. Note that $E_T \delta(E - H) = \delta(E/E_T - H/E_T)$ is dimensionless. We divide by $N!$ because we do not distinguish among configurations where we simply interchange particle labels. The two configurations shown in Fig. 1.8 are therefore to be treated as equivalent. In the direct integration over all q_N we count a particular configuration $N!$ times. Finally, we divide the integral by h^{3N}, where h is Planck's constant [31], to make the number of configurations dimensionless (remember that $h \approx$ [energy] $\times$ [time] $\approx$ [length] $\times$ [momentum]). This then specifies how we count the number of available configurations. We should expect, and should verify, that observables should not depend on $\hbar$ or the speed of light in the classical regime. The ensemble average of some observable A, which is a function of the phase-space coordinates, is given by

$$\langle A \rangle = \int \frac{dq_N}{h^{3N} N!} E_T \delta(E - H) P(q_N) A(q_N) \tag{82}$$

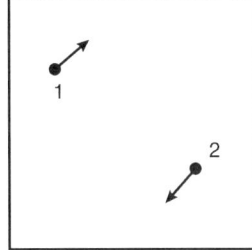 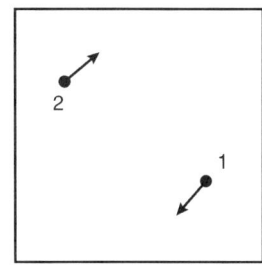

FIGURE 1.8 Physical systems are invariant under relabeling of particles.

where $P(q_N)$ is the appropriate probability distribution. To save some writing it is convenient to introduce the notation

$$T_r^M PA \equiv \int \frac{dq_N}{h^{3N} N!} E_T \delta(E - H) P(q_N) A(q_N) \tag{83}$$

so that T_r^M is the *microcanonical trace* or sum over all allowable configurations. The entropy is given in this case by

$$S(E, N, V) = -k T_r^M P \ln P \tag{84}$$

and we have the normalization condition on the probability distribution

$$T_r^M P = 1 \tag{85}$$

and the definition

$$Z_M(E, N, V) = T_r^M \cdot 1. \tag{86}$$

In our definition of the entropy, $S(E, N, V)$, energy is fixed by the δ-function constraint, N is fixed by construction, and V is fixed by a geometrical constraint. The equilibrium configuration is given by that probability distribution where we have a maximum entropy consistent with fixed E, N, and V. We therefore want to find the probability distribution P that maximizes S. We do this by comparing two different probability distributions P_0 and P_1 which are both properly normalized,

$$T_r^M P_0 = T_r^M P_1 = 1, \tag{87}$$

and are positive definite. For reasons that will become clear as we proceed, we want to consider the quantity $-\ln(P_1/P_0)$. Using the identity

$$-\ln x \leq \frac{1}{x} - 1 \tag{88}$$

32 GENERAL PRINCIPLES OF STATISTICAL MECHANICS

(see Appendix D), valid for $x \geq 0$ and letting $x = P_1/P_0$, we obtain

$$-\ln \frac{P_1}{P_0} \leq \frac{P_0}{P_1} - 1. \tag{89}$$

Multiply this inequality on both sides by kP_1 (which is positive definite) and take the microcanonical trace to obtain

$$-kT_r^M P_1 \ln \frac{P_1}{P_0} \leq kT_r^M P_1 \left(\frac{P_0}{P_1} - 1\right). \tag{90}$$

Note, however, that we can rewrite the right-hand side as

$$kT_r^M P_1 \left(\frac{P_0}{P_1} - 1\right) = k(T_r^M P_0 - T_r^M P_1) = 0, \tag{91}$$

and Eq. (90) reduces to

$$-kT_r^M P_1 \ln P_1 + kT_r^M P_1 \ln P_0 \leq 0. \tag{92}$$

We see then that the entropy associated with an arbitrary probability distribution P_1, given by

$$S(P_1) = -kT_r^M P_1 \ln P_1, \tag{93}$$

is bounded:

$$S(P_1) \leq -kT_r^M P_1 \ln P_0. \tag{94}$$

If we choose P_0 to be that distribution that is independent of the phase-space coordinates (i.e., a constant) on the energy-invariant subspace, then

$$-kT_r^M P_1 \ln P_0 = -k \ln P_0 T_r^M P_1 = -k \ln P_0. \tag{95}$$

However, the entropy associated with P_0 is

$$\begin{aligned} S(P_0) &= -kT_r^M P_0 \ln P_0 = -k \ln P_0 T_r^M P_0 \\ &= -k \ln P_0 \end{aligned} \tag{96}$$

so that

$$S(P_1) \leq S(P_0). \tag{97}$$

We see then that P_0 is the distribution that maximizes S since for any other distribution P_1, S is smaller or equal. We have the very simple result that the equilibrium probability distribution for the microcanonical ensemble is a simple constant that we can determine from the normalization condition Eq. (87), or

$$P_0 = \frac{1}{T_r^M} = \frac{1}{Z_M(E, N, V)}. \tag{98}$$

Therefore, P_0 is inversely proportional to the total number of configurations allowed for given values of E, N, and V. This has the physical interpretation that all configurations for a given E, N, and V are *equally* probable.

Notice that our result for the equilibrium probability distribution in the microcanonical ensemble agrees with our earlier conclusions that a mixing system relaxes to a final state such that averages can be computed with a constant invariant ensemble on the hypersurface characterized by the isolating integrals (here only the energy).

The entropy for the equilibrium state is

$$\begin{aligned} S(E, N, V) &= -k \ln P_0 \\ &= k \ln Z_M(E, N, V). \end{aligned} \tag{99}$$

This relation between entropy and the number of available configurations, known as *Boltzmann's principle*, was first published by Planck [32] in 1906.

1.9.2 Ideal Gas

The simplest example where we can construct the explicit dependence of the entropy S on E, N, and V is the ideal or noninteracting gas. In this case the interaction potential is zero and the Hamiltonian is given by

$$H = \sum_{i=1}^{N} \frac{\mathbf{p}_i^2}{2m}. \tag{100}$$

The entropy is just the logarithm of the microcanonical partition function, which is given by Eq. (80). In evaluating this quantity, it is convenient to use the fact that the δ-function is the derivative of the stepfunction,

$$\delta(x - y) = \frac{\partial}{\partial x} \theta(x - y), \tag{101}$$

where the step function is defined as

$$\theta(x) = \begin{cases} 1 & x > 0 \\ \frac{1}{2} & x = 0 \\ 0 & x < 0 \end{cases} \tag{102}$$

and write

$$Z_M = E_T \frac{\partial}{\partial E} \int_V \frac{dq_N}{h^{3N}N!} \theta(E - H). \tag{103}$$

Consider then the integral

$$I_N = \int_V \frac{dq_N}{h^{3N}N!} \theta(E - H), \tag{104}$$

which gives the number of states with energy less than E. Since H is independent of the spatial coordinates $\mathbf{r}_1, \mathbf{r}_2, \ldots, \mathbf{r}_N$, we can immediately do the space integrals to obtain

$$I_N = \frac{V^N}{h^{3N}N!} \int d^3p_1 d^3p_2 \cdots d^3p_N \theta\left(E - \sum_{i=1}^{N} \frac{\mathbf{p}_i^2}{2m}\right). \tag{105}$$

If we then change to the dimensionless variables

$$\mathbf{x}_i = \frac{\mathbf{p}_i}{(2mE)^{1/2}}, \tag{106}$$

we obtain

$$I_N = \frac{V^N}{h^{3N}N!} (2mE)^{3N/2} C_{3N} \tag{107}$$

where

$$C_d \equiv \int dx_1 dx_2 \cdots dx_d \, \theta\left(1 - \sum_{i=1}^{d} x_i^2\right) \tag{108}$$

is the volume of a d-dimensional sphere of unit radius. We then have, since C_d is independent of E,

$$Z_M = \frac{\partial}{\partial E} \left[\frac{E_T V^N}{h^{3N}N!} (2mE)^{3N/2} C_{3N}\right]$$

$$= \frac{V^N}{h^{3N}N!} E_T \frac{3N}{2E} (2mE)^{3N/2} C_{3N}. \tag{109}$$

The determination C_{3N} is carried out in Appendix E with the result

$$C_{3N} = \frac{\pi^{3N/2}}{\Gamma(3N/2 + 1)} \tag{110}$$

where Γ is the usual gamma function. The total number of configurations for an ideal gas in the microcanonical ensemble is then given by

$$Z_M = \left[\frac{V(2mE\pi)^{3/2}}{h^3}\right]^N \frac{3NE_T}{N!2E\Gamma(3N/2+1)} \tag{111}$$

and the entropy is given by

$$S(E,N,V) = k\ln\left\{\left[\frac{V(2mE\pi)^{3/2}}{h^3}\right]^N \frac{3NE_T}{N!2E\Gamma(3N/2+1)}\right\}. \tag{112}$$

Inspection of Eq. (112) shows that S is linear in N and is an *extensive variable*. An extensive variable is one that increases linearly with the size of the system for large systems. E and V are clearly also extensive variables. Since we are interested in the case of very large N, we want to take the large-N limit. We must, however, exercise some care. If our system is uniform, then in the large-N limit, we expect the particle density

$$n = \frac{N}{V} \tag{113}$$

and the energy per particle

$$\epsilon = \frac{E}{N} \tag{114}$$

to be finite and non zero. The correct *thermodynamic limit* [33] is to eliminate E and V in terms of ϵ, n, and N and then take the large-N limit. In this limit the entropy per particle is the appropriate quantity to determine:

$$\begin{aligned}
s &= \lim_{N\to\infty} \frac{S(N\epsilon, N/n, N)}{N} \\
&= \lim_{N\to\infty}\left\{k\ln\left[\frac{N}{nh^3}(2\pi mN\epsilon)^{3/2}\right] + \frac{k}{N}\ln\left[\frac{3NE_T}{2\epsilon NN!\Gamma(3N/2+1)}\right]\right\} \\
&= \lim_{N\to\infty}\left\{k\ln\left[\frac{N}{nh^3}(2\pi mN\epsilon)^{3/2}\right] - \frac{k}{N}\ln N! - \frac{k}{N}\ln\Gamma\left(\frac{3N}{2}+1\right)\right\}.
\end{aligned}$$

If we again use Stirling's formula, Eq. (64), remembering that $\Gamma(n+1) = n!$ and that $\lim_{N\to\infty} N^{-1}\ln N \to 0$, we find that

$$s = k\ln\left[\frac{e^{5/2}}{nh^3}\left(\frac{4\pi m\epsilon}{3}\right)^{3/2}\right] \tag{115}$$

which relates the *intensive* variables s, n, and ϵ. Note that this result (known as the Sackur–Tetrode equation [34]) is independent of the scale we used to measure the total energy E_T. This equation, properly interpreted, contains all available thermodynamic information about an ideal gas system. We are not yet properly situated to extract all of this information. To gain this perspective we move on to a different description of equilibrium statistical mechanics.

1.10 OPEN ENSEMBLES

1.10.1 Local Conservation Laws

a. Subsystems. In Section 1.9 we discussed, within the Boltzmann paradigm, an *isolated* system at rest with a fixed energy and particle number. It is for truly isolated systems that we run into the challenging problems of ergodic theory. As we have emphasized above, the concept of imperfectly prepared ensembles and small perturbations at the boundary lead to a randomization of the microvariables. If our system is truly isolated, we do not have these perturbations at the boundary. In the Langevin paradigm one deals more realistically with boundary effects. Suppose that our system of interest, labeled 1 in Fig. 1.9, is immersed in a much larger system (labeled 2). We suppose that the walls of 1 are such that energy and particles can be transferred between 1 and 2. System 1 is very different from the isolated system we treated before. It is clear that the energy of system 1, E_1, is not conserved and similarly, the number of particles in system 1, N_1, changes with time. Thus we see that the idea of the $6N_1 - 7$ microvariables being conserved (or even being constructed) is ridiculous since N_1 itself is changing with time. However, things are not as hopeless as they seem. Although E_1 and N_1 are *not* conserved, the *average* values of E_1 and N_1 are reproducible and time independent in equilibrium. This is simply due to the fact that *in equilibrium*, as much energy will pass into 1 as will leave. Otherwise, system 1 will heat up or cool down at the expense of the rest of the system. Similarly, as many particles, on average, will enter 1 as will leave it.

b. Densities. A key point in this development is the notion of a *local* conservation law. Thus far we have talked about *global* constants of the motion

$$\frac{d}{dt}\Phi_i(q) = 0. \tag{116}$$

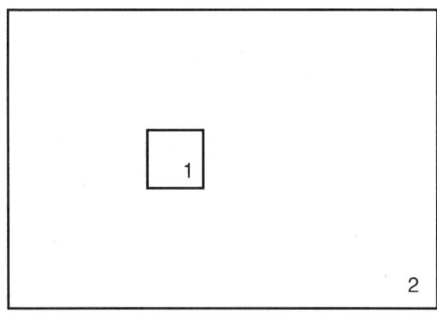

FIGURE 1.9 Open subsystem (1) embedded in a very large heat bath (2).

Clearly, conservation of energy, momentum, and particle number for an isolated system can be written in this form

$$\frac{dE}{dt} = 0 \qquad (117)$$

$$\frac{dN}{dt} = 0 \qquad (118)$$

$$\frac{d\mathbf{P}}{dt} = 0. \qquad (119)$$

One of the properties that distinguishes such quantities as energy, momentum, and particle number is that their definition is not coupled into the boundary conditions in a problem. Looking at Fig. 1.10, we can, of course, count up the total number of particles N in the volume V. Similarly, at some time t we can count up the particles N' in volume V'. If we divide V up into a set of volumes V_α such that

$$V = \sum_\alpha V_\alpha \qquad (120)$$

we have

$$N = \sum_\alpha N_\alpha. \qquad (121)$$

We have the property of additivity for the total number of particles. Similarly, for the energy and momentum, we can write

$$E = \sum_\alpha E_\alpha \qquad (122)$$

$$\mathbf{P} = \sum_\alpha \mathbf{P}_\alpha \qquad (123)$$

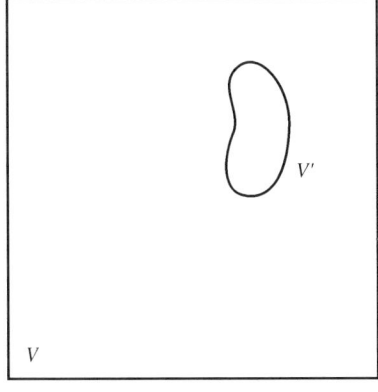

FIGURE 1.10 Volume V divided up into subsystems like V'.

where E_α is the energy [35] and $\mathbf{P}_\alpha$ the momentum associated with a volume V_α. This additivity property separates the macrovariables from the microvariables. In the limit, as we let $V_\alpha \to 0$, the sum over α goes over to an integral over the total volume V and N_α goes over to a set of δ-functions located at the positions of the particles

$$N = \int_V d^3x \sum_{i=1}^N \delta(\mathbf{x} - \mathbf{r}_i). \tag{124}$$

It is natural to introduce the local particle density $n(\mathbf{x})$ whose integral gives the total number of particles:

$$N = \int_V d^3x\, n(\mathbf{x}). \tag{125}$$

Comparing the last two equations, we can identify the microscopic definition of the particle density:

$$n(\mathbf{x}) = \sum_{i=1}^N \delta(\mathbf{x} - \mathbf{r}_i). \tag{126}$$

We can introduce the energy density $\epsilon(\mathbf{x})$ and momentum density $\mathbf{g}(\mathbf{x})$ via

$$E = \int_V d^3x\, \epsilon(\mathbf{x}) \tag{127}$$

and

$$\mathbf{P} = \int_V d^3x\, \mathbf{g}(\mathbf{x}) \tag{128}$$

where

$$\mathbf{g}(\mathbf{x}) = \sum_{i=1}^N \mathbf{p}_i \delta(\mathbf{x} - \mathbf{r}_i) \tag{129}$$

$$\epsilon(\mathbf{x}) = \sum_{i=1}^N \epsilon_i \delta(\mathbf{x} - \mathbf{r}_i) \tag{130}$$

and ϵ_i and $\mathbf{p}_i$ are the energy and momentum of the ith particle. The angular momentum density is given by

$$\mathbf{l}(\mathbf{x}) = \sum_{i=1}^N (\mathbf{r}_i \times \mathbf{p}_i)\delta(\mathbf{x} - \mathbf{r}_i) = \mathbf{x} \times \mathbf{g}(x). \tag{131}$$

c. Continuity Equations. Let us consider the time evolution of the particle density. Using the chain rule for differentiation, we obtain [36]

$$\frac{\partial n(\mathbf{x},t)}{\partial t} = \sum_{i=1}^{N} \frac{\partial}{\partial t} \delta(\mathbf{x} - \mathbf{r}_i(t))$$

$$= -\sum_{i=1}^{N} \frac{d\mathbf{r}_i(t)}{dt} \cdot \nabla_x \delta(\mathbf{x} - \mathbf{r}_i(t))$$

$$= -\nabla_x \cdot \frac{\mathbf{g}(\mathbf{x},t)}{m} \qquad (132)$$

since

$$\mathbf{p}_i(t) = m \frac{d}{dt} \mathbf{r}_i(t). \qquad (133)$$

The physical interpretation of this result is that particles leave a region by flowing out of it, not by being created or destroyed. It is less obvious, but true, that the other conserved densities also satisfy such a continuity equation:

$$\frac{\partial}{\partial t} (\text{density}) = -\nabla \cdot (\text{current}). \qquad (134)$$

We show in Appendix F that local conservation of momentum [37] can be expressed in the form

$$\frac{\partial g_\alpha(\mathbf{x},t)}{\partial t} = -\sum_\beta \nabla_x^\beta \sigma_{\alpha\beta}(\mathbf{x},t) \qquad (135)$$

where the microscopic expression for the stress tensor $\sigma_{\alpha\beta}$ is given in Appendix F by Eq (2609). Similarly, conservation of energy [38] can be expressed in the form

$$\frac{\partial \epsilon(\mathbf{x},t)}{\partial t} = -\nabla_x \cdot \mathbf{J}_\epsilon(\mathbf{x},t) \qquad (136)$$

where the microscopic expression for the energy current $\mathbf{J}_\epsilon$ is given by Eq. (2615).

An important distinguishing characteristic between microvariables and (conserved) macrovariables is that we can identify local variations and flows associated with macrovariables and such a concept is nonsensical for microvariables. Since we cannot create energy in a region, it must, in a rather smooth manner, flow in or flow out. Thus we can understand how, on average, it makes sense to identify an energy and particle number associated with the subvolume 1.

40 GENERAL PRINCIPLES OF STATISTICAL MECHANICS

It is appropriate here to quote [39] from the preface of Landau and Lifshitz, *Statistical Physics*: "We do not share the view, which one encounters sometimes, that statistical physics is the least well-founded branch of theoretical physics (as regards its basic principles). We believe that the difficulties are created artificially, because the problems are often not stated sufficiently rationally. If one talks from the very beginning about the statistical distribution for small parts of a system (subsystems) and not for a closed system as a whole, then one avoids the whole question of the ergodic or similar hypotheses, which are not really essential for physical statistics."

1.10.2 Open Boundaries

Systems where we allow for a coupling to the outside world are called *open systems*. There are two particular systems of interest.

1. *Canonical ensemble*. In the canonical ensemble [40] the system of interest has a weak thermal coupling to the external world. Thus in Fig. 1.9 system 1 can exchange energy but not particles with system 2. The number of particles in system 1 is held fixed. The condition determining the equilibrium probability distribution for system 1 is that the entropy S be a maximum consistent with the average energy $\langle H_1 \rangle$ and the total number of particles $\langle N_1 \rangle$ being held fixed.

2. *Grand canonical ensemble*. In the grand canonical ensemble [41] the system of interest has a weak coupling to the external world, allowing energy and particle transport. The condition determining the equilibrium probability distribution is that the entropy S be a maximum consistent with the average energy $\langle H_1 \rangle$ and the average number of particles $\langle N_1 \rangle$ being held fixed.

We discuss the equivalence of these different formulations of statistical mechanics as we proceed.

1.10.3 Introduction of Temperature and Chemical Potential

As an important application of our entropy principle in the case of open systems, consider the situation where we have two separate systems (see Fig. 1.11) characterized by energies E_1 and E_2, volumes V_1 and V_2, and particle numbers N_1 and N_2. Since the macrovariables and the entropy are additive, one has for the combined [42] system

$$E_T = E_1 + E_2 \tag{137}$$

$$N_T = N_1 + N_2 \tag{138}$$

$$V_T = V_1 + V_2 \tag{139}$$

$$S_T = S_1 + S_2. \tag{140}$$

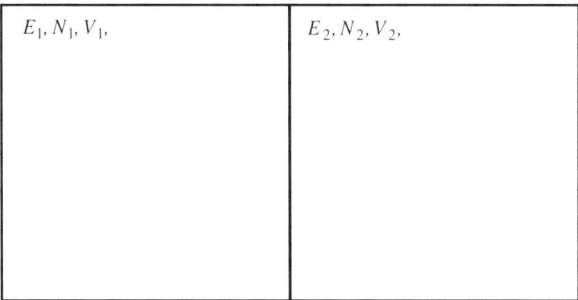

FIGURE 1.11 Two macroscopic systems in thermal communication across a common wall.

where $S_1 = S_1(N_1, V_1, E_1)$ and $S_2 = S_2(N_2, V_2, E_2)$. Initially, the two systems are thermally isolated and there is no exchange of energy or particles between the two systems. At sometime we remove the thermal barrier between the two systems and they are allowed to exchange energy—they are thermally coupled. We also assume that the volumes of the two systems are fixed and there is no particle exchange (N_1 and N_2 are fixed). Consider then the variation in the total entropy due to heat exchange between the two systems as they attempt to equilibrate:

$$\delta S_T = \left(\frac{\partial S_1}{\partial E_1}\right)_{N_1, V_1} \delta E_1 + \left(\frac{\partial S_2}{\partial E_2}\right)_{N_2, V_2} \delta E_2. \tag{141}$$

Since the total energy is conserved,

$$\delta E_2 = -\delta E_1, \tag{142}$$

and

$$\delta S_T = \left[\left(\frac{\partial S_1}{\partial E_1}\right)_{N_1, V_1} - \left(\frac{\partial S_2}{\partial E_2}\right)_{N_2, V_2}\right] \delta E_1. \tag{143}$$

If the system is in the process of moving to equilibrium, S_T is increasing:

$$\delta S_T = \left[\left(\frac{\partial S_1}{\partial E_1}\right)_{N_1, V_1} - \left(\frac{\partial S_2}{\partial E_2}\right)_{N_2, V_2}\right] \delta E_1 > 0. \tag{144}$$

If $\delta E_1 > 0$, energy is flowing from body 2 to body 1 and

$$\left(\frac{\partial S_1}{\partial E_1}\right)_{N_1, V_1} > \left(\frac{\partial S_2}{\partial E_2}\right)_{N_2, V_2}. \tag{145}$$

Clearly, for each body, the quantity

$$\left(\frac{\partial S_i}{\partial E_i}\right)_{N_i, V_i} \equiv \frac{1}{T_i} \tag{146}$$

is important. Inserting this notation into Eq. (145) gives

$$\frac{1}{T_1} > \frac{1}{T_2} \tag{147}$$

or $T_2 > T_1$. We then have the statement that heat flows from body 2 with greater T to body 1 with lower T. We recognize this as the fundamental property of *temperature* [43]. Heat flows from a hotter body to a cooler body. In equilibrium $\delta S_T = 0$ and

$$\frac{1}{T_1} = \frac{1}{T_2} \tag{148}$$

or, for two bodies to be in thermal equilibrium, they must be at the same temperature:

$$T_1 = T_2. \tag{149}$$

Let us next return to our system of two coupled systems which are now assumed to be held at the same common temperature $T_1 = T_2 = T$. We assume that initially there is a partition between the two systems and there is no exchange of particles. At some time we remove a partition to reveal a boundary formed by a membrane that preserves the volumes of the two systems but allows for exchange of particles. After the system has returned to equilibrium, the variation in the total entropy due to removing the partition is given by

$$\delta S_T = \left(\frac{\partial S_1}{\partial E_1}\right)_{N_1, V_1} \delta E_1 + \left(\frac{\partial S_1}{\partial N_1}\right)_{E_1, V_1} \delta N_1$$
$$+ \left(\frac{\partial S_2}{\partial E_2}\right)_{N_2, V_2} \delta E_2 + \left(\frac{\partial S_2}{\partial N_2}\right)_{E_2, V_2} \delta N_2. \tag{150}$$

Conservation of total energy and total number of particles gives

$$\delta E_1 + \delta E_2 = 0 \tag{151}$$
$$\delta N_1 + \delta N_2 = 0 \tag{152}$$

and Eq. (150) reduces to

$$\delta S_T = \left(\frac{1}{T_1} - \frac{1}{T_2}\right)\delta E_1 + \left[\left(\frac{\partial S_1}{\partial N_1}\right)_{E_1, V_1} - \left(\frac{\partial S_2}{\partial N_2}\right)_{E_2, V_2}\right]\delta N_1. \tag{153}$$

Since the temperatures are equal, we obtain

$$\delta S_T = \left[\left(\frac{\partial S_1}{\partial N_1}\right)_{E_1,V_1} - \left(\frac{\partial S_2}{\partial N_2}\right)_{E_2,V_2}\right]\delta N_1. \tag{154}$$

If the system is approaching equilibrium, then

$$\delta S_T = \left[\left(\frac{\partial S_1}{\partial N_1}\right)_{E_1,V_1} - \left(\frac{\partial S_2}{\partial N_2}\right)_{E_2,V_2}\right]\delta N_1 > 0. \tag{155}$$

If particles flow from system 2 to 1, then $\delta N_1 > 0$ and

$$\left(\frac{\partial S_1}{\partial N_1}\right)_{E_1,V_1} > \left(\frac{\partial S_2}{\partial N_2}\right)_{E_2,V_2}. \tag{156}$$

It is conventional to define the chemical potential μ_i [44] as

$$\left(\frac{\partial S_i}{\partial E_i}\right)_{N_i,V_i} = -\frac{\mu_i}{T_i}. \tag{157}$$

Then following the same logic as for the temperature, we have the condition for equilibrium,

$$\mu_1 = \mu_2 \tag{158}$$

and in approaching equilibrium,

$$-\frac{\mu_1}{T} > -\frac{\mu_2}{T} \tag{159}$$

which reduces to

$$\mu_1 < \mu_2. \tag{160}$$

Thus we see that particles flow from regions of high chemical potential to regions of low chemical potential. Usually, we think of particles flowing from regions of high density to regions of low density. In an ideal gas, where μ increases with the density, our intuition is correct. In more complicated situations, where one could, for example, have phase coexistence, it is the chemical potential and not the density that controls the flow.

44 GENERAL PRINCIPLES OF STATISTICAL MECHANICS

Similar arguments can be made concerning volume changes where we introduce the pressure:

$$\left(\frac{\partial S}{\partial V}\right)_{E,N} = \frac{p}{T}. \tag{161}$$

Then, if $p_1 > p_2$, the volume of system 1 will increase at the expense of system 2. Thus the high-pressure system will expand into the low-pressure system. For the two systems to be in mechanical and thermal equilibrium, the two pressures must be equal. Later we discuss in more detail this identification of the pressure.

This then completes the set of derivatives of the entropy with respect to its natural variables in the case of a simple fluid:

$$\left(\frac{\partial S}{\partial E}\right)_{N,V} = \frac{1}{T} \equiv k_B F_E \tag{162}$$

$$\left(\frac{\partial S}{\partial \overline{N}}\right)_{E,V} = -\frac{\mu}{T} \equiv k_B F_N \tag{163}$$

$$\left(\frac{\partial S}{\partial V}\right)_{E,N} = \frac{p}{T} \equiv k_B F_V \tag{164}$$

where k_B is Boltzmann's constant defined below. If we adopt the notation $X_E = E$, $X_N = N$, $X_V = V$, these equations can be written in the form

$$\frac{\partial S}{\partial X_i} = k_B F_i. \tag{165}$$

These equations giving the intensive variables in terms of the macrovariables are called *equations of state*. We can work these out explicitly for the case of an ideal gas starting with the Sackur–Tetrode equation for the entropy (with $k = k_B$)

$$S = Nk_B \ln\left[\frac{e^{5/2}}{nh^3}\left(\frac{4\pi m\epsilon}{3}\right)^{3/2}\right] \tag{166}$$

and taking the derivatives. We easily find that

$$\left(\frac{\partial S}{\partial E}\right)_{N,V} = \frac{1}{T} = \frac{3k_B}{2\epsilon}, \tag{167}$$

$$\left(\frac{\partial S}{\partial N}\right)_{E,V} = -\frac{\mu}{T} = -k_B \ln n\lambda^3 \tag{168}$$

where it is natural to introduce the length $\lambda = (3h^2/4\pi m\epsilon)^{1/2}$, and

$$\left(\frac{\partial S}{\partial V}\right)_{E,N} = \frac{p}{T} = nk_B \tag{169}$$

which can be reexpressed as equations of state relating intensive quantities:

$$\epsilon = \tfrac{3}{2} k_B T \tag{170}$$

$$\mu = k_B T \ln n\lambda^3 \tag{171}$$

and

$$p = nk_B T. \tag{172}$$

Equation (172) is the usual ideal gas law [45] relating pressure, density, and temperature. Practically speaking, by working in a dilute system, we can use this expression to define a temperature scale [46]. This is one way to determine Boltzmann's constant to be $k_B = 1.280 \times 10^{16}$ ergs/degree. This thermometer can then be used to calibrate other systems.

1.10.4 Grand Canonical Ensemble

Let us now discuss the grand canonical ensemble (GCE) in some detail. We will come back to the canonical ensemble later. We note here that the grand canonical ensemble is the one used typically in modern statistical mechanics. This is because the conditions of energy and particle exchange with the environment are more realistic than the *isolation* associated with the other ensembles.

The first point in treating the GCE is to enumerate the possible configurations. In this case we must allow for all possible numbers of particles. Averages are then of the form

$$\langle A \rangle = \sum_{N=0}^{\infty} \int \frac{dq_N}{h^{3N} N!} P_N(q_N) A(q_N) \tag{173}$$

where the probability distribution $P_N[q_N]$ is now labeled by the number of particles. We can rewrite Eq. (173) in a more compact form,

$$\langle A \rangle = \text{Tr}_{cl} PA \tag{174}$$

where Tr stands for *trace* and the subscript cl stands for *classical* (as opposed to the quantum mechanical trace we discuss later). The probability distribution for the GCE is determined by maximizing the entropy

$$S = -k_B \text{Tr}_{cl} P \ln P \tag{175}$$

subject to the constraints

$$\text{Tr}_{cl} P = 1 \tag{176}$$

$$\text{Tr}_{cl} PH = E \tag{177}$$

and
$$\mathrm{Tr}_{cl}\, PN = \bar{N} \tag{178}$$

where E and $\bar{N}$ are the average energy and particle number. We can take these constraints into account using the method of Lagrange multipliers. We discuss the basic method in Appendix G. In the case we treat here we need one Lagrange multiplier for each constraint. We then introduce the functional of the probability distribution:

$$\begin{aligned} W[P] &= S - k_B f_0 \,\mathrm{Tr}_{cl} P - k_B f_E \,\mathrm{Tr}_{cl}\, PH - k_B f_N \,\mathrm{Tr}_{cl}\, PN \\ &= -k_B \mathrm{Tr}_{cl}\, P(\ln P + f_0 + f_E H + f_N N) \end{aligned} \tag{179}$$

where f_0, f_E, and f_N are the Lagrange multipliers (we include a factor of k_B in the multipliers for convenience). The conditions determining the probability distribution appropriate to the equilibrium state corresponds to finding the maximum of $W[P]$ with respect to P with *no* restriction. Looking for the extrema of W gives

$$\delta W = -k_B \mathrm{Tr}_{cl} \delta P(\ln P + 1 + f_0 + f_E H + f_N N) = 0 \tag{180}$$

where δP is an arbitrary variation of the probability distribution. The vanishing of δW for arbitrary δP dictates that

$$\ln P = -(1 + f_0) - f_E H - f_N N. \tag{181}$$

This gives immediately the *grand probability distribution*

$$P_G = e^{-(1+f_0+f_E H+f_N N)}. \tag{182}$$

We see that in this case the probability distribution depends on the Hamiltonian and the number of particles. We can eliminate the Lagrange multiplier f_o by using the normalization condition, Eq. (176),

$$1 = \mathrm{Tr}_{cl} P_G = e^{-(1+f_o)} \mathrm{Tr}_{cl} e^{-(f_E H + f_N N)} \tag{183}$$

so

$$e^{1+f_0} = \mathrm{Tr}_{cl} e^{-(f_E H + f_N N)} \equiv Z_G \tag{184}$$

where Z_G is the partition function in the GCE. Then we can write the probability distribution in the form

$$P_G = \frac{e^{-(f_E H + f_N N)}}{Z_G}. \tag{185}$$

The other two constraint equations, Eqs. (177) and (178), can be written

$$E = \frac{\text{Tr}_{\text{cl}} H e^{-(f_E H + f_N N)}}{Z_G} \quad (186)$$

and

$$\bar{N} = \frac{\text{Tr}_{\text{cl}} N e^{-(f_E H + f_N N)}}{Z_G}. \quad (187)$$

We still need to show that our extremum is a maximum. As a first step we use Eq. (185) to determine the entropy associated with our solution:

$$\begin{aligned} S[P_G] &= -k_B \text{Tr}_{\text{cl}} P_G \ln P_G \\ &= -\frac{k_B}{Z_G} \text{Tr}_{\text{cl}} e^{-(f_E H + f_N N)} (-f_E H - f_N N - \ln Z_G) \\ &= k_B (f_E E + f_N \bar{N} + \ln Z_G). \end{aligned} \quad (188)$$

Consider the entropy for a general probability distribution P',

$$S[P'] = -k_B \text{Tr}_{\text{cl}} P' \ln P' \quad (189)$$

where P' must satisfy the constraints given by Eqs. (176), (177), and (178). It should be clear that we can now use the inequality, Eq. (88), that we used in treating the microcanonical case, to obtain

$$S[P'] \leq -k_B \text{Tr}_{\text{cl}} P' \ln P_G. \quad (190)$$

Using Eq. (185) for P_G in Eq. (190) gives

$$\begin{aligned} S[P'] &\leq -k_B \text{Tr}_{\text{cl}} P' (-\ln Z_G - f_E H - f_N N) \\ &= k_B (\ln Z_G \text{Tr}_{\text{cl}} P' + f_E \text{Tr}_{\text{cl}} P' H + f_N \text{Tr}_{\text{cl}} P' N) \\ &= k_B (\ln Z_G + f_E E + f_N \bar{N}). \end{aligned} \quad (191)$$

However, we recognize from Eq. (188) that the right-handside of Eq. (191) is just the entropy for the equilibrium state and for any P',

$$S[P'] \leq S[P_G]. \quad (192)$$

Thus we see that P_G indeed maximizes the entropy consistent with the appropriate constraints and is *the* grand canonical ensemble probability distribution.

48 GENERAL PRINCIPLES OF STATISTICAL MECHANICS

Since the distribution function P_G depends on the Lagrange multipliers f_E and f_N, we need to discuss their significance. It is clear that the equations for the average energy and particle number, given by Eqs. (186) and (187), are of the form $E = E(f_E, f_N, V)$ and $\bar{N} = \bar{N}(f_E, f_N, V)$. We can, in principle, solve these two equations for the two unknowns f_E and f_N to obtain $f_E = f_E(E, \bar{N}, V)$ and $f_N = f_N(E, \bar{N}, V)$. In carrying out this inversion we will come to realize that f_E and f_N have their own direct physical interpretation. As a first step it is useful to introduce the thermodynamic potential [47]

$$\Omega \equiv -\frac{1}{f_E} \ln Z_G, \qquad (193)$$

so that the entropy, given by Eq. (188), can be written as

$$S = k_B(-f_E \Omega + f_E E + f_N \bar{N}). \qquad (194)$$

Ω has a number of useful properties. It is easy to show, using Eqs. (184), (186), and (187), that the energy and average number of particles can be obtained as derivatives of Ω:

$$E = \frac{\partial}{\partial f_E}(f_E \Omega) \qquad (195)$$

and

$$\bar{N} = \frac{\partial}{\partial f_N}(f_E \Omega). \qquad (196)$$

The interpretations of f_E and f_N follow from a discussion of the variation of the entropy, $S(E, \bar{N}, V)$, as functions of E and $\bar{N}$, where V is held constant. In this variation we treat f_E and f_N as functions of E and $\bar{N}$, so

$$df_E = \left(\frac{\partial f_E}{\partial E}\right)_{\bar{N}} dE + \left(\frac{\partial f_E}{\partial \bar{N}}\right)_E d\bar{N} \qquad (197)$$

$$df_N = \left(\frac{\partial f_E}{\partial E}\right)_{\bar{N}} dE + \left(\frac{\partial f_E}{\partial \bar{N}}\right)_E d\bar{N} \qquad (198)$$

and we have, remembering that $\Omega = \Omega(f_E, f_N, V)$,

$$dS = k_B \left[-\frac{\partial(f_E \Omega)}{\partial f_E} df_E - \frac{\partial(f_E \Omega)}{\partial f_N} df_N + E\, df_E + f_E\, dE + \bar{N}\, df_N + f_N\, d\bar{N} \right]. \qquad (199)$$

Using Eqs. (195) and (196), we can cancel the coefficients of df_E and df_N to obtain the nice result

$$dS = k_B(f_E\, dE + f_N\, d\bar{N}). \qquad (200)$$

Clearly, we can identify the Lagrange multipliers with the intensive parameters F_i introduced in the microcanonical ensemble:

$$k_B f_E = \left(\frac{\partial S}{\partial E}\right)_{\bar{N},V} = k_B F_E = \frac{1}{T} \qquad (201)$$

and

$$k_B f_N = \left(\frac{\partial S}{\partial \bar{N}}\right)_{E,V} = k_B F_N = -\frac{\mu}{T}. \qquad (202)$$

It is conventional to define

$$F_E = \frac{1}{k_B T} \equiv \beta \qquad (203)$$

and

$$F_N = -\frac{\mu}{k_B T} \equiv -\beta\mu \equiv \alpha. \qquad (204)$$

Starting with Eq. (194), and using Eqs. (195) and (196), it is easy to see that:

$$\left(\frac{\partial S}{\partial V}\right)_{E,N} = k_B \left(\frac{\partial}{\partial V}(-F_E \Omega)\right)_{F_E,F_N}$$

$$= -\frac{1}{T}\left(\frac{\partial \Omega}{\partial V}\right)_{\alpha,\beta}. \qquad (205)$$

Using Eq. (161) to evaluate the derivative of the entropy, canceling common factors, we obtain the expression for calculating the pressure in the grand canonical ensemble:

$$p = -\left(\frac{\partial \Omega}{\partial V}\right)_{\alpha\beta}. \qquad (206)$$

We can go further. It is clear that Ω is an extensive variable and can be written, in the thermodynamic limit, in the form

$$\Omega = V g(\alpha, \beta). \qquad (207)$$

Inserting this result back into Eq. (206) gives

$$p = -\frac{\partial}{\partial V} Vg(\alpha, \beta) = -g(\alpha, \beta) \tag{208}$$

or

$$pV = -\Omega. \tag{209}$$

If we use this expression for the thermodynamic potential in Eq. (194), the entropy can then be written as the sum of terms:

$$TS = pV + E - \mu \bar{N}. \tag{210}$$

We therefore have a relationship [48] between the extensive $(S, V, E, \bar{N})$ and intensive (T, p, μ) variables. Using the notation introduced earlier, we see that Eq. (210) can be written in the form

$$S = \sum_{i=E,N,V} F_i X_i. \tag{211}$$

Let us summarize our results for the grand canonical ensemble: The average of some microscopic quantity $A_N[q_N]$ is given in the grand canonical ensemble by

$$\langle A \rangle = \sum_{N=0}^{\infty} \int_V \frac{dq_N}{h^{3N} N!} P_G[q_N] A[q_N]. \tag{212}$$

The grand probability distribution is given by

$$P_G = e^{-\beta(H - \mu N - \Omega)} \tag{213}$$

where the thermodynamic potential is given by

$$\Omega = -\beta^{-1} \ln Z_G, \tag{214}$$

and Z_G is the grand canonical partition function [49]

$$Z_G = \sum_{N=0}^{\infty} \int_V \frac{dq_N}{h^{3N} N!} e^{-\beta(H - \mu N)} \tag{215}$$

which is a natural function of $\beta = (k_B T)^{-1}$, where T is the absolute temperature, $\alpha = -\beta \mu$, where μ is the chemical potential, and V is the volume. The average energy, particle number, pressure, and entropy are given by

$$E = \frac{\partial}{\partial \beta}(\beta \Omega) \tag{216}$$

$$\bar{N} = \frac{\partial}{\partial \alpha}(\beta \Omega) \tag{217}$$

$$pV = -\Omega \tag{218}$$

$$TS = E - \mu N + pV. \tag{219}$$

1.10.5 Ideal Gas in the Grand Canonical Ensemble

We computed the entropy per particle in the microcanonical ensemble for an ideal gas with the result given by the Sackur–Tetrode law, Eq. (115). How does this result compare with the results from the grand canonical ensemble? We can determine the entropy per particle by first explicitly calculating the grand partition function Z_G, taking the logarithm to obtain the grand potential Ω, taking derivatives, using Eqs. (216) and (217), to obtain E and $\bar{N}$ and then plugging these results back into Eq. (219) to obtain the entropy. We will pick up useful experience working in the GCE along the way.

First we must calculate the grand partition function for the Hamiltonian

$$H_0 = \sum_{i=1}^{N} \frac{\mathbf{p}_i^2}{2m}. \tag{220}$$

We find directly that in contrast to the complexity of the calculation in the microcanonical ensemble, the grand partition function is given by

$$\begin{aligned}
Z_G &= \sum_{N=0}^{\infty} \int_V \frac{dq_N}{N! h^{3N}} e^{-\beta(H_0 - \mu N)} \\
&= \sum_{N=0}^{\infty} \frac{V^N e^{\beta \mu N}}{N! h^{3N}} \left(\int d^3 p \, e^{-\beta p^2 / 2m} \right)^N \\
&= \sum_{N=0}^{\infty} \frac{1}{N!} \left[\frac{V e^{\beta \mu} (2\pi m \beta^{-1})^{3/2}}{h^3} \right]^N \\
&= \exp\left(\frac{V e^{\beta \mu}}{\lambda^3} \right)
\end{aligned} \tag{221}$$

where λ is the thermal de Broglie wavelength [50]

$$\lambda = \hbar \left(\frac{2\pi \beta}{m} \right)^{1/2}. \tag{222}$$

The thermodynamic potential is given by

$$\beta\Omega = -\ln Z_G = -V\frac{e^{\beta\mu}}{\lambda^3}.\tag{223}$$

We derive from this the average particle number by taking the derivative ($\alpha = -\beta\mu$),

$$\bar{N} = \frac{\partial}{\partial\alpha}(\beta\Omega) = \frac{Ve^{\beta\mu}}{\lambda^3},\tag{224}$$

so the particle density is given by

$$n = \frac{e^{\beta\mu}}{\lambda^3}.\tag{225}$$

The average energy for an ideal gas is given by

$$E = \frac{\partial}{\partial\beta}(\beta\Omega) = -Ve^{-\alpha}\frac{\partial}{\partial\beta}\lambda^{-3}$$

$$= \frac{3}{2}\beta^{-1}\frac{Ve^{\beta\mu}}{\lambda^3}.\tag{226}$$

Looking back at Eq. (224), giving the average number of particles, we see that the average total energy can be expressed as

$$E = \tfrac{3}{2}\beta^{-1}\bar{N} = \tfrac{3}{2}k_BT\bar{N},\tag{227}$$

and the average energy per particle is given by

$$\epsilon = \tfrac{3}{2}k_BT.\tag{228}$$

Using Eq. (218), the pressure is given by

$$\beta p = -\frac{1}{V}\beta\Omega = \frac{e^{\beta\mu}}{\lambda^3}.\tag{229}$$

Since the particle density is given by Eq. (225), we regain the ideal gas law

$$p = \beta^{-1}n = nk_BT.\tag{230}$$

We can next determine the entropy per particle starting with Eq. (219). We find on dividing by $\bar{N}$ that

$$T\frac{S}{\bar{N}} = \epsilon - \mu + \frac{pV}{\bar{N}}$$

$$= \epsilon - \mu + \beta^{-1} = \tfrac{5}{3}k_BT - \mu.\tag{231}$$

This gives the entropy per particle in terms of the temperature and chemical potential. If we are to compare with the calculation in the microcanonical ensemble, we must express the entropy per particle in terms of the energy per particle and the particle density. Clearly, from Eq. (228), $\beta^{-1} = 2\epsilon/3$, and from Eq. (225), the chemical potential is given by

$$\mu = \tfrac{2}{3}\epsilon \ln n\lambda^3 \tag{232}$$

where we can express the de Broglie wavelength in the form

$$\lambda = h\left(\frac{3}{4\pi m\epsilon}\right)^{1/2}. \tag{233}$$

Inserting these results back into Eq. (231) and canceling a factor of ϵ, we obtain

$$\frac{S}{N} = k_B\left(\frac{5}{2} - \ln n\lambda^3\right) = k_B \ln \frac{e^{5/2}}{n\lambda^3}. \tag{234}$$

Looking back to Eq. (115) we see, after introducing the de Broglie wavelength, that the two expressions agree and the entropy per particle for an ideal gas is the same in both the microcanonical and the grand canonical ensemble.

1.11 FLUCTUATIONS

Let us consider the obvious contrast between the microcanonical and the grand canonical ensembles. In one case we have a fixed energy, and in the other case we have a distribution of energies with a fixed average. Let us consider the dispersion in the energy

$$\Delta_E \equiv \frac{[\langle (H-E)^2\rangle]^{1/2}}{E} \tag{235}$$

where, as above, $E = \langle H \rangle$. Clearly, there is zero dispersion in the microcanonical ensemble. In the GCE the average energy is given by

$$E = \frac{\text{Tr}_{cl} H e^{-\beta(H-\mu N)}}{Z_G}. \tag{236}$$

Taking the derivative of this result with respect to β, holding $\alpha = -\beta\mu$ constant, gives

$$\begin{aligned}
\left(\frac{\partial E}{\partial \beta}\right)_\alpha &= -\frac{1}{Z_G}\text{Tr}_{cl} H^2 e^{-\beta H - \alpha N} + \text{Tr}_{cl} H e^{-\beta H - \alpha N}\frac{\partial}{\partial \beta}Z_G^{-1} \\
&= -\langle H^2\rangle + \langle H\rangle^2 \\
&= -\langle (H-E)^2\rangle.
\end{aligned} \tag{237}$$

Since we know that the average energy E is an extensive variable, we can write for a large system:

$$E(\beta, \alpha, V) = V\epsilon(\beta, \alpha) \tag{238}$$

where ϵ is a function of β and α alone. Using Eq. (238) to evaluate the left-hand side of Eq. (237) leads to the expression for the dispersion

$$\begin{aligned}\Delta_E &= \frac{1}{V\epsilon(\beta,\alpha)}\left[V\left(-\frac{\partial\epsilon(\beta,\alpha)}{\partial\beta}\right)\right]^{1/2}\\ &= \frac{1}{\sqrt{V}}\frac{1}{\epsilon}\left(-\frac{\partial\epsilon}{\partial\beta}\right)^{1/2}.\end{aligned} \tag{239}$$

In the thermodynamic limit where V and $\bar{N}$ go to infinity with fixed density $n(=\bar{N}/V)$, the dispersion in the energy goes to zero as $1/\sqrt{N}$. The distribution of energy states is very sharply peaked about the average value for large $\bar{N}$. Clearly, the same arguments follow for the distribution of the average number of particles. The fact that the dispersion in total energy goes as $1/\sqrt{N}$ in a large system means that the energy H is equal to the average energy up to negligibly small correction terms. In this sense the microcanonical and canonical ensembles are equivalent. In Problem 1.30 it is shown that in the thermodynamic limit in the canonical and grand canonical ensembles,

$$\langle f(H)\rangle = f(E)[1 + \mathcal{O}(N^{-1/2})] \tag{240}$$

where f is an arbitrary function. Since we have trivially in the microcanonical ensemble that

$$\langle f(H)\rangle = f(E) \tag{241}$$

we see that any function of the Hamiltonian, $f(H)$, in the GCE will equal that calculated in the microcanonical ensemble in the thermodynamic limit. It should be clear that similar considerations hold for the number of particles. Although it is true that the various ensembles give the same average values for functions of H and N, they will not give the same results for the averages of fluctuations in these variables. It may not yet be obvious, but averages of fluctuations such as $\langle (H-E)^2 \rangle$ and $\langle (N-\bar{N})^2 \rangle$ are interesting and measurable. In a situation where we are dealing with fluctuations it is clear that it is more natural to use one of the open ensembles.

1.12 CANONICAL ENSEMBLE

We have skipped over the canonical ensemble [40] where the volume V and particle number N are held fixed and the average energy $E = \langle H \rangle$ is held constant. We could, of course, use our information-theoretic approach to find the probability distribution. This is left as an exercise (see Problem 1.40). Here we want to show how a microcanonical treatment of a combined system (see Fig. 1.9) consisting of a system of interest (1) immersed in and weakly interacting with a large system (2) leads to a canonical ensemble probability distribution for system 1. We assume that system 1 has particle number N_1 and volume V_1, while system 2 has N_2 particles in a volume V_2. System 2 is much bigger than 1, so $N_2 \gg N_1$. We assume that the combined system is isolated and can be described by a microcanonical ensemble. The condition that the two systems be weakly interacting requires that the Hamiltonian for the combined system be of the form

$$H_T = H_1 + H_2 \tag{242}$$

where H_2 is the Hamiltonian for system 1 and depends only on the degrees of freedom of system 1 and similarly for H_2. Of course, there must be some interaction between the two systems, but we assume that this occurs at the surface between the two systems. This surface interaction involves $\bar{N}_1^{(d-1)/d}$ atoms in d dimensions and the associated surface energy is down by a factor $\bar{N}_1^{-1/d}$ compared to H_1 and can therefore be neglected for large $\bar{N}_1$.

The probability distribution for the combined microcanonical system is

$$P_T = \frac{1}{Z_M^T} \tag{243}$$

where

$$Z_M^T(E) = \mathrm{Tr}_{\mathrm{cl}}^T \delta(E - H_T) E_T \tag{244}$$

is the microcanonical partition function for the combined system, E is the energy of the total system, and $E_T = (N_1 m_1 + N_2 m_2)c^2$. $\mathrm{Tr}_{\mathrm{cl}}^T$ corresponds to a sum over the phase-space coordinates of both systems and can be written

$$\mathrm{Tr}_{\mathrm{cl}}^T = \mathrm{Tr}_{\mathrm{cl}}^{N_1} \mathrm{Tr}_{\mathrm{cl}}^{N_2} \tag{245}$$

where, for example, $\mathrm{Tr}_{\mathrm{cl}}^{N_1}$ is the sum over the degrees of freedom of system 1:

$$\mathrm{Tr}_{\mathrm{cl}}^{N_1} = \int \frac{dq_{N_1}}{N_1! h^{3N_1}} \cdot \tag{246}$$

The probability distribution for system 1 alone is given by summing P_T over all the phase-space coordinates of system 2 consistent with the δ-function constraint on the total energy:

$$P_1 = \frac{\mathrm{Tr}_{\mathrm{cl}}^{N_2} \delta(E - H_T) E_T}{Z_M^T}. \tag{247}$$

P_1 is properly normalized since

$$\mathrm{Tr}_{\mathrm{cl}}^{N_1} P_1 = \frac{\mathrm{Tr}_{\mathrm{cl}}^{N_1}\mathrm{Tr}_{\mathrm{cl}}^{N_2}\delta(E-H_T)E_T}{Z_M^T}$$

$$= \frac{\mathrm{Tr}_{\mathrm{cl}}^{N}\delta(E-H_T)E_T}{Z_M^T} = \frac{Z_M^T}{Z_M^T} = 1. \qquad (248)$$

Returning to Eq. (247), we note that

$$\mathrm{Tr}_{\mathrm{cl}}^{N_2}\delta(E-H_T)E_T = \mathrm{Tr}_{\mathrm{cl}}^{N_2}\delta(E-H_1-H_2)E_T$$

$$= \frac{Z_M^{(2)}(E-H_1)E_T}{E_T^{(2)}} \qquad (249)$$

where

$$Z_M^{(2)}(E) = \mathrm{Tr}_{\mathrm{cl}}^{N_2}\delta(E-H_2)E_T^{(2)}, \qquad (250)$$

and $E_T^{(2)} = N_2 m_2 c^2$. Similarly, the partition function for the total system, Eq. (244), can be written in the form

$$Z_M^T(E) = \mathrm{Tr}_{\mathrm{cl}}^{N_1}\mathrm{Tr}_{\mathrm{cl}}^{N_2}\delta(E-H_1-H_2)E_T$$

$$= \frac{\mathrm{Tr}_{\mathrm{cl}}^{N_1} Z_M^{(2)}(E-H_1)E_T}{E_T^{(2)}}. \qquad (251)$$

We see that $Z_M^{(2)}(E-H_1)$ is the microcanonical partition function for system 2 at energy $E-H_1$. The probability distribution for the subsystem 1 can then be written

$$P_1 = \frac{Z_M^{(2)}(E-H_1)}{\mathrm{Tr}_{\mathrm{cl}}^{N_1} Z_M^{(2)}(E-H_1)}. \qquad (252)$$

Using the method of stationary phase [51], we can show in the thermodynamic limit (see Appendix H) that

$$Z_M(E) = \frac{e^{\beta E + W_2(\beta)} E_T}{[2\pi \partial^2 W_2(\beta)/\partial \beta^2]^{1/2}}[1 + \mathcal{O}(N_2^{-1/2})] \qquad (253)$$

where

$$W_2(\beta) = \ln \mathrm{Tr}_{\mathrm{cl}}^{(2)} e^{-\beta H_2}. \qquad (254)$$

and the parameter β is defined implicitly by

$$\beta = \frac{1}{k_B}\left(\frac{\partial S_2}{\partial E_2}\right)_{N_2,V_2} \tag{255}$$

where S_2 and β are, respectively, the entropy and inverse temperature of the heat bath corresponding to system 2. The result given by Eq. (253) can also be applied to the quantity $Z_M(E - H_1)$. We have

$$Z_M^{(2)}(E - H_1) = \frac{e^{\beta(E-H_1)+W_2(\beta)}E_T}{[2\pi\partial^2 W_2(\beta)/\partial\beta^2]^{1/2}}[1 + \mathcal{O}(N_2^{-1/2})] \tag{256}$$

where we have used

$$\frac{1}{k_B}\frac{\partial S_2(E - H_1, N_2, V_2)}{\partial(E - H_1)} = \frac{1}{k_B}\frac{\partial S_2(E_2, N_2, V_2)}{\partial E_2}[1 + \mathcal{O}(N_1/N_2)^{1/2}]$$

$$= \beta(1 + \mathcal{O}(N_1/N_2)^{1/2}] \tag{257}$$

which follows from the requirement that $N_2 \gg N_1$. Inserting Eq. (256) into our expression, Eq. (252), for P_1, we obtain

$$P_1 = \frac{e^{\beta(E-H_1)+W_2(\beta)}(2\pi\partial^2 W_2/\partial\beta^2)^{1/2}}{(2\pi\partial^2 W_2/\partial\beta^2)^{1/2}\mathrm{Tr}_{\mathrm{cl}}^{N_1}\exp[\beta(E - H_1) + W_2(\beta)]} \tag{258}$$

which, after canceling several common factors, takes the simple form

$$P_1 = \frac{e^{-\beta H_1}}{\mathrm{Tr}_{\mathrm{cl}}^{N_1}e^{-\beta H_1}}. \tag{259}$$

This is the same probability distribution that we would obtain by applying our information-theoretic approach to a single system. We will use the notation

$$P_c = \frac{e^{-\beta H}}{Z_c} \tag{260}$$

and

$$Z_c = \mathrm{Tr}_{\mathrm{cl}}^N e^{-\beta H} \tag{261}$$

is the canonical ensemble partition function. The main point from the analysis above is that the coordinates of the large reservoir do *not* appear in P_c. The only remnant of the bath is that system 1 is at the same temperature T as the bath.

In the canonical ensemble the fundamental variables are β, N, and V and

$$P_c = P_c(\beta, N, V). \tag{262}$$

In this case the entropy of system 1 can be written

$$\begin{aligned}
S &= -k_B \mathrm{Tr}^N_{\mathrm{cl}} P_c \ln P_c \\
&= -k_B \mathrm{Tr}^N_{\mathrm{cl}} P_c \ln \frac{e^{-\beta H}}{Z_c} \\
&= -k_B \mathrm{Tr}^N_{\mathrm{cl}} P_c (-\beta H - \ln Z_c) \\
S &= \frac{E}{T} + k_B \ln Z_c.
\end{aligned} \tag{263}$$

In analogy with the GCE case, it is convenient to introduce a *potential*

$$\beta F \equiv -\ln Z_c. \tag{264}$$

In this case F is the Helmholtz free energy [52] and, using Eqs. (263) and (264), it is related to the energy and entropy by

$$F = E - TS. \tag{265}$$

We note that just as $\beta\Omega$ generated the dependent variables $E, \bar{N}$, and p by differentiation in the GCE, we have

$$E = \langle H \rangle = \frac{\partial}{\partial \beta}(\beta F) \tag{266}$$

in the canonical ensemble. One can also show (see Problem 1.52) that

$$p = -\left(\frac{\partial F}{\partial V}\right)_{T,N} \tag{267}$$

$$\mu = \left(\frac{\partial F}{\partial N}\right)_{T,V}.$$

Thus in the canonical ensemble the independent variables are T, N, and V and the dependent variables E, μ, and p are derived as derivatives of the Helmholtz free energy F.

1.13 SYMMETRY AND EQUILIBRIUM ENSEMBLES

The ideas of symmetry and invariance principles were introduced earlier when we discussed conservation laws. Here we want to investigate the nature of these

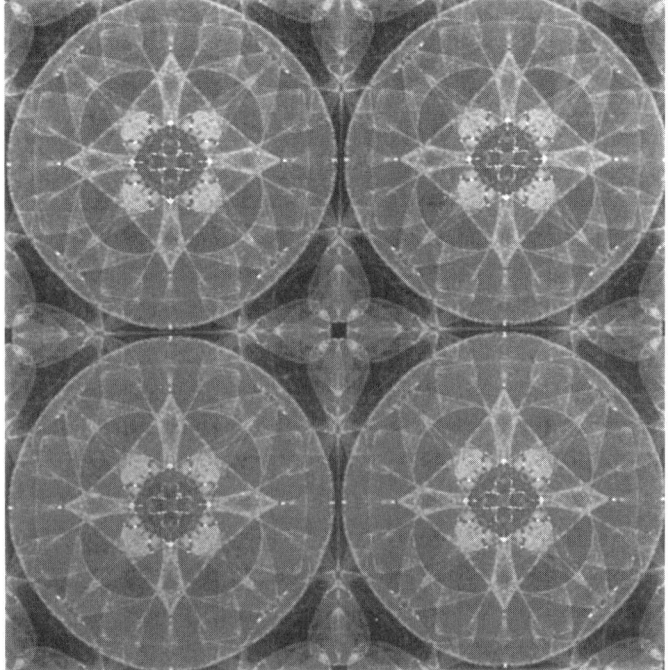

FIGURE 1.12 Symmetrical figure. (From I. Stewart and M. Golubitsky, *From Fearful Symmetry*, Penguin, New York, 1993.)

symmetry principles in more detail. The ideas we develop will then have immediate application when we study mechanical force and pressure in statistical mechanical systems. Let us begin with a more elementary discussion of symmetry and its consequences in statistical mechanics.

Consider the pattern shown in Fig. 1.12. The most obvious aspect of this picture is that it is pretty. One element contributing to the appealing nature of the figure is its symmetry. We know in general that symmetry is very important in developing a sense of aesthetics [53]. Symmetry is also crucially important in statistical mechanics in particular and physics in general [54]. To understand why the concept of symmetry is so central in physics, we must find a way to measure it. This appears difficult at first sight since symmetry seems such a qualitative notion. The first step in quantifying our idea of symmetry is to establish which, in comparing two situations, has *more* symmetry.

It should be clear that in a broad sense symmetry is associated with the notion of patterns. Among other things we notice that Fig. 1.12 has as an element a repeated figure. If we think more abstractly and replace each figure with a point, we obtain, repeating the structure, a periodic square lattice. It is natural to think of a periodic lattice structure as highly symmetric. We can, however, look at this situation from a different perspective. As we discuss in more detail below, symmetry is associated with the equivalence of points in space. Configurations that distinguish between different points have a lower symmetry. Thus a physical system with an imposed lattice, where there are *special* positions associated with the lattice sites, is less symmetric than one with no imposed lattice and where all points are equivalent.

The symmetry of a system is governed by the number of transformations on that system that leaves the system unchanged. A very simple example is the bilateral symmetry of human beings. In this case, if one exchanges points equidistant from a symmetry plane the human shape is left unchanged. This *inversion transformation* is an example of a transformation in space. Other examples are rotations and translations. Thus the objects shown in Fig. 1.13 are invariant under rotations of integer multiples of 60°. The periodic structure in Fig. 1.14 is invariant under translations by an integer multiple of the lattice spacing along one of two axis. The symmetry of an object in space is given in a quantitative sense as the collection (group) of all translations, rotations, and inversions that leave the object unchanged. It is in this sense that homogeneous space, which is invariant under all such transformations, is the most symmetric system of all.

Suppose that we put a diatomic molecule on each lattice site as shown in Fig. 1.15. In this case the symmetry of the situation is a bit more complicated. Because of the orientation of the molecules, the lattice sites are no longer equivalent. The introduction of the molecular axis seems to have introduced a high degree of asymmetry into the problem. The position of each atom picks out a special point in space. Instead of this static snapshot, suppose that we take a dynamic point of view and allow the molecules to rotate in time. If we think of a photograph exposed over a time long compared to the average rotation time of a molecule, we will obtain the picture shown in Fig. 1.16. Here we seem to have restored the symmetry within each cell. Thus it appears that the average behavior of a collection of particles will have higher symmetry than any particular configuration of that system. Implicit in this statement is the very important idea that the symmetry of a system, after averaging, depends on the thermodynamic state of the system.

Let us investigate this idea more carefully. Consider the role of temperature in determining the symmetry of the set of dumbbells [55]. Starting with the system, corresponding to Fig. 1.16, suppose that we lower the temperature to a value below a phase transition where the dipoles appear to be aligned in a particular direction and the system, on average, looks like Fig. 1.17. This system is of lower symmetry than in Fig. 1.16 because now the rotational symmetry at each site is broken. Although each site is equivalent, there is now a special direction in the problem. Now suppose that we raise the temperature of the system beginning with the state shown in Fig. 1.17. We eventually regain the phase where the dipoles will rotate freely and the rotational symmetry will be restored as in Fig. 1.16. If we continue to raise the temperature, we approach the melting temperature. The molecules will then have enough kinetic energy to wonder away from their lattice sites and the solid melts. What is the symmetry of this situation? Is it of higher or lower symmetry than the solid? Again the answer to this question depends on whether we look at a typical microscopic configuration of the system or the average over a sequence of configurations. If we take a static point of view and look at a solid and a fluid at a particular time, it is clear that the solid has higher symmetry. However, if we take the

FIGURE 1.13 Symmetrical figure. (From I. Stewart and M. Golubitsky, *From Fearful Symmetry*, Penguin, New York, 1993.)

62   GENERAL PRINCIPLES OF STATISTICAL MECHANICS

FIGURE 1.14 Symmetrical figure. (From I. Stewart and M. Golubitsky, *From Fearful Symmetry*, Penguin, New York, 1993.)

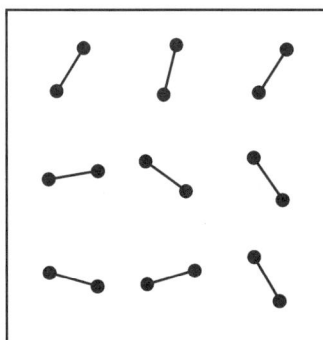

FIGURE 1.15 Schematic snapshot of a set of diatomic molecules on a lattice at high temperatures.

dynamic point of view and ask about the average properties, we obtain a quite different result. If we observe the average number of particles n_v in a box of volume $v = L^3$ centered at position **R**, we see something quite different for the solid and the fluid. In the solid the average number of particles in a box is strongly tied to the position of the box relative to the lattice sites. n_v is a periodic function of **R**. In the fluid, however, over time, the particles will roam (see Fig. 1.18) throughout the available volume. No particular position is special. In this case the average density n

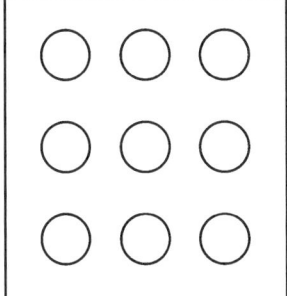

FIGURE 1.16 Time average of Fig. 1.15 at high temperatures.

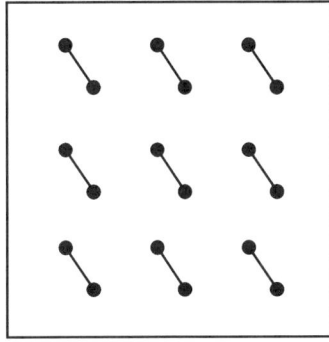

FIGURE 1.17 Broken-symmetry low-temperature state for diatomic molecules.

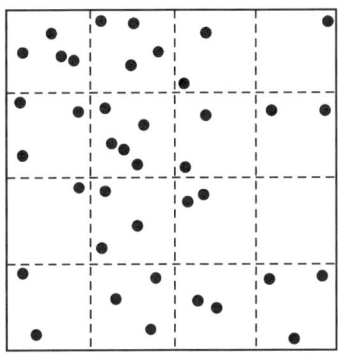

FIGURE 1.18 Spatial symmetry of a fluid. One observes a varying number of particles in each cell as a function of time.

is independent of **R**. We see then that from the point of view of average properties, the fluid system is extremely symmetric—all positions are equivalent! In the solid, cells associated with each lattice site are equivalent, but points within a cell are inequivalent. Thus the solid is less symmetric [56] than the fluid.

The fluid phase is the most symmetric state of matter. All positions and directions are, on average, equivalent. This leads us to a general principle: As one cools a system, it will tend toward states with lower symmetry. An excellent example of this principle is the evolution of the early universe [57]. At the earliest of times after the big bang, the force laws of nature were highly symmetric: Gravity, the

64 GENERAL PRINCIPLES OF STATISTICAL MECHANICS

electromagnetic force, and the weak and strong nuclear forces were all equivalent. As the universe expanded and cooled, it developed regions of broken symmetry where the forces are inequivalent. Indeed, this symmetry breaking took place several times as the various forces differentiated themselves.

There is a clear connection between the tendency of a system to be, on average, uniform in space and the construction of the equilibrium canonical ensembles. At high enough temperatures, there is sufficient energy for particles to move throughout the available space and establish a state that is stationary in time and uniform in space. These invariances are what is needed to establish the relevance of local conservation of energy and total momentum in understanding open systems and establishing the structure of the canonical ensembles.

The degree of symmetry of a system is tested by checking its invariance under an appropriate operation. The simplest operations are those that test whether various positions and directions are equivalent. If we translate the system in space through some distance l, does the system look the same? If we rotate the system by some angle θ, does the system look the same? Is the system invariant under these operations? It is this collection of invariances that we use to quantify the symmetry of a system. The equivalence of positions in space is tested by translational invariance. The equivalence of directions in space is tested by rotational invariance.

The machinery for applying these transformations which test for symmetry is developed in Appendix B. Here we summarize the results. Each transformation can be associated with an observable G, which generates the transformations [58], and an evolution operator $U_G(s)$, which carries out the transformation on some observable A:

$$A'(s) = U_G(s)A \qquad (269)$$

where s is a parameter associated with the transformation and A' is the transformed variable, which depends on s. The simplest example corresponds to translation in space, say along the z-direction. Such translations are generated by the total momentum P_z, and in the classical case the evolution operator is given by

$$U_{P_T^z} = \exp\left(z \sum_{i=1}^{N} \frac{\partial}{\partial r_i^z}\right) \qquad (270)$$

where $\mathbf{r}_i$ is the position of the ith particle. Then, for any reasonably well behaved function of the particles positions, translation along the z-direction a distance z is given explicitly by

$$f(\mathbf{r}_i + z\hat{z}) = \exp\left(z \sum_{i=1}^{N} \frac{\partial}{\partial r_i^z}\right) f(\mathbf{r}_i). \qquad (271)$$

We see that this is simply an elaborate way of writing a Taylor series expansion in powers of z. Similar developments exist for translations in time, where the

Hamiltonian H is the generator of the translation, $U_H(t)$, and rotations in space where the total angular momentum L_T is the generator of the transformation, $U_{L_T}(\theta)$.

The important point here is to look at the equilibrium average of the transformed observable

$$\langle A'(s) \rangle = \langle U_G(s) A \rangle. \tag{272}$$

We can write the right-hand side in the form

$$\langle U_G(s) A \rangle = \operatorname{Tr} P_E U_G(s) A \tag{273}$$

where P_E is a general equilibrium probability distribution. A key point is that one can, through a series of integration by parts [59], arrange for the *transpose* of U_G to act on the probability distribution:

$$\operatorname{Tr} P_E U_G(s) A = \operatorname{Tr} A U_G(-s) P_E. \tag{274}$$

The equilibrium probability distribution is invariant under the transformation if

$$U_G(-s) P_E = P_E. \tag{275}$$

If this is true, one has the *invariance principle*

$$\langle A'(s) \rangle = \langle A \rangle. \tag{276}$$

This is our main result in this section.

How do we know if the distribution is invariant and Eq. (275) holds? As is discussed in some detail in Appendix B, in a classical system the distribution is invariant if the generator of the symmetry G has zero Poisson bracket with the probability distribution:

$$\{G, P_E\} = 0. \tag{277}$$

For general variables A and B of the phase-space coordinates q_N, the Poisson bracket is defined for a system of classical particles by

$$\{A, B\} = \sum_{i,\alpha} \left(\frac{\partial A}{\partial r_i^\alpha} \frac{\partial B}{\partial p_i^\alpha} - \frac{\partial A}{\partial p_i^\alpha} \frac{\partial B}{\partial r_i^\alpha} \right) \tag{278}$$

where i labels the particles and α is a vector component label. In the case of the classical grand canonical ensemble,

$$P_E = P_G = \frac{1}{Z_G} e^{-\beta(H - \mu N)}, \tag{279}$$

and we see that the condition of invariance amounts to having a zero Poisson bracket [60] between the generator of the symmetry and the Hamiltonian

$$\{G, H\} = 0. \tag{280}$$

For the simplest case of point particles with central two-body forces, the generators G, which have zero Poisson bracket with the Hamiltonian, are the Hamiltonian itself $(G \to H)$, which implies time-translational invariance $(G \to \mathbf{P}_T =$ the total momentum), which implies translational invariance in space, and $(G \to \mathbf{L}_T =$ total angular momentum), which implies rotational invariance in space [22].

Let us now look at the practical consequences of these symmetry principles given by Eq. (276). Consider first-time translational invariance where the generator is the Hamiltonian and the parameter s the amount translated in time. Suppose that the observable A depends on time only through the phase-space coordinates, $A = A(q_N(t))$. Then the symmetry principle given by Eq. (276) takes the form

$$\langle A(q_N(t+s)) \rangle = \langle A(q_N(t)) \rangle. \tag{281}$$

Clearly, since the left-hand side depends on the parameter s and the right-hand side does not, $\langle A(q_N(t)) \rangle$ must be independent of time:

$$\frac{d}{dt} \langle A(q_N(t)) \rangle = 0. \tag{282}$$

An interesting set of quantities that give information about the internal dynamics of a many-body system in equilibrium are time correlation functions such as

$$C_{AB}(t, t') = \langle A(t) B(t') \rangle \tag{283}$$

where A and B are observables, for example the number density, which depend on time only through the phase-space coordinates. We can apply our symmetry principle to this quantity and obtain

$$\langle A(t+s) B(t'+s) \rangle = \langle A(t) B(t') \rangle. \tag{284}$$

Since s is arbitrary, we can choose $s = -t'$, with the result

$$C_{AB}(t, t') = \langle A(t-t') B(0) \rangle = C_{AB}(t-t'), \tag{285}$$

and the average depends only on the time difference.

Turn next to translational invariance in space. In this case it is interesting to consider densities of the form

$$\rho(\mathbf{x}) = \sum_{i=1}^{N} \rho_i \delta(\mathbf{x} - \mathbf{r}_i) \tag{286}$$

where ρ_i could be the charge or mass of particle i. We assume that ρ_i does not depend on $\mathbf{r}_i$. In this case the translation depends on a vector $\mathbf{s}$ and the invariance principle reads

$$\left\langle \sum_{i=1}^{N} \rho_i \delta(\mathbf{x} - \mathbf{r}_i - \mathbf{s}) \right\rangle = \langle \rho(\mathbf{x}) \rangle \quad (287)$$

or

$$\langle \rho(\mathbf{x} - \mathbf{s}) \rangle = \langle \rho(\mathbf{x}) \rangle. \quad (288)$$

Again, since $\mathbf{s}$ is arbitrary, $\langle \rho(\mathbf{x}) \rangle$ must be independent of $\mathbf{x}$!

If we have correlation functions in space, then, following the same analysis as for the case of time-translational invariance, we obtain

$$C_{AB}(\mathbf{x}, \mathbf{x}') = \langle A(\mathbf{x}) B(\mathbf{x}') \rangle = \langle A(\mathbf{x} - \mathbf{x}') B(0) \rangle = C_{AB}(\mathbf{x} - \mathbf{x}'). \quad (289)$$

Notice that if we preserve the full translational invariance, we cannot describe a solid. In a solid the particle densities are periodic in space (as discussed in detail in Chapter 6)

$$\langle n(\mathbf{x}) \rangle = \langle n(\mathbf{x} + \mathbf{R}) \rangle \quad (290)$$

where $\mathbf{R}$ is a lattice vector and $\langle n(\mathbf{x}) \rangle$ varies with position within a cell. This has the significance that the grand canonical ensemble (and all other ensembles that preserve translational invariance) cannot be used to describe solids. Instead, one must use a restricted ensemble. This idea of a restricted ensemble indicates that there are, in these broken symmetry states, additional macrovariables which must be specified. More about this below.

Turning to the case of rotational invariance we obtain the symmetry principle

$$\langle A \rangle = \langle A'(\boldsymbol{\theta}) \rangle \quad (291)$$

where we rotate all phase-space coordinates, which are vectors, through an angle θ about an axis in the $\hat{\theta}$-direction. This result has several consequences, depending on the nature of the observable A. Suppose that A is a vector, A_α. A typical example is the total momentum. Then, by definition,

$$A'_\alpha(\boldsymbol{\theta}) = \sum_\beta M_{\alpha\beta}(\boldsymbol{\theta}) A_\beta \quad (292)$$

where the $M_{\alpha\beta}(\boldsymbol{\theta})$ are *rotation matrices* [61]. The invariance principle, Eq. (276), gives in this case

$$\langle A_\alpha \rangle = \sum_\beta M_{\alpha\beta}(\boldsymbol{\theta}) \langle A_\beta \rangle \quad (293)$$

for general θ. It is clear that the general solution to this equation is

$$\langle A_\alpha \rangle = 0. \tag{294}$$

The equilibrium average of a vector is zero for systems with rotational invariance. It is left as an exercise (see Problem 1.48) to discuss the case of averages of products of vectors.

Let us turn to the cases where at lower temperatures we have broken symmetries. The most obvious case of broken translational symmetry associated with going from a fluid to a solid is somewhat involved. Instead, we can consider the simpler case of the breaking of rotational invariance in going from a paramagnet to a ferromagnet. Let us consider the case where the Hamiltonian, H, in the absence of any applied external magnetic fields, is rotationally invariant and has zero Poisson bracket with the total magnetization $\mathbf{M}_T$,

$$\{H, \mathbf{M}_T\} = 0. \tag{295}$$

Assuming that $\mathbf{M}_T$ is proportional to the total angular momentum in the system, this result tells us that the Hamiltonian is rotationally invariant. If we have a canonical ensemble, where the probability distribution, P_E, depends only on the Hamiltonian and shares the same symmetries as the Hamiltonian, we find

$$\{P_E, \mathbf{M}_T\} = 0, \tag{296}$$

and the probability distribution is rotationally invariant. Using our invariance principle, leading to Eq. (294), we have immediately that the average total magnetization must be zero:

$$\langle \mathbf{M}_T \rangle = 0 \tag{297}$$

in this ensemble. This is not the physical situation in a ferromagnet where, essentially by definition,

$$\langle \mathbf{M}_T \rangle = \mathbf{M}_0 \neq 0. \tag{298}$$

In this case a physical choice for the appropriate restricted ensemble is to assume that the probability distribution depends on a new Hamiltonian,

$$H_T = H - \mathbf{B} \cdot \mathbf{M}_T \tag{299}$$

where we have included a Zeeman term, proportional to the externally applied magnetic field $\mathbf{B}$, which breaks the rotational invariance in the system. The physically relevant analysis in the ferromagnetic phase is to do the calculation in the presence of a small field $\mathbf{B}$, take the thermodynamic limit to obtain $\mathbf{M}_0$, which will be aligned along $\mathbf{B}$. For sufficiently small $\mathbf{B}$ one finds that M_0^2 is independent of B^2. If one starts with **B** = 0, one always finds that $M_0 = 0$. Thus the small **B** and thermodynamic limits do not commute in the ferromagnetic phase. This is discussed in detail in Chapter 5.

Summarizing our results: The symmetry of a physical system can be quantified in terms of the set of invariance conditions satisfied by the system. The average behavior of a collection of particles is more symmetric than a particular generic configuration of this system. The symmetry associated with the average behavior of a large system depends on the thermodynamic state of the system. The classic example is that of a simple ferromagnet. At sufficiently high temperatures the system has no net magnetization, is paramagnetic, and has full rotational symmetry. At lower temperatures, where the system becomes a ferromagnet, the system develops a net magnetization along a particular direction, and the system is not rotationally invariant.

1.14 MECHANICAL FORCES: SOLIDS AND LIQUIDS

1.14.1 Shear Forces

It is clear that there must be a close relationship between thermal and mechanical equilibrium. Intuitively we understand that in equilibrium the pressure of a gas must be balanced by the force exerted at the walls of a container. However, the general question of the application of mechanical force to a macroscopic object is more subtle than one might, at first, imagine. The subtle points are tied up with the requirements of equilibrium. This can be appreciated by considering the following simple experiment. Suppose that we have the sample of fluid shown in Fig. 1.19a. We then take a piston, as shown in Fig. 1.19b, and very slowly apply it to the fluid as shown in Fig. 1.19c. Of course, this does not appear to be a very dramatic experiment since the piston simply sinks into the fluid. The final state of the fluid is unchanged by the motion of the piston if the piston moves very slowly. Only the shape of the container is changed. Consider next a related experiment (see Fig. 1.20), where we replace the fluid with a solid. In this case the piston has a very different effect. After making contact with the face of the solid, it will, for pressures that are not too large, act to compress and deform the entire solid. Thus the state of the solid will be altered since the average distance between atoms along the z-axis will be changed by an amount proportional to the distance Δz over which the solid is compressed. You should consider how you would go about calculating, given the magnitude of the applied force, the size and shape of the compression of the solid. This is addressed in Chapter 6.

We can draw a significant conclusion from these simple experiments: The response of a macroscopic object to an applied macroscopic force depends on the thermodynamic state of the object. If we are to quantify this statement, we need to identify what is fundamentally different between the fluid and the solid. Focus on the region in Fig. 1.19c near the edge of the piston as shown in Fig. 1.21a. The fluid directly beneath the piston is free to flow in the *xy*-plane. Thus the increase in

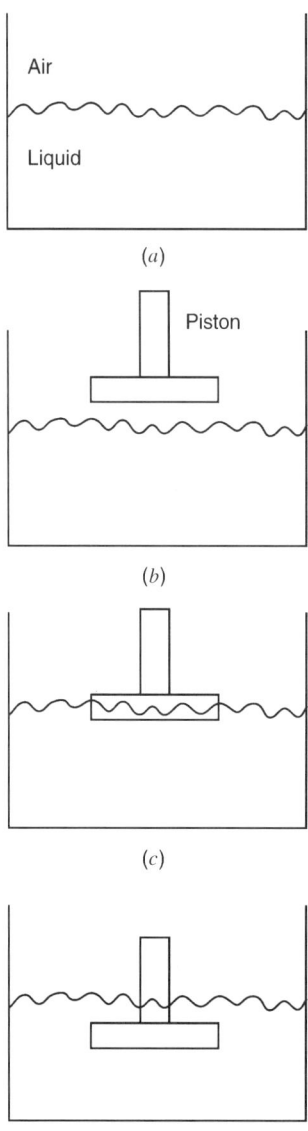

FIGURE 1.19 Schematic of a piston moving into a fluid.

pressure directly below the piston can be relieved by flow of fluid out of that region, which in turn forces fluid up around the side of the piston. The piston is unable to generate an opposing force outside the region below the piston. In the case of a solid, the situation is different (see Fig. 1.21b). The force applied directly below the piston is transmitted along the solid to the point displaced a distance Δx along the x-axis and the force F'_z acts to compress the solid even though it is not beneath the piston. A localized force will cause a large-scale *deformation* in the solid. This leads us to the

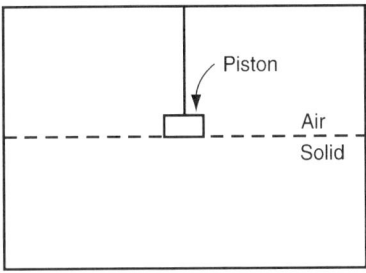
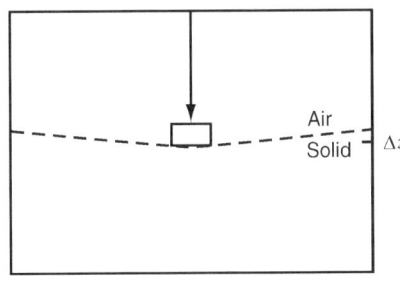

FIGURE 1.20 Schematic of a piston moving into a solid. This situation is investigated quantitatively in Chapter 6.

concept of shear forces. When we apply a force F_z along the z-direction we elicit a response force in the solid, which is a function of x. We have a finite shear if $dF_z(x)/dx$ is nonzero. The fundamental difference [62] between solids and fluids is that solids can sustain (i.e., that is transmit into the bulk) shears applied at a surface, while fluids cannot transmit static shears applied at a surface.

We can get some feeling for the concept of shear through the following example. Apply opposing forces along the boundaries of a sample as shown in Fig. 1.22a. These forces have no influence at all on the equilibrium state of a fluid. In a solid one obtains the sheared configuration shown in Fig. 1.22b. As we shall see, the fundamental ramification of this result is that the thermodynamic state of a solid depends on the vector nature of the forces applied to it.

Similarly, if we apply a force in the xy-plane at $z = 0$ which generates a torque about the z-axis, do we generate a torque in the system for values of $z > 0$? Again this is true for solids and liquid crystals, but not for ordinary fluids. As discussed in Volume 2, liquid crystals are unusual in that they can communicate torques applied at a surface long distances into the sample; they cannot, however, transmit applied shears since they are able to flow to *relieve* the shear.

The question then arises of how to incorporate these physical ideas into a statistical mechanical treatment. We shall proceed with an approach that allows us to treat both fluids and solids at the same time and allows us to see clearly the difference between the two.

1.14.2 Distortions

Let us assume that we apply a force to a macroscopic system at its surface. We can think of this in terms of movement (compression, rotation, sliding, etc.) of the walls

72 GENERAL PRINCIPLES OF STATISTICAL MECHANICS

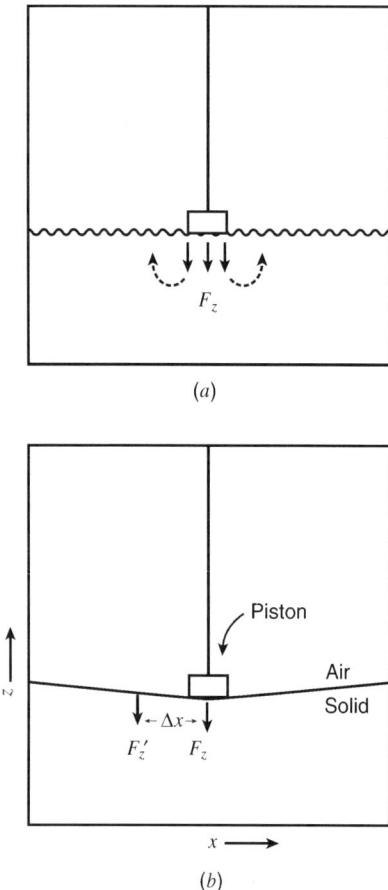

FIGURE 1.21 (*a*) Shear flow as a piston moves into a fluid; (*b*) sustained shear as piston moves into a solid.

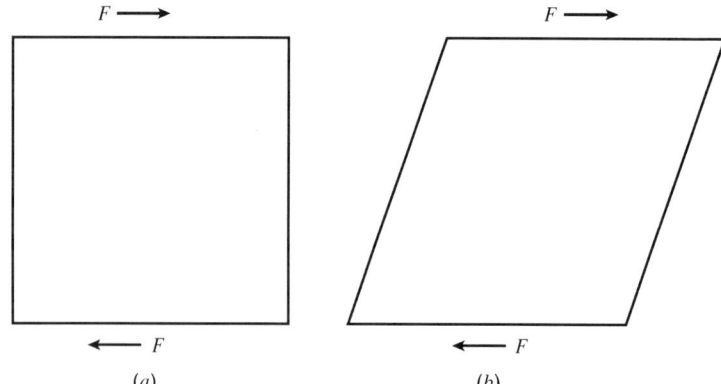

FIGURE 1.22 (*a*) Application of a shearing force to a system; (*b*) result of the application of a symmetric shearing force to a solid.

surrounding the system. We assume that the system was in thermal equilibrium before shifting the walls and returns to thermal equilibrium after our mechanical disturbance of the surface. Are these *distortions* of the walls communicated to the bulk of the system? If this distortion is communicated to the bulk of the system, the particle positions [63] must reflect this distortion. Thus a distortion at the walls will be accompanied in the bulk by a change in the phase-space coordinates of the form

$$r_i^\alpha \to r_i^{\prime\alpha} = r_i^\alpha + \sum_\beta U_{\alpha\beta} r_i^\beta \qquad (300)$$

$$p_i^\alpha \to p_i^{\prime\alpha} = p_i^\alpha - \sum_\beta U_{\beta\alpha} p_i^\beta \qquad (301)$$

where the matrix $U_{\alpha\beta}$ is known [64] as the *strain tensor*. In general, the strain tensor may depend on the position in the sample. Here we assume that the strain tensor is a constant in space. Clearly, in a fluid, such a strain at the surface will not be carried into the bulk. The situation is quite different for a solid. If **R** corresponds to a lattice point in the undistorted solid (see Fig. 1.23), the strain at the surface leads to a distorted set of lattice points:

$$R_\alpha' = R_\alpha + \sum_\beta U_{\alpha\beta} R_\beta. \qquad (302)$$

To gain some feeling for the nature of the strain tensor, consider some special cases. Suppose that the strain tensor has only one nonzero diagonal element:

$$U_{\alpha\beta} = \delta_{\alpha\beta} \delta_{\alpha,z} U_z. \qquad (303)$$

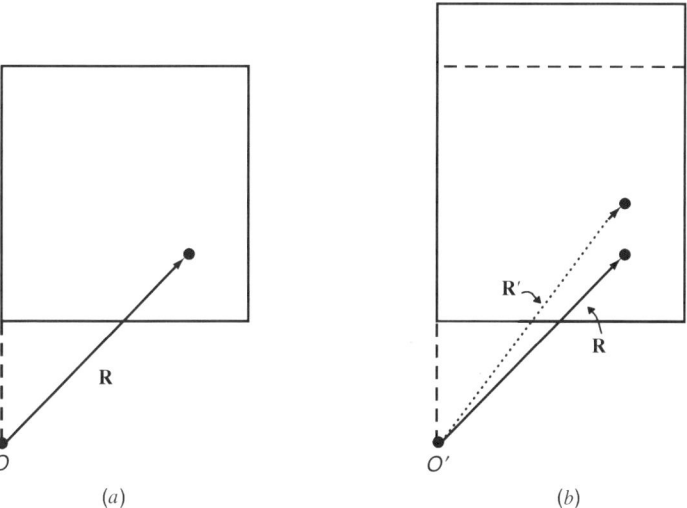

FIGURE 1.23 (*a*) Points on an undistorted solid are denoted by **R**; (*b*) equivalent points on the distorted solid are denoted by **R'**.

Then Eq. (302) reduces to

$$R'_x = R_x \tag{304}$$
$$R'_y = R_y \tag{305}$$
$$R'_z = (1 + U_z)R_z. \tag{306}$$

The distortion in this case corresponds to a uniaxial compression ($U_z < 0$) or elongation ($U_z > 0$) along the z-axis (see Fig. 1.24). In these cases one has that the length of the sample along the z-direction has changed from L to

$$L' = (1 + U_z)L. \tag{307}$$

Thus U_z gives the compression or expansion per unit length in the system

$$U_z = \frac{L' - L}{L} = \frac{\Delta z}{L}. \tag{308}$$

Consider next the case where we have a diagonal strain tensor of the form

$$U_{\alpha\beta} = \delta_{\alpha\beta}U_0. \tag{309}$$

Then Eq. (302) becomes

$$R'_\alpha = R_\alpha(1 + U_0) \tag{310}$$

and each direction is compressed or expanded by the same amount. This is called a *hydrostatic compression* or *elongation* and corresponds to a volume change with no shape change.

More generally, the change in an infinitesimal volume element due to a distortion is given by

$$d^3r' = J d^3r \tag{311}$$

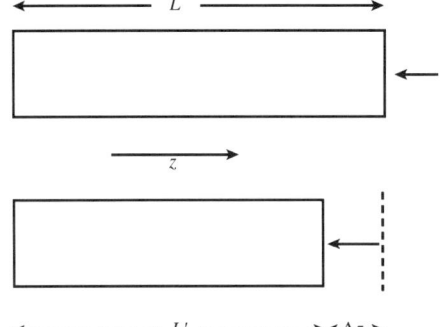

FIGURE 1.24 Schematic of a uniaxial compression of a rod.

where the Jacobian relating the two-coordinate systems is given by

$$J = \det\left(\frac{\partial r'_\alpha}{\partial r_\beta}\right) \tag{312}$$

$$= \det(\delta_{\alpha\beta} + U_{\alpha\beta}). \tag{313}$$

It is easy to show for small values of the strain tensor that

$$J = 1 + \sum_\alpha U_{\alpha\alpha} \equiv 1 + \operatorname{Tr} U \tag{314}$$

plus higher-order terms in the strain tensor. If we then integrate over space we have that the new system volume V' is related to the old undistorted volume V by

$$V' = (1 + \operatorname{Tr} U)V. \tag{315}$$

The change in the volume due to a distortion is given by

$$\delta V = V' - V = (\operatorname{Tr} U)V. \tag{316}$$

Next consider an off-diagonal component of the strain tensor,

$$U_{\alpha\beta} = \epsilon \delta_{\alpha x} \delta_{\beta y}, \tag{317}$$

so that Eq. (302) reads

$$R'_x = R_x + \epsilon R_y \tag{318}$$
$$R'_y = R_y \tag{319}$$
$$R'_z = R_z. \tag{320}$$

This change in coordinates corresponds to the situation shown in Fig. 1.25a, where shearing force is applied along a face and the solid deforms into the rhombus shown in Fig. 1.25b. In this case the y-component is unchanged, while from the geometry, the x-component is given by

$$R'_x = R_x + R_y \frac{\Delta x}{L} \tag{321}$$

and we identify the parameter ϵ in Eq. (317) as

$$\epsilon = \frac{\Delta x}{L} = \tan\theta. \tag{322}$$

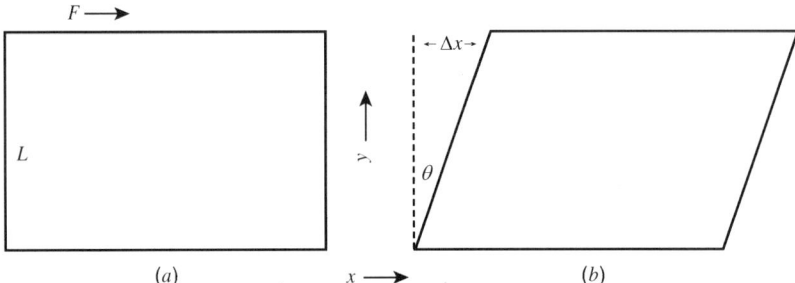

FIGURE 1.25 Schematic of an unsymmetrical shear in a solid.

A key question for us is how the Hamiltonian changes when we apply a force that causes a small distortion in a system. We can conveniently address this question using transformation theory. A distortion can be viewed as a transformation on our system of phase-space coordinates with a generator

$$S = \sum_{i=1}^{N} \sum_{\alpha\beta} p_i^\alpha U_{\alpha\beta} r_i^\beta. \tag{323}$$

The transformed phase-space coordinates are given by the Poisson bracket relations

$$\delta r_i^\alpha = \{r_i^\alpha, S\} = \sum_\beta U_{\alpha\beta} r_i^\beta \tag{324}$$

$$\delta p_i^\alpha = \{p_i^\alpha, S\} = -\sum_\beta U_{\beta\alpha} p_i^\beta. \tag{325}$$

Note that this set of equations is equivalent to the set given by Eqs. (300) and (301). This means that the change in any function, A, of the phase-space coordinates is given, using the chain rule for differentiation, by

$$\begin{aligned} \delta A &= \sum_{i=1}^{N} \sum_\alpha \left(\frac{\partial A}{\partial r_i^\alpha} \delta r_i^\alpha + \frac{\partial A}{\partial p_i^\alpha} \delta p_i^\alpha \right) \\ &= \sum_{i=1}^{N} \sum_\alpha \left(\frac{\partial A}{\partial r_i^\alpha} \{r_i^\alpha, S\} + \frac{\partial A}{\partial p_i^\alpha} \{p_i^\alpha, S\} \right) \\ &= \sum_{i=1}^{N} \sum_\alpha \left(\frac{\partial A}{\partial r_i^\alpha} \frac{\partial S}{\partial p_i^\alpha} - \frac{\partial A}{\partial p_i^\alpha} \frac{\partial S}{\partial r_i^\alpha} \right) \\ &= \{A, S\}. \end{aligned} \tag{326}$$

MECHANICAL FORCES: SOLIDS AND LIQUIDS 77

Let us apply this result to the case where the function A is the Hamiltonian in the system. The change in the Hamiltonian due to a distortion is given by

$$\delta H = \{H, S\}. \tag{327}$$

Note, however, that we can write the generator given by Eq. (323) in terms of the momentum density defined by Eq. (129):

$$S = \sum_{\alpha\beta} \int d^3x\, g_\alpha(\mathbf{x}) U_{\alpha\beta} x_\beta. \tag{328}$$

The change in the Hamiltonian due to the distortion is then given by

$$\delta H = \sum_{\alpha\beta} \int d^3x\, x_\beta U_{\alpha\beta} \{H, g_\alpha(\mathbf{x})\}. \tag{329}$$

We can go further if we remember the continuity equation for the momentum density, Eq. (135), can be written in the form

$$\frac{\partial g_\alpha(\mathbf{x})}{\partial t} = \{g_\alpha(\mathbf{x}), H\} = -\sum_\beta \nabla_x^\beta \sigma_{\alpha\beta}(\mathbf{x}) \tag{330}$$

where $\sigma_{\alpha\beta}(\mathbf{x})$ is the stress tensor. The change in the Hamiltonian due to the application of a distortion is then given by

$$\begin{aligned}
\delta H &= \sum_{\alpha\beta} \int d^3x\, x_\beta U_{\alpha\beta} \sum_\gamma \nabla_x^\gamma \sigma_{\alpha\gamma}(\mathbf{x}) \\
&= -\sum_{\alpha\beta\gamma} \int d^3x\, x_\beta U_{\alpha\beta} \sigma_{\alpha\gamma}(\mathbf{x}) \nabla_x^\gamma x_\beta \\
&= -\int d^3x \sum_{\alpha,\beta} U_{\alpha\beta} \sigma_{\alpha\beta}(\mathbf{x})
\end{aligned} \tag{331}$$

where in the second line we have carried out an integration by parts and assumed that the stress tensor vanishes outside the sample. The change in the Hamiltonian is proportional to an integral over the stress tensor times the strain tensor. This is the change of the energy of the system due to the motion of the walls. The stress tensor is constructed such that $\sum_\beta \sigma_{\alpha\beta}\, dA_\beta$ is the αth component of the force on the surface $d\mathbf{A}$. Conceptually, we think of some imposed stresses, as reflected by the stress tensor, causing a distortion in the system reflected in the strain tensor. For sufficiently weak stresses, the stress and strain tensors are proportional and related by a generalized form of Hooke's law. We return to this point in detail in Chapter 6.

78 GENERAL PRINCIPLES OF STATISTICAL MECHANICS

It turns out that one can always construct the stress tensor to be symmetric ($\sigma_{\alpha\beta} = \sigma_{\beta\alpha}$). This means, looking at Eq. (331) and using the symmetry of $\sigma_{\alpha\beta}$, that it is only the symmetric part of the strain tensor,

$$\tilde{U}_{\alpha\beta} = \tfrac{1}{2}(U_{\alpha\beta} + U_{\beta\alpha}) \tag{332}$$

which enters into the change in the Hamiltonian when applying a distortion. This leads to the result that rotations, which are associated with antisymmetric contributions to $U_{\alpha\beta}$, do not contribute to $\tilde{U}_{\alpha\beta}$. $\tilde{U}_{\alpha\beta}$ is a superposition of compressions and shears. Henceforth, unless stated otherwise, we assume that the strain tensor, $U_{\alpha\beta}$, is symmetric.

The symmetrized version of the strain tensor for the shearing distortion given by Eq. (317) goes over to

$$U_{\alpha\beta} = \epsilon'(\delta_{\alpha x}\delta_{\beta y} + \delta_{\alpha y}\delta_{\beta x}), \tag{333}$$

and implies the transformation

$$R'_x = R_x + \epsilon' R_y \tag{334}$$
$$R'_y = R_y + \epsilon' R_x. \tag{335}$$

If we take the equilibrium average of Eq. (331), we obtain the change in the average energy,

$$\delta E = -\sum_{\alpha,\beta} V_0 U_{\alpha\beta} \langle \bar{\sigma}_{\alpha\beta} \rangle \tag{336}$$

where

$$\langle \bar{\sigma}_{\alpha\beta} \rangle = \frac{1}{V_0} \int d^3x \langle \sigma_{\alpha\beta}(\mathbf{x}) \rangle \tag{337}$$

is the average stress tensor and V_0 the undistorted volume. Since we can make $U_{\alpha\beta}$ arbitrarily small, we can rewrite Eq. (336) as a differential equation:

$$\frac{\partial E}{\partial (V_0 U_{\alpha\beta})} = -\langle \bar{\sigma}_{\alpha\beta} \rangle. \tag{338}$$

This relation can be treated as the fundamental thermodynamic definition [65] of the average stress tensor. It also says that the energy depends on the strain tensor. We shall return to this important result below.

1.14.3 External Forces

It is instructive here to look at this problem from a slightly expanded point of view. Suppose that we look at the combined system of the sample plus the walls containing the sample. The effects of a wall are represented by an external force $\mathbf{F}_E(\mathbf{r}_i)$ which acts on particle i when it is near the wall [66]. We assume that this external force is generated by a potential $U_E(\mathbf{x})$ such that

$$F_E^\alpha(\mathbf{r}_i) = -\frac{\partial}{\partial r_i^\alpha} U_E(\mathbf{r}_i). \tag{339}$$

The Hamiltonian for the combined system is given by

$$H_T = H + H_E, \tag{340}$$

where the external contribution can be written as

$$H_E = \sum_{i=1}^{N} U_E(\mathbf{r}_i) = \int d^3x\, n(\mathbf{x}) U_E(\mathbf{x}). \tag{341}$$

Suppose that we investigate the behavior of the distorted system in the time-independent equilibrium state. Then, for any observable $A[q]$, we have

$$\frac{d\langle A\rangle}{dt} = 0. \tag{342}$$

If we choose A to be the momentum density, then, using Eq. (330), with $H \to H_T$, we find that the condition for mechanical stability, Newton's law, is given by

$$\frac{\partial}{\partial t}\langle g_\alpha(\mathbf{x})\rangle = -\sum_\beta \nabla_\beta \langle \sigma_{\alpha\beta}(\mathbf{x})\rangle + \langle F_E^\alpha(\mathbf{x})\rangle = 0 \tag{343}$$

where the average external force is given by

$$\langle \mathbf{F}_E(\mathbf{x})\rangle = \langle\{\mathbf{g}(\mathbf{x}), H_E\}\rangle = -\langle n(\mathbf{x})\rangle \nabla_x U_E(\mathbf{x}), \tag{344}$$

and is small except near the boundary. Thus the gradient of the average stress tensor is balanced by the applied external force. This means that the average stress tensor is essentially uniform except at the boundary. It is interesting to integrate Eq. (343) over a pillbox of cross-sectional area A and depth $2h$, as shown in Fig. 1.26. The pillbox has a volume $V_p = 2Ah$. A wall constraining the system cuts through the pillbox and we assume, over the area A, is flat, perpendicular to the x-direction, and

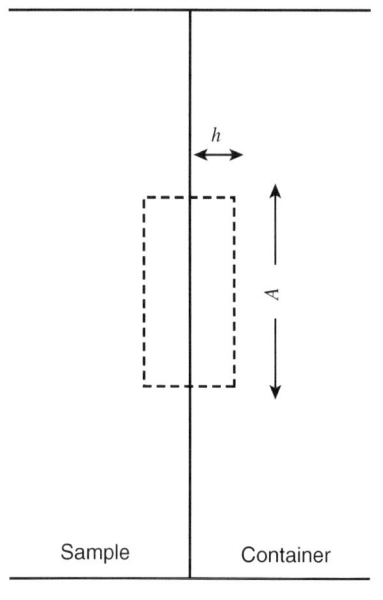

FIGURE 1.26 Pill box geometry for integration over Eq. (343).

located at position $x = a$. We then obtain

$$\int_{V_p} d^3x \langle F_E^\alpha(\mathbf{x}) \rangle = \int_{V_p} d^3x \sum_\beta \nabla_\beta \langle \sigma_{\alpha\beta}(\mathbf{x}) \rangle. \tag{345}$$

In this pillbox we can write the average stress tensor in the form

$$\langle \sigma_{\alpha\beta}(\mathbf{x}) \rangle = \bar{\sigma}_{\alpha\beta} \theta(a - x), \tag{346}$$

since it must vanish at the wall and remain zero for $x > a$. The associated force is given by

$$\sum_\beta \nabla_\beta \langle \sigma_{\alpha\beta}(\mathbf{x}) \rangle = \sum_\beta \nabla_\beta \bar{\sigma}_{\alpha\beta} \theta(a - x) = \sum_\beta \bar{\sigma}_{\alpha\beta} \left(-\delta(x - a) \delta_{\beta,x} \right)$$
$$= -\bar{\sigma}_{\alpha x} \delta(x - a). \tag{347}$$

The integral over the pillbox gives

$$-\int_{V_p} d^3x \, \bar{\sigma}_{\alpha x} \delta(x - a) = -\bar{\sigma}_{\alpha x} A = \int_{V_p} d^3x \langle F_E^\alpha(\mathbf{x}) \rangle \equiv \bar{F}_E^\alpha, \tag{348}$$

or

$$\bar{\sigma}_{\alpha x} = -\frac{\bar{F}_E^\alpha}{A}. \qquad (349)$$

As discussed above, $\sigma_{\alpha\beta}$ is the αth component of the force on unit area perpendicular to the axis in the β-direction. Thus σ_{xx} corresponds to a force in the x-direction applied along a plane perpendicular to x (see Fig. 1.27). Thus this stress produces a uniaxial strain. The stress σ_{xy} corresponds to a force in the x-direction applied in a plane perpendicular to $\hat{y}$. Clearly, the stress σ_{xy} generates a pure shear strain (see Fig. 1.28).

In a solid we can have forces applied in the y and z directions which correspond to shear. In fluids the system can sustain only forces in the x-direction, $\bar{\sigma}_{xx}$ is just the pressure, and Eq. (349) reduces to

$$p = -\frac{\bar{F}_E^x}{A}. \qquad (350)$$

If we multiply Eq. (343) by $x_\gamma U_{\alpha\gamma}$, sum over γ, and integrate over $\mathbf{x}$, we obtain the balance between work done on the system and the work done by the system:

$$\langle \delta H \rangle = -\int d^3x\, U_{\alpha\beta} \langle \sigma_{\alpha\beta} \rangle = -\langle \delta H_E \rangle = \int d^3x\, U_{\alpha\beta} x_\alpha \langle F_E^\beta(\mathbf{x}) \rangle. \qquad (351)$$

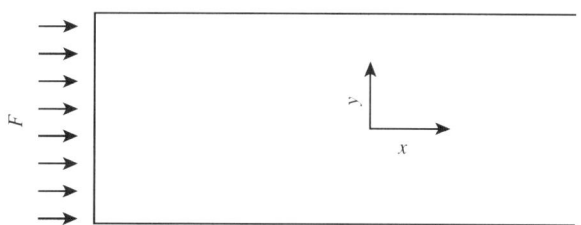

FIGURE 1.27 Schematic of uniaxial stress applied to a solid.

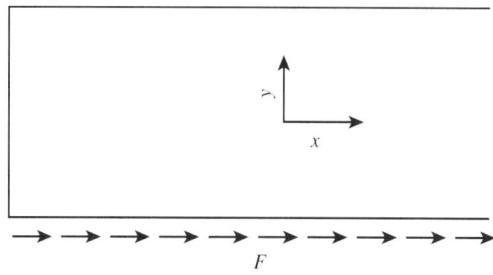

FIGURE 1.28 Schematic of shear stress applied to a solid.

This equation shows the reaction of the system of interest to the work performed by the forces at the surface. This force balance equation can be written in the form

$$\int d^3x\, U_{\alpha\beta}\left[\langle\sigma_{\alpha\beta}(\mathbf{x})\rangle + x_\alpha \langle F_E^\beta(\mathbf{x})\rangle\right] = 0. \tag{352}$$

1.14.4 Fluids

Starting from Newton's laws for a classical system interacting with a pair potential, one can show (see Appendix F) that the momentum density satisfies a continuity equation with a current given by the stress tensor which has the explicit form

$$\sigma_{\alpha\beta}(\mathbf{x}) = \sum_{i=1}^{N} \frac{p_i^\alpha p_i^\beta}{m} \delta(\mathbf{x} - \mathbf{r}_i) - \frac{1}{4}\int_{-1}^{+1} ds \int d^3r\, \frac{r_\alpha r_\beta}{r}\frac{\partial V(r)}{\partial r}$$

$$\times \sum_{i\neq j=1}^{N} \delta\left(\mathbf{x} + \frac{s+1}{2}\mathbf{r} - \mathbf{r}_i\right)\delta\left(\mathbf{x} + \frac{s-1}{2}\mathbf{r} - \mathbf{r}_j\right). \tag{353}$$

$V(r)$ is the pair potential for two particles separated by a distance $\mathbf{r}$. This rather unpleasant-looking result for the stress tensor has the compensating virtue that it shows that the stress tensor can be constructed explicity and is independent of the thermodynamic state of the system.

As shown in detail in Appendix F, if we take the average of the stress tensor and the system has full translational and rotational symmetry (it is a fluid), we easily find that the average stress tensor is given by

$$\langle \sigma_{\alpha\beta}(\mathbf{x})\rangle = p\delta_{\alpha\beta} \tag{354}$$

where p, which is to be identified with the pressure, is given explicitly by [67]

$$p = \frac{2}{3}\epsilon_K - \frac{1}{6}\int d^3r\, r\, \frac{\partial V(r)}{\partial r} n^2 g(r) \tag{355}$$

where ϵ_K is the kinetic energy per unit volume and the *radial distribution function*, $g(r)$, is defined by

$$n^2 g(r) = \left\langle \sum_{i\neq j=1}^{N} \delta(\mathbf{r} - \mathbf{r}_i)\delta(\mathbf{r}_j) \right\rangle \tag{356}$$

where n is, as usual, the particle density.

We should pause at this point and convince ourselves that the pressure we have introduced corresponds to our physical intuition about pressure. This is most conveniently carried out in the low-density limit, where the effects of the interaction

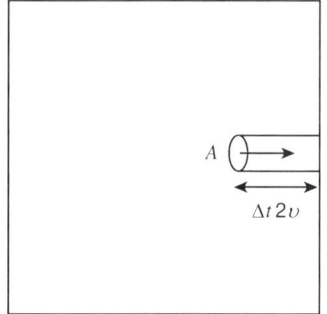

FIGURE 1.29 Geometry for pressure estimate. Particle of interest, at time t, is entering the tube shown in the figure.

potential can be ignored and we can use the kinetic theory of gases [68]. Our intuitive notion of pressure is that it is the average force per unit area. Thus we measure the pressure by putting a wall into a fluid or looking at the walls surrounding the fluid. The force on the wall is due to collision of the fluid particles with the wall as shown in Fig. 1.29. We can then estimate the pressure using the following qualitative argument. Assume that the particles hitting the wall on average have a velocity $\mathbf{v}$ before the collision. We also assume, on average, that the magnitude of the velocity is not changed during the collision (elastic collision) but that the momentum is changed by $\Delta \mathbf{p} = -2m\mathbf{v}$. The mean force exerted on the wall is, by Newton's laws, just equal to the mean rate of change of momentum of the wall. Hence the mean force on the wall can be obtained simply by multiplying the average momentum $2m\mathbf{v}$ gained by the wall per collision by the mean number of collisions per unit time with the wall. The number of collisions with the wall of area A in time Δt is given by the number of particles in the volume $Av \Delta t$ times the fraction $\frac{1}{6}$ with a velocity component in the direction of the wall. The pressure is just this number times the force per collision $2mv/\Delta t$ divided by the area:

$$p = \frac{1}{A}\left(\frac{n}{6} A v \Delta t\right) \frac{2mv}{\Delta t}$$
$$= \frac{1}{3} nmv^2 = \frac{2}{3}\frac{K}{V} = \frac{2}{3}\epsilon_K \qquad (357)$$

where ϵ_K is the average kinetic energy per unit volume and this is in agreement with Eq. (355) in the low-density limit.

The conclusion that the stress tensor is diagonal in an isotropic fluid has immediate consequences. Consider Eq. (352), which in the fluid reduces to

$$\int d^3x \sum_{\alpha,\beta} U_{\alpha\beta}[p\delta_{\alpha\beta} + x_\alpha \langle F_E^\beta(\mathbf{x})\rangle] = 0. \qquad (358)$$

We immediately see that the pressure *cannot* balance shear forces which correspond to off-diagonal components of α and β. Thus, as we expected from physical arguments, fluids cannot equilibrate (sustain or oppose) applied shear forces. Instead, they flow to eliminate the shear forces and then equilibrate. Thus, ultimately, these shear forces are ineffective in determining the equilibrium state of a fluid.

Let us return to the expression for the change in energy, given by Eq. (336), due to a distortion in the case of a fluid. Equation (336) in this case reduces to

$$\delta E = -\int d^3 x\, p \sum_\alpha U_{\alpha\alpha} = -p \sum_\alpha U_{\alpha\alpha} V_0. \tag{359}$$

However, we showed earlier that a volume change due to a distortion is given by

$$\delta V = V' - V = \sum_\alpha U_{\alpha\alpha} V_0, \tag{360}$$

so the change in energy due to the distortion can be written as

$$\delta E = -p\, \delta V. \tag{361}$$

This is our most important result concerning the application of mechanical force to a fluid. The change in energy of the fluid depends only on volume changes, not on rotations and shears. Thus the state does not generally depend on the $U_{\alpha\beta}$, only on hydrostatic compressions and expansions $\sum_\alpha U_{\alpha\alpha} = \delta V/V$.

This has the immediate consequence, setting $U_{\alpha\beta} = \delta_{\alpha\beta} U$ (which corresponds to a dilation of the system) in Eq. (352), that

$$\begin{aligned} pV &= -\tfrac{1}{3} \int d^3 x \sum_\alpha x_\alpha \langle F_E^\alpha(\mathbf{x}) \rangle \\ &= -\tfrac{1}{3} \left\langle \sum_{i=1}^N \mathbf{r}_i \cdot \mathbf{F}_E(\mathbf{r}_i) \right\rangle \end{aligned} \tag{362}$$

which is commonly known as the *virial theorem* [69]. The forces at the boundary balance the pressure of the fluid.

1.14.5 Solids

This picture for fluids must be contrasted with that for solids. In a fluid the system flows in such a way as to produce an isotropic pressure at any interface. Change the wall and the system will rapidly readjust to produce a uniform pressure at the new wall. The nature of any applied forces at the surface is essentially irrelevant as long as they provide walls. The stress tensor is, to a large degree, determined by the kinetic energy and collisions of particles with the walls. Thus we can understand how a gas expands to fill all available space. The average stress tensor is a function only of the density and temperature of the system.

In a solid, things are quite different. A solid can be freestanding *without* walls. Thus in a solid the notion of pressure and the stress tensor is qualitatively different from a fluid. Applied forces at the surface induce distortions in the solid which are communicated into the bulk of the system. Thus the internal stress depends on the nature of this distortion, although it is independent of the manner in which the distortion was produced. The upshot of this discussion is that in a solid the thermodynamic state of the system requires specification of all macroscopic distortions or strains. This means that we must include the strain field $V_0 U_{\alpha\beta}$ in our list of macrovariables characterizing a solid. As discussed in detail in Volume 2, the fundamental reason *why* the strain field must be included in our list of macrovariables is tied to the notion [70] of Nambu–Goldstone modes and broken continuous symmetry. Thus this macrovariable, unlike the energy, particle number, or total momentum, is not tied directly to being a conserved quantity.

The case of liquid crystals is intermediate between a liquid and a solid. These systems can flow to relieve shears, but the rotation of a wall can couple to the orientation of the large molecules in the system and can be communicated into the bulk system. In this case changes in energy are associated with such rotations. Thus we must include the *director field*, which characterizes these orientations when constructing the thermodynamics and the statistical mechanics.

1.14.6 Adiabatic Processes

We are not yet finished with our discussion of the role of mechanical forces acting on systems in equilibrium. We need to explore in a bit more depth the expressions giving the change in energy due to a change in strain:

$$dE = -\sum_{\alpha\beta} \langle \bar{\sigma}_{\alpha\beta} \rangle d(V_0 U_{\alpha\beta}) \qquad (363)$$

or, in a fluid,

$$dE = -p\,dV. \qquad (364)$$

We can understand such *mechanical* changes as occurring due to changes in the Hamiltonian describing the system. To see how this is applicable to the case of a change in volume, we can, instead of rigid boundaries, introduce an external potential that confines the system within some well-defined volume. This potential is then part of the system. In one dimension we can write

$$U_E = \sum_{i=1}^{N} [u(x_i - a) + u(-x_i)] \qquad (365)$$

where, for example,

$$u(x) = u_0 e^{x/l_0} \qquad (366)$$

and l_0 is taken to be a microscopically small length characterizing the transition layer into the solid creating the wall. The total Hamiltonian for this system can be written in the form

$$H_T = H + U_E, \qquad (367)$$

and depends on the parameter a, which gives the length of the system and is macroscopically adjustable. We can change the volume of the system by changing a, which amounts to moving the face of a piston.

Let us consider the situation where we change the energy through a *slow* variation in a parameter, such as a, appearing in the Hamiltonian. Such a process is referred to as *adiabatic*.

If the Hamiltonian $H(q_N, a)$ depends on the parameter a, which changes with time [71], the change in the Hamiltonian is given by

$$\frac{dH}{dt} = \frac{\partial H}{\partial a} \frac{da}{dt}. \qquad (368)$$

Assuming that this variation in a occurs over a time range $0 < t < \tau$, we obtain the change in energy

$$dE = \int_0^\tau dt \frac{dH}{dt} = \int_0^\tau dt \frac{\partial H}{\partial a} \frac{da}{dt}. \qquad (369)$$

Assuming a very slow constant rate, $da/dt = \gamma$, we have the change in a over the time τ, $da = \gamma \tau$, and

$$\frac{da}{dt} = \frac{da}{\tau}. \qquad (370)$$

Inserting this result back into Eq. (369), we have

$$dE = \int_0^\tau dt \frac{\partial H}{\partial a} \frac{da}{\tau}. \qquad (371)$$

We can take the large τ limit and obtain, for a mixing system,

$$\lim_{\tau \to \infty} \frac{1}{\tau} \int_0^\tau dt \frac{\partial H}{\partial a} = \left\langle \frac{\partial H}{\partial a} \right\rangle. \qquad (372)$$

The change in the energy due to this slow process is given by

$$dE = \left\langle \frac{\partial H}{\partial a} \right\rangle da, \qquad (373)$$

and follows the change in the Hamiltonian.

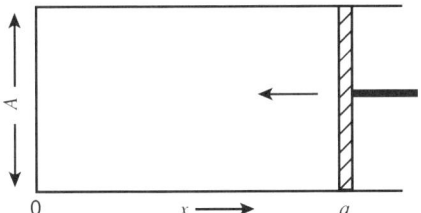

FIGURE 1.30 Piston adiabatically compressing a system with cross-sectional area A.

As a specific example, consider the case of a piston compressing a cylinder as shown in Fig. 1.30. This container of cross section A is closed by a piston at position $x = a$. The volume of the system of interest is aA. The piston interacts with the fluid in the cylinder with a potential energy

$$U_E = \sum_{i=1}^{N} u(x_i - a) \qquad (374)$$

where x_i is the x-component of the position $\mathbf{r}_i$ of particle i, and, for simplicity, we take the other boundaries to be rigidly fixed. We are interested in the change in the total Hamiltonian, H, including U_E describing the piston, due to the change in a. We have

$$\frac{\partial H}{\partial a} = \sum_{i=1}^{N} \frac{\partial}{\partial a} u(x_i - a)$$

$$= -\sum_{i=1}^{N} \frac{\partial}{\partial x_i} u(x_i - a) = F_x^E \qquad (375)$$

where F_x^E is the x-component of the *external* force acting on the particles. From our analysis in the preceding section, we showed that, after taking averages, in a fluid

$$\langle F_x^E \rangle = -pA. \qquad (376)$$

Combining the average of Eq. (375) with Eq. (376), we obtain

$$\left\langle \frac{\partial H}{\partial a} \right\rangle = -pA. \qquad (377)$$

Since the change of volume is given by $dV = A\,da$, Eq. (377) can be rewritten as

$$\left\langle \frac{\partial H}{\partial V} \right\rangle = -p. \qquad (378)$$

88 GENERAL PRINCIPLES OF STATISTICAL MECHANICS

If the process is adiabatic, Eq. (373) tells us that

$$-p = \left\langle \frac{\partial H}{\partial V} \right\rangle = \frac{\partial E}{\partial V}. \tag{379}$$

For a more general distortion we can write

$$\frac{\partial E}{\partial (V_0 U_{\alpha\beta})} = -\langle \bar{\sigma}_{\alpha\beta} \rangle. \tag{380}$$

Thus application of mechanical force to a statistical mechanical system falls within that class of perturbations where we change the system by changing a parameter in the Hamiltonian.

We now need to look at how the entropy changes when we very slowly change a parameter in the Hamiltonian. It is useful to point out at this stage that the original definition of the entropy for the microcanonical ensemble

$$S = k_B \ln[Z_M(E)], \tag{381}$$

where Z_M is the number of states corresponding to an energy E, can for large systems, be replaced by

$$S = k_B \ln[\Sigma_M(E)] \tag{382}$$

where

$$\Sigma_M(E) = \int \frac{dq_N}{h^{3N} N!} E_T \theta(E - H) \tag{383}$$

is the number of states with energy less than E. One easily finds that

$$Z_M = \frac{\partial \Sigma_M(E)}{\partial E}. \tag{384}$$

This replacement of Z_M by Σ_M in S is justified since one can show (see Problem 1.51) that

$$\lim_{N \to \infty} \frac{1}{N}[\ln \Sigma_M - \ln Z_M] = \lim_{N \to \infty} \frac{1}{N} \ln \left\langle \frac{2}{3N}(E - U) \right\rangle = 0 \tag{385}$$

where U is the potential energy.

Consider now the entropy given by Eq. (382) where we assume that the Hamiltonian depends on a parameter a and we allow for the total energy to vary with a:

$$S(a) = k_B \ln \Sigma_M(a) \tag{386}$$

where

$$\Sigma_M(a) = \int \frac{dq_N}{h^{3N} N!} E_T \theta(E(a) - H(a)). \tag{387}$$

MECHANICAL FORCES: SOLIDS AND LIQUIDS   89

Now look at the change in the entropy corresponding to a change in a parameter a:

$$\begin{aligned} dS(a) &= k_B \frac{1}{\Sigma_M(a)} \int \frac{dq_N}{h^{3N} N!} E_T \delta(E(a) - H(a)) \left[dE - \frac{\partial H(a)}{\partial a} da \right] \\ &= k_B \frac{1}{\Sigma_M(a)} \left[Z_M(a) dE - Z_M(a) \left\langle \frac{\partial H(a)}{\partial a} \right\rangle da \right] \\ &= k_B \frac{Z_M(a)}{\Sigma_M(a)} \left[dE - \left\langle \frac{\partial H(a)}{\partial a} \right\rangle da \right]. \end{aligned} \quad (388)$$

If the process is slow, the variation of the average energy is given by Eq. (373), and Eq. (388) reduces to

$$dS(a) = 0. \quad (389)$$

Thus in an adiabatic process the entropy is held constant. We also see that $\Sigma_M(a)$ is invariant. When we very slowly change a parameter in the Hamiltonian, we do not change the number of available states, only their relative position in phase space.

1.14.7 Pressure, Thermodynamic Derivatives, and Equations of State

The main result from the preceding section is that the thermodynamic derivatives given by Eqs. (379) and (380) should be evaluated at constant entropy:

$$\left(\frac{\partial E}{\partial (V_0 U_{\alpha\beta})} \right)_{S,N} = -\langle \bar{\sigma}_{\alpha\beta} \rangle \quad (390)$$

or, for fluids,

$$\left(\frac{\partial E}{\partial V} \right)_{S,N} = -p. \quad (391)$$

We can write these equations in another form which fits firmly into the rest of our development. To save writing we assume that we have a fluid, but the final results clearly hold for a solid.

In our treatments of the microcanonical and grand canonical ensembles, we established that the derivative of the entropy with respect to its natural variables, the energy and particle number, are related to the intensive variables, the temperature and chemical potential. What about the derivative of the entropy with respect to its other natural variable the volume? In our earlier discussion of the entropy principle, we made the identification of the pressure via

$$\left(\frac{\partial S}{\partial V} \right)_{E,N} = \frac{p}{T}. \quad (392)$$

Using the methods of thermodynamic transformation theory introduced in the next chapter we can easily show that Eqs. (391) and (392) are equivalent. Thus Eq. (392) follows from Eq. (391).

1.15 NOSÉ DYNAMICS

As we have gone along in this chapter we have moved in perspective from the Boltzmann paradigm of *clean* isolated systems to the Langevin paradigm of open systems interacting with their environment. Thus we have moved from the microcanonical to the canonical ensembles. When we consider dynamics, the usual Newtonian mechanics for isolated systems conserves energy and is naturally compatible with the the microcanical ensemble with an associated probability distribution given by Eq. (98). In many circumstances it is more natural to think of dynamics which, instead of a precise conservation of energy, corresponds to a dynamics with a fixed temperature. Thus one looks for dynamics that are compatible with the canonical distribution function. One can obtain [72] these equations of motion, which are consistent with a fixed temperature, if one couples the system of interest to a collective variable that plays the role of a heat bath.

It seems physically plausible that a fixed-temperature ensemble requires controlling the fluctuations in the kinetic energy, K, whose average can be used to define the temperature [73] in the canonical ensemble. Suppose that we modify Newton's equation governing a set of N particles with coordinates $q_i(t)$ by adding a damping term proportional to the velocity $\dot{q}_i(t)$:

$$\ddot{q}_i(t) = \frac{F_i(q_i(t))}{m} - \zeta(t)\dot{q}_i(t) \qquad (393)$$

where F_i is the usual force on particle i and for simplicity we suppress the vector index. Thus, appropriate to an approach within the Langevin paradigm, we have a type of Langevin equation. However, unlike the Langevin equation [74], we can maintain a set of reversible equations if we assume that the damping coefficient $\zeta(t)$ satisfies the equation of motion

$$\dot{\zeta}(t) = \frac{1}{Q}(2K - Xk_BT) \qquad (394)$$

where $K = \frac{1}{2}\sum_i m\dot{q}_i^2(t)$ is the kinetic energy, Q a positive damping coefficient, X a measure of the total number of degrees of freedom (to be determined below), and the system is held at temperature T. Since in equilibrium we require that $\langle\dot{\zeta}(t)\rangle = 0$, we must choose

$$2\langle K \rangle = \langle X \rangle k_BT. \qquad (395)$$

Since we easily find [73] that in equilibrium at temperature T,

$$\langle K \rangle = \tfrac{3}{2} N k_B T, \tag{396}$$

we have that we must arrange

$$\langle X \rangle = 3N. \tag{397}$$

Analyzing the equilibrium probability distribution associated with Eqs. (393) and (394) is not straightforward since these equations cannot be derived directly from a Hamiltonian. Fortunately, these equations can be transformed to a Hamiltonian form through a few changes of variables. The key point is that we introduce a conjugate coordinate and momentum for the bath. The first step is to define the momentum of the bath via

$$p_s = Q\zeta. \tag{398}$$

Next we introduce a variable $s(t)$, defined by

$$\frac{ds}{dt} = \frac{s p_s}{Q}, \tag{399}$$

and a new particle momentum variable,

$$p_i = m s \dot{q}_i. \tag{400}$$

The equations of motion, Eqs. (393) and (394), are then equivalent to

$$\dot{p}_i = s F_i(q) \tag{401}$$

and

$$\dot{p}_s = \sum_i \frac{p_i^2}{m s^2} - X k_B T. \tag{402}$$

These equations can be put into canonical form if we change the time variable from t to τ, where $d\tau/dt = s(\tau)$. In terms of the new time variable we obtain the new equations of motion

$$\frac{dq_i}{d\tau} = \frac{p_i}{m s^2} \tag{403}$$

$$\frac{dp_i}{d\tau} = F_i(q) \tag{404}$$

$$\frac{ds}{d\tau} = \frac{p_s}{Q} \tag{405}$$

and

$$\frac{dp_s}{d\tau} = \sum_i \frac{p_i^2}{ms^3} - X \frac{k_B T}{s}. \tag{406}$$

We see then that $s(\tau)$ plays the role of a bath coordinate. These equations of motion follow in the usual way from the Hamiltonian,

$$H = V + \sum_i \frac{p_i^2}{2ms^2} + X k_B T \ln(s) + \frac{p_s^2}{2Q} \tag{407}$$

where V is the usual potential energy associated with the coordinates q_i. The equilibrium probability distribution for the full phase space, bath plus particles, $\{q_i, p_i, s, p_s\}$, is the microcanonical equilibrium probability distribution

$$P[q_i, p_i, s, p_s] = \frac{\delta(E - H)}{Z_M} \tag{408}$$

where the normalizing partition function is given by

$$Z_M = \frac{1}{N!} \int \prod_i \frac{d^3 q_i d^3 p_i}{(2\pi\hbar)^3} ds\, dp_s \delta(E - H). \tag{409}$$

Clearly, we are interested in the probability distribution governing the system of particles alone. This is given by the reduced probability distribution,

$$P[q_i, P_i] = \int ds\, dp_s P[q_i, p_i, s, p_s] \delta\left(P_i - \frac{p_i}{s}\right). \tag{410}$$

It is left to Problem 1.53 to show that reduced probability distribution $P[q_i, P_i]$ is canonical and $X = 3N + 1$. This result means that the canonical ensemble average

$$\langle A \rangle_c = \text{Tr}_c\, P[q_i, p_i] A(q_i, p_i) \tag{411}$$

is equivalent to the time average

$$\langle A \rangle_c = \lim_{\tau \to \infty} \frac{1}{\tau - t_0} \int_{t_0}^{\tau} d\tau'\, A\left(\frac{q_i(\tau'), p_i(\tau')}{s(\tau')}\right) \tag{412}$$

where $q_i(\tau), p_i(\tau)$, and $s(\tau)$ are generated by the four equations of motion listed above. For further discussion, see the papers by Nosé [72] and Hoover [75].

1.16 ENSEMBLES AND INDEPENDENT VARIABLES

Let us end this chapter with a summary of the ensembles discussed and their structure. Thus far, in the case of simple fluids, we have considered seven variables: E, N, V, T, p, S, and μ. One of our main results is that for simple fluids there are only three independent variables. The other four variables can be expressed in terms of the three independent variables by taking derivatives of a potential that can be determined by a microscopic calculation. These potentials for the various ensembles are listed in Table 1.2. In Table 1.3 we list the fundamental independent variables for the three ensembles and the quantities we need to calculate. For example, in the microcanonical ensemble, it is the entropy that is determined by a microscopic calculation, and the intensive variables p, T, and μ are calculated by taking derivatives as indicated in the table. Once we have the potential relevant to a given ensemble as a function of the independent variables in that ensemble, the other variables are given as derivatives of the potential with respect to the independent variables.

We have discussed in some detail in this chapter how we can characterize the equilibrium state. We have focused on a fluid system for definiteness. The ideas we developed, however, are quite general. The macrostate of a system is characterized by the conserved variables, plus variables governing any macroscopic constraints (e.g., a fixed volume or a distortion in a solid) [76]. The equilibrium state is that state where the macrovariables are known and the entropy is maximized consistent with these constraints.

TABLE 1.2 Potentials, Partition Function, and Probability Distributions for the Canonical Ensembles

Ensemble	Potential	Partition Function	Probability
Microcanonical	$S = k_B \ln Z_M$	$Z_M = \text{Tr}_{\text{cl}}^N \delta(E - H)$	$\dfrac{1}{Z_M}$
Canonical	$F = -\beta^{-1} \ln Z_C$	$Z_C = \text{Tr}_{\text{cl}}^N e^{-\beta H}$	$\dfrac{e^{-\beta H}}{Z_C}$
Grand canonical	$\Omega = -\beta^{-1} \ln Z_G$	$Z_G = \text{Tr}_{\text{cl}} e^{-\beta(H-\mu N)}$	$\dfrac{e^{-\beta(H-\mu N)}}{Z_G}$

TABLE 1.3 Independent Variables, Denoted by G, and Derived Variables, Indicated by Derivatives of a Potential, in the Canonical Ensembles

Ensemble	E	V	N	T^{-1}	μ	p	TS
Microcanonical	G	G	G	$\dfrac{\partial S}{\partial E}$	$-T\dfrac{\partial S}{\partial N}$	$T\dfrac{\partial S}{\partial V}$	$\beta^{-1} \ln Z_M$
Canonical	$\dfrac{\partial(\beta F)}{\partial \beta}$	G	G	G	$\dfrac{\partial F}{\partial N}$	$-\dfrac{\partial F}{\partial V}$	$E - F$
Grand canonical	$\dfrac{\partial(\beta \Omega)}{\partial \beta}$	G	$-\dfrac{\partial(\beta \Omega)}{\partial \alpha}$	G	G	$-\dfrac{\partial \Omega}{\partial V}$	$E - \mu N - \Omega$

REFERENCES

[1] P. W. Anderson ["More is different", *Science* **177**, 393 (1972)], discusses the ideas of broken symmetry and rigidity that we will introduce here as we go along. These ideas permeate many of the important problems of interest in statistical mechanics in the second half of the twentieth century.

[2] See S. G. Brush, *Statistical Physics and the Atomic Theory of Matter: From Boyle and Newton to Landau and Onsager*, Princeton University Press, Princeton, N. J., 1983, Sec. 1.8 and 1.9; also, P. S. Epstein, *Textbook of Thermodynamics*, Wiley, New York, 1937, p. 27.

[3] One of the goals of statistical mechanics is to answer questions of the type: If we push the pencil with a certain force, how far is the surface of the solid depressed along its surface? This particular question is addressed in some detail in Chapter 6.

[4] The use of molecular dynamics was pioneered in the late 1950s by B. J. Alder and T. E. Wainwright, *J. Chem. Phys.* **31** (1960). They studied the dynamics of hard disks (two dimensions) and hard spheres. See also, for example, B. J. Alder and T. E. Wainwright, *Sci. Am.* **200**, 113 (Oct. 1959). Early work on continuous potentials was pioneered by A. Rahman, *Phys. Rev. A* **13**, 405 (1964). This technique has now been extended to a wide variety of physical systems. For reviews, see J. J. Erpenbeck and W. W. Wood, "Molecular dynamics techniques for hard-core systems", in *Statistical Mechanics*, Part B: *Time Dependent Processes*, ed. B. J. Berne, Plenum Press, New York, 1977; and J. Kushick and B. J. Berne, "Molecular dynamics methods: continuous potentials," in *Statistical Mechanics*, Part B: *Time Dependent Processes*, ed. B. J. Berne, Plenum Press, New York, 1977. For an introduction at an elementary level, see H. Gould and J. Tobochnik, *An Introduction to Computer Simulation Methods*, Addison-Wesley, Reading, Mass., 1988, Chap. 6 of Part 1. For a discussion of the calculation of specific heats and compressibilities using molecular dynamics see P. S. Y. Cheung, *Mol. Phys.* **33**, 519 (1977).

[5] James Clerk Maxwell coined the name *statistical mechanics* in *Cambridge Philos. Soc. Trans.* **12**, 547 (1879).

[6] A general reference on probability theory is W. Feller, *An Introduction to Probability Theory and Its Applications*, Wiley, New York, 1968. From the point of view of statistical mechanics, see L. E. Reichl, *A Modern Course in Statistical Physics*, University of Texas Press, Austin, Texas, 1980, Chap. 5; and P. Resibois and M. DeLeener, *Classical Kinetic Theory of Fluids*, Wiley, New York, 1977.

[7] Ensemble theory (and indeed the term *ensemble*) was introduced by J. Williard Gibbs in his famous book on statistical mechanics published in 1902 J.W. Gibbs, *Elementary Principles in Statistical Mechanics*, Ox Bow Press, Woodbridge, Conn., 1981; see the preface.

[8] Clearly, there is the underlying assumption that this limit exists. This puts limitations on the nature of the ensembles under investigation.

[9] The statistics of coin flipping can be explored in detail by visiting the applet located on the Web at *http://stp.clarku.edu/java/CoinTossing.html*.

[10] As will be made clear as we proceed, the basic ideas that support statistical mechanics are independent of the particular system we study and whether we use classical or quantum mechanics. We develop the basics in terms of classical fluids since they are the easiest realistic system to conceptualize.

[11] The term Γ-*space*, which stands for *gas*, was introduced by P. Ehrenfest and T. Ehrenfest, *Enzyklopädie Mathematischen Wissenschaften*, Bd. IV, Teil 32, Leipzig, 1911.

[12] There is a philosophical question at this point of considerable importance. From the point of view of nineteenth-century physics, we live in a clockwork universe. Consider the famous statement by Laplace: "We ought then to regard the present state of the universe as the effect of its anterior state and as the cause of the one which is to follow. Given for one instant an intelligence which could comprehend all the forces by which nature is animated and the respective situation of the beings who compose it—an intelligence sufficiently vast to submit these data to analysis—it would embrace in the same formula the movements of the greatest bodies of the universe and those of the lightest atom; for it, nothing would be uncertain and the future, as the past, would be present to its eyes." Pierre-Simon Marquis de Laplace, *A Philosophical Essay on Probabilities*, translated from the 6th French edition by F. W. Truscott and F. L. Emory, New York, 1917, p. 4.

Thus it is, from this point of view, possible to precisely prepare a system in a specific configuration q_N^l. The implication is: if one can prepare the system in this state once, we can do it again. Our point of view here allows for the possibility that in the sample preparation there are uncertainties that preclude precise specification of the state. Such uncertainties may, for example, be associated with boundary effects that differ very slightly from one preparation of the system to the next.

[13] See, for example, H. Goldstein, *Classical Mechanics*, 2nd ed., Addison-Wesley, Reading, Mass., 1980. This is just a restatement of the point of view of Laplace quoted in Ref. 12: that if we know the configuration of the universe, everything is determined.

[14] For an introduction to the physics of liquid crystals, see P. G. de Gennes and J. Proust, *The Physics of Liquid Crystals*, 2nd ed. Oxford University Press, New York, 1993.

[15] For an introduction to polymer physics, see P. G. deGennes, *Scaling Concepts in Polymer Physics*, Cornell University Press, Ithaca, N.Y., 1979.

[16] The statistical mechanics of harmonic solids is discussed in Chapter 6.

[17] A model for binary alloys is discussed in Chapter 2.

[18] Magnetic models are discussed in some detail in Chapter 5.

[19] J. Perron, *Atoms*, Ox Bow Press, Woodridge, Conn., 1990.

[20] Ludwig Boltzmann, one of the central figures in the development of statistical mechanics, was caught between the *clockwork* model of the universe held by many physicists of his time and the inherently statistical nature of the laws of thermodynamics. Early on, he stated the ergodic theorem in its loosest form, that every disturbed system will approach thermal equilibrium as its end state if left alone, and then proceeded to attempt to prove it generally. Boltzmann's first significant paper, "The mechanical meaning of the second law of thermodynamics," 1866 shows his intentions. He had a number of subsequent attempts: L. Boltzmann, *Wien. Ber.* **58**, 517 (1868); **63**, 679 (1871); **76**, 373 (1877); *J. Math.* **100**, 201 (1887). He did not succeed completely since, as we shall see, he had set an impossible goal. He did, however, make impressive progress, including his famous work on kinetic theory in which he introduced his kinetic equation and his H-theorem. Although this work is still powerful within its range of applicability (dilute fluids), Boltzmann did not recognize its limitations.

These early efforts by Boltzmann created much confusion and conflict, much of the conflict arising between the need of one part of the community to see things from a mechanistic point of view and those who did not trust the microscopic–atomic viewpoint. Boltzmann's efforts plunged him into a series of bitter controversies.

In the 1870s Boltzmann and his Viennese colleague Josef Loschmid entered into a bitter debate over the *reversibility paradox*. In 1874, William Thomson (Lord Kelvin) called attention to the apparent contradiction between the irreversibility of thermodynamics and the reversibility of Newtonian mechanics. Newton's laws say that one should be able to return to any initial state merely by reversing the molecular velocities. See J. Loschmidt, "Über den Zustand des Wärmgleichgewichtes eines Systemes von Korpern mit Rücksicht auf fie Schwerkraft," *Sitzungsberichte, K. Akademie der Wissenschaften*, Wien, *Math. Naturwiss. Kl.*, **73**, 128–142 (1865); and William Thomson, "The kinetic theory of the dissipation of energy." *Proc. R. Soc. Edinburgh* **8**, 325–334 (1874).

A second bitter controversy involving Boltzmann broke out after Poincaré's work, H. Poincaré, *Acta Math.* **13**, 1 (1890), on celestial mechanics. A theorem due to Poincaré says that every such system eventually returns arbitrarily close to its initial state. The time involved is enormous for a macroscopic system. Boltzmann estimated a typical Poincaré period for a system of size $100\,cm^3$ to be $10^{10^{10}}$ s. Ernest Zermelo used the recurrence theorem to attack the mechanistic world view. He held that the second law of thermodynamics is an absolute truth, and any mechanical explanation of the second law must be excluded by Poincaré's recurrence theorem. See E. Zermelo, *Ann. Phys. Ser. 3* **57**, 485–494; **59**, 793–801 (1896). For more details and reprints, see S. Brush, *Kinetic Theory*, 2 vols., Pergamon Press, Elmsford, N.Y., 1965 and 1966. These controversies cast a broad shadow on the basic foundations of statistical mechanics and laid the ground work for the work in ergodic theory carried out in the twentieth century.

[21] The Toda Hamiltonian is discussed in Appendix C.

[22] The connection between a constant of the motion and an invariance principle is sometimes called *Noether's principle*. E. Noether, *Nach. Ges. Wiss. Goettingen* **2**, 235 (1918). The general principle is discussed by H. Goldstein, *Classical Mechanics*, 2nd ed., Addison-Wesley, Reading, Mass., 1980.

[23] The concept of a *generic variable* was introduced by J. W. Gibbs, *Elementary Principles in Statistical Mechanics*, Ox Bow Press, Woodbridge, Conn., 1981, p. 187.

[24] These ideas are clear in the case of a system where a set of large particles is suspended in a solvent of smaller particles. In this case, while the particle trajectories are not reproducible, the average speed and root-mean-square displacement are meaningful.

[25] The origins of the concept of thermodynamic equilibrium are long and deep and tied to the evolution of the ideas of energy conservation and temperature. Significant contributions came from work on biology, technology, and engineering.

[26] These ideas are very general but assume that the equilibration times are short compared with times over that one carries out experiments. There are systems that relax very slowly to equilibrium, and their treatment on experimental time scales requires inclusion in the analysis of additional *quenched variables*. There are a number of systems where, because of the slowness of some subset of variables, full equilibration is problematic over observationally relevant time scales. This is particularly true for solid systems at low temperatures, where some atoms may hop around much more slowly than others.

[27] The basic ideas of information theory go back to the seminal work of C. E. Shannon, "The mathematical theory of communication," *Bell Syst. Tech. J.* **27**, 379–424, 623–657 (1948); and N. Wiener, *Cybernetics*, Wiley, New York, 1948, within the context of

communication theory. The application to statistical physics is due to E. T. Jaynes, "Information theory and statistical mechanics," *Phys. Rev.* **106**, 620; **108**, 171 (1957). A complete introduction is given by L. Brillouin, *Science and Information Theory*, Academic Press, New York, 1956. Books in which this material is discussed from a point of view close to that given here are those by A. Katz, *Principles of Statistical Mechanics: The Information Theory Approach*, W. H. Freeman, San Francisco, 1967; and H. S. Robertson, *Statistical Thermophysics*, Prentice Hall, Englewood Cliffs, N.J., 1993.

[28] The term *entropy* was introduced by R. Clausius, *Ann. Phys. Ser. 2* **125**, 353 (1865). See also "On several convenient forms of the fundamental equations of the mechanical theory of heat," in *The Mechanical Theory of Heat*, ed. T. A. Hirst, Van Voorst, London, 1867. Entropy is taken from the Greek word τροπη, meaning *transformation* and was chosen to have a resemblance to the word *energy*. The concept of entropy was introduced 11 years earlier, before the name was coined, also by Clausius: R. Clausius, *Ann. Phys. Ser. 2* **93**, 481 (1854); *Philos. Mag. Ser. 4* **12**, 81 (1856). For further discussion see S. G. Brush, *Kinetic Theory*, Vol. II, Pergamon Press, Elmsford, N. Y., 1966, p. 8.

[29] Building on the earlier work of Carnot (1824), the second law was developed by R. Clausius, *Ann. Phys. (Leipzig)* **79**, 368, 500(1850); and W. Thomson, *Trans. R. Soc. Edinburgh* **20**, 261 (1853). Carnot's work is translated in *The Second Law of Thermodynamics*, ed. J. Keston, Dowden, Hutchinson and Ross, Stroudsburg, Pa., 1976.

[30] The microcanonical ensemble was introduced and named by J. W. Gibbs, *Elementary Principles in Statistical Mechanics*, Ox Bow Press, Woodbridge, Conn., 1981, p. 115. He developed the theory of the microcanonical ensemble with a phase density $\approx e^{-(H-E)^2/\sigma^2}$ and carried out the limit process $\sigma \to 0$ only at the end of the argument. See A. Münster, *Statistical Thermodynamics*, Springer-Verlag, Berlin, 1969, p. 95. For a more detailed treatment than we give here of the microcanonical ensemble, see T. Caugin and J. Ray, *Phys. Rev. A* **37**, 247 (1988). J. H. Ray and H. W. Graben, *Mol. Phys.* **43**, 1293 (1981), discuss direct calculation of fluctuation formulas in the microcanonical ensemble.

[31] One should check that classical observables do not depend on the particular choice for $h = 2\pi\hbar$.

[32] Boltzmann's principle,

$$S = k_B \ln Z_M \tag{413}$$

is engraved on Boltzmann's tombstone in Vienna. Quoting A. Sommerfeld, *Thermodynamics and Statistical Mechanics*, Academic Press, New York, 1964: "So it stands carved out on Boltzmann's memorial in the Central cemetery in Vienna, floating in the clouds over his majestic bust." Remarkably, Boltzmann himself never wrote down the equation in this form. This was first done by Planck (e.g., in the first edition of his "*Vorlesunger über die Theorie der Wärmestralung*," 1906). The constant k_B was also introduced by Planck and not by Boltzmann. Boltzmann only referred to the proportionality between S and the logarithm of the probability of a state. The designation of Boltzmann's principle was advocated by Einstein for the reverse relation,

$$W = e^{S/k_B}, \tag{414}$$

which determined W from the empirically known S. A. Einstein, *Ann. Phys. (Leipzig)* **22**, 180 (1907).

[33] The limit $N \to \infty$ while $n = N/V$, $\epsilon = E/N$, and $s = S/N$ are held fixed is known as the *thermodynamic limit*. This limit is appropriate for standard stable macroscopic systems. There are some *abnormal* materials for which a different limit is appropriate. These are the systems with long-range interactions (gravitational and coulombic forces) where the potential energy does not increase linearly with the volume of the system. See P. T. Landberg, "Is equilibrium always an entropy maximum?" *J. Stat. Phys.* **35**, 159 (1984). In Chapter 3, when we discuss harmonically trapped systems, we find an interesting deviation from the conventional thermodynamic limit.

[34] O. Sackur, *Ann. Phys.* **36**, 958 (1911); **40**, 67 (1913); H. Tetrode, *Ann. Phys.* **38**, 434 (1912).

[35] In the case of systems with long-range interactions the identification of an energy E_α for a given volume is not so obvious. With care, such separation can be achieved.

[36] The continuity equation expressing local conservation of mass was introduced by d'Alembert according to L. M. Milne-Thomson, *Theoretical Hydrodynamics*, Macmillan, New York, 1960.

[37] The momentum equation for an ideal gas, called *Euler's equation*,

$$\frac{\partial \mathbf{v}}{\partial t} + (\mathbf{v} \cdot \nabla)\mathbf{v} = -\frac{1}{\rho}\nabla p \tag{415}$$

was first obtained by L. Euler in 1755, "Principes généraux du movement des fluids," *Hist. Acad. Berlin* (1755). If one adds dissipative terms one obtains the Navier–Stokes equation: C. L. M. H. Navier, "Mémoire sur les lois du mouvement des fluides," *Mém. Acad. Sci.* **6**, 389 (1822); G. G. Stokes, "On the theories of the internal friction of fluids in motion," *Cambridge Trans.* **8**, 287 (1845). For further discussion see L. D. Landau and E. M. Lifshitz, *Fluid Mechanics*, Pergamon Press, Oxford, 1959.

[38] The development of the local statement of conservation of energy is tied up with the development of the heat equation by Joseph Fourier, "Théorie de la propagation de la chaleur dans les solides," 1807, presented to the Institut de France, published in *Joseph Fourier* by I. Grattan-Guiness with R. R. Ravetz, MIT. Press, Cambridge, Mass., 1972. See also *Théorie analytique de la chaleur, The Analytical Theory of Heat* Didot Paris; English translation by A. Freeman, London, 1978.

[39] From the preface to the first English edition of L. D. Landau and E. M. Lifshitz, *Statistical Physics*, 2nd ed., Addison-Wesley, Reading, Mass., 1969.

[40] The canonical ensemble was introduced by J. W. Gibbs, *Elementary Principles in Statistical Mechanics*, Ox Bow Press, Woodbridge, Conn., 1981, p. 32.

[41] The grand canonical ensemble and its name were introduced by J. W. Gibbs, *Elementary Principles in Statistical Mechanics*, Ox Bow Press, Woodbridge, Conn., 1981, p. 189.

[42] We assume that any interaction term in the Hamiltonian is small (proportional to the number of atoms near the surface of the subsystem) and can be neglected in the thermodynamic limit.

[43] The definition of temperature, using the relation

$$\frac{1}{T} = \frac{\partial S}{\partial E},$$

is related to the identification of an absolute temperature scale. Historically, this came through the study of cyclic reversible processes. Suppose one has that the heat input is ΔQ_1 during an isothermal process at temperature T_1 and that the heat input in a second and final isothermal process, at T_2 is ΔQ_2. These two isothermal legs of the process are connected by two adiabatic ($\Delta Q = 0$) processes. The sequence of four processes defines a cycle.

A reversible engine operating between two heat reservoirs with temperatures T_1 and T_2 has a *heat ratio*

$$\eta = -\frac{\Delta Q_1}{\Delta Q_2}.$$

One can show (see, e.g., R. Becker, *Theory of Heat*, Springer-Verlag, New York, 1967, p. 19) that this ratio is independent of the reversible engine and that the type of reservoir η depends only on the temperatures of the reservoirs,

$$\eta = \eta(T_1, T_2).$$

This statement can be used to identify a temperature scale. It is straightforward to show for a sequence of two reversible engines attached by a heat bath at temperature T_2 that

$$\eta(T_1, T_2)\eta(T_2, T_3) = \eta(T_1, T_3).$$

This equation has the general solution

$$\eta(T_1, T_2) = \frac{f(T_1)}{f(T_2)}$$

where f is a general function of T. The simplest choice for f is simply $f = T$ and

$$\frac{\Delta Q_1}{\Delta Q_2} = -\frac{T_1}{T_2}.$$

The choice $f = T$ establishes a particular temperature scale known as the *absolute scale*.

Note that we have the result for a reversible cycle

$$\frac{\Delta Q_1}{T_1} + \frac{\Delta Q_2}{T_2} = 0$$

This means that the sum of variations of the quantity

$$\Delta S = \sum_i \frac{\Delta Q_i}{T_i} = 0 \tag{416}$$

in a reversible process. This implies that the variable S is a *function of state*. Thus S has the same value at a point in the thermodynamic plane independent of how one arrived at this point. This establishes the nature of entropy as a thermodynamic variable. Holding all other thermodynamic variables fixed, one can identify ΔQ_1 with the change in energy ΔE_1 associated with a change in ΔS_1, and one is led to $\partial S/\partial E = T^{-1}$. Gibbs was apparently the one to establish this. See the discussion in A. Sommerfeld, *Thermodynamics and Statistical Mechanics*, Academic Press, New York, 1964.

[44] The chemical potential was introduced by J. W. Gibbs, "On the equilibrium of heterogeneous substances," *Trans. Conn. Acad.*, **3**, 108 (1876); 343 (1878); reprinted in J. W. Gibbs, *The Collected Works*, Longmans, New York, 1931.

[45] The ideal gas law combined Boyle's law, $V \approx 1/p$, and Gay-Lussac's law $V \approx T$, to obtain $V \approx T/p$ or $pV/T \approx$ constant. For a discussion of the history, see G. Holton, *Introduction to Concepts and Theories in Physical Science*, Princeton University Press, Princeton, N.J., 1985.

[46] W. Thomson, *Proc. Cambridge Philos. Soc.* **1**, 69 (1848). While the nature of the absolute temperature scale was discussed in Ref. 43, the actual implementation of a temperature scale is an important and continuing technical problem. For a discussion, see D. Roller, *The Early Development of the Concepts of Temperature and Heat*, Harvard University Press, Cambridge, Mass., 1950.

[47] The thermodynamic potential was introduced by J. W. Gibbs, *Elementary Principles in Statistical Mechanics*, Ox Bow Press, Woodbridge, Conn., 1981, p. 201 and earlier.

[48] The general nature of the Euler relation given by Eq. (211) is discussed in Chapter 2, where we find that the simplicity of the expression is not an accident.

[49] R. H. Fowler, *Proc. Cambridge Philos. Soc.* **34**, 382 (1938), coined the term *grand partition function*.

[50] The meaning of this intrinsically quantum mechanical length is discussed in detail in Chapter 3.

[51] The stationary phase approach in statistical physics has been emphasized by C. S. Darwin and R. H. Fowler, *Philos. Mag.* **44**, 450, 823 (1922); **45**, 1 (1923); and R. H. Fowler, *Statistical Mechanics*, 2nd ed., Cambridge University Press, Cambridge, Mass., 1936. Discussion of the Darwin and Fowler method is given by K. Huang, *Statistical Mechanics*, 2nd ed. Wiley, New York, 1987.

[52] H. von Helmholtz, "Sitzber," *Kgl. Preuss. Akad. Wiss.* **1**, 22 (1882). "The thermodynamics of chemical process", communicated to the Berlin Academy on Feb. 2 and July 27, 1882.

[53] J. R. Newmann, *The World of Mathematics*, Vol. 1, Simon and Schuster, New York, 1956, p. 672, in his introduction to Hermann Weyl's retirement lecture on symmetry comments: "Every day symmetry carries the meaning of balance, proportion, harmony, regularity of form." Weyl went on to say that "Symmetry, as wide or as narrow as you may define its meaning, is one idea by which man through the ages has tried to comprehend and create order, beauty, and perfection."

[54] As we shall come to understand in detail as we go forward, the importance of symmetry in physics is fundamental. Symmetry arises in physics as one considers various transformations that take one physical system into another. An invariance principle, a transformation that leaves a physical system unchanged, implies a conservation law. See E. P. Wigner, "Symmetry and conservations laws," *Proc. Natl. Acad. Sci. (U.S.A.)* **51**, 956 (1964). See also Ref. 22.

[55] This example does not present a realistic physical picture of how systems become ferroelectric.

[56] We say that a solid has broken translational symmetry. Rigidity is a consequence of broken translational symmetry, and rigidity is easily verifiable experimentally. As quoted by P. W. Anderson: "when you kick a stone, no doubt remains in your mind that it possesses a property that we call rigidity, which we recognize as the very epitome of broken translational invariance." From *Basic Notions of Condensed Matter Physics*, Benjamin/Cummings, Menlo Park, Calif., 1984, p. 9.

[57] The universe at the time of the big bang, it is believed, was extremely symmetric. Indeed, at these earliest of times, when the universe was at an extraordinarily high temperature, all the forces of nature—gravity, electromagnetic, and weak and strong nuclear forces—were part of a single type of force. As the universe expanded and cooled there was symmetry breaking, which led to the establishment of the very different forces that we observe today. For a popular introduction to the evolution of the early universe, see *The*

First Three Minutes, by S. Weinberg, Bantam Books, New York, 1977. For a more technical discussion, see S. Weinberg, *Gravitation and Cosmology*, Wiley, New York, 1972.

[58] Transformation theory and the analysis of symmetry is more familiar in a quantum mechanical setting in terms of unitary operators and commutators.

[59] This manipulation is somewhat more recognizable in quantum mechanics, where we can write transformations in terms of unitary operators. The generalization of U_G in quantum mechanics is known as a *superoperator*.

[60] The role of Poisson brackets in classical systems in analogous to the use of commutators in quantum mechanics. These points are discussed in some detail in Appendix B.

[61] Rotation matrices are discussed in L. Schiff, *Quantum Mechanics*, 3rd ed., McGraw-Hill, New York, 1968.

[62] In simple terms a solid is rigid. One can pick up a solid, whereas one cannot pick up a fluid.

[63] In the final section of Appendix B we indicate how these ideas can be formalized within transformation theory. Our analysis here follows that of Professor Paul Martin, lecture notes, Harvard University, 1968–1969.

[64] In the case of solids this development forms the basis for standard theory of elasticity of solids. This is discussed in some detail in Chapter 6.

[65] See the additional discussion in Chapters 2 and 6.

[66] The boundaries can be treated carefully by assuming a potential which increases rapidly as a particle approaches a wall. One then has that the particle density vanishes rapidly as one approaches the wall.

[67] This expression for the pressure is discussed by J. E. Mayer and M. G. Mayer, *Statistical Mechanics*, 2nd ed., Wiley, New York, 1977, p. 298, who identify it as "the equation for the virial of Clausius."

[68] The kinetic theory of gases has an old, important, and controversial history. Robert Boyle published work in 1660 and 1662 which established for fixed temperature the quantitative result PV = constant. Boyle's law was apparently proposed first by the British scientists Henry Power and Richard Towncley on the basis of experiments begun in 1653. They did not publish their results until 1660 and 1661, when they were passed on to Boyle, who only then realized that his experiments of 1660 could be fit by *Boyle's law*. The Swiss physicist Daniel Bernoulli in 1738 published a prophetic kinetic model for gases which led to the result that PV/T is constant for a sample of gas. This work was 100 years ahead of its time. D. Bernoulli, *Hydrodynamics*, 1738. See S. Brush, *Kinetic Theory*, Vol. 1, Pergamon Press, Elnsford, N. Y., 1965 for further discussion.

[69] As discussed by Sir James Jeans, *The Dynamical Theory of Gases*, Dover, New York, 1954, R. Clausius, *Philos. Mag.*, Aug. 1870, derived a virial theorem for what he called the *virial*:

$$\frac{1}{2}\sum \mathbf{r} \cdot \frac{\partial V(r)}{\partial \mathbf{r}}. \qquad (417)$$

[70] The important role of Nambu–Goldstone modes in modern statistical mechanics is discussed in Volume 2. The important underlying concept is again symmetry. Background on the general significance of these broken symmetry modes is given by P. W. Anderson, *Basic Notions of Condensed Matter Physics*, Benjamin/Cummings, London,

1984; see also, J. Goldstone, *Nuovo Cimento* **19**, 154, (1961); Y. Nambu, *Phys. Rev.* **117**, 648 (1960); and P. W. Anderson, *Phys. Rev.* **112**, 1900 (1958).

[71] P. Ehrenfest, *Ann. Physics* **51**, 327 (1916). For further discussion of the adiabatic theorem, see A. Münster, *Statistical Thermodynamics*, Springer-Verlag, Berlin, 1969, p. 51; R. Becker, *Theory of Heat*, Springer-Verlag, New York, 1967, p. 129; and L. D. Landau and E. M. Lifshitz, *Statistical Physics*, 3rd ed., Pergamon Press, Oxford, 1980, p. 38.

[72] S. Nosé, *J. Chem. Phys.* **81**, 511 (1984); *Mol. Phys.* **52**, 255 (1984). For a generalization, see J. Jellinek and R. S. Berry, *Phys. Rev. A* **38**, 3069 (1988).

[73] This result is established for interacting classical systems in Chapter 4.

[74] The conventional Langevin equation for a particle moving with velocity **v** through a viscous medium is given by

$$\frac{d\mathbf{v}}{dt} = -\gamma \mathbf{v} + \boldsymbol{\eta}$$

where γ is a damping term and $\boldsymbol{\eta}$ is thermal noise with an autocorrelation proportional both to γ and the temperature of the bath in which the particle moves.

[75] W. G. Hoover, *Phys. Rev. A* **31**, 1695 (1985); **37**, 252 (1988).

[76] In the case of a system with a broken continuous symmetry, one must also include new variables in the description. The clearest example is that of a solid where the broken translational symmetry leads to the need to include the complete strain tensor among the list of macrovariables.

PROBLEMS

1.1. A bus has 9 seats facing forward and 8 facing backward. In how many ways can 7 passengers be seated if 2 refuse to ride facing forward and 3 refuse to ride facing backward?

1.2. A die is loaded so that even numbers occur three times as often as odd numbers. If the die is thrown 9 times, what is the probability that an odd number occurs 5 times?

1.3. In a system governed by Poisson statistics, the probability of finding n events of one type out of $N(\gg n)$ total events is

$$P_N(n) = \frac{a^n e^{-a}}{n!}$$

where a is a constant. Calculate, in the limit $N \to \infty$, the average value of n. Calculate the *fluctuations* in n:

$$\langle (\delta n)^2 \rangle = \langle (n - \langle n \rangle)^2 \rangle.$$

1.4. Consider a system where the events A and B exhaust all possibilities, so that $P(A \cup B) = 1$. Also assume that $P(A) = \frac{2}{3}$ and $P(B) = \frac{2}{3}$.

(a) Find $P(A \cap B)$.

(b) Are A and B independent?

(c) Find $P(B|A)$.

1.5. A random walker trying to walk a straight line has an equally likely chance of going forward or backward a distance of one unit with each step. After N steps, not necessarily large, what is his mean displacement $\langle x \rangle$, and what is the variance $\langle x^2 \rangle - \langle x \rangle^2$? What is the meaning of variance in this case?

1.6. Show that $P(B \cup C)$ and $P(B \cap C)$ are related by
$$P(B \cup C) = P(B) + P(C) - P(B \cap C).$$

1.7. Consider a system with two states, 1 and 2, with the probability of being in state 1 given by
$$P_1 = \frac{h_1}{h_1 + h_2}$$
and the probability of being in state 2 given by
$$P_2 = \frac{h_2}{h_1 + h_2},$$
where the positive numbers h_1 and h_2 are assumed known. Determine the fluctuation and dispersion for the stochastic variable A which takes on the values $A_1 = +1$ and $A_2 = -1$ in the two states. Discuss the limit $h_1 \gg h_2$.

1.8. The coin-flipping example discussed in the text can be generalized to a system with n equally probable outcomes, and one can consider the stochastic variable A which takes on values
$$A_l = \cos \frac{2\pi l}{n}$$
for $l = 1, 2, \ldots, n$. Notice that coin flipping corresponds to the case $n = 2$ where $A_1 = -1$ (tails) and $A_2 = 1$ (heads).
(a) Find the average value of A for this system.
(b) Determine the fluctuation in A as a function of n.
(c) What difference does it make to these results if we change the definition of A to
$$A_l = \cos\left(\frac{2\pi l}{n} + \phi\right)$$
where ϕ is a constant? This type of system is called an *n-state clock model*. Why?

1.9. Consider an n-state system where the states, labeled $l = 1, 2, 3, \ldots, n$, are equally probable. In this system there are two stochastic variables of interest:
$$A_l = A_0 \cos \frac{2\pi l}{n}$$

104   GENERAL PRINCIPLES OF STATISTICAL MECHANICS

and

$$B_l = B_0 \frac{l}{n}$$

where A_0 and B_0 are constants. Determine the correlation between A and B as a function of n.

1.10. A container of volume V contains N molecules.

(a) Assuming that there is no correlation between the locations of the various molecules, compute the probability $P(n, v)$ that a region of volume v (inside the container) contains exactly n molecules. Define $p = v/V$.

(b) Show that $\langle n \rangle = Np$ and

$$\sqrt{\langle (\delta n)^2 \rangle} = [Np(1-p)]^{1/2}.$$

(c) Show that if n and N are large, $P(n, v)$ is a Gaussian.

(d) Show that, if, in addition, $p \ll 1$, the function $P(n, v)$ is a Poisson distribution

$$P(n) = e^{-\langle n \rangle} \frac{\langle n \rangle^n}{n!}.$$

1.11. Suppose that we observe the spin of a proton at regular time intervals Δt and the spin can point either up or down. We can therefore associate the spin with a stochastic variable σ which can be $+1$ (up spin) or -1 (down spin). Suppose we know that the probability that the spin will flip during any interval Δt is

$$P_F(\sigma) = \frac{\alpha}{2}(1 - \sigma H)$$

where α is a positive parameter (< 1), and $H^2 < 1$. Find the average value of σ. (*Hint:* The key is to find the probability $P[\sigma, t]$ that σ has a particular value at time t. The probability at time $t + \Delta t$ is related to the probability at time t by

$$P[\sigma, t + \Delta t] = [1 - P_F(\sigma)]P[\sigma, t] + P_F(-\sigma)P[-\sigma, t].$$

Explain why this equation is correct.)

1.12. Suppose that for a system of N particles we define a vector $\mathbf{q} = (r_1^x, r_1^y, r_1^z, \ldots, p_N^x, p_N^y, p_N^z)$ which has the equation of motion

$$\dot{q}_i = F_i(\mathbf{q}).$$

Show using Hamiltonian's equations that

$$\sum_{i=1}^{6N} \nabla_q^i F_i(\mathbf{q}) = \nabla_q \cdot \mathbf{F}(q) = 0.$$

This equation is characteristic of conservative (as opposed to dissipative) systems.

Lorenz developed the following system of equations to describe the interrelations of temperature variation and convective motion:

$$\frac{dx}{dt} = -\sigma x + \sigma y$$
$$\frac{dy}{dt} = -xz + rx - y$$
$$\frac{dz}{dt} = xy - bz$$

where σ, r, and b are constants. Is this system of equations conservative?

1.13. A single harmonic oscillator, described by the Hamiltonian

$$H = \frac{p^2}{2m} + \frac{k}{2}x^2,$$

is ergodic. Show this by explicitly computing the time average

$$\overline{x^2(t)} = \lim_{T \to \infty} \frac{1}{T - t_0} \int_{t_0}^{T} dt\, x^2(t)$$

and the ensemble average $\langle x^2(t) \rangle$ with respect to the appropriate invariant probability distribution and showing that they are equal.

1.14. Discuss the ergodic properties of the simple pendulum as a function of energy.

1.15. Consider two uncoupled harmonic oscillators described by a Hamiltonian

$$H = p_1^2 + p_2^2 + \omega_1^2 x_1^2 + \omega_2^2 x_2^2.$$

We know that two constants of the motion are the individual energies of the oscillators

$$\Phi_1 = p_1^2 + \omega_1^2 x_1^2$$
$$\Phi_2 = p_2^2 + \omega_2^2 x_2^2.$$

Suppose that $\omega_2 = n\omega_1$, where n is an integer. Show that the third integral of the motion is a multivalued function with n branches. You need not determine the function explicitly. Determine Φ_3 explicitly for the case $n = 2$.

1.16. A system consists of two independent subsystems, with configurations c and c'. Thus the probability of a combined configuration (c, c') is simply the product of two independent probabilities: $p_c q_{c'}$. Show that the entropy of the combined system is the sum of those of the two subsystems.

1.17. Derive and investigate the validity of *Stirling's formula*:

$$n! = \sqrt{2\pi n}\, n^n e^{-n} \quad \text{for} \quad n \gg 1.$$

Using the integral representation for the factorial

$$n! = \int_0^{+\infty} x^n e^{-x} dx,$$

and noting that $x^n e^{-x}$ is sharply peaked about a particular value of x for very large n, derive Sterling's formula and the leading correction for large n. That is, find α_0 defined by

$$n! = \sqrt{2\pi n}\, n^n e^{-n} \left[1 + \frac{\alpha_0}{n} + \mathcal{O}\left(\frac{1}{n^2}\right)\right].$$

If $N = 15$, what error does one make when using Stirling's formula?

1.18. Consider a situation where we have two choices. Suppose we know that the probability of choice 1 is equal to that for choice 2. What information S_1 do we obtain if we perform an experiment distinguishing between the two possibilities? Suppose we know that the probability of choice 1 is $\frac{1}{3}$ before the experiment. What information S_2 do we obtain from the experiment? Is $S_1 > S_2$, $S_1 = S_2$ or $S_2 > S_1$? How could you know which of these cases holds before carrying out the detailed calculation? Explain!

1.19. Consider a situation where there are a total of n possible configurations. If P_i is the probability that the system is the ith configuration ($i = 1, 2, ..., n$), the missing information or degree of randomness is defined by

$$S = -k \sum_{i=1}^{n} P_i \ln P_i$$

where k is a constant. Find the probability distribution P_i that maximizes S.

1.20. Consider a system with 2^n possible configurations, each having equal probability. A given realization of the system is one of these configurations: c. One way to quantify the information needed to determine the configuration is the minimal number of yes–no questions we must ask in order to determine c. Accordingly, we ask is the occupied configuration on the left half? This eliminates one half. For the remaining half we ask is it in the left quarter?... And so on.

(a) How many guesses are required using this scheme?

(b) What is the entropy of this system using base 2 logarithms?

1.21. Evaluate the binomial probability distribution given by Eq. (2406) for the maximum value $n_1 = n_{\max} = Np$ for large N. Assume that n_1 is approximately continuous and that one can use the Stirling formula.

1.22. Consider an Ising model of noninteracting spins $\sigma(\mathbf{R}) = \pm 1$ where there are a total of N lattice sites labeled by $\mathbf{R}$.

(a) Assuming that the system is symmetric between the up $=\uparrow= +1$ and down $=\downarrow= -1$ states, compute the entropy/missing information for this system.

(b) If we assume that there is an applied magnetic field, the probability of obtaining an up or down configuration is biased by the field. Assuming that we have the probabilities for each site,

$$P[\uparrow] = \frac{e^H}{e^H + e^{-H}}$$

$$P[\downarrow] = \frac{e^{-H}}{e^H + e^{-H}}$$

where H is proportional to the applied magnetic field, compute the entropy or missing information for this system.

(c) Compare the two calculations.

1.23. Consider a 10,000-letter message constructed from the 27-letter (including a space) alphabet. First construct the missing information by assuming equal a priori probabilities. Calculate the missing information per letter. Use the following table

Symbol	Probability	Symbol	Probability	Symbol	Probability
Wordspace	0.2	S	0.052	YW	0.012
E	0.105	H	0.047	G	0.011
T	0.072	D	0.035	B	0.0105
O	0.0654	L	0.029	V	0.008
A	0.063	C	0.023	K	0.003
N	0.059	FU	0.0225	X	0.002
I	0.055	M	0.021	JQZ	0.001
R	0.054	P	0.0175		

to construct the information per letter, taking into account a priori probabilities of the individual letters. What is the expected relationship between these two values for the missing information per letter?

1.24. In the limit of large N and large pN, the binomial distribution approaches a Gaussian distribution. This follows from a series of steps. First show that the binomial distribution $P_N(n_1)$ exhibits a maximum at $n_1 = \langle n_1 \rangle = Np$. Thus for $p > 0$, the average value $\langle n_1 \rangle$ grows with N. For $N \gg 1$, the values of n_1 in the region of $\langle n_1 \rangle$ will be very large relative to integer changes in n_1 which can occur. Therefore, in the region of the maximum we can treat n_1 as a continuous variable. Next use Stirling's formula to find $P_N(n_1)$ for large N and n_1. This involves expanding $\ln[\sqrt{2\pi N} P_N(n_1)]$ about its maximum value. Show that we have a Taylor series expansion about $n_1 = Np$ of the form

$$\ln[\sqrt{2\pi N} P_N(n_1)] = \sum_{s=0}^{\infty} \frac{A_s}{s!}(n_1 - Np)^s$$

where the coefficients are given by

$$A_s = \frac{d^s}{dn_1^s} \ln[\sqrt{2\pi N}\, P_N(n_1)]\big|_{n_1=Np}$$

where

$$A_1 = 0$$

$$A_2 = -\frac{1}{Npq}$$

and in general

$$|A_s| = \frac{1}{(Npq)^{s-1}}.$$

Remembering that the standard deviation for the binomial distribution is $\sigma_N = \sqrt{Npq}$, show that we can write $P_N(n_1)$ in the form

$$P_N(n_1) = \frac{1}{\sqrt{2\pi}\,\sigma_N} \exp\left[-\frac{1}{2}\frac{(n_1 - \langle n_1 \rangle)^2}{\sigma_N^2}\right].$$

1.25. Compute the total number of states for an ideal gas ($H = \sum_{i=1}^{N} \mathbf{p}_i^2/2m$) in the microcanonical ensemble where we hold both the energy E and the total momentum, $\mathbf{P} = \sum_{i=1}^{N} \mathbf{p}_i$, fixed. That is, determine

$$Z_M(E, N, V, \mathbf{P}_0) = \int_V \frac{dq_N}{h^{3N} N!} E_T \delta(E - H) P_T^3 \delta(\mathbf{P}_0 - \mathbf{P})$$

where $P_T = mcN$. Determine the entropy per particle in the thermodynamic limit. Discuss the role of the total momentum and identify the variable $\partial S/\partial \mathbf{P}_0$ conjugate to $\mathbf{P}_0$.

1.26. Consider a classical, ideal, monatomic gas confined by a spherical box with perfect reflecting walls. In this case, the total angular momentum, $\mathbf{J}$, is conserved, so dynamical motion is confined to the $(6N-4)$-dimensional surface, S', in Γ defined by $H = E$ and $\mathbf{J} = \mathbf{J}_0$. If the motion is appropriately ergodic on S', the gas will be described by the modified microcanonical ensemble $\delta(H - E)\delta(\mathbf{J} - \mathbf{J}_0)\,d^{3N}q\,d^{3N}p$. Find the probability distribution $P(\mathbf{v})$ that a typical particle has velocity $\mathbf{v}$.

1.27. The angular momentum density for a fluid system is defined by

$$\mathbf{l}(\mathbf{x}, t) = \sum_{i=1}^{N} \mathbf{r}_i(t) \times \mathbf{p}_i(t) \delta(\mathbf{x} - \mathbf{r}_i(t)).$$

Show that $\mathbf{l}(\mathbf{x}, t)$ satisfies a continuity equation and that the associated current can be expressed in terms of the stress tensor.

1.28. The microscopic expressions for the particle, momentum, and energy densities for an ideal gas of N particles are given by

$$n(\mathbf{x},t) = \sum_{i=1}^{N} \delta(\mathbf{x} - \mathbf{r}_i(t))$$

$$\mathbf{g}(\mathbf{x},t) = \sum_{i=1}^{N} \mathbf{p}_i(t)\delta(\mathbf{x} - \mathbf{r}_i(t))$$

$$\epsilon(\mathbf{x},t) = \sum_{i=1}^{N} \frac{\mathbf{p}_i^2(t)}{2m}\delta(\mathbf{x} - \mathbf{r}_i(t)).$$

Derive the associated continuity equations for each. We know that the averages of these equations must be locally satisfied for the existence of thermal equilibrium. Show that this is indeed the case for the equations you derived above.

1.29. Using the information-theoretic approach developed in the text, find the equilibrium probability distribution in the case where the average energy E, average number particles $\bar{N}$, volume V, average total momentum $\mathbf{P}_0$, average total angular momentum $\mathbf{L}_0$, and average generator of Galilean transformations $\mathbf{G}_0$ are known. Identify, and relate where possible, the associated Lagrange multipliers.

1.30. Consider a function $f(x)$ that possesses a power series expansion

$$f(x) = \sum_{n=0}^{\infty} f_n x^n.$$

Show in the grand canonical ensemble that, if one assumes that

$$\langle H \rangle = V f(\beta, \alpha)$$

then,

$$\langle f(H) \rangle = f(\langle H \rangle)$$

for large volume V.

1.31. Determine the specific heat at constant volume of an ideal gas in a cylindrical box of radius R and length L that is rotating about its axis with angular velocity ω.

1.32. (a) A classical gas of N particles at temperature T is confined by a central potential (rather than a box) of the form $V(r) = ar^b$, where a and b are constants. Calculate the expected energy E and the specific heat C of the gas.
(b) A cubic box of side L of a classical gas of N particles of mass m is placed in

a uniform gravitational field of strength g. (The box is oriented so that the gravitational field is normal to a pair of faces.) The gas is in thermal equilibrium at temperature T. Calculate the average number density, $n(\mathbf{x})$, of the gas and the specific heat (at constant volume) of the gas.

1.33. (a) Write down the general expression for the probability that a system in the grand canonical ensemble has exactly N particles.

(b) Show that for a classical ideal gas the expression found in part (a) becomes a Poisson distribution. Calculate the root-mean-square value of $\langle(\delta N)^2\rangle$ and show that it agrees with the value found directly.

1.34. Consider the case where an external nonlocal potential field $U(\mathbf{x})$ acts on an ideal gas. We then have the Hamiltonian

$$H = \sum_{i=1}^{N} \frac{p_i^2}{2m} + \int d^3x\, U(\mathbf{x}) n(\mathbf{x}).$$

A particular case is the gravitational potential $U(\mathbf{x}) = mgz$. Starting from the stationarity condition satisfied by the equilibrium average of the continuity equation for the momentum density, find the average particle density and pressure. Assume that you are given the value of the particle density at a particular point in space $\mathbf{x}_0$, $n_0 = \langle n(\mathbf{x}_0)\rangle$.

1.35. Consider an ensemble in which the walls of each system are flexible and heat conducting. Each system is described by N, T, and p (pressure). Let the average values of the total energy and the volume be E and $\bar{V}$. Following the lines of our treatment of the grand canonical ensemble:

(a) Show that the partition function is

$$Z(T, N, p) = \mathrm{Tr}\, e^{-\beta(H+pV)}$$

Give the proper definition of the trace.

(b) Show that the characteristic *potential* for this ensemble,

$$G = -k_B T \ln Z(T, N, p)$$

is the Gibbs free energy.

(c) Derive an expression for the volume fluctuations in this ensemble. Express your answer in terms of isothermal compressibility.

1.36. Discuss the dispersion of energy and particle number

$$\Delta_E = \frac{\sqrt{\langle (H-E)^2\rangle}}{E}$$

$$\Delta_N = \frac{\sqrt{\langle (N-\langle N\rangle)^2\rangle}}{\langle N\rangle}$$

in the microcanonical, canonical, and grand canonical ensembles. Calculate Δ_E and Δ_N in the three ensembles in the case of an ideal classical gas, and contrast your results.

1.37. The mass m of a (classical body) is measured by hanging it on a spring with spring constant K in a uniform gravitational field g using the formula $m = Kx/g$, where x is the displacement from equilibrium. Suppose that the displacement readout system is perfect, so that the only source of error in the measurement is the buffeting of the mass by air molecules at temperature T. Assuming that only one reading of the displacement is taken, what uncertainty Δm should be assigned to the measurement of m?

1.38. Consider a simple classical model for a magnet where we have a lattice of N sites and on each site i there is a magnetic moment $\mathbf{m}_i$ which is free to take on any values. We assume that the Hamiltonian governing these moments is given by

$$H = \sum_{i=1}^{N} \frac{\alpha_0}{2} \mathbf{m}_i^2$$

where α_0 is a material parameter and the total magnetization is given by

$$\mathbf{M}_T = \sum_{i=1}^{N} \mathbf{m}_i.$$

(a) Working in the canonical ensemble, where we introduce the Lagrange multipliers β and $-\beta \mathbf{H}$ where $\mathbf{H}$ is the applied magnetic field, compute the average total magnetization. Look at the behavior of this quantity for large and small applied magnetic fields. Do your results make physical sense?

(b) Find the entropy for this system in terms of its natural variables N, E, and $\langle \mathbf{M}_T \rangle$.

(Note: Try not to confuse the scalar Hamiltonian H with the applied external field $\mathbf{H}$.)

1.39. Consider a collection of N noninteracting classical particles each with mass m. This system is in equilibrium at temperature T and with a net average total momentum $\mathbf{P}_0$. Working in the canonical ensemble, find the equilibrium probability distribution governing this system. Use this result to compute the average energy current

$$\mathbf{J}_E = \sum_{i=1}^{N} \mathbf{v}_i \tfrac{1}{2} m v_i^2,$$

where $\mathbf{v}_i$ is the velocity of particle i, in terms of N, m, T, and $\mathbf{P}_0$.

1.40. Find the canonical ensemble probability distribution using our information-theoretic approach.

1.41. A classical gas of N particles in one spatial dimension is confined by a box to the coordinate range $0 \leq q \leq L$. A logarithmic potential is applied, so that the Hamiltonian of the system is

$$H = \sum_{i=1}^{N} \left(\frac{p_i^2}{2m} + V_0 \ln(1 - q_i) \right).$$

The gas is in thermal equilibrium at temperature T. What is its expected average energy E?

1.42. A classical particle moves in one dimension in a potential $V(x) = \lambda x^4$. It is in thermodynamic equilibrium with a heat reservoir at temperature β^{-1}. What are its mean energy and specific heat?

1.43. Consider a classical ideal gas of N particles with total energy E in a box of volume V. Imagine that at some randomly chosen time, we suddenly insert a partition that divides the box into two equal volumes, labeled 1 and 2. On average, of course, the number of particles, N_1, in the first volume will be $N/2$, and the energy E_1 in that volume will be $E/2$.

(a) Estimate the probability that N_1 is nearly $N/2$ but $E_1 = aE/2$, where a differs from unity by more than 1 percent.

(b) Suppose that an energy fluctuation of 1% is observed [i.e., $E_1 = 1/2(1.01)E$]. How many particles N_1 would you expect to find in the first half of the box?

1.44. Determine the set of symmetry operations that leave the pattern in Fig.1.12 unchanged.

1.45. Show that the operator $U_{P_T^z}(z_0)$, defined by Eq. (270), does indeed translate coordinates of the system a distance z_0 along the z-axis.

1.46. Consider a set of classical magnetic moments $S_\alpha(\mathbf{R})$ $(\alpha = x, y, z)$ defined on a cubic lattice and located at sites $\mathbf{R}$. The Hamiltonian describing this system is given by

$$H = -\frac{1}{2} \sum_\alpha \sum_\mathbf{R} \sum_{\mathbf{R}'} J_\alpha(|\mathbf{R} - \mathbf{R}'|) S_\alpha(\mathbf{R}) S_\alpha(\mathbf{R}')$$

where $J_\alpha(|\mathbf{R} - \mathbf{R}'|)$ is a function of the distance between magnetic moments. Assume that the fundamental Poisson brackets between the moments are given by

$$\{S_\alpha(\mathbf{R}), S_\beta(\mathbf{R}')\} = \sum_\gamma \epsilon_{\alpha\beta\gamma} S_\gamma(\mathbf{R}) \delta_{\mathbf{R},\mathbf{R}'},$$

where $\epsilon_{\alpha\beta\gamma}$ is the usual antisymmetric tensor. Work out explicitly the equation of motion satisfied by $S_\alpha(\mathbf{R}, t)$:

$$\frac{\partial}{\partial t} S_\alpha(\mathbf{R}, t) = \{S_\alpha(\mathbf{R}, t), H\}.$$

What conditions must be satisfied by the J_α's for components of the total magnetization

$$M_\alpha = \sum_R S_\alpha(R)$$

to be conserved? Assume that the total magnetization is proportional to the total angular momentum in the system. Conservation of a component of the total magnetization implies what invariance principle?

1.47. Consider the equilibrium-averaged correlation function,

$$C(\mathbf{x} - \mathbf{x}') = \langle n(\mathbf{x})n(\mathbf{x}')\rangle,$$

where the particle density is given by

$$n(\mathbf{x}) = \sum_{i=1}^{N} \delta(\mathbf{x} - \mathbf{r}_i).$$

For a system with translational and rotational invariance, show that

$$C(\mathbf{x} - \mathbf{x}) = C(|\mathbf{x} - \mathbf{x}'|).$$

1.48. Consider the average of the product of two vectors $\mathbf{A}$ and $\mathbf{B}$ of the form

$$K_{\alpha\beta} = \langle A_\alpha B_\beta\rangle \delta_{\alpha\beta}.$$

If the underlying probability distribution is rotationally invariant, use the invariance principle discussed in the text to show that

$$K_{\alpha\beta} = \tfrac{1}{3}\langle \mathbf{A} \cdot \mathbf{B}\rangle \delta_{\alpha\beta}.$$

1.49. For an isotropic solid the stress and strain tensors are related for modest strains by Hooke's law:

$$\sigma_{\alpha\beta} = -B \sum_l U_{ll}\delta_{\alpha\beta} - 2\mu\left(U_{\alpha\beta} - \tfrac{1}{3}\delta_{\alpha\beta}\sum_l U_{ll}\right)$$

where $B(>0)$ is the bulk modulus and $\mu(>0)$ is the shear modulus. Consider a rectangular rod of length L aligned along the z-axis with a square cross section of side a. Assume that a pressure p is applied along the z-axis. Determine the amount of transverse expansion of the rod in terms of p, B, μ, L, and a.

1.50. Show that the expression for the pressure given by Eq. (355),

$$p = \frac{2}{3}\epsilon_K - \frac{1}{6}\int d^3r\, r\frac{\partial V(r)}{\partial r} g(r)n^2,$$

can be rewritten as

$$pV = \left\langle \sum_{i=1}^{N} \frac{\mathbf{p}_i^2}{3m} \right\rangle - \frac{1}{6} \left\langle \sum_{i \neq j=1}^{N} |\mathbf{r}_i - \mathbf{r}_j| \frac{\partial V(|\mathbf{r}_i - \mathbf{r}_j|)}{\partial |\mathbf{r}_i - \mathbf{r}_j|} \right\rangle.$$

1.51. Starting with the rather general Hamiltonian

$$H = \sum_{i=1}^{N} \frac{\mathbf{p}_i^2}{2m} + U(\mathbf{r}_1, \mathbf{r}_2, \ldots, \mathbf{r}_N),$$

show that the number of states with energy less than E is given by

$$\Sigma_M = \int d^3 r_1 d^3 r_2 \cdots d^3 r_N A_N (E - U)^{3N/2} \theta(E - U)$$

and identify A_N, which is independent of E. Show that in the thermodynamic limit the entropies per particle, which follow from the two definitions for the entropy,

$$S_\Sigma \equiv k_B \ln \Sigma_M$$

and

$$S = k_B \ln Z_M,$$

where Z_M is the microcanonical partition function defined in the text, are equivalent.

1.52. Use the results obtained in the microcanonical ensemble

$$\frac{p}{T} = \left(\frac{\partial S}{\partial V} \right)_{E,N}$$

$$\frac{\mu}{T} = -\left(\frac{\partial S}{\partial N} \right)_{E,V},$$

to show that in the canonical ensemble,

$$p = -\left(\frac{\partial F}{\partial V} \right)_{T,N}$$

$$\mu = \left(\frac{\partial F}{\partial N} \right)_{T,V}$$

where F is the Helmholtz free energy:

$$F = E - TS.$$

1.53. Starting with Eq.(410) with H given by Eq. (407) and integrating over s and p_s holding $P_i = p_i/s$ fixed, show that the probability distribution $P[q_i, P_i]$ is canonical and that $X = 3N + 1$.

1.54. In the microcanonical ensemble for an interacting classical fluid, show that the temperature can be expressed in terms of the average of the inverse kinetic energy via

$$k_B T = \frac{1}{(3N/2 - 1)} \frac{1}{\langle K^{-1} \rangle}.$$

1.55. In the microcanonical ensemble for an interacting classical fluid, show that the specific heat at constant volume can be written in the form

$$\frac{1}{C_V} = \frac{2}{3k_B N} \left[1 - \left(\frac{3N}{2} - 1 \right) \langle K \rangle \langle K^{-1} \rangle \right]$$

where K is the kinetic energy.

1.56. In the microcanonical ensemble, show that

$$\left\langle p_i^\alpha \frac{\partial H}{\partial p_i^\alpha} \right\rangle = \left\langle r_i^\alpha \frac{\partial H}{\partial r_i^\alpha} \right\rangle = \frac{1}{d \ln \Sigma_M(E)/dE}$$

for all i and α and where $\Sigma_M(E)$ is the number of states with energy less that E.

1.57. Show that the potential-energy contribution to the energy density in an interacting simple fluid, $\epsilon_p(\mathbf{x})$, has a part that is quadratic in the phase-space density $f(\mathbf{x}, \mathbf{p})$.

1.58. The equation of motion for the phase-space density is given by Eq. (2603). Show that the potential-energy contribution in this equation,

$$\int d^3 x_1 d^3 p d^3 p_1 [\nabla_x^\alpha V(\mathbf{x} - \mathbf{x}_1)] f_2(\mathbf{x}, \mathbf{p}, \mathbf{x}, \mathbf{p}_1),$$

can be written in the form

$$\int d^3 x' \, \nabla_x^\alpha V(\mathbf{x} - \mathbf{x}') n_2(\mathbf{x}, \mathbf{x}') = \frac{1}{4} \sum_\beta \nabla_x^\beta \int d^3 r \int_{-1}^{+1} ds \, \frac{r_\beta r_\alpha}{r} \frac{\partial V(r)}{\partial r}$$

$$\times n_2 \left(\mathbf{x} + \frac{(s+1)}{2} \mathbf{r}, \mathbf{x} + \frac{(s-1)}{2} \mathbf{r} \right).$$

Explain why this is of interest.

1.59. Show that the energy-density current defined by the continuity equation for the energy density, Eq. (2597), can be written in the form

$$J_\epsilon^\alpha(\mathbf{x}) = \int d^3 p \left[\frac{p^2}{2m} + \int d^3 r \frac{1}{2} V(\mathbf{x} - \mathbf{r}) n(\mathbf{r}) \right] p_\alpha f(\mathbf{x}, \mathbf{p})$$

$$- \frac{1}{4} \int_{-1}^{+1} ds \int d^3 r \int d^3 p d^3 p' \sum_\beta \frac{r_\alpha r_\beta}{r} \frac{\partial V(r)}{\partial r} (p_\beta + p'_\beta)$$

$$\times f_2 \left(\mathbf{x} - \mathbf{r} \frac{s+1}{2}, \mathbf{p}, \mathbf{x} - \frac{\mathbf{r}(s-1)}{2}, \mathbf{p}' \right).$$

2 Principles of Thermodynamics

2.1 INTRODUCTION

Several ideas developed in Chapter 1 are very general and independent of the microscopic details or structure of the physical system we studied. We were able to articulate an *entropy principle* [1]: The entropy of a system in equilibrium is a maximum consistent with all macroscopic constraints on the system. We also saw how for a given choice of independent variables (e.g., the macrovariables E, N, and V) we could microscopically define and in principle compute one of the potentials S, F, or Ω. We then showed how various other quantities of physical interest could be obtained from these potentials by taking derivatives. These ideas are put on a more general footing in *thermodynamics*.

The basic assumption in thermodynamics is that there generally exists a potential (the entropy) which is a function of the macrovariables. One can then (1) extend and apply the entropy principle to systems with constraint conditions that are more complicated than those discussed in Chapter 1, and (2) systematize the interrelationship between different ensembles (different choices of independent variables) for the same physical system.

Because thermodynamics correlates information associated with macrovariables, it treats systems on length and time scales long compared to any intrinsic length or time scales in the system. An example of such an intrinsic length scale is the mean free path in a fluid or the lattice spacing in a crystal. For the simplest systems, such as noble gases, this means a treatment of uniform bulk behavior such as specific heats. It does not include spatial correlations on interatomic distances as measured, for example, in x-ray scattering and discussed in Chapter 4. There are more complicated systems, such as solids and dielectric materials, where there are long-range interactions and where there are spatial variations in the system associated with, for example, forces applied at the boundaries.

At some level thermodynamics is a subfield [2] of statistical mechanics. One needs the methods of statistical mechanics to identify the macrovariables and explicitly determine the potentials. However, given the existence of the macrovariables and a potential, the results of thermodynamics ride above the details of the particular microscopic statistical mechanical development. Thermodynamics allows us to make rather general statements about the properties of large systems independent of the details of the system investigated. The formal structure of the thermodynamics can be quite similar for apparently very different physical systems once one has [3] the proper macrovariables.

118 PRINCIPLES OF THERMODYNAMICS

Our discussion of thermodynamics will be broken into three basic parts. In the first section we discuss the formal foundations of thermodynamics and the related transformation theory which allows us to relate different ensembles. In the middle part of this chapter we discuss the application of the entropy principle to a variety of physical situations (phase transitions, mixtures, chemical reactions) from a general point of view. A key assumption [4] in thermodynamics is that we *know* the equations of state. Thus to apply the thermodynamic ideas developed in the first two portions of this chapter in detail, we must construct the appropriate equations of state. In the last third of this chapter we discuss approximate equations of state for three different physical systems: Isotropic magnets, simple mixtures and classical fluids. We show how to extract useful information from these equations of state.

Our discussion of thermodynamics will be rather incomplete. We do not discuss the very important application of thermodynamic ideas to the area of energy [5] use and efficiency. The associated discussion of heat cycles, refrigerators, and so on, is, of course, very important from an engineering point of view.

2.2 GENERAL POSTULATES OF THERMODYNAMICS

We start in this section with a formal statement of the basic principles of thermodynamics. We should keep in mind, however, that these postulates are simply a restatement of many of the general principles developed in Chapter 1.

We begin our formulation of the general principles of thermodynamics by making a list of the macrovariables in our system: $(X_0, X_1, \ldots, X_t)$. In our simple fluid example the macrovariables are E, N, and V. We develop thermodynamics for very general systems. Thus we could add to this list the number of particles for different chemical species in a mixture of gases, the magnetization in a magnetic system, the strain tensor in a solid, or as we mentioned earlier, the linear and angular momentum. We use the convention that $X_0 = E$, the total energy. Recall that macrovariables are extensive and additive when combining subsystems; therefore, $X_i = X_i^{(1)} + X_i^{(2)}$, where, for example, $X_i^{(2)}$ is the value of the parameter in the subsystem 2 and X_i is its value in the combined system.

The basic postulates [6] of thermodynamics are:

1. The equilibrium state of a system is characterized by a specification of its macrovariables (i.e., the X's).
2. The entropy is a function of the macrovariables

$$S = S(X_0, X_1, \ldots, X_t) \qquad (418)$$

with the property that the equilibrium values of the macrovariables are those that maximize S consistent with any applied constraints. The entropy is postulated to have the following mathematical properties:

   (a) S is a homogeneous first-order function of the macrovariables. That is, if all the macrovariables of the system are multiplied by a constant λ, the entropy is multiplied by the same constant,

$$S(\lambda X_i) = \lambda S(X_i). \tag{419}$$

(b) S is a continuous, differentiable function and can be inverted uniquely, to give

$$E \equiv X_0 = E(S, X_1, \ldots, X_t) \tag{420}$$

$$\left(\frac{\partial S}{\partial E}\right)_{X_1, X_2, \ldots X_t} > 0. \tag{421}$$

3. All possible thermodynamic information about a system is contained in the fundamental equation

$$S = S(E, X_1, \ldots, X_t). \tag{422}$$

A few comments are in order to tie these postulates to our work in Chapter 1. The idea of the macrostate being specified by the macroscopic variables, the X_i's, is one that we have emphasized before. Again we expect the choice of the X's to depend on the symmetries of the Hamiltonian and possibly, as in solids, includes variables, such as the strain tensor $U_{\alpha\beta}$, associated with broken symmetry.

We have to say more about the extremum principle for the entropy in order to relate it back to our principle of maximum missing information. We first have to discuss what we mean by an internal constraint. Let us suppose that two simple systems are contained within a closed cylinder, separated from each other by an internal piston [7]. Initially, system 1 has energy, volume, and particle number $E^{(1)}$, $V^{(1)}$, and $N^{(1)}$, respectively. Similarly, system 2 is specified by $E^{(2)}$, $V^{(2)}$, and $N^{(2)}$. We assume that the cylinder walls and the piston are rigid, impermeable to matter, there is no heat flow, and the position of the piston is firmly fixed. Each system is closed so the macrovariables can be specified for each system separately. If we now free the piston, it will seek a new position. The removal of the constraint will result in the onset of some spontaneous process that will eventually result in a new equilibrium state. Similarly, if the adiabatic coating is stripped from the piston so that heat can flow between the two systems, there will be a redistribution of energy between the two systems. Again, if the holes are punched in the piston, there will be a redistribution of matter (and also energy) between the two systems. Our postulate says, in each case, that the new state will be the one with maximum entropy consistent with the remaining constraints. If we remove the piston altogether, punch enough holes, the only constraints left (assuming that the subsystems are originally filled with the same type of material) are that the total energy, volume, and number of particles are held fixed for the combined system.

Postulate 2 is completely consistent with our earlier statements about the missing information. A macroscopic constraint on a system is a known piece of information. If we remove the constraint and a chaotic process follows, we have lost information about the state of the system. The missing information must increase.

120   PRINCIPLES OF THERMODYNAMICS

The mathematical properties we give for the entropy follow directly from our statistical mechanical discussion. In particular, the homogeneity property follows directly from the result (for simple fluids)

$$TS = E - \mu N + pV, \tag{423}$$

which we extend below to the more general set of variables (X_i).

Postulate 3 follows once we recall that the entropy serves as the potential in the microcanonical ensemble from which we can obtain temperature, chemical potential, and pressure and, consequently, knowing the functional interrelationships among these parameters gives complete thermodynamic information. In some sense this is a definition of what we mean by thermodynamic information.

2.3   FIRST-DERIVATIVE INFORMATION

2.3.1   Conjugate Forces

Although the entropy contains all available thermodynamic information, it is not directly experimentally observable. Experimentally, we can only measure derivatives of S. Since S is a function of the independent variables X_i, it is natural to consider the first derivatives:

$$dS = \sum_{i=0}^{t} \left(\frac{\partial S}{\partial X_i}\right) dX_i \tag{424}$$

and define

$$k_B F_{X_i} = \frac{\partial S}{\partial X_i} \tag{425}$$

so

$$dS = \sum_{i=0}^{t} k_B F_{X_i} dX_i. \tag{426}$$

The F_{X_i}'s, called generalized *forces* [8], are said to be *conjugate* to the X_i's. Note that the F_{X_i}'s are intensive (i.e., they do not scale with the size of a system because they are the *ratio* of two extensive quantities). We have immediately, from Chapter 1, that

$$k_B F_E = \left(\frac{\partial S}{\partial E}\right)_{X_1,\ldots,X_t} = \frac{1}{T} = k_B \beta \tag{427}$$

where T is the temperature, for fluids with $x_1 = V$, the volume, we have

$$k_B F_V = \left(\frac{\partial S}{\partial V}\right)_{X_0,X_2,\ldots,X_t} = \frac{p}{T} \tag{428}$$

where p is the pressure, and if we choose $X_2 = N$, the average number of particles, then

$$k_B F_N = \left(\frac{\partial S}{\partial N}\right)_{X_0, X_1, X_3, \ldots, X_t} = \frac{-\mu}{T} = k_B \alpha. \tag{429}$$

For a solid we must include the stain tensor $V_0 U_{\alpha\beta}$ as a member of the set of macrovariables, and from the discussion in Chapter 1,

$$k_B F_{U_{\alpha\beta}} = \frac{\partial S}{\partial (V_0 U_{\alpha\beta})} = \frac{\langle \sigma_{\alpha\beta} \rangle}{T} \tag{430}$$

where $\langle \sigma_{\alpha\beta} \rangle$ is the average stress tensor. Another important example occurs in magnetic problems, where the magnetization **M** must be included in the list of extensive variables. The conjugate variable in this case is

$$k_B F_{M_\alpha} = \frac{\partial S}{\partial M_\alpha} = -\frac{H_\alpha}{T} \tag{431}$$

where H_α is the αth component of the applied magnetic field. Our differential relation for the entropy, Eq. (426), can be written

$$dS = \frac{dE}{T} + k_B \sum_{i=1}^{t} F_{X_i} \, dX_i \tag{432}$$

and reduces in the case of a fluid to

$$dS = \frac{dE}{T} + p\frac{dV}{T} - \mu\frac{dN}{T}. \tag{433}$$

2.3.2 Energy Representation

The formulation of thermodynamics given above, which centers on the entropy as the fundamental *potential*, is not the only possibility [9]. Indeed, historically, this role was played by the energy and it was the search for the macrovariable associated with heat that led to the identification of entropy [10] as an independent thermodynamic variable. The change of independent variables from $X = \{E, X_1, \ldots, X_t\}$ to $Y = \{S, X_1, \ldots, X_t\}$ is not difficult. The first step is to rewrite Eq. (432) in the form

$$dE = T\,dS - k_B T \sum_{i=1}^{t} F_{X_i}\, dX_i. \tag{434}$$

If we define the conjugate forces in this description as

$$g_{Y_i} = \frac{\partial E}{\partial Y_i}, \tag{435}$$

holding all other Y_j fixed, then Eq. (434) takes the compact form

$$dE = \sum_{i=0}^{t} g_{Y_i} dY_i. \qquad (436)$$

The new conjugate forces are now given in terms of the old by

$$g_0 = g_S = T \qquad (437)$$

and for $i \neq 0$,

$$g_{Y_i} = -k_B T F_{X_i}. \qquad (438)$$

We have then, in particular,

$$g_V = -k_B T F_V = -p \qquad (439)$$
$$g_N = \mu \qquad (440)$$
$$g_{U_{\alpha\beta}} = -\langle \sigma_{\alpha\beta} \rangle \qquad (441)$$

and

$$g_{M_i} = H_i. \qquad (442)$$

Scrutinizing the g_{Y_i}, we see that they are physically more fundamental than the F_{X_i}. Indeed, they have the direct interpretation in terms of applied forces and their own historic symbols! A change in the energy for a system must be accounted for in terms of changes in the associated macrovariables Y. Each term contributing to dE in Eq. (436) can be associated with a type of work: $g_V dV = -p dV$ is the mechanical work, $g_N dN = \mu dN$ is the chemical work, and $T\,dS$ is the heat flow. In a magnetic system we have the magnetic work $\mathbf{H} \cdot d\mathbf{M}$. Equation (436) is just a statement of conservation of energy and is conventionally known as the *first law of thermodynamics* [11].

It is clear that we need some nomenclature for dealing with these various descriptions. If the fundamental potential is the entropy, we say that we are in the *entropy representation*. If the fundamental potential is the energy, we say we are in the *energy representation*. Within a given representation, as we change independent variables, we say that we change ensembles.

2.3.3 Euler Equation

The derivative equations, defined by Eq. (425), give $t + 1$ *equations of state*, one for each conjugate force F_{X_i}. Thus for an ideal gas, where the entropy is given by the Sackur–Tetrode formula, we have

$$\frac{\partial S}{\partial E} = \frac{1}{k_B T} = \frac{3N}{2E} \qquad (443)$$

$$\frac{\partial S}{\partial N} = -\frac{\mu}{T} = -k_B \ln\left(n\lambda^3\right) \qquad (444)$$

and

$$\frac{\partial S}{\partial V} = \frac{p}{T} = \frac{Nk_B}{V}. \qquad (445)$$

In statistical mechanics the main goal, within the context of thermodynamic information [12], is to calculate the equations of state from the partition function. Traditionally, before the advent of statistical mechanics, thermodynamics assumed that we have experimental knowledge of the equations of state. Given the equations of state, we can construct the entropy or *fundamental equation* using the Euler [13] equation:

$$S(X_i) = k_B \sum_{j=0}^{t} F_{X_j} X_j. \qquad (446)$$

Euler's equation follows from the homogeneity property given by Eq. (419). Differentiate Eq. (419) with respect to λ and set $\lambda = 1$. We find, using the chain rule for differentiation, that

$$\begin{aligned} S(X_i) &= \left[\frac{\partial}{\partial \lambda} S(\lambda X_i) \right]_{\lambda=1} \\ &= \left[\sum_i \frac{\partial}{\partial Z_i} S(Z_i) \frac{\partial Z_i}{\partial \lambda} \right]_{\lambda=1} \end{aligned} \qquad (447)$$

where $Z_i = \lambda X_i$. We then have

$$S(X_i) = \sum_{i=0}^{t} \frac{\partial S(X_i)}{\partial X_i} X_i, \qquad (448)$$

which, on using Eq.(425), leads directly to Eq.(446). It turns out that we can obtain the entropy from a knowledge of one fewer than all the equations of state. Using the Euler equation, Eq.(446), we obtain

$$\frac{dS}{k_B} = \sum_i X_i \, dF_{X_i} + \sum_i F_{X_i} \, dX_i. \qquad (449)$$

If we use Eq. (426) to substitute for dS/k_B, we can cancel the term proportional to the dF_{X_i} and obtain the Gibbs–Duhem relation [14]

$$\sum_{i=0}^{t} X_i \, dF_{X_i} = 0. \qquad (450)$$

Given t equations of state, we can find the other equation of state up to an additive constant by using the Gibbs–Duhem relation.

It is easy to reproduce these arguments in the energy representation and obtain

$$E(Y) = \sum_{i=0}^{t} g_i Y_i, \qquad (451)$$

and

$$\sum_{i=0}^{t} Y_i \, dg_i = 0. \qquad (452)$$

For a simple fluid these take the form

$$E = TS - pV + \mu N \qquad (453)$$

and

$$S \, dT - V \, dp + N \, d\mu = 0. \qquad (454)$$

2.4 LEGENDRE TRANSFORMS

2.4.1 Introduction

In our discussion of statistical mechanics we noted that the various ensembles corresponded to different choices of independent variables with different potentials, such as S, F, Ω, and E. In the microcanonical ensemble the extensive parameters play the roles of mathematically independent variables, while the intensive parameters arise as derived concepts. This situation is in direct contrast to the practical situation in the laboratory. The experimentalist typically finds that the intensive parameters are more easily measured and controlled and therefore is likely to think of the intensive parameters as operationally independent variables and of the extensive parameters as operationally derived quantities. The extreme instance of this situation is provided by the conjugate variables entropy and temperature. No practical instruments exist for the measurement and control of entropy, whereas thermometers and thermostats, for the measurement and control of the temperature, are obviously very common. We therefore want to recast our general mathematical formulation such that intensive parameters replace extensive parameters as independent variables. This reformulation allows us to transform from one ensemble to another. It should be noted that these transformations to new ensembles are purely a matter of convenience. All thermodynamic information is included in any particular ensemble [15].

We are interested, for example, in the situation where, in the energy representation, we change independent variables from the entropy to the temperature. Thus we start, ignoring at this point the other extensive variables, with the equations

$$E = E(S) \qquad (455)$$

and

$$T = \frac{dE(S)}{dS}. \tag{456}$$

We want to go over to a description where we treat temperature as the independent variable. A very convenient procedure for carrying out such a change of variables is called a *Legendre transformation* [16]. If we have the set of equations

$$y = y(x) \tag{457}$$

and

$$p = \frac{dy(x)}{dx}, \tag{458}$$

where $y(x)$ is a function of the independent variable x, a Legendre transformation amounts to a mapping $(x, y) \rightarrow (p, \psi)$, where p is the new independent variable and ψ, defined by

$$\psi(p) = y(x(p)) - px(p), \tag{459}$$

is the new dependent variable and we must invert Eq. (458) to obtain $x = x(p)$. If we look at the derivative of the new dependent variable, we obtain

$$\frac{d\psi(p)}{dp} = \frac{dy(x)}{dx}\frac{dx}{dp} - x - p\frac{dx}{dp} = -x, \tag{460}$$

and the old independent variable x is now given by

$$x(p) = -\frac{d\psi(p)}{dp}. \tag{461}$$

Returning to our problem of interest in thermodynamics and Eqs. (455) and (456), we can make the identifications $x \rightarrow S$, $y \rightarrow E$, $p \rightarrow T$. Then, after a Legrendre transformation, the *fundamental potential* is no longer the energy but the Helmholtz free energy,

$$F[T] = E - TS, \tag{462}$$

which corresponds to the new dependent variable ψ in the example above. The old independent variable, the entropy, is now a derived quantity,

$$S = -\frac{\partial F[T]}{\partial T}. \tag{463}$$

If we work in the entropy representation where $S = S(E)$, then $T^{-1} = \partial S(E)/\partial E$ is the natural intensive variable and the appropriate Legendre-transformed potential is given by

$$S_1[T^{-1}] \equiv S - T^{-1}E, \tag{464}$$

where the old dependent variable is now given by

$$E = -\frac{\partial S_1[T^{-1}]}{\partial (T^{-1})}. \tag{465}$$

$S_1[T^{-1}]$ is clearly related to the Helmholtz free energy by

$$F = -TS_1(T^{-1}). \tag{466}$$

We can express the energy as a function of temperature in two equivalent ways. First, we have in the energy representation

$$E = F + TS = F - T\frac{\partial F[T]}{\partial T}, \tag{467}$$

while in the entropy representation

$$E = -\frac{\partial}{\partial (T^{-1})} S_1[T^{-1}] = -\frac{\partial}{\partial (T^{-1})} \left(-T^{-1}F[T]\right) \tag{468}$$

$$= \frac{\partial}{\partial \beta}(\beta F), \tag{469}$$

which is the result we found in the canonical ensemble. We can now see that a change of ensemble from the microcanonical to the canonical ensemble corresponds to a Legendre transformation from (S, E) to $(S_1[T^{-1}], T^{-1})$.

2.4.2 Application in Classical Mechanics

There is a more familiar situation where we have seen the usefulness of Legendre transformations. This is in the change from Hamiltonian to Lagrangian mechanics. In Hamiltonian mechanics the independent variables are the coordinates q and the momentum p and the central object of investigation is the Hamiltonian. For a particle of mass m in a potential $V(q)$, the Hamiltonian is given by

$$H = \frac{p^2}{2m} + V(q), \tag{470}$$

and Hamilton's equations read

$$\dot{q} = \frac{\partial H}{\partial p} \tag{471}$$

$$\dot{p} = -\frac{\partial H}{\partial q}. \tag{472}$$

The transformation to Lagrangian mechanics corresponds to a change of independent variables from p to $\dot{q}$ via a Legendre transformation. The Lagrangian is the transformed dependent function,

$$L(\dot{q}, q) = -H + \dot{q}p. \tag{473}$$

We can then work out the derivatives of the Lagrangian. First we have

$$\left(\frac{\partial L}{\partial \dot{q}}\right)_q = -\left(\frac{\partial H}{\partial p}\right)_q \left(\frac{\partial p}{\partial \dot{q}}\right)_q + p + \dot{q}\left(\frac{\partial p}{\partial \dot{q}}\right)_q \tag{474}$$
$$= p$$

after using Eq. (471). We see in the Lagrangian formulation that the conjugate momentum is a derived quantity. Next we have

$$\left(\frac{\partial L}{\partial q}\right)_{\dot{q}} = -\left(\frac{\partial H}{\partial p}\right)_q \left(\frac{\partial p}{\partial q}\right)_{\dot{q}} - \left(\frac{\partial H}{\partial q}\right)_p + \dot{q}\left(\frac{\partial p}{\partial q}\right)_{\dot{q}}$$
$$= -\left(\frac{\partial H}{\partial q}\right)_p. \tag{475}$$

Inserting Eqs. (474) and (475) back into Eq. (472), the equation of motion in the Lagrangian formulation is given by

$$\frac{d}{dt}\left(\frac{\partial L}{\partial \dot{q}}\right) = \frac{\partial L}{\partial q}. \tag{476}$$

2.4.3 Multiple Legendre Transformations

Let us now return to thermodynamics and the general role of Legendre transformations. In general, a partial Legendre transformation can be made by replacing the set of variables $X_0, \ldots X_s$ by $F_0, \ldots, F_s$ and the Legendre-transformed potential is given by [17]

$$S_s[F_0, \ldots, F_s] \equiv \tilde{S} - \sum_{j=0}^{s} F_j X_j. \tag{477}$$

Notice that in this section we simplify the notation by using F_j for F_{X_j} and $\tilde{S} = S/k_B$. In this case the $s+1$ transformed variables, $i = 0, 1, \ldots, s$, are functions of the form

$$X_i = X_i(F_0, \ldots, F_s, X_{s+1}, \ldots, X_t). \tag{478}$$

Then, differentiating Eq. (477) with respect to the untransformed variables X_i for $i = s+1, \ldots, t$ gives

$$\frac{\partial}{\partial X_i} S_s[F_0, \ldots, F_s] = \frac{\partial}{\partial X_i}\left[\tilde{S}(X) - \sum_{j=0}^{s} F_j X_j\right]$$

$$= \frac{\partial \tilde{S}}{\partial X_i} + \sum_{j=0}^{s} \frac{\partial \tilde{S}}{\partial X_j}\frac{\partial X_j}{\partial X_i} - \sum_{j=0}^{s} F_j \frac{\partial X_j}{\partial X_i}$$

$$= F_i. \tag{479}$$

For the Legendre-transformed variables, for $i = 0, \ldots, s$, we have

$$\frac{\partial}{\partial F_i} S_s[F_0, \ldots, F_s] = \sum_{j=0}^{s} \frac{\partial \tilde{S}}{\partial X_j}\frac{\partial X_j}{\partial F_i} - X_i - \sum_{j=0}^{s} F_j \frac{\partial X_j}{\partial F_i}$$

$$= -X_i. \tag{480}$$

These results are summarized by the differential relation

$$dS_s[F_0, \ldots, F_s] = \sum_{i=0}^{s}(-X_i)\,dF_i + \sum_{i=s+1}^{t} F_i\,dX_i. \tag{481}$$

If we start with the energy as the basic potential with the associated dependent variables $Y_i = (S, X_1, \ldots, X_t)$, the Legendre transformation is to the new set of variables $g_0, \ldots, g_s, Y_s+1, \ldots, Y_t$ and the new potential is

$$E_s[g_0, \ldots, g_s] = E - \sum_{i=0}^{s} g_i Y_i. \tag{482}$$

Clearly, in complete analogy to Eq. (481), we obtain

$$dE_s[g_0, \ldots, g_s] = \sum_{i=0}^{s}(-Y_i)\,dg_i + \sum_{i=s+1}^{t} g_i\,dY_i. \tag{483}$$

As we see below, both the entropy and energy descriptions are useful.

2.4.4 Examples

a. Gibbs Ensemble. As a simple example, consider the case where we choose temperature, pressure, and particle number as independent variables. This is typical in the laboratory. Since the intensive variables T and p occur naturally as conjugate variables using the energy as the fundamental potential, the appropriate Legendre-transformed function is

$$\begin{aligned} G \equiv E_2[T,p] &= E - g_S S - g_V V \\ &= E - TS + pV \\ &= \mu N. \end{aligned} \quad (484)$$

The last step follows from Euler's relation, Eq. (453). We then have directly from Eq. (483) that

$$\frac{\partial G}{\partial T} = -S \quad (485)$$

$$\frac{\partial G}{\partial p} = V \quad (486)$$

$$\frac{\partial G}{\partial N} = \mu. \quad (487)$$

b. Grand Canonical Ensemble. In the grand canonical ensemble, and from a theoretical point of view, the natural set of variables are β, α, and V or F_E, F_N, and V, and one works in the entropy representation with the twice Legendre-transformed potential

$$\begin{aligned} S_2[\alpha,\beta] &= \frac{S}{k_B} - F_E E - F_N N \\ &= \frac{1}{k_B}\left(S - \frac{E}{T} + \frac{\mu N}{T}\right) \\ &= \beta p V = -\beta \Omega \end{aligned} \quad (488)$$

where in the next-to-last step we again used Euler's equation. We obtain directly from Eq. (481) that

$$E = -\frac{\partial S_2}{\partial \beta} = \frac{\partial(\beta \Omega)}{\partial \beta} \quad (489)$$

$$N = -\frac{\partial S_2}{\partial \alpha} = \frac{\partial(\beta \Omega)}{\partial \alpha} \quad (490)$$

and

$$\beta p \frac{\partial S_2}{\partial V} = -\frac{\partial(\beta \Omega)}{\partial V} \quad (491)$$

as we found directly when working in the GCE.

c. Low Temperature Behavior. Consider the case where we are given the Helmholtz free energy in the energy representation as a power series in temperature:

$$E_1[T] = F[T] = V\left(a_0 + b_0 T + \frac{c_0}{2}T^2\right) \tag{492}$$

where a_0, b_0, and c_0 are material parameters. Since we work in the energy representation, we have from Eq. (483), that the entropy is given by

$$S = -\frac{\partial F[T]}{\partial T} = -V(b_0 + c_0 T). \tag{493}$$

The energy and free energy are related by Eq. (462), so we can construct the energy as a function of the entropy by inserting $T = (-1/c_0)(s + b_0)$ in $E = F + TS$, where $s = S/V$, to obtain

$$\frac{E}{V} = a_0 - \frac{1}{2c_0}(s + b_0)^2. \tag{494}$$

This completes the Lengendre transformation from $F[T]$ to $E[S]$. Let us now consider the physical interpretation of the parameters a_0, b_0, and c_0. If we express the energy in terms of the temperature, we obtain

$$\frac{E}{V} = a_0 - \frac{c_0}{2}T^2. \tag{495}$$

Clearly, we can identify $a_0 V$ as the ground-state energy, and the specific heat is given by

$$C_V = \left(\frac{\partial E}{\partial T}\right)_V = -V T c_0. \tag{496}$$

Since the specific heat is positive, we see that c_0 must be negative.

While a_0 is associated with the ground-state energy and c_0 with the low-temperature contribution to the specific heat, what about the parameter b_0? This quantity is interesting since it does *not* appear in the expression for the energy. Its interpretation is simple. It is the zero-temperature contribution to the entropy. According to *Nernst's theorem*, sometimes called the *third law of thermodynamics* [18], the degeneracy of the ground state Σ_{GS} is not extensive:

$$\lim_{T \to 0} s = \lim_{N \to \infty} \frac{S_{GS}}{N} = \lim_{N \to \infty} \frac{1}{N} k_B \ln \Sigma_{GS} = 0. \tag{497}$$

We see then that we should expect $b_0 = 0$. From a classical point of view, b_0 is an arbitrary constant in the entropy which can be chosen to be zero.

2.5 SECOND-DERIVATIVE INFORMATION

2.5.1 Physical Observables

In many experiments we do not measure the *first-derivative* information, the equations of state. We actually measure second-derivative information. To make this clearer, let us choose a particular thermodynamic ensemble. We choose p and T as our independent variables, and we will keep N fixed in this part of the discussion. The thermodynamic potential in this ensemble is the Gibbs potential G. The first derivatives of G, for fixed N, are given by Eqs. (485) and (486):

$$S = -\left(\frac{\partial G}{\partial T}\right)_{p,N} \tag{498}$$

$$V = \left(\frac{\partial G}{\partial p}\right)_{T,N}. \tag{499}$$

The second derivatives $\partial^2 G/\partial T^2$, $\partial^2 G/\partial T \partial p$, $\partial^2 G/\partial p^2$ are experimentally measurable and contain interesting information. Using the first derivatives, we see that

$$\frac{\partial^2 G}{\partial T^2} = -\left(\frac{\partial S}{\partial T}\right)_p = -\frac{C_p}{T} \tag{500}$$

where C_p is the specific heat at constant pressure,

$$\frac{\partial^2 G}{\partial T \partial p} = \left(\frac{\partial V}{\partial T}\right)_p \equiv \alpha_T V \tag{501}$$

where α_T is the coefficient of thermal expansion, and

$$\frac{\partial^2 G}{\partial p^2} = \left(\frac{\partial V}{\partial p}\right)_T \equiv -V \kappa_T \tag{502}$$

and κ_T is the isothermal compressibility. C_p, α_T, and κ_T are directly experimentally measurable. In Problem 2.4 you are asked to work out these three quantities for an ideal gas.

For fixed N, C_p, κ_T, and α form a complete set of second derivatives. Any second derivatives for another set of independent variables (still holding N fixed) can be expressed [19] in terms of C_p, α, and κ_T. As a concrete example, consider the specific heat at constant volume

$$C_V = \left(\frac{\partial E}{\partial T}\right)_{V,N} = T\left(\frac{\partial S}{\partial T}\right)_{V,N}. \tag{503}$$

If we assume that we are working in the ensemble where T, V, and N are the natural independent variables and the Helmholtz free energy is the appropriate potential, we

132   PRINCIPLES OF THERMODYNAMICS

can obtain the average energy, using Eq. (467). Inserting this result into the first form in Eq. (503), we find that

$$C_V = \frac{\partial}{\partial T}\left(F - T\frac{\partial F}{\partial T}\right)$$
$$= -T\left(\frac{\partial^2 F}{\partial T^2}\right)_{N,V}. \qquad (504)$$

We see then that C_V is a second derivative in this ensemble. Since C_V assumes constant N, we should, if the assertions above are correct, be able to express it in terms of the complete set C_p, α, κ_T (and possibly V and T). As a prelude to this derivation, we discuss certain mathematical properties of systems of partial derivatives.

2.5.2   Maxwell Relations

We may ask why in evaluating $\partial^2 G/\partial T\, \partial p$ in Eq. (501) we evaluated the derivative with respect to pressure first. Since the order of differentiation is interchangeable, we could write

$$\frac{\partial^2 G}{\partial T\, \partial p} = \frac{\partial}{\partial p}\frac{\partial G}{\partial T} = -\left(\frac{\partial S}{\partial p}\right)_T. \qquad (505)$$

Comparing Eqs. (501) and (505), we obtain the identity

$$\left(\frac{\partial V}{\partial T}\right)_p = -\left(\frac{\partial S}{\partial p}\right)_T = \alpha_T V. \qquad (506)$$

This is an example of a Maxwell's relation [20] relating cross derivatives. In the more general case, a particular thermodynamic potential $S_s[F_0,\ldots,F_s]$ or $E_s[g_0,\ldots,g_s]$, expressed in terms of its $(t+1)$ natural variables, has $t(t+1)/2$ separate pairs of mixed second derivatives. Thus each potential yields $t(t+1)/2$ Maxwell relations. For the thermodynamic potential $G(p,T,N)$ and $t=2$, we see that there are $2(3)/2 = 3$ Maxwell's relations. In addition to Eq. (506), we have

$$-\left(\frac{\partial S}{\partial N}\right)_{p,T} = \left(\frac{\partial \mu}{\partial T}\right)_{p,N} \quad (T,N) \qquad (507)$$

and

$$\left(\frac{\partial V}{\partial N}\right)_{T,p} = \left(\frac{\partial \mu}{\partial p}\right)_{T,N} \quad (p,N). \qquad (508)$$

In the more general case the Maxwell relations read

$$\frac{\partial X_j}{\partial F_k} = \frac{\partial X_k}{\partial F_j} \quad \text{if} \quad j,k \leq s \tag{509}$$

$$\frac{\partial X_j}{\partial X_k} = -\frac{\partial F_k}{\partial F_j} \quad \text{if} \quad j \leq s \quad \text{and} \quad k > s \tag{510}$$

$$\frac{\partial F_j}{\partial X_k} = \frac{\partial F_k}{\partial X_j} \quad \text{if} \quad j,k > s. \tag{511}$$

The Maxwell relations are very useful for relating various thermodynamic derivatives which show up in calculations or in data interpretation. They connect certain second derivatives within the same ensemble.

2.5.3 Jacobians and Transformations

If we want to express C_V in terms of C_p, α, and κ_T we need the mathematical apparatus for changing ensembles [21]. One approach is to use the mathematical identities

$$\left(\frac{\partial X}{\partial Y}\right)_Z = 1 \bigg/ \left(\frac{\partial Y}{\partial X}\right)_Z \tag{512}$$

$$\left(\frac{\partial X}{\partial Y}\right)_Z = \left(\frac{\partial X}{\partial W}\right)_Z \bigg/ \left(\frac{\partial Y}{\partial W}\right)_Z \tag{513}$$

$$\left(\frac{\partial X}{\partial Y}\right)_Z = -\left(\frac{\partial Z}{\partial Y}\right)_X \bigg/ \left(\frac{\partial Z}{\partial X}\right)_Y. \tag{514}$$

A more general method of manipulation of thermodynamic derivatives is based on the mathematical properties of Jacobians [22]. Jacobians should be familiar from the standard theory of coordinate transformations. In Appendix I we discuss the case of two variables. Here we simply state the relevant properties of Jacobians we need. Proofs of the general theorems can be found in texts on differential calculus. If $u, v, \ldots, w$ are functions of $x, y, \ldots, z$, the Jacobian is defined as

$$\frac{\partial(u,v,\ldots,w)}{\partial(x,y,\ldots,z)} = \det \begin{bmatrix} \frac{\partial u}{\partial x} & \frac{\partial u}{\partial y} & \cdots & \frac{\partial u}{\partial z} \\ \frac{\partial v}{\partial x} & \frac{\partial v}{\partial y} & \cdots & \frac{\partial v}{\partial z} \\ \cdot & \cdot & \cdots & \cdot \\ \cdot & \cdot & \cdots & \cdot \\ \frac{\partial w}{\partial x} & \frac{\partial w}{\partial y} & \cdots & \frac{\partial w}{\partial z} \end{bmatrix}$$

The property that makes the Jacobian particularly useful in thermodynamic applications is

$$\left(\frac{\partial u}{\partial x}\right)_{y,\ldots,z} = \frac{\partial(u, y, \ldots, z)}{\partial(x, y, \ldots, z)}. \tag{515}$$

Consequently, a thermodynamic derivative of interest can be expressed in terms of Jacobians, and the well-developed properties of Jacobians can then be used. The properties of use to us are

$$\frac{\partial(u, v, \ldots, w)}{\partial(x, y, \ldots, z)} = -\frac{\partial(v, u, \ldots, w)}{\partial(x, y, \ldots, z)} \tag{516}$$

which follows from the definition in terms of a determinant, a generalization of the chain rule for derivatives,

$$\frac{\partial(u, v, \ldots, w)}{\partial(x, y, \ldots, z)} = \frac{\partial(u, v, \ldots, w)}{\partial(r, s, \ldots, t)} \frac{\partial(r, s, \ldots, t)}{\partial(x, y, \ldots, z)} \tag{517}$$

and

$$\frac{\partial(u, v, \ldots, w)}{\partial(x, y, \ldots, z)} = 1 \bigg/ \frac{\partial(x, y, \ldots, z)}{\partial(u, v, \ldots, w)}. \tag{518}$$

Some simple examples using these results are discussed in the problems at the end of the chapter. The results quoted above should suffice for most standard thermodynamic applications. The problem of reduction of a given derivative to, for example C_p, α, and κ_T, is equivalent to the transformation to T, p, and N as independent parameters.

2.5.4 Example: Expressing $C_V = C_V(C_p, \alpha_T, \kappa_T)$

Let us go back to our example of C_V. We have from Eq. (503)

$$C_V = T\left(\frac{\partial S}{\partial T}\right)_V = T\frac{\partial(S, V)}{\partial(T, V)}$$
$$= T\frac{\partial(S, V)}{\partial(T, p)} \bigg/ \frac{\partial(T, V)}{\partial(T, p)} \tag{519}$$

which effects our transformation to T and p as independent variables. We have next, using Eqs. (515) and (502),

$$\frac{\partial(T, V)}{\partial(T, p)} = \left(\frac{\partial V}{\partial p}\right)_T = -V\kappa_T \tag{520}$$

so

$$C_V = -\frac{T}{V\kappa_T}\frac{\partial(S,V)}{\partial(T,p)}. \tag{521}$$

Writing out the remaining Jacobian explicitly,

$$\frac{\partial(S,V)}{\partial(T,p)} = \left(\frac{\partial S}{\partial T}\right)_p \left(\frac{\partial V}{\partial p}\right)_T - \left(\frac{\partial S}{\partial p}\right)_T \left(\frac{\partial V}{\partial T}\right)_p, \tag{522}$$

the various terms can be identified with observables using Eqs. (500), (501), (502), and the Maxwell relation given by Eq. (506), with the result

$$\frac{\partial(S,V)}{\partial(T,p)} = \frac{C_p}{T}(-\kappa_T V) + (\alpha_T V)^2. \tag{523}$$

Inserting Eq. (523) back into Eq. (521) leads to the standard result

$$\begin{aligned}C_V &= -\frac{T}{V\kappa_T}\left[-C_p\frac{\kappa_T V}{T} + (\alpha_T V)^2\right] \\ &= C_p - T\alpha_T^2 V/\kappa_T. \end{aligned} \tag{524}$$

Those who have derived this result in other ways will appreciate the usefulness of this approach. Since C_V, C_p, α_T, and κ_T can be expressed as second derivatives, we can rewrite Eq. (524) in the form

$$\frac{\partial^2 F}{\partial T^2} = \frac{\partial^2 G}{\partial T^2} - \left(\frac{\partial^2 G}{\partial T\, \partial p}\right)^2 \left(\frac{\partial^2 G}{\partial p^2}\right)^{-1}. \tag{525}$$

This shows how the second derivatives in the Gibbs ensemble form a complete set.

2.6 FLUCTUATIONS AND STABILITY

2.6.1 Introduction

We have not yet made full use of our entropy principle. We can begin to probe its power by considering the following rather general situation. Assume that there are r subsystems immersed in an enormous heat bath (see Fig. 2.1). We assume that the bath and the r subsystems can be characterized by the set of extensive parameters X_i^α. $\alpha = 1$ to r ranges over all the subsystems and $\alpha = B$ corresponds to the extensive variable for the bath. i labels the different extensive variables (e.g., E, N, V, etc.). We assume that the total system is isolated and the extensive variables

136 PRINCIPLES OF THERMODYNAMICS

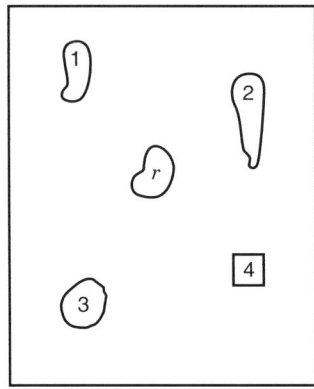

FIGURE 2.1 Schematic of r macroscopic subsystems embedded in a very large heat bath.

for the combined system

$$X_i^T = X_i^B + \sum_{\alpha=1}^{r} X_i^\alpha \tag{526}$$

are fixed (conserved). The entropy for the combined system is

$$S^T = S^B + \sum_{\alpha=1}^{r} S^\alpha \tag{527}$$

where

$$S^\alpha = S^\alpha[X_i^\alpha]. \tag{528}$$

Consider now the situation where the X_i^α are slightly removed from their equilibrium values, which, in this section only, we denote by $\bar{X}_i^\alpha$. Let us expand S^T in the deviations from equilibrium $S^T = \bar{S}^T + \delta S^T + \delta^2 S^T + \cdots$. δS^T corresponds to first-order variations in the X_i^α and $\delta^2 S^T$ to the second-order variations. Clearly, from the schematic in Fig. 2.2, the condition that S_T be a maximum is

$$\delta S^T = 0 \tag{529}$$

while

$$\delta^2 S^T < 0. \tag{530}$$

These conditions are extremely powerful.

Consider first Eq. (529). We have that

$$\delta S^B = k_B \sum_i F_i^B \, \delta X_i^B \tag{531}$$

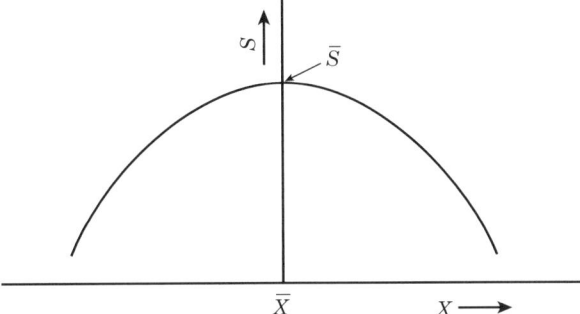

FIGURE 2.2 Schematic of the entropy as a function of one of its arguments near the maximum corresponding to the equilibrium state.

and

$$\delta S^\alpha = k_B \sum_i F_i^\alpha \, \delta X_i^\alpha \qquad (532)$$

where the F_i^α's are the generalized forces for the ith extensive variable in the αth subsystem. We should remember that the F_i^α's are to be evaluated in equilibrium. Since X_i^T is a constant,

$$\delta X_i^T = 0 = \delta X_i^B + \sum_\alpha \delta X_i^\alpha \qquad (533)$$

and Eq. (529) becomes

$$\frac{\delta S^T}{k_B} = 0 = \sum_i F_i^B \, \delta X_i^B + \sum_{i,\alpha} F_i^\alpha \, \delta X_i^\alpha$$

$$= \sum_{i,\alpha} (-F_i^B + F_i^\alpha) \, \delta X_i^\alpha. \qquad (534)$$

Since this must hold for arbitrary variations of the X_i^α's, we obtain

$$F_i^\alpha = F_i^B. \qquad (535)$$

This means that in equilibrium the intensive variables of all the subsystems must equal those of the bath *and* each other!

We next consider the second variations:

$$\delta^2 S^B = \sum_{i,j} \left(\frac{\partial^2 S^B}{\partial X_i^B \partial X_j^B} \right)_{X^B = \bar{X}^B} \delta X_i^B \delta X_j^B \qquad (536)$$

and

$$\delta^2 S^\alpha = \sum_{i,j} \left(\frac{\partial^2 S^\alpha}{\partial X_i^\alpha \partial X_j^\alpha} \right)_{X^\alpha = \bar{X}^\alpha} \delta X_i^\alpha \delta X_j^\alpha, \tag{537}$$

so Eq. (530) requires that

$$\delta^2 S^T = \sum_{i,j} (M_{ij}^B \delta X_i^B \delta X_j^B + \sum_\alpha M_{ij}^\alpha \delta X_i^\alpha \delta X_j^\alpha) < 0 \tag{538}$$

where we define the matrices

$$M_{ij}^B = \left(\frac{\partial^2 S^B}{\partial X_i^B \partial X_j^B} \right)_{X^B = \bar{X}^B} \tag{539}$$

and

$$M_{ij}^\alpha = \left(\frac{\partial^2 S^\alpha}{\partial X_i^\alpha \partial X_j^\alpha} \right)_{X^\alpha = \bar{X}^\alpha}. \tag{540}$$

Using Eq. (533) to eliminate δX_i^B, Eq. (538) can be rewritten as

$$\delta^2 S^T = \sum_{i,j} \sum_{\alpha,\beta} (M_{ij}^B + \delta_{\alpha,\beta} M_{ij}^\alpha) \delta X_i^\alpha \delta X_j^\beta < 0. \tag{541}$$

We then note that M_{ij}^B, given by Eq. (539), goes as $1/V^B$, where V^B is the volume of the bath. Similarly, $M_{ij}^\alpha \approx 1/V^\alpha$, where V^α is the volume of the αth subsystem. In the limit $V^B \gg V^\alpha$, where the volume of the bath is very large compared to the subsystems, we have

$$\delta^2 S^T = \sum_{i,j} \sum_\alpha M_{ij}^\alpha \delta X_i^\alpha \delta X_j^\alpha < 0. \tag{542}$$

There is one slightly subtle point associated with the inequality given by Eq. (542). This arises due to the Gibbs–Duhem relation satisfied by each subsystem

$$\sum_j X_j^\alpha dF_{X_j}^\alpha = 0. \tag{543}$$

Taking the derivative with respect to X_i^α, this can be written as

$$\sum_j X_j^\alpha \frac{\partial F_{X_j}^\alpha}{\partial X_i^\alpha} = 0. \tag{544}$$

However the matrix M^α, defined by Eq. (540), can be written as

$$M^\alpha_{ij} = \frac{\partial F^\alpha_{X_j}}{\partial X_i}, \quad (545)$$

and Eq. (544) gives the constraint

$$\sum_j X^\alpha_j M^\alpha_{ij} = 0. \quad (546)$$

Let us write (suppressing the α-index)

$$X_i = X x_i, \quad (547)$$

where X is any one of the extensive variables, say $X \equiv X_k$. Then the x_i are intensive densities and $x_k = 1$. If we substitute

$$\delta X_i = x_i \, \delta X + X \, \delta x_i \quad (548)$$

into Eq. (542), we obtain, after some rearrangement,

$$\delta^2 S^T = \sum_{i,j} M_{ij} [X^2 \, \delta x_i \, \delta x_j + 2X \, \delta X x_i \, \delta x_j + (\delta X)^2 x_i x_j]. \quad (549)$$

However, Eq. (546) can be written in the form, for each subsystem α,

$$X \sum_j x_j M_{ij} = 0 \quad (550)$$

and, restoring the label α,

$$\delta^2 S^T = \sum_\alpha (X^\alpha)^2 \sum_{i,j} M^\alpha_{ij} \, \delta x^\alpha_i \, \delta x^\alpha_j < 0. \quad (551)$$

Notice, that $\delta x^\alpha_k = 0$, the kth component does not contribute to the sum.

From a mathematical point of view, the inequality given by Eq. (551) means that the eigenvalues associated with the symmetric matrix

$$M^\alpha_{ij} \equiv \frac{\partial^2 S^\alpha}{\partial X^\alpha_i \partial X^\alpha_j} \quad (552)$$

must be negative. If there are $t+1$ extensive variables, this leads to t stability conditions.

We can reexpress these results in the energy representation. Suppressing the α label for simplicity and holding X_t fixed, we first note that $\delta^2 S$, given by Eq. (542), can be rewritten in the form

$$\delta^2 S = k_B \sum_{i=0}^{t-1} \delta F_i \delta X_i, \tag{553}$$

since the fluctuation in the intensive variables can be written as

$$k_B \delta F_i = k_B \sum_{j=0}^{t-1} \frac{\partial F_i}{\partial X_j} \delta X_j = \sum_{j=0}^{t-1} \frac{\partial^2 S}{\partial X_i \partial X_j} \delta X_j. \tag{554}$$

In complete analogy with the development in the entropy representation, we have in the energy representation, with fixed Y_t,

$$\delta E = \sum_{i=0}^{t-1} g_i \delta Y_i \tag{555}$$

and

$$\delta^2 E = \sum_{i=0}^{t-1} \sum_{j=0}^{t-1} \delta Y_j \delta Y_i \frac{\partial^2 E}{\partial Y_i \partial Y_j} = \sum_{i=0}^{t-1} \delta g_i \delta Y_i. \tag{556}$$

How is $\delta^2 E$ related to $\delta^2 S$? We can answer this question by reexpressing $\delta^2 S$, given by Eq. (553), in terms of natural variables in the energy representation. The intensive variables g_{Y_i} are related to the intensive variables for the entropy representation F_i by Eqs. (437) and (438). For $i \neq E$ the extensive variables are equivalent, $Y_i = X_i$. If we express Eq. (553) in terms of the g's and Y's, we obtain

$$\begin{aligned}
\delta^2 S &= k_B \delta\left(\frac{1}{k_B g_S}\right) \delta E + k_B \sum_{i=1}^{t-1} \delta\left(-\frac{g_i}{k_B g_S}\right) \delta Y_i \\
&= -\frac{1}{g_S^2} \delta g_S \delta E + \sum_{i=1}^{t-1} \left(\frac{g_i}{g_S^2} \delta g_S - \frac{\delta g_i}{g_S}\right) \delta Y_i \\
&= -\frac{1}{g_S^2} \delta g_S \delta E + \frac{1}{g_S^2} \delta g_S \delta E - \frac{\delta g_S}{g_S^2} g_S \delta S - \frac{1}{g_S} \sum_{i=1}^{t-1} \delta g_i \delta Y_i \\
&= -\frac{1}{g_S} \delta g_S \delta S - \frac{1}{g_S} \sum_{i=1}^{t-1} \delta g_i \delta Y_i = -\frac{1}{g_S} \sum_{i=0}^{t-1} \delta g_i \delta Y_i
\end{aligned}$$

and finally, recalling Eq. (556),

$$\delta^2 S = -\frac{\delta^2 E}{T}. \tag{557}$$

It then follows that we can replace the problem where we maximize the entropy with the problem where we minimize the energy

$$\delta E^T = 0 \tag{558}$$

$$\delta^2 E^T > 0. \tag{559}$$

If we change independent variables from the Y_i to the mixed set $\{g_0, ..., g_s, Y_{s+1}, ..., Y_t\}$, the fluctuations of the dependent variables are given in this case by

$$\delta Y_i = \sum_{j=0}^{s} \frac{\partial Y_i}{\partial g_j} \delta g_j + \sum_{j=s+1}^{t-1} \frac{\partial Y_i}{\partial Y_j} \delta Y_j \qquad 0 \leq i \leq s \tag{560}$$

$$\delta g_i = \sum_{j=0}^{s} \frac{\partial g_i}{\partial g_j} \delta g_j + \sum_{j=s+1}^{t-1} \frac{\partial g_i}{\partial Y_j} \delta Y_j \qquad s \leq i \leq t. \tag{561}$$

Inserting these results into Eq. (556), identifying

$$\frac{\partial Y_i}{\partial g_j} = -\frac{\partial^2 E_s}{\partial g_i \partial g_j} \tag{562}$$

$$\frac{\partial g_i}{\partial Y_j} = \frac{\partial^2 E_s}{\partial Y_i \partial Y_j}, \tag{563}$$

and noting that the cross-terms vanish because for $i \leq s$ and $j > s$,

$$\frac{\partial Y_i}{\partial Y_j} + \frac{\partial g_j}{\partial g_i} = -\frac{\partial^2 E_s}{\partial g_i \partial Y_j} + \frac{\partial^2 E_s}{\partial Y_j \partial g_i} = 0, \tag{564}$$

we obtain in the energy representation

$$\delta^2 E = \sum_{ij=0}^{s} \delta g_i \delta g_j \left(-\frac{\partial E_s}{\partial g_i \partial g_j} \right) + \sum_{ij=s+1}^{t-1} \delta Y_i \delta Y_j \frac{\partial^2 E_s}{\partial Y_i \partial Y_j} > 0, \tag{565}$$

and similarly, in the entropy representation,

$$\delta^2 S = \sum_{ij=0}^{s} \delta F_i \delta F_j \left(-\frac{\partial S_s}{\partial F_i \partial F_j} \right) + \sum_{ij=s+1}^{t-1} \delta X_i \delta X_j \frac{\partial^2 S_s}{\partial Y_i \partial Y_j} < 0. \tag{566}$$

We see then that if we hold the $s + 1$ intensive variables constant, we have

$$\delta^2 E = \sum_{ij=s+1}^{t-1} \delta Y_i \delta Y_j \frac{\partial^2 E_s}{\partial Y_i \partial Y_j} = \delta^2 E_s > 0, \tag{567}$$

and the problem reduces to minimizing the Legendre-transformed quantities E_s while holding their natural intensive variables fixed:

$$\delta E_s^T = 0 \tag{568}$$
$$\delta^2 E_s^T > 0. \tag{569}$$

As an important example, in a problem where we hold the temperature constant, we minimize the free energy.

2.6.2 Single-Component Subsystem

We discuss these general results first in the case of a single subsystem: $r = 1$. We assume for definiteness that this subsystem is a fluid characterized by extensive variables E, N, and V. The bath is characterized by the extensive variables E^B, N^B, and V^B. The equilibrium conditions, given by Eq. (535), reduce in this case to

$$\frac{1}{T} = \frac{1}{T^B} \tag{570}$$

$$\frac{p}{T} = \frac{p_B}{T_B} \tag{571}$$

and

$$-\frac{\mu}{T} = -\frac{\mu_B}{T_B}. \tag{572}$$

This simply says that the bath and subsystem will be in equilibrium only if all their conjugate intensive variables are equal. Clearly, it makes sense that the systems must be at the same temperature, pressure, and chemical potential.

Let us move on to the stability conditions $\delta^2 S = -\delta^2 E/T < 0$ for a single-component system. From the form of Eqs. (565) and (566), we obtain simplicity if we choose independent variables that mix intensive and extensive variables. For fixed N there are two possibilities. First we choose as independent variables T and V, where the associated Legendre-transformed energy potential

$$E_1[T] = E - TS = F \tag{573}$$

is just the Helmholtz free energy. We can immediately write Eq. (565) in the form

$$\delta^2 E = (\delta T)^2 \left(-\frac{\partial^2 F}{\partial T^2}\right) + (\delta V)^2 \frac{\partial^2 F}{\partial V^2} > 0. \tag{574}$$

Since T and V can be varied independently, we have the stability conditions,

$$-\frac{\partial^2 F}{\partial T^2} = \frac{1}{T} C_V > 0 \tag{575}$$

and

$$\frac{\partial^2 F}{\partial V^2} = -\left(\frac{\partial p}{\partial V}\right)_{N,T} = \frac{1}{V\kappa_T} > 0. \tag{576}$$

Thus for a thermodynamically stable system the specific heat and isothermal compressibilities *must* be positive.

The second choice of mixed independent variables is (S, p), where the associated energy potential

$$E_1[p] = E + pV \equiv W \tag{577}$$

is the *enthalpy* [23]. In this case Eq. (565) reads

$$\delta^2 E = (\delta p)^2 \left(-\frac{\partial^2 W}{\partial p^2}\right) + (\delta S)^2 \frac{\partial^2 W}{\partial^2 S} > 0 \tag{578}$$

and this results in the stability conditions,

$$-\frac{\partial^2 W}{\partial p^2} = -\left(\frac{\partial V}{\partial p}\right)_{S,N} > 0 \tag{579}$$

$$\frac{\partial^2 W}{\partial^2 S} = -\left(\frac{\partial T}{\partial S}\right)_p = \frac{T}{C_p} > 0. \tag{580}$$

In analogy with the isothermal compressibility given by Eq. (502), the adiabatic compressibility κ_s is defined by

$$V\kappa_s = -\left(\frac{\partial V}{\partial p}\right)_{S,N}, \tag{581}$$

and Eq. (579) requires that

$$\kappa_s > 0. \tag{582}$$

Since κ_s is a second derivative with fixed N, we should be able to express it in terms of C_p, α, and κ_T. It is left as an exercise (Problem 2.6) to show that

$$\kappa_s = \kappa_T \frac{C_V}{C_p}. \tag{583}$$

Since we showed earlier that

$$C_p - C_V = \frac{TV\alpha^2}{\kappa_T}, \tag{584}$$

we can make a statement stronger than Eq. (580), $C_p > C_V > 0$.

144   PRINCIPLES OF THERMODYNAMICS

It is worth noting that the stability condition $C_V > 0$ follows directly in a statistical mechanical description. If we work in the canonical ensemble, then

$$C_V = \left(\frac{\partial E}{\partial T}\right)_{N,V} = -\frac{1}{k_B T^2}\left(\frac{\partial E}{\partial \beta}\right)_{N,V}$$
$$= \frac{1}{k_B T^2}\langle(H-E)^2\rangle \geq 0. \qquad (585)$$

2.6.3 Phase Equilibrium

a. Phase Coexistence. Thus far we have been discussing homogeneous systems in equilibrium. There are circumstances where a system is separated into two homogeneous parts that are in contact and are in different thermodynamic states. The simplest example is of a cylinder filled with the gas and liquid phases of the same substance, as shown in Fig. 2.3. This is a perfectly good equilibrium situation. How can we understand the coexistence of phases? We can identify this situation with our general development of r subsystems in a heat bath by treating the two phases as separate subsystems. Thus the gas phase is characterized by extensive variables E_G, V_G, and N_G, and the liquid phase is characterized by E_L, V_L, and N_L. In this case we allow for exchange of heat with a surrounding heat bath which has a rigidly fixed volume and does not exchange particles with the two subsystems of interest. We obtain immediately that the conditions of equilibrium, Eq. (535), are in this case

$$F_i^1 = F_i^2 = F_i^B \qquad (586)$$

or

$$T_G = T_L = T_B \qquad (587)$$
$$p_G = p_L \qquad (588)$$
$$\mu_G = \mu_L. \qquad (589)$$

Thus the temperature, pressures, and chemical potentials of the subsystems must be the same. The conditions on the temperature and pressure are physically obvious, and the equality of the chemical potentials guarantees that there is no net particle

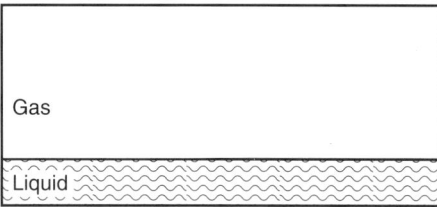

FIGURE 2.3 Schematic of gas–liquid phase separation under the influence of gravity.

flow between the two systems. The condition given by Eq. (589) relating the chemical potentials has nontrivial and nonobvious content. If we work in the Gibbs ensemble (p, T, N), we can write

$$\mu_G(p_G, T_G, N_G) = \mu_L(p_L, T_L, N_L), \tag{590}$$

which, after enforcing the other conditions for coexistence, reads

$$\mu_G(p, T, N_G) = \mu_L(p, T, N_L). \tag{591}$$

It is clear that in the thermodynamic limit $N_G, N_L \to \infty$ that the intensive variables μ_G and μ_L are finite and independent of N_G and N_L and we have

$$\mu_G(p, T) = \mu_L(p, T). \tag{592}$$

This condition tells us that two phases cannot be in equilibrium with each other at all pressures and temperature. Indeed, Eq. (592) fixes $p = p(T)$ or $T = T(p)$ in the region of coexistence. A phase equilibrium curve in a pressure versus temperature diagram is shown schematically in Fig 2.4. On this line liquid and gas can coexist. The points on either side of the curve will be the completely homogeneous phases of either pure gas or liquid. We have seen previously, in the case of the ideal gas, that we can find an equation of state of the form $p = p(n, T)$ when n is the density. This is quite general. In a single-phase region we can invert this equation to obtain $n = n(p, T)$. However, since $p = p(n, T)$ is in general a nonlinear equation in n, one can have multiple solutions for $n = n(p, T)$. In particular, in a two-phase region there are two solutions for the same p and T, $n_G = n_G(p, T)$ and $n_L = n_L(p, T)$. If we plot

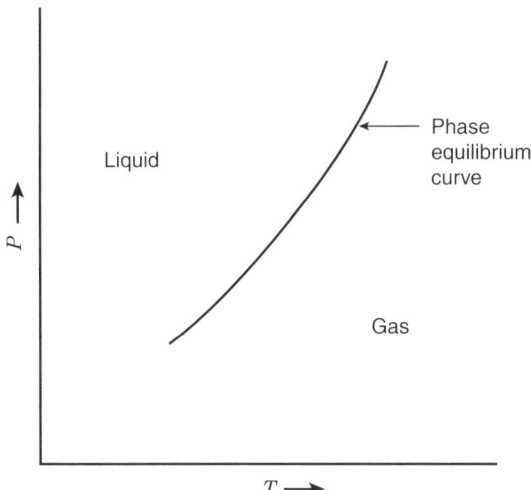

FIGURE 2.4 Schematic of liquid–gas phase coexistence curve in the pressure versus temperature plane.

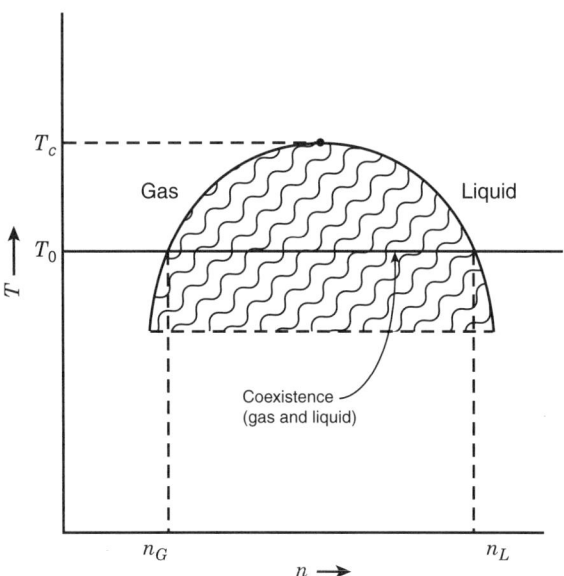

FIGURE 2.5 Schematic of liquid–gas phase coexistence curve in the temperature and density plane. At a given temperature T_0 one has two values of density, n_G and n_L. T_c marks the critical point.

temperature versus density, the states in which two phases coexist occupy a region (hatched in Fig. 2.5) and not just a line as in the p versus T plot. The reason is that the density, unlike pressure, is not the same for the two phases.

Let us investigate the situation shown in Fig. 2.5. Assume that we are at some fixed temperature T_0, with a sufficiently low density so that we are in the gas phase. If we then decrease the volume but hold the temperature and total number of particles constant, the density will increase and the pressure ($\approx nk_B T_0$) will increase. Eventually, the pressure will reach the value p_0 where $\mu_G(p_0, T_0) = \mu_L(p_0, T_0)$ and the phases can coexist. At this point the density of the gas will have the value n_G shown in Fig. 2.5. If we further compress the system at constant temperature, we find that the system can support regions with different densities. This is reflected in the existence of two solutions when we invert the equation of state $p = p(n, T)$ to obtain $n_G(p, T)$ and $n_L(p, T)$. If we slowly compress the system while maintaining a constant temperature, we must, of course, extract heat from the system. In this portion of the phase diagram the pressure will *not* increase. Instead, the system becomes inhomogeneous, *phase separates*, and some gas is converted into liquid. If we define the average density $n = N/V$ for the entire system, then in the region $n_G < n < n_L$,

$$N = n_G V_G + n_L V_L \tag{593}$$

$$V = V_G + V_L \tag{594}$$

where V_G and V_L are the volumes occupied by gas and liquid, respectively. We then move from $n = n_G$ to $n = n_L$ by decreasing V while holding N and T constant. Thus V_G will decrease while V_L will increase until $V = V_L$ and $n = n_L$.

The process just described is an example of a *phase transition*. In this case we have gone from the gas to the liquid phase. This process involved a jump in a thermodynamic parameter. In this case the difference in density between the gas and liquid phases corresponding to the same temperature, pressure, and particle number can be substantial: $n_L(p, T) \gg n_G(p, T)$. Phase transformations involving a discontinuous change in a macrovariable are called first-order phase transitions.

b. Clausius–Clapeyron Equation. The gas–liquid phase transition is, as we discussed above, accompanied by the transfer of heat called the *latent heat* of transition. Since the process (the phase transformation) is at constant p and T the heat change is $dQ = T \, dS|_{p,T}$. It is conventional to consider the heat change per particle $dq = dQ/N$ in going from $n_L = N/V_L$ to $n_G = N/V_G$ at fixed pressure and temperature:

$$q = \int_{n_L}^{n_G} T d\left(\frac{S}{N}\right)\Big|_{p,T} = T[s_G - s_L] \tag{595}$$

where s is the entropy per particle.

We can learn something about q using our stability arguments. We showed above that the condition for equilibrium for a system constrained to have fixed values $g_0, ..., g_s, Y_{s+1}, ..., Y_t$ is that the associated energy potential $E[g_0, ..., g_s]$ be a minimum. In the case of interest here where p, T, and N are the independent variables, it is the Gibbs potential $G = \mu N$ which must be a minimum. If we carry out a calculation of G for fixed T and p, we find, just as for the density, that there are two possible functions, G_G and G_L. If we assume that we have a fixed number of particles, we have a chemical potential $\mu_G = G_G/N$ for the gas phase and $\mu_L = G_L/N$ for the liquid phase. We plot μ_G and μ_L as a function of T (for fixed p) in Fig. 2.6. What is *the* chemical potential for this system? It is simply the smallest between μ_L and μ_G at any temperature. Again we use our generalized entropy principle. We see that for $T < T_0$, the system *wants* to be a liquid since $\mu_L < \mu_G$. For $T > T_0$ the system will *choose* to be a gas since $\mu_G < \mu_L$. Thus particles in a liquid droplet will flow away into the gas region because of the lower chemical potential. At $T = T_0$, $\mu_G = \mu_L$ and the phases can coexist. Clearly, from Fig. 2.6 we obtain at $T = T_0$,

$$\left(\frac{\partial \mu_L}{\partial T}\right)_{p,N} > \left(\frac{\partial \mu_G}{\partial T}\right)_{p,N} \tag{596}$$

(these are negative slopes). Since we know, using Eq. (485), that

$$s = \frac{S}{N} = -\frac{1}{N}\left(\frac{\partial G}{\partial T}\right)_{p,N} = -\left(\frac{\partial \mu}{\partial T}\right)_{p,N}, \tag{597}$$

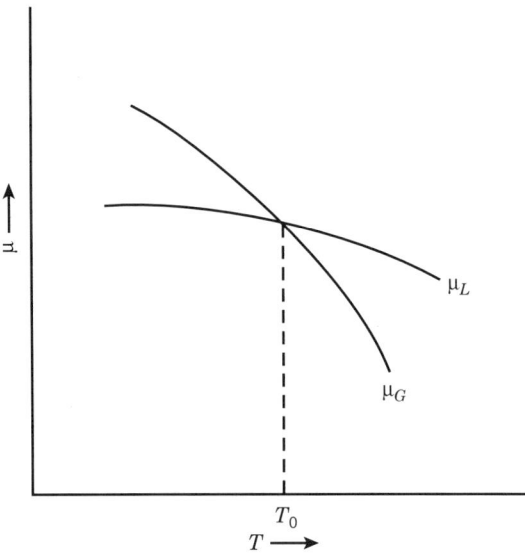

FIGURE 2.6 Schematic plot of chemical potentials for liquid (μ_L) and gas (μ_G) phases versus temperature for fixed pressure. Coexistence occurs at temperature T_o, where $\mu_L = \mu_G$.

we obtain

$$s_L < s_G. \tag{598}$$

The entropy per particle is higher in the high-temperature gas phase. Thus, as we would expect intuitively, heat must be added to the system in going from a low- to a high-temperature phase, and

$$q = T[s_G - s_L] > 0. \tag{599}$$

Conversely, if we go from a high-temperature phase to a low-temperature phase, as in our example above where we compressed the system, $q < 0$ and we must draw heat out of the system.

We can derive a relationship between the pressure and temperature in the coexistence region by differentiating $\mu_G(p,T) = \mu_L(p,T)$ with respect to T:

$$\left(\frac{\partial \mu_G}{\partial T}\right)_{p,N_G} + \left(\frac{\partial \mu_G}{\partial p}\right)_{T,N_G}\frac{dp}{dT} = \left(\frac{\partial \mu_L}{\partial T}\right)_{p,N_L} + \left(\frac{\partial \mu_L}{\partial p}\right)_{T,N_L}\frac{dp}{dT}. \tag{600}$$

Using Eq. (597) and

$$\left(\frac{\partial \mu}{\partial p}\right)_{N,T} = \frac{1}{N}\left(\frac{\partial G}{\partial p}\right)_{N,T} = \frac{V}{N} = \frac{1}{n} \tag{601}$$

we find that

$$\frac{dp}{dT} = \frac{-s_L + s_G}{1/n_G - 1/n_L}. \qquad (602)$$

If $v_G = 1/n_G$ is the specific volume and we use Eq. (599), we obtain the Clausius–Clapeyron equation [24]:

$$\frac{dp}{dT} = \frac{q}{T(v_G - v_L)}. \qquad (603)$$

This relation governs the coexistence curve $p = p(T)$. Although our discussion has focused on the coexistence of a liquid and a gas, the same ideas apply to the case of a solid–liquid coexistence. In Fig. 2.7 we show the phase diagram for neon. We see that the solid–liquid coexistence curve is very steep compared to the liquid–gas line of coexistence.

c. Triple Point. Thus far we have been discussing the coexistence of two phases. Can we have the coexistence of more than two phases? Clearly, the equilibrium of three phases of the same substance is governed by the equation

$$T_1 = T_2 = T_3 \equiv T \qquad (604)$$
$$p_1 = p_2 = p_3 \equiv p \qquad (605)$$

and

$$\mu_1 = \mu_2 = \mu_3. \qquad (606)$$

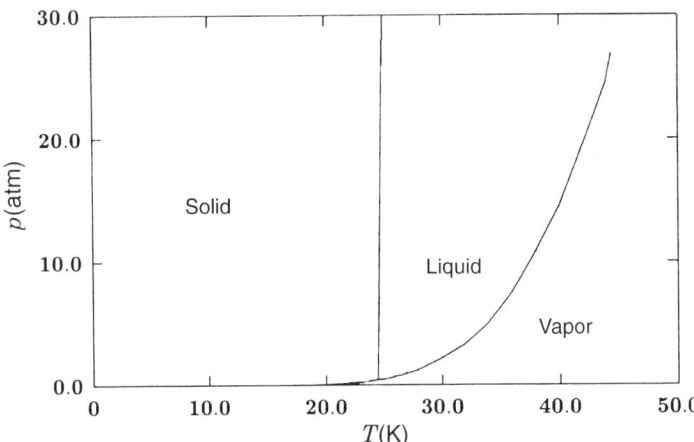

FIGURE 2.7 Equation of state for neon. (Data from G. A. Cook, ed., *Argon, Helium and the Rare Gases*, Vol. 1, Interscience, New York, 1961.)

TABLE 2.1 Triple-Point Parameters for Noble Gases

Gas	T_{tp}^0 (K)	p_{tp}(atm)
Ne	24.55	0.43
Ar	83.78	0.68
Kr	115.95	0.72
Xe	161.30	0.81

Source: Data from Ref. 25.

This last equation can be written

$$\mu_1(p,T) = \mu_2(p,T) = \mu_3(p,T). \tag{607}$$

These give two equations for two unknowns, and their solutions are specific pairs of values of p and T. The states in which three phases are present simultaneously (called *triple points*) in the (p,T) diagram are represented by isolated points which are the points of intersection of the equilibrium curves of each pair of phases (see Fig. 2.7). The triple-point parameters for the noble gases is given in Table 2.1.

The equilibrium of more than three phases of a one-component substance is clearly impossible.

d. Critical Point. Let us return to Fig. 2.5. Notice that as we increase temperature from a value T_0 to the value T_c, the jump $\Delta n = n_L - n_G$ decreases. Eventually, as $T \to T_c$, $\Delta n \to 0$ and we lose the distinction between liquid and gas. This point where $\Delta n \to 0$ is called a *critical point* and the associated phase transition near this point is called a *continuous* (no jump in the density) or (for historical reasons) *second-order phase transition* [26]. If we return to the phase diagram for neon given by Fig. 2.7 we see that the critical point lies at the end of the liquid–gas coexistence line. The values of the critical point parameters for the noble gases are given in Table 2.2.

As we discuss in more detail below, a critical point is special beyond being the end of a coexistence line. If we approach the critical point from high temperatures, we find (see Fig. 2.16) that the isotherms develop an inflection point. This means that

TABLE 2.2 Critical-Point Parameters for Noble Gases

Gas	T_c^0 (K)	p_c(atm)	ρ_c (g·cm^{-3})	ρ_{tps} (g·cm^{-3})[a]
He	6.06	2.26	0.0693	–
Ne	44.45	26.9	0.484	1.444
Ar	150.85	48.3	0.536	1.623
Kr	209.35	54.3	0.908	2.826
Xe	289.74	57.64	1.10	3.540

Source: Data from Ref. 25.
[a] Values of the mass density for the solid phase at the triple point are included for contrast.

$\partial p/\partial n$ is becoming small, or more dramatically, the isothermal compressibility is blowing up as we approach the critical point. This is discussed in much more detail below.

2.6.4 Mixtures

a. Macrovariables. We turn now to some more complicated thermodynamic systems. We first discuss mixtures of various substances. In the case of m different substances we have the macrovariables $E, V, N_1, N_2, \ldots, N_m$ and the associated set of intensive variables $T, p, \mu_1, \mu_2, \ldots, \mu_m$. Such systems are considerably more complicated than single-component systems because of the additional processes of mixing and unmixing (or phase separation), which correspond to new phases in addition to the standard one-component phase transitions. Thus we can have very complicated phase diagrams, as discussed, for example, by Landau and Lifshitz [27]. We discuss the phase-separation process further below, but let us begin here with an elementary discussion of mixing.

b. Mixing. Suppose that we have two cylinders filled with two different gases, A and B (see Fig. 2.8). Suppose also that the gases A and B are at the same temperature. Let us then remove the partition separating them. The resulting process is a mixing together of the two fluids. Since A and B are different, it is clear that in this process we lose information about the system (we knew initially that all of A was in one cylinder and B in the other) and therefore the missing information, the entropy, must increase. This means that the entropy of mixing must be positive:

$$\Delta S > 0. \tag{608}$$

This picture is correct as long as the particles are different. Consider the situation where the particles in the two cylinders are the same and we have the same density in

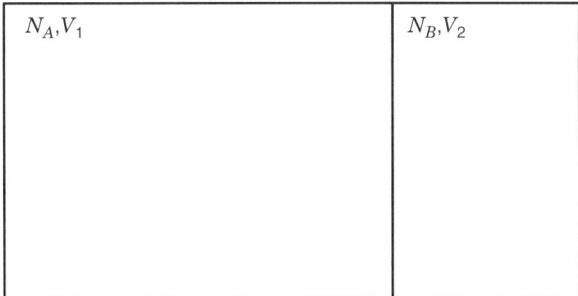

FIGURE 2.8 Container 1 with volume V_1 borders container 2 with volume V_2. The two containers are thermally coupled and maintained at temperature T. The combined system is prepared such that there are N_A particles of type A in container 1 and N_B particles of type B in container 2.

each cylinder. If we then remove the partition, we lose no information. The gas remains at the same temperature, pressure, and density, and

$$\Delta S = 0. \tag{609}$$

In Problem 2.25 you are asked to prove this result for a *mixture* of ideal gases. It is absolutely crucial in obtaining this result to include the factor $N!$ in our counting of configurations in statistical mechanics. This factor arose naturally in our discussion because we did not distinguish among the various particles. If we do not include the $N!$, the entropy of mixing for identical particles at the same density and temperature will not be zero. Historically, the theoretical difficulty of finding $\Delta S = 0$ for identical mixtures is known as *Gibbs paradox* [28]. The $N!$ factor solves the Gibbs paradox.

c. Ideal Mixtures. Let us consider the simple case of noninteracting mixtures in the canonical ensemble. Suppose that in the more general case, we have s components and that the constituents of the ith component have N_i number of particles and mass m_i. Clearly, the canonical partition function in this case is given by

$$Z(T, V, N_1, N_2, ..., N_s) = \prod_{i=1}^{s} \frac{1}{N_i!} \left(\frac{V}{\lambda_i^3} \right)^{N_i} \tag{610}$$

where λ_i is the thermal de Broglie wavelength for species i,

$$\lambda_i = \hbar \left(\frac{2\pi\beta}{m_i} \right)^{1/2}. \tag{611}$$

The free energy is then given by

$$F(T, V, N_i) = -\beta^{-1} \ln Z$$
$$= \beta^{-1} \sum_{i=1}^{s} \left[\left(\ln N_i! - N_i \ln \frac{V}{\lambda_i^3} \right) \right]. \tag{612}$$

Using Stirling's formula for $\ln N_i!$, we easily obtain

$$F(T, V, N_i) = \beta^{-1} \sum_{i=1}^{s} N_i \ln \frac{N_i \lambda_i^3}{Ve}. \tag{613}$$

One can then find the average energy,

$$E = \frac{\partial}{\partial \beta}(\beta F) = \frac{3}{2} k_B T \sum_{i=1}^{s} N_i, \tag{614}$$

the pressure

$$p = -\frac{\partial F}{\partial V} = \frac{\beta^{-1}}{V}\sum_{i=1}^{s} N_i, \qquad (615)$$

and the chemical potentials

$$\mu_i = \frac{\partial F}{\partial N_i} = \beta^{-1}\ln\frac{N_i \lambda_i^3}{V} = \beta^{-1}\ln n_i \lambda_i^3 \qquad (616)$$

where $n_i = N_i/V$ is the density of species i.

d. Osmotic Pressure. An interesting physical situation involving mixtures arises when two solutions of the same substance in the same solvent but with different solute concentrations c_1 and c_2 are separated by a semipermeable membrane. In this case the solvent molecules can pass through the membrane but the solute molecules cannot and there will be a pressure difference across the membrane called the osmotic pressure. We can understand the existence of this pressure jump from our entropy principle.

We can characterize our problem in terms of the variables E_i, V_i, N_i, and N_i', in each compartment, where N_i and N_i' are the number of solvent and solute particles in compartment $i = 1$ or 2. The concentrations are defined by $c_i = N_i'/(N_i + N_i')$. We have the constraints in the problem that the V_i, N_i', $N = N_1 + N_2$, and $E = E_1 + E_2$ are held constant. There is energy flow and particle flow by the solvent.

If we look at fluctuations in the E_i and N_i (since the V_i and N_i' are fixed) we have, following our earlier analysis for one component, the equilibrium condition

$$\delta S = \left(\frac{\partial S_1}{\partial E_1} - \frac{\partial S_2}{\partial E_2}\right)\delta E_1 + \left(\frac{\partial S_1}{\partial N_1} - \frac{\partial S_2}{\partial N_2}\right)\delta N_1 = 0. \qquad (617)$$

As before, these give the equilibrium conditions,

$$T_1 = T_2 \equiv T, \qquad (618)$$

while the chemical potentials for the solvents in the two compartments satisfy

$$\mu_1 = \mu_2. \qquad (619)$$

For a set of ideal gas solvents we have, using Eq. (616) in Eq. (619),

$$\beta^{-1}\ln(n_1\lambda_1^3) = \beta^{-1}\ln(n_2\lambda_2^3). \qquad (620)$$

Since the solvent has the same mass in both compartments, $\lambda_1 = \lambda_2$, and we have the equilibrium condition

$$n_1 = n_2 = n. \qquad (621)$$

The pressures in the compartments are given by Eq. (615). For compartment 1,

$$p_1 = \beta^{-1}\frac{N_1 + N_1'}{V_1} = \frac{\beta^{-1}n}{1 - c_1}, \qquad (622)$$

where the second expression follows by using $N_i' = c_i N_i/(1 - c_i)$. From the symmetry of the problem the pressure of the second compartment is given by

$$p_2 = \frac{\beta^{-1}n}{1 - c_2}. \qquad (623)$$

The pressure difference between the two components is given by

$$\Delta p = p_1 - p_2 = \frac{\beta^{-1}n(c_1 - c_2)}{(1 - c_1)(1 - c_2)}. \qquad (624)$$

In the limit of small concentrations the osmotic pressure jump we obtain is given by

$$\Delta p = \beta^{-1}n(c_1 - c_2) \qquad (625)$$

which is known as *van t'Hoff's formula* [29].

2.6.5 Chemical Reactions

Let us finish our general discussion of the entropy principle with a brief analysis of the conditions for chemical equilibrium. Consider a homogeneous system (consisting of a single phase) which contains m different kinds of molecules. Let us designate the chemical symbols of these molecules by $B_1, B_2, ..., B_m$. Assume that there exists the possibility of one chemical reaction occurring between these molecules whereby molecules can be transformed into each other. These chemical reactions are characterized by a reaction equation,

$$\sum_{i=1}^{m} b_i B_i = 0. \qquad (626)$$

As an example, for a system consisting of $H_2 = B_1$, $O_2 = B_2$, and $H_2O = B_3$, we have the reaction

$$2H_2 + O_2 \rightarrow 2H_2O \qquad (627)$$

which corresponds to $b_1 = 2, b_2 = 1$, and $b_3 = -2$. A second example, from high-energy physics, is pair production $\gamma \rightarrow e^+ + e^-$. The conditions for equilibrium in these cases follow from our general development. First note that the extensive variables in this case are $E, V, N_1, ..., N_m$. However, variations in the $N_i's$ are *not* independent. If we change N_i, then N_j will change by an amount governed by the chemical equation

$$\frac{\delta N_i}{\delta N_j} = \frac{b_i}{b_j}. \tag{628}$$

If we work in the Gibbs ensemble, the fundamental potential is

$$G = G(T, p, N_1, ..., N_m). \tag{629}$$

The equilibrium condition is that G is a minimum with respect to the variation of the uncontrolled parameters. In this case we cannot control the individual N_i. We can then demand that G be a minimum with respect to changes in any one of the N_i

$$\left(\frac{\partial G}{\partial N_i}\right)_{T,p} = \left(\frac{\partial G}{\partial N_i}\right)_{T,p,N_i \neq N_j} + \sum_{j \neq i} \left(\frac{\partial G}{\partial N_j}\right)_{T,p,N_k \neq N_j} \frac{\partial N_j}{\partial N_i} = 0. \tag{630}$$

Defining the chemical potential for each species,

$$\mu_i = \left(\frac{\partial G}{\partial N_i}\right)_{T,p,N_j \neq N_i}, \tag{631}$$

and using Eq. (628), we obtain the basic equation

$$\sum_{i=1}^{m} \mu_i b_i = 0. \tag{632}$$

As a practical example, consider the case of reacting ideal gases. We know that the chemical potential for the rth component of our reacting system is given by

$$\mu_r = \beta^{-1} \ln(n_r \lambda_r^3) \quad r = 1, ..., m \tag{633}$$

where the thermal de Broglie wavelength of the rth component with mass m_r is given by

$$\lambda_r = \hbar \left(\frac{2\pi\beta}{m_r}\right)^{1/2}. \tag{634}$$

Our equilibrium condition, Eq. (632), becomes

$$0 = \sum_{r=1}^{m} b_r \beta^{-1} \ln(n_r \lambda_r^3). \tag{635}$$

This can be rewritten, after exponentiation, as

$$1 = \prod_{r=1}^{m} (n_r \lambda_r^3)^{b_r}. \tag{636}$$

It is conventional to write this in the form

$$\prod_{r=1}^{m}(n_r)^{b_r} = K \qquad (637)$$

where

$$K = \prod_{r=1}^{m}(\lambda_r)^{-3b_r} \qquad (638)$$

is called the *equilibrium constant*. Equation (637), known [30] as the *law of mass action*, gives a relationship between the densities of particles in terms of the masses of the particles. An important example of its use is discussed in Chapter 4.

Thus far in this chapter we have been discussing thermodynamics from a general point of view. It should be clear, however, that we require knowledge of the equations of state if we are to go further. In the rest of this chapter we investigate the types of information we can extract from model equations of state. These equations of state are phenomenological in nature. They are constructed here, to a large degree, using physical arguments. Derivations of these equations of states require going back to a statistical mechanical point of view. Such developments are carried out in subsequent chapters.

2.7 LANDAU THEORY OF PHASE TRANSITIONS

2.7.1 Order Parameters

The basic properties of phase transitions can be understood using a phenomenological approach due to Landau [31]. The key idea, introduced in the 1930s, is that there is a variable, called the *order parameter*, which can be used to characterize each phase transition. Although the actual variables corresponding to the order parameters for a pair of physical systems may be very different, the statistical properties of the order parameter as a function of temperature may be quite similar for the two systems. One uses symmetry or other general arguments in a problem to identify the order parameter as that variable which has zero thermodynamic average at high temperatures but *spontaneously* develops a nonzero value below some critical temperature T_c. The choice or name, order parameter, comes from the expectation that systems develop order at low temperatures. The identification of the order parameter is simplified in those systems that undergo symmetry breaking for temperatures less than some critical value. Then the vanishing of the order parameter in the high-temperature phase reflects a symmetry respected. The nonzero values of the order parameter in the low-temperature phase reflects a broken symmetry.

As an important example, consider an ideal isotropic ferromagnet. Such systems, in the absence of an external magnetic field, possess rotational symmetry at high temperatures. A magnetic moment **μ(R)** located at lattice site **R** is equally likely to point in any direction. The order parameter in this system is the total magnetization,

$$\mathbf{M} = \sum_{\mathbf{R}} \boldsymbol{\mu}(\mathbf{R}). \tag{639}$$

For high temperatures the rotational symmetry of the system is respected, the individual moments can point in any direction, and on average,

$$\langle \mathbf{M} \rangle = 0 \tag{640}$$

and we are in the paramagnetic phase shown in Fig. 2.9. For temperatures below the Curie [32] temperature, T_c, the system develops a nonzero average magnetization and the system is ferromagnetic.

Order parameters can be identified in a wide variety of systems. Sometimes, identification of the order parameter for a particular class of systems is obvious, as in a ferromagnet. It took 50 years to identify the correct order parameter associated with the onset, in certain systems, of superconductivity [33]. In the case of continuous transitions or critical points, as discussed in Chapter 5 and Volume 2, it is possible to develop a general theory that is widely applicable. The ideas developed

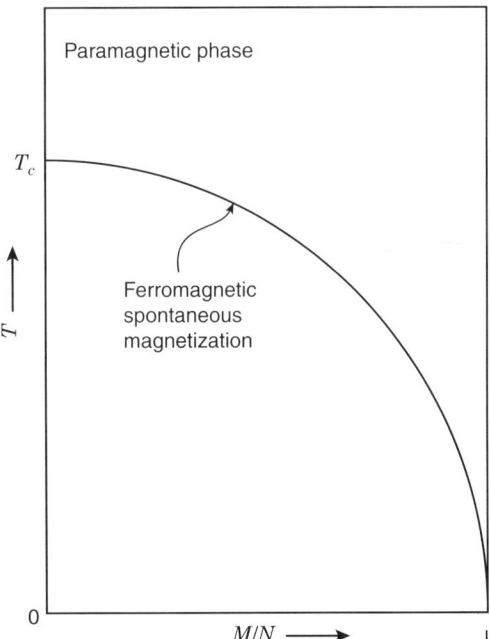

FIGURE 2.9 Ferromagnetic phase diagram in zero external field. For $T > T_c$, the Curie temperature, the average magnetization per spin is zero, indicating the paramagnetic phase. For $T < T_c$ one has a nonzero average magnetization per spin, which saturates at zero temperature at the fully aligned value μ.

in Landau theory can be extended to give a quantitative description of critical phenomena in fluids, magnetic systems, superfluids, liquid crystals, superconductors, ferroelectric systems, and others! An important property of the order parameter is it symmetry structure. Is it a scalar (such as concentration) or a vector (such as the magnetization in an isotropic ferromagnet)? Another important property of the order parameter is whether or not it is conserved. We have obvious conserved order parameters, such as the concentration in a binary mixture. Important physical examples where the order parameter is not conserved, discussed in Volume 2, are superfluids and superconductors. In some cases, such as the magnetization in a ferromagnet, the order parameter is one of the macrovariables characterizing the high-temperature phase. Typically, however, as discussed below, it corresponds to a new variable in addition to the X_i.

Consider the case of an antiferromagnet [34] where, unlike ferromagnets where spins want to align ($\uparrow\uparrow\uparrow\uparrow$) at low temperatures, the spins want to antialign ($\uparrow\downarrow\uparrow\downarrow$). It is clear that the total magnetization is *not* the order parameter for an antiferromagnet, since

$$\langle \mathbf{M} \rangle = \left\langle \sum_{\mathbf{R}} \boldsymbol{\mu}(\mathbf{R}) \right\rangle = 0. \tag{641}$$

The order parameter in this case is the *staggered magnetization*, whose definition depends on the nature of the lattice on which the system orders. For a two-dimensional square lattice it is given by

$$\mathbf{N} = \sum_{\mathbf{R}} (-1)^{n_x+n_y} \boldsymbol{\mu}(\mathbf{R}), \tag{642}$$

where $\mathbf{R} = n_x \hat{x} + n_y \hat{y}$. At low temperatures the staggered field will add up to give a nonzero value. Clearly, while the total magnetization is conserved in an isotropic system, it is clear that the staggered magnetization is *not* conserved.

It is also important to identify the variable thermodynamically conjugate to the order parameter. Examples of such conjugate fields that we have already encountered are the applied magnetic field **H** in the magnetic case (which is conjugate to the magnetization) and the chemical potential in a fluid (which is conjugate to the number of particles).

We now develop a simple theory to describe phase transitions based on the simplest assumptions of symmetry and analyticity. This *generic* theory must be modified to correspond to a particular realistic system when applied to first-order phase transitions, as discussed in the two examples at the end of this chapter. As explained in Volume 2, the Landau theory serves as the beginning step in a general development of complete theory of critical phenomena.

2.7.2 Free Energy

We assume that we have identified the *total* order parameter Ψ for a system. To keep the development as simple as possible, we assume a scalar order parameter. In the ferromagnetic case this corresponds to treating an anisotropic *Ising model*. We also introduce the order parameter density $\psi = \Psi/V$. The macrovariables in this problem are E, V, and Ψ, and the entropy satisfies the fundamental equation

$$S = S(E, \Psi, V). \tag{643}$$

If we choose to work in terms of an energy potential, with independent variables T and Ψ, the thermodynamic potential of interest is the Helmholtz free energy,

$$F[T] = E - TS. \tag{644}$$

The intensive variable conjugate to Ψ is the applied external field

$$h = \frac{\partial F[T]}{\partial \Psi}. \tag{645}$$

If we introduce the free-energy density

$$f[T] = \frac{F[T]}{V}, \tag{646}$$

we have the equation of state

$$h = \frac{\partial f[T]}{\partial \psi}. \tag{647}$$

If we make a Legendre transformation to

$$F[T, h] = F[T] - h\Psi \tag{648}$$

or, in terms of intensive densities,

$$f[T, h] = f[T] - h\psi, \tag{649}$$

then, as usual,

$$\psi = -\frac{\partial f[T, h]}{\partial h}. \tag{650}$$

The order parameter susceptibility is defined by

$$\chi = \left(\frac{\partial \psi}{\partial h}\right)_T = -\frac{\partial^2 f[T, h]}{\partial h^2}, \tag{651}$$

and is, again, a second derivative of a free energy, which is experimentally important. For a ferromagnet, χ is the magnetic susceptibility, which is directly observable.

Notice that the inverse order-parameter susceptibility can be obtained using

$$\chi^{-1} = \left(\frac{\partial h}{\partial \psi}\right)_T = \frac{\partial^2 f[T]}{\partial \psi^2}. \tag{652}$$

If $f[T,h]$ depends on a range of possible values of the order parameter, *the equilibrium solution corresponds to those values of the order parameter that minimize $f[T,h]$ with respect to ψ*. The condition for an extremum,

$$\frac{\partial f[T,h]}{\partial \psi} = 0, \tag{653}$$

gives back the equation of state, Eq. (647). Since the free energy $f[T,h]$ is to be minimized, the second derivative gives

$$\frac{\partial^2 f[T,h]}{\partial \psi^2} = \frac{\partial h}{\partial \psi} = \chi^{-1} > 0 \tag{654}$$

and the order-parameter susceptibility must be positive for a minimum free energy and thermodynamic stability.

2.7.3 Critical Points

Earlier we discussed the critical point for the liquid–gas system. In that case we have to fix both the temperature and pressure (or chemical potential) to a particular value to find the critical point. More generally, to locate a critical point we have to adjust the temperature and the conjugate field h. In the liquid–gas case, the order parameter is the density and the conjugate field is the pressure. More generally, once we adjust the conjugate field to its critical value, one can find an order parameter ψ whose average is zero [35] for $T > T_c$ and is small for T slightly less than T_c. It turns out that one can assume a power law dependence

$$\psi \approx (T_c - T)^\beta \qquad T \leq T_c, \tag{655}$$

where β is called a *critical index* [36]. The critical point is a point of marginal thermodynamic stability where

$$\chi^{-1} = \frac{\partial^2 f[T]}{\partial \psi^2} = 0. \tag{656}$$

Thus the order-parameter susceptibility blows up at the critical point.

2.7.4 Effective Free Energy

Suppose that we carry out a statistical mechanical calculation of the free energy as a function of the order parameter ψ. In general, when we construct the free energy as a function of the order parameter, we will obtain some effective free-energy density $f_E[T]$. Why it is called an effective free energy is discussed below. Let us construct a model f_E. The first step in our analysis is to expand in a Taylor series in ψ:

$$f_E[T] = \sum_{n=0}^{\infty} \frac{a_n(T)}{n!} \psi^n \tag{657}$$

where the temperature-dependent coefficients are given by

$$a_n(T) = \left(\frac{\partial^n f_E[T]}{\partial \psi^n}\right)_{\psi=0}. \tag{658}$$

We assume, due to inversion symmetry, that $f_E[T]$ is even under $\psi \to -\psi$, and therefore the a_n vanish for odd n. As the simplest model let us assume that we can truncate f_E at the lowest nontrivial order. As we will see below, this requires us to keep at least terms of fourth order and we consider

$$f_E[T] = a_0 + \frac{a_2}{2}\psi^2 + \frac{a_4}{4!}\psi^4. \tag{659}$$

If we restrict our analysis to the region near the critical point where the order parameter is small, we should be able [37] to truncate our power series expansion with a few powers of ψ. In Problem 2.26 you are asked to study the case where the ψ^6 term is also kept.

If we introduce the conventional notation, $a_2 = r$ and $a_4 = 6u$, the effective free energy is given by

$$f_E[T] = a_0 + \frac{r}{2}\psi^2 + \frac{u}{4}\psi^4. \tag{660}$$

We suppose, as is justified in some detail below, that r is linear in the temperature,

$$r = r_0\left(\frac{T}{T_0} - 1\right), \tag{661}$$

and r_0, T_0, and u are positive constants assumed to be independent of temperature. We also assume that the temperature dependence of a_0 is sufficiently weak that we can take it to be a constant, and without loss of generality, we can set this constant to zero. What information about the associated physical system can we extract from our model effective free energy? We look first at the simplest situation where the free energy is to be minimized without constraints. We note immediately that overall thermodynamic stability requires that the coefficient $u > 0$. If $u < 0$, the thermodynamic preferred state that minimizes f_E would be $\psi^2 \to \infty$. This, of course, is unphysical.

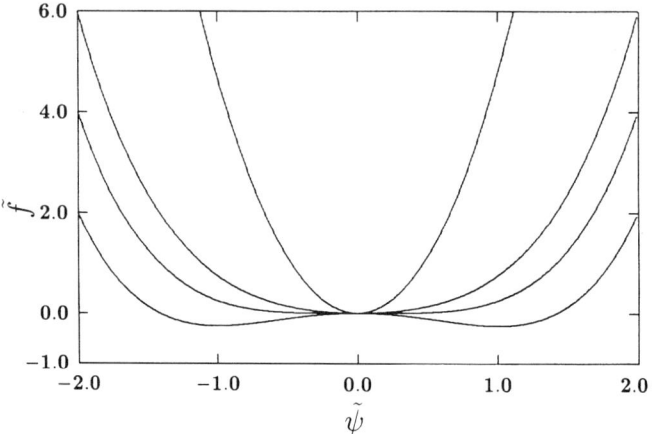

FIGURE 2.10 Scaled Landau free energy versus scaled order parameter for various scaled temperatures. From bottom to top, $T/T_0 = 0, 1, 2,$ and 10.

We refer to $f_E[T]$ as an effective free-energy density since it is clearly *not* convex [38] for $T < T_0$. This is seen in Fig. 2.10, where we plot the scaled effective free-energy density $\tilde{f}$, defined by

$$f_E = \frac{r_0^2}{u}\tilde{f}, \qquad (662)$$

versus the scaled order parameter $\tilde{\psi}$, defined by

$$\psi = \sqrt{\frac{r_0}{u}}\tilde{\psi}. \qquad (663)$$

For $T > T_0$, $\tilde{f}$ is convex, but not for $T < T_0$.

Given Eq. (659) for $f_E[T]$ and using Eq. (647), we obtain the equation of state

$$h = r\psi + u\psi^3. \qquad (664)$$

If the natural independent variables are T and h and the order parameters is allowed to fluctuate, the equilibrium values of ψ are those that *minimize* the free energy

$$\begin{aligned} f_E[T,h] &= f_E[T] - h\psi \\ &= \frac{1}{2}r\psi^2 + \frac{u}{4}\psi^4 - h\psi. \end{aligned} \qquad (665)$$

The idea here is that we can construct the physically convex free-energy density f by minimizing f_E with respect to ψ. If we have the effective free-energy density $f_E[T,h]$

with no further constraints, the equilibrium value of ψ is that value which minimizes f_E,

$$\frac{\partial f_E[T,h]}{\partial \psi} = 0. \tag{666}$$

This reduces back to the equation of state given by Eq. (664). Let us investigate how this works in detail.

2.7.5 Zero Conjugate Field

Let us consider first the case with zero conjugate field $h = 0$, which includes the critical point in our model. The equation of state is then given by

$$r\psi + u\psi^3 = 0. \tag{667}$$

We see that this equation has multiple solutions for ψ:

$$\psi = 0, \quad \pm\sqrt{\frac{-r}{u}}. \tag{668}$$

How do we decide among them? For $r > 0$ only the $\psi = 0$ solution is physically acceptable since ψ is real. For $r < 0$ and $T < T_0$, all three solutions are physically meaningful. We choose among these solutions by comparing the associated values of the free energy:

$$f_E = \begin{cases} 0 & \text{for } \psi = 0 \\ -\dfrac{r^2}{4u} & \text{for } \psi = \pm\sqrt{\dfrac{-r}{u}} \equiv \psi_\pm. \end{cases} \tag{669}$$

Clearly, the $\psi \neq 0$ solutions give a lower free energy, and $f_E = -r^2/4u$ is the equilibrium free-energy density for this model with $r < 0$. We also have that there are two *degenerate* equilibrium states or phases, ψ_+ and ψ_-. We do not know which phase the system chooses without knowing how the system was prepared. The energy due to the ordering is called the *condensation energy*.

Clearly, the important variable in the problem is $r = r_0(T/T_0 - 1)$. When r changes sign we go from the paramagnetic to the ferromagnetic regime, as shown in Fig. 2.9. The point $r = 0$ corresponds to the *critical point* and we identify, $T_c = T_0$, as the critical temperature. The average order parameter in a zero conjugate field is given by

$$\psi = \begin{cases} 0 & T > T_c \\ \pm\sqrt{\dfrac{r_0(T_c - T)}{u^0}} & T < T_c. \end{cases} \tag{670}$$

In Fig. 2.11 we plot the average magnitude of the order parameter, ψ_+, and the physical free energy as functions of temperature. Notice that for $T > T_c$ both are zero for all temperatures in this range.

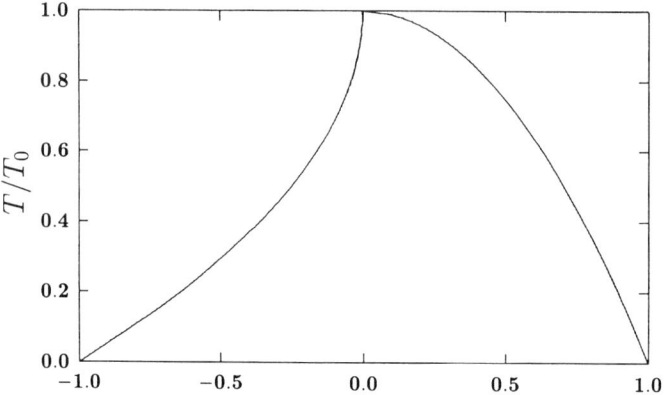

FIGURE 2.11 Scaled temperature versus scaled free energy, $f_E/f_0 (f_0 = r_0^2/4u)$, on the left-hand side and versus scaled order parameter, $\psi_+/\psi_0 (\psi_0 = \sqrt{r_0/u})$, on the right-hand side.

Moving from the average order parameter to the order-parameter susceptibility, we have, using Eq. (652),

$$\chi^{-1} = \frac{\partial^2 f_E[T]}{\partial \psi^2} = r + 3u\psi^2, \tag{671}$$

or

$$\chi = \frac{1}{r + 3u\psi^2}. \tag{672}$$

Inserting the equilibrium expressions for the physical values for the average values of the order parameter, we obtain

$$\chi = \begin{cases} \dfrac{1}{r_0(T - T_c)} & T > T_c \\ \dfrac{1}{2r_0(T_c - T)} & T < T_c, \end{cases} \tag{673}$$

and we see that χ *diverges* as $T \to T_c$.

The behavior near a critical point, where for a zero conjugate field,

$$\psi \approx (T_c - T)^\beta \tag{674}$$

and

$$\chi \approx |T - T_c|^{-\gamma} \tag{675}$$

is general and seen in a wide variety of materials. We found here, for Landau theory, that $\beta = \frac{1}{2}$ and $\gamma = 1$. The values observed for these critical indices and their origins for different systems are discussed in Chapter 5 and, more generally, in Volume 2.

2.7.6 Nonzero Conjugate Field

Let us turn to the situation where $h \neq 0$. In this case the applied field breaks the *up–down* symmetry and we obtain a unique solution to the equation of state, given by Eq. (664), for all temperatures. For $T > T_c$ and very small h, Eq. (664) reduces to

$$\psi = \frac{h}{r}. \tag{676}$$

This solution must be modified as we approach the critical temperature, where we obtain

$$\psi = \left(\frac{h}{u}\right)^{1/3}. \tag{677}$$

So $\psi \to 0$ as $h \to 0$ at the critical point, and again this is power-law behavior with an exponent of $1/\delta = \frac{1}{3}$. Note that the divergence in the susceptibility is cut off by a nonzero magnetic field since, setting $r = 0$ and using Eq. (677) for ψ, Eq. (672) reduces to

$$\chi^{-1}(T = T_c, h) = 3u\left(\frac{h}{u}\right)^{2/3}. \tag{678}$$

For $T < T_c$, the analysis of the equation of state is more involved, but the essential point is that the order parameter is a continuous nonzero function of temperature with two branches ($h \lessgtr 0$). In terms of the dimensionless order parameter defined by

$$\psi = \sqrt{\frac{|r|}{u}}\phi, \tag{679}$$

the equation of state can be written as

$$H = -\phi + \phi^3 \tag{680}$$

where the dimensionless conjugate field is defined by

$$H = \frac{h}{|r|}\sqrt{\frac{u}{|r|}}. \tag{681}$$

The order parameter susceptibility is still given by Eq. (672) and reduces, in this notation, to

$$\chi^{-1} = |r|(3\phi^2 - 1). \tag{682}$$

For small $H (> 0)$ we have, to linear order in H,

$$\phi = 1 + \frac{H}{2} + \cdots, \tag{683}$$

$$\chi^{-1} = \frac{|r|}{u}(2 + 3H) > 0, \tag{684}$$

and we have thermodynamic stability. For large H, $\phi \approx H^{1/3}$ and $\chi^{-1} = |r|3H^{2/3}$, in agreement with Eq. (678).

Our model leads to a very simple phase diagram. For $h \neq 0$ we have no phase transition as we lower the temperature from high to low values but there are two branches corresponding to $h \gtrless 0$. For $h = 0$ we have $\psi = 0$ for high temperatures $T > T_c$. For $T < T_c$, there is competition between two degenerate *ordered* phases, where the average value of the order parameter is nonzero.

2.7.7 Phase Separation and a Conservation Law

Suppose that we have a system governed by the same effective free-energy density as in the preceding section, Eq. (660), but now the order parameter is a conserved quantity. This means that the amount of ψ in the system is fixed:

$$\Psi = \Psi_0 \tag{685}$$

and if the system is spatially uniform, the average value of the order parameter has the constrained value

$$\psi = \psi_0 = \Psi_0/V. \tag{686}$$

We control the value of Ψ_0 by how *much* of ψ we put into the sample. An example of a conserved order parameter is the concentration in a binary mixture of different particles.

In the case where we have a conservation law, we must minimize $f_E[T]$ subject to the constraint that the average value of ψ is fixed. As usual, this constraint requires introducing the Lagrange multiplier h and minimizing the free energy $f_E[T, h]$. The minimization condition reduces to the equation of state

$$r\psi_0 + u\psi_0^3 = h \tag{687}$$

which just gives $h = h(\psi_0)$. In the case of a binary mixture h is related to the chemical potentials. The inverse susceptibility is given by

$$\chi^{-1} = r + 3u\psi^2. \tag{688}$$

If $r > 0$ and $\psi = \psi_0$, then $\chi^{-1} > 0$ and as in the case where the order parameter is not conserved, nothing very interesting happens as a function of r, u, and ψ_0.

Let us focus on the more interesting case $r < 0$, where the free-energy density, eliminating h using Eq. (687), takes the form

$$f_E[T, h] = \frac{|r|}{2}\psi_0^2 - \frac{3u}{4}\psi_0^4 \tag{689}$$

and from Eq. (688), the inverse susceptibility is given by

$$\chi^{-1} = -|r| + 3u\psi_0^2. \tag{690}$$

Clearly, the uniform solution $\psi = \psi_0$ breaks down for $r + 3u\psi_0^2 < 0$. If $-r_0 + 3u\psi_0^2 < 0$, there is a temperature where this solution becomes thermodynamically unstable. This temperature, $T_s = T_s(\psi_0)$, is called the *spinodal temperature*. For lower temperatures one has lost the uniform solution all together. What does nature do in this case?

It turns out that in the region $r < 0$ and $\psi_0^2 < |r|/u$, there is another set of solutions to our problem. For this new set of *inhomogeneous solutions* the system breaks up into macroscopic volumes of two competing degenerate phases—we have phase coexistence. The system finds these solutions by locally fixing the conjugate field $h = 0$ and minimizing the effective free energy given by Eq. (660). The two solutions are just those degenerate phases found in the preceding section for the system without a constraint. Thus the system organizes itself such that it breaks up into domains where

$$\psi = \psi_\pm = \pm\sqrt{\frac{|r|}{u}}. \tag{691}$$

The conservation law is satisfied by having just enough of ψ_+ and ψ_- so that Eq. (685) is satisfied. Thus if α is the fraction of the system in the $+$ phase and β is the fraction of the system in the $-$ phase, we require that

$$\psi_0 = \alpha\psi_+ + \beta\psi_- \tag{692}$$

and

$$1 = \alpha + \beta. \tag{693}$$

In the present case these degenerate solutions are symmetric, and we have $\psi_- = -\psi_+$, so

$$\psi_0 = (2\alpha - 1)\psi_+ \tag{694}$$

and

$$\alpha = \frac{1}{2}\left(1 + \frac{\psi_0}{\psi_+}\right) \tag{695}$$

$$\beta = \frac{1}{2}\left(1 - \frac{\psi_0}{\psi_+}\right). \tag{696}$$

168 PRINCIPLES OF THERMODYNAMICS

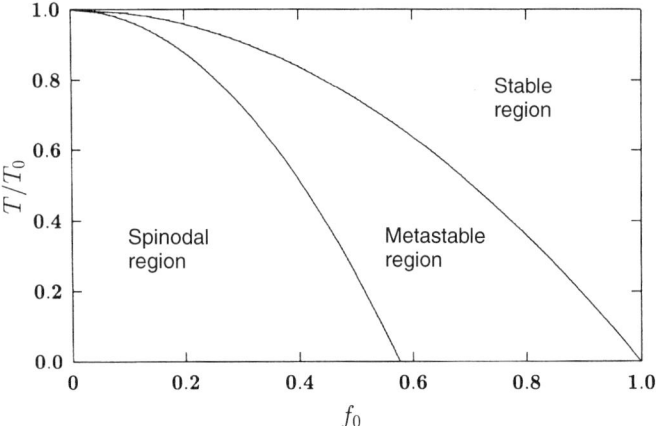

FIGURE 2.12. Equation of state for a system with a conserved order parameter. The scaled temperature is plotted versus $f_0 = \psi_0 u/r_0$, where ψ_0 is the conserved density. The metastable region is separated from the stable region by the coexistence curve.

The phase diagram for this system is shown in Fig. 2.12. How do we know that nature picks out this phase-separated system compared to the homogeneous system in the range above the spinodal line separating the spinodal and metastable regions in Fig. 2.12? We must compare the free energies for these two solutions: the homogeneous solution and the inhomogeneous solution. For the homogeneous solution we have

$$f_E^H[T,h] = \frac{\psi_0^2}{2}\left[|r| - \frac{3}{2}u\psi_0^2\right], \tag{697}$$

while for the inhomogeneous solution we have, $f_E^+[T,h] = f_E^-[T,h]$ and

$$f_E^I[T,h] = \alpha f_E^+[T,h] + \beta f_E^-[T,h] = -\frac{r^2}{4u}. \tag{698}$$

Comparing these two results, we obtain

$$f_E^H[T,h] - f_E^I[T,h] = \frac{\psi_0^2}{2}\left[|r| - \frac{3}{2}u\psi_0^2\right] + \frac{r^2}{4u}. \tag{699}$$

If we write this in terms of $\psi_+^2 = |r|/u$, we have

$$f_E^H[T,h] - f_E^I[T,h] = \frac{u}{4}\left[\psi_+^4 - 3\psi_0^4 + 2\psi_+^2\psi_0^2\right]. \tag{700}$$

If we write $\psi_0^2 = \psi_+^2 f$, then

$$f_E^H[T,h] - f_E^I[T,h] = \frac{u}{4}\psi_+^4[(1+3f)(1-f)] \geq 0 \tag{701}$$

if $0 \leq f \leq 1$. Thus the system picks out the phase-separated solution for the entire range $r < 0$ and $\psi_0^2 \leq \psi_+^2$. In the regime between the spinodal line and the coexistence curve,

$$-3u\psi_0^2 < r < 0, \qquad (702)$$

the homogenerous solution is a metastable solution. It is possible to prepare the system in this state, but it will eventually *nucleate* over to the thermodynamically stable inhomogeneous phase-separated state. For $f > 1$ in Eq. (701) we see that the homogeneous solution is stable.

2.8 PHASE SEPARATION IN BINARY ALLOYS

Let us now move on to an example of phase separation that is a bit more realistic than our generic or simplified magnetic example of the preceding section. Here we discuss a simple model for a binary alloy.

We consider a crystalline solid mixture in a system with N lattice sites. Each site is occupied by either an atom of type A or an atom of type B. For simplicity, we do not allow for vacancies, an unoccupied lattice site, and require that the number of atoms N_A and B atoms N_B add to N:

$$N = N_A + N_B. \qquad (703)$$

Let us assume, as will be true at sufficiently high temperature, that the probability that a given site is occupied by an A or B atom is independent of the group of atoms occupying neighboring sites. It is then clear that the probability of a particular site $\mathbf{R}_i$ being occupied by an A atom is

$$P_A^i = \frac{N_A}{N} \equiv c, \qquad (704)$$

while the probability that a site is occupied by a B atom is

$$P_B^i = \frac{N_B}{N} = 1 - c. \qquad (705)$$

Thus c is the *concentration* of A atoms. Going back to our original definition of entropy, we find immediately that the entropy for this system is given by

$$S = -k_B \sum_i \sum_{\alpha = A \text{ or } B} P_\alpha^i \ln P_\alpha^i. \qquad (706)$$

Assuming that the probability of occupation is given [39], on average, by Eq. (704), the entropy reduces to

$$S = -k_B N [c \ln c + (1-c) \ln(1-c)]. \qquad (707)$$

Since $0 < c < 1$, the entropy is positive.

Suppose that there is an interaction potential energy U_{AA} between two neighboring A atoms. Similarly, there is an interaction U_{BB} for two neighboring B atoms and an interaction U_{AB} between unlike neighbors. Consider a lattice where each site has z nearest neighbors. Then, on average, the interaction energy of an A atom with its environment is given by

$$U_A = z[cU_{AA} + (1-c)U_{AB}]. \tag{708}$$

Similarly for B atoms,

$$U_B = z[cU_{AB} + (1-c)U_{BB}], \tag{709}$$

and the average energy for the combined system is given by

$$E = \frac{N}{2}z(cU_A + (1-c)U_B), \tag{710}$$

where the total number of pairs is $Nz/2$. The $\frac{1}{2}$ comes from correcting for double counting. Then the approximate effective free energy is of the form

$$F_E = E - TS = N\left[\frac{z}{2}(c^2 U_{AA} + 2c(1-c)U_{AB} + (1-c)^2 U_{BB})\right.$$
$$\left. + k_B T(c \ln c + (1-c)\ln(1-c))\right]$$
$$\equiv N f_E(c) \tag{711}$$

where f_E is the free-energy density, which is a function of the concentration c. If there are no constraints, the observed value of the concentration is given by that value of c that minimizes the free energy. In this case we have several constraints on the system reflecting conservation laws. We have that the total number of A and B atoms are conserved separately. We handle these constraints, as usual, by introducing chemical potentials, Lagrange multipliers, for each species and defining the effective Gibbs free energy

$$G_E(c) = N[f_E(c) - \mu_1 c - \mu_2(1-c)]. \tag{712}$$

However, since $N = N_A + N_B$ is fixed, we require that

$$\mu_1 = -\mu_2 = \mu \tag{713}$$

and

$$G_E(c) = N[f_E(c) - \mu(2c-1)] \equiv N g_E(c). \tag{714}$$

Our goal is to understand the phase structure associated with this model.

To understand the physics in this example in the most straightforward fashion, it is useful to define the new variable

$$M = 2c - 1 \tag{715}$$

or

$$c = \tfrac{1}{2}(1 + M). \tag{716}$$

In terms of this variable, the Gibbs free energy can be written in the form

$$g_E(c) = f_0 - (\Delta + \mu)M + E_0 M^2$$
$$+ \frac{k_B T}{2}[(1 - M)\ln(1 - M) + (1 + M)\ln(1 + M)] \tag{717}$$

where we have defined the quantities:

$$f_0 = \frac{z}{8}(U_{AA} + U_{BB} + 2U_{AB}) + k_B T \ln\left(\frac{1}{2}\right) \tag{718}$$

$$\Delta = \frac{z}{4}(U_{BB} - U_{AA}) \tag{719}$$

and

$$E_0 = \frac{z}{8}(U_{AA} + U_{BB} - 2U_{AB}). \tag{720}$$

A key observation is that the only asymmetry between A and B particles is through the term proportional to Δ. This term can be removed by shifting the definition of the chemical potential

$$\mu \to \mu - \Delta \tag{721}$$

and obtaining

$$g_E(M) = f_0 - \mu M + E_0 M^2$$
$$+ \frac{k_B T}{2}[(1 - M)\ln(1 - M) + (1 + M)\ln(1 + M)]. \tag{722}$$

In this case we see that the Gibbs free energy is invariant under the symmetry operation $M \to -M$ and $\mu \to -\mu$. In terms of the original concentration variables, we have that there is a symmetry about $c = \tfrac{1}{2}$. There is no physical reason for this symmetry in this particular system. However, working with a model with this up–down symmetry simplifies the analysis considerably.

In equilibrium the Gibbs free energy is a minimum consistent with the concentration

$$c_0 = \tfrac{1}{2}(1 + M_0). \tag{723}$$

Thus we look for solutions to

$$\left.\frac{\partial g_E(c)}{\partial c}\right|_{c=c_0} = 0, \qquad (724)$$

or equivalently,

$$\left.\frac{\partial g_E(M)}{\partial M}\right|_{M=M_0} = -\mu + 2E_0 M_0 + \frac{k_B T}{2} \ln\frac{1+M_0}{1-M_0} = 0. \qquad (725)$$

This equation can be solved for the chemical potential and put back into the Gibbs free energy to obtain, after some cancellations,

$$g_E(M) = f_0 - E_0 M_0^2 + \frac{k_B T}{2}\ln(1-M_0^2). \qquad (726)$$

For most of the phase diagram for this system the chemical potential is given by Eq. (725) as

$$\mu = \frac{k_B T}{2}\ln\frac{1+M_0}{1-M_0} + 2E_0 M_0. \qquad (727)$$

We are also interested in the compressibility for the system, which is defined in terms of the second derivative of the free energy with respect to concentration:

$$\frac{\partial c}{\partial \mu} \equiv \chi = \left(\frac{\partial^2 g_E}{\partial c^2}\right)^{-1} \qquad (728)$$

or

$$\chi^{-1} = \frac{1}{4}\frac{\partial^2 g_E}{\partial M^2} = \frac{1}{4}\left[2E_0 + \frac{k_B T}{2}\left(\frac{1}{1+M} + \frac{1}{1-M}\right)\right] \qquad (729)$$

which must be positive for thermodynamic stability.

Let now consider the simpler case where $E_0 > 0$, which corresponds to the situation where the energy required to have like neighboring particles is higher than the energy to have unlike particles as neighbors. In this case the system wants to mix A and B particles, and μ and χ are rather unremarkable functions of c_0 and T. As c_0 ranges from 0 to 1 and M_0 ranges between -1 and 1, μ ranges smoothly from $-\infty$ to ∞. χ vanishes as $c \to 0$ and 1 and has a single peak at $c = \frac{1}{2}$ or $M = 0$, where

$$\chi = \frac{4}{k_B T + 2E_0}. \qquad (730)$$

Thus this peak increases as temperature decreases. Over the entire temperature range, as we expect from general thermodynamic arguments, $\chi > 0$ and

$\partial^2 g/\partial c^2 > 0$. Thus for $E_0 > 0$, the system is in a single mixed phase for all temperatures. The situation for $E_0 < 0$ is quite different. In this case it is favorable for the system to phase separate at low temperatures into A- and B- rich macroscopic domains. As a first indication of what we are up against, consider the compressibility in this case, where Eq. (729) takes the form

$$\chi^{-1} = \frac{1}{4(1-M^2)} \left[k_B T - 2|E_0|(1-M^2) \right]. \qquad (731)$$

For a given concentration $M_0 = 2c_0 - 1$, there is a temperature T_s, given by

$$k_B T_s = 2|E_0|(1-M_0^2), \qquad (732)$$

below which the system is unstable. The highest temperature for which this instability occurs is when $c_0 = \frac{1}{2}$ of $M_0 = 0$ and

$$k_B T_c = 2|E_0|. \qquad (733)$$

For temperatures $T > T_c$ the system is in a single mixed phase with compressibility given by Eq. (731).

Suppose that we look in the vicinity of $T = T_c$ and $M_0 \approx 0$. In this case we can expand the free energy in powers of M and obtain

$$g_E(M) = f_0 - HM + \frac{r}{2}M^2 + \frac{u}{4}M^4 + \cdots \qquad (734)$$

and to facilitate the comparison with the Landau theory, we have introduced the quantities

$$\frac{r}{2} = \frac{1}{2}(k_B T - 2|E_0|) \qquad (735)$$

$$\frac{u}{4} = \frac{k_B T}{12} \qquad (736)$$

and $H = \mu$. We see immediately that our free energy here, Eq. (734), is identical in form to Eq. (665) in the Landau theory. Clearly, we have a critical point for $H = \mu = 0$ and $T = T_c$. The properties near the critical point are clearly just those given by the Landau theory.

Note that there is a precise connection between M in the current problem and the order parameter in the Landau theory. We also connect the variable H here with the external conjugate field in the Landau theory. The order parameter susceptibility is proportional to the compressibility. The divergence of χ for $H = 0$ as $T \to T_c$ has the same critical indices as in the Landau theory. Thus, for example, Eq. (731) leads to the exponent $\gamma = 1$.

Let us now focus on the case of $T < T_c$. It is crucial to realize, as in our treatment of the generic problem, one picks up a new set of physical solutions to the problem. In the case $T < T_s$, we know that some new solutions must develop since the

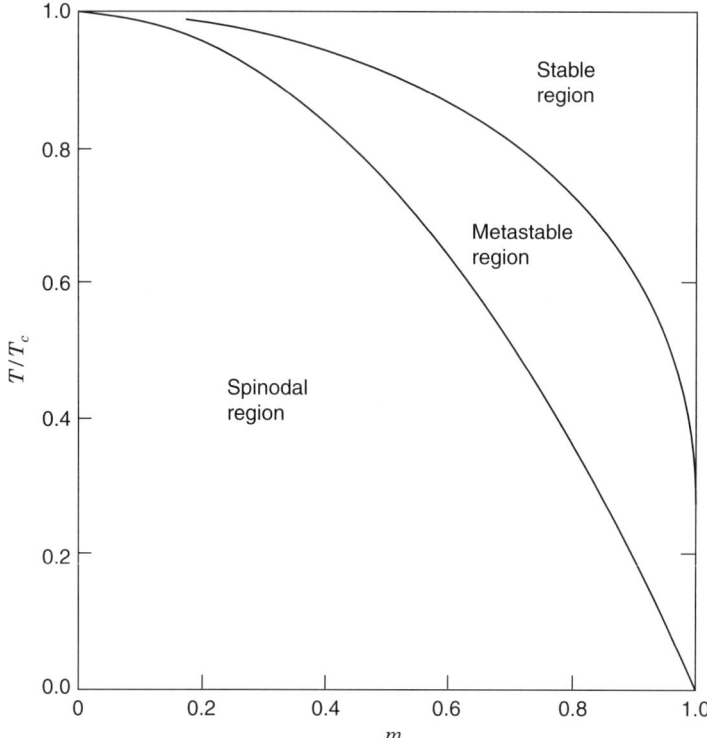

FIGURE 2.13 Equation of state for binary solid mixture. The scaled temperature is plotted versus the reduced concentration m. The metastable region is separated from the stable region by the coexistence curve $m(T)$. The metastable region is separated from the spinodal region by the spinodal curve $M_s(T)$.

continuation of the mixed-phase solution is unstable. For a given temperature $T < T_c$, we can define the spinodal line corresponding to concentration:

$$M_s^2 = 1 - \frac{k_B T}{2|E_0|} \tag{737}$$

as shown in Fig. 2.13. To see that there are new solutions for $T < T_c$, let us look at the equation of state obtained from minimizing the Gibbs free energy:

$$\mu = \frac{k_B T}{2} \ln \frac{1+M}{1-M} - 2|E_0|M. \tag{738}$$

For $T > T_c$ this equation simply gives $\mu = \mu(M_0)$. As one approaches the critical point, one sees that $\mu \to 0$. Below T_c one finds that there are three solutions to Eq. (738). The two solutions of interest to us here are those with $\mu = 0$, $M = \pm m$, and m satisfies

$$2|E_0|m = \frac{k_B T}{2} \ln \frac{1+m}{1-m}. \tag{739}$$

We easily find that there are symmetric solutions to Eq. (739) for positive and negative m which together trace out the *coexistence curve* shown in Fig. 2.13. For a given temperature and $M_0^2 < M_s^2$, concentrations inside the spinodal curve, these solutions are the only solutions available and thus the system must select them. How does the system maintain the conservation law? Since there is a fixed amount of M_0, the system organizes itself such that it divides into macroscopic phases of m and $-m$. The amounts of m and $-m$ phases are determined by the fixed amount of M_0. This situation is very similar to that treated in the preceding section. Thus if the fraction of the system that is in state m is α and the amount that is in state $-m$ is β, we require that

$$M_0 = \alpha m + \beta(-m) \qquad (740)$$

and

$$1 = \alpha + \beta. \qquad (741)$$

We can solve this set of equations to obtain

$$\alpha = \frac{1}{2}\left(1 + \frac{M_0}{m}\right). \qquad (742)$$

What happens in the region between the coexistence curve and the spinodal curve shown in Fig. 2.13? In this case we need to compare the Gibbs free energy for the mixed state with that for the phase-separated state. In the phase-separated case the free energy is given by

$$g_P = g(m) = g(-m) = |E_0|m^2 + \frac{k_B T}{2}\ln(1 - m^2). \qquad (743)$$

This must be compared with the mixed-state expression

$$g_M = |E_0|M_0^2 + \frac{k_B T}{2}\ln(1 - M_0^2). \qquad (744)$$

We plot $g_P/|E_0|$ and $g_M/|E_0|$ versus T/T_c in Fig. 2.14. We see that the phase-separated phase has the lower free energy inside the coexistence curve. Analytically, we can see this as follows. We can write

$$g_M - g_P = |E_0|(M_0^2 - m^2) + \frac{k_B T}{2}\ln\frac{1 - M_0^2}{1 - m^2}. \qquad (745)$$

If we remember the inequality, Eq. (88),

$$\ln x \geq 1 - \frac{1}{x}, \qquad (746)$$

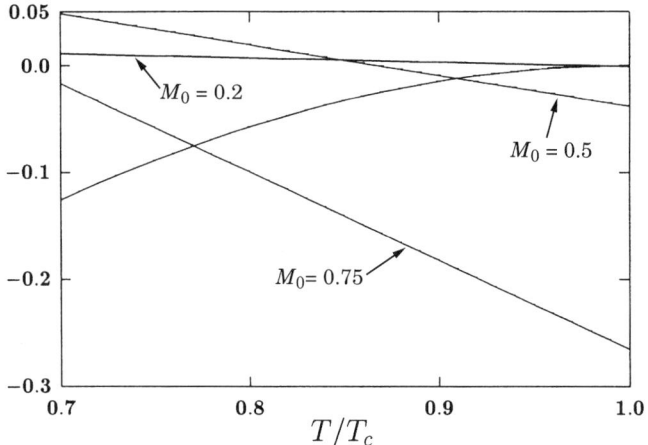

FIGURE 2.14 Free energy comparisons determining the equation of state for a binary mixture. $\tilde{g}_p = g_p/|E_0|$ is the free energy for the phase-separated state given by Eq. (743), with m determined by the solutions to Eq. (739). This free energy is independent of M_0 and represented by the line with increasing slope in the figure. $\tilde{g}_M = g_M/|E_0|$ is the free energy for the mixed state and is given by Eq. (744), shown for the labeled values of M_0. The equilibrium state for a given value of M_0 is the lower of g_p and g_M for a given temperature. The interesection of g_p and g_M, for a given M_0, fixes a point on the coexistence curve $m(T)$.

we can write, after some minor rearrangement,

$$g_M - g_P \geq \frac{1}{2}\left(\frac{m^2 - M_0^2}{1 - M_0^2}\right)\left[k_B T - 2|E_0|(1 - M_0^2)\right]. \tag{747}$$

The last factor on the right-hand side of Eq. (747) is positive if $M_0^2 \geq M_s^2$. Thus the mixed phase has a higher Gibbs free energy than the phase-separated phase,

$$g_M - g_P \geq 0 \tag{748}$$

in the region of Fig. 2.13, where $m^2 \geq M_0^2 \geq M_s^2$. Thus the phase-separated solution is the physically selected solution for all M_0 and T under the coexistence curve.

At low temperatures it is easy to show that the solution for m is given by

$$m = 1 - 2e^{-|E_0|/2T} + \cdots \tag{749}$$

and the compressibility by

$$\chi = \frac{8}{k_B T} e^{-|E_0|/2T}. \tag{750}$$

In more general systems where there is asymmetry in the problem that cannot be absorbed into the chemical potential, treatment of the phase-separated state is more involved. If one has coexisting phases c_1 and c_2, we require equal chemical potentials and (equivalent to pressure) Gibbs free energy:

$$\mu(c_1) = \mu(c_2) \tag{751}$$
$$g(c_1) = g(c_2). \tag{752}$$

2.9 van der WAALS EQUATION OF STATE

2.9.1 Motivation

Our next example of a phenomenological equation of state is the most famous. In the nineteenth century, before the advent of statistical mechanics, the lack of a method for determining the equation of state of an interacting fluid severely restrained progress in thermodynamics. The van der Waals phenomenologically motivated equation of state [40], proposed in 1873, led not only to great progress in understanding the physics of fluids, but pushed forward the development of statistical mechanics itself. This push arose from the desire to derive the van der Waals equation of state from first principles. In Chapter 4 we discuss the extension of statistical mechanics to nonideal gases and the derivation of the van der Waals equation of state.

It is conventional to write the van der Waals equation in the form

$$\left(p + \frac{a}{V_m^2}\right)(V_m - b) = RT \tag{753}$$

where p is the pressure, R the gas constant, V_m the specific volume per mole:

$$V_m = \frac{L}{n} \tag{754}$$

where L is Avogadro's number, n the density, and a and b are positive constants peculiar to a given system. We will find it more convenient to write Eq. (753) in terms of the density, and obtain

$$(p + \bar{a}n^2)(1 - \bar{b}n) = nk_BT \tag{755}$$

where we have introduced

$$\bar{a} = \frac{a}{L^2} \tag{756}$$

and

$$\bar{b} = \frac{b}{L} \tag{757}$$

and

$$k_B = \frac{R}{L} \tag{758}$$

is Boltzmann's constant. If we set $\bar{a}$ and $\bar{b}$ equal to zero, we recover the ideal gas law. van der Waals argued that there should be two modifications of the ideal gas law, due to intermolecular interactions (see Fig. 2.15). The first modification is due to the harsh repulsive part of the interaction at short distances. Particles can get no closer together than a distance σ, which would be the diameter of billiard ball–like particles. This means that particles are excluded from the volume occupied by other particles. The available volume is therefore less than the volume of the container by the amount V_{ex}, where V_{ex} is just the volume excluded per particle $\bar{b}$ times the total number of particles:

$$V_{\text{ex}} = \bar{b}N. \tag{759}$$

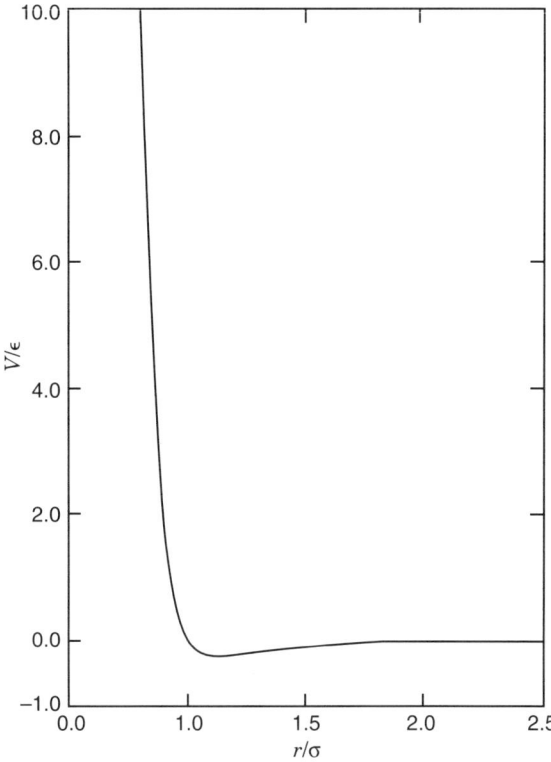

FIGURE 2.15 Form of pair potential V for simple single-component systems with short-range interactions. ϵ is some characteristic energy and σ a characteristic distance (discussed below and in Chapter 4).

The effective density is given then by

$$n_{\text{eff}} = \frac{N}{V - V_{\text{ex}}} = \frac{N}{V - \bar{b}N}$$
$$= \frac{n}{1 - \bar{b}n}. \tag{760}$$

We are then to replace n by n_{eff} in the ideal gas law to obtain

$$p = n_{\text{eff}} k_B T = \frac{n}{1 - \bar{b}n} k_B T. \tag{761}$$

We have not yet included the effects of the attractive part of the potential. The basic point is that an attractive force tends to pull the particles together. This clearly reduces the pressure they exert on the walls. van der Waals assumed that this reduction in pressure could be taken into account by adding a term $p_{\text{att}} = -\bar{a}n^2$ to the equation above. The n^2 dependence of p_{att} comes from realizing that the effects of attraction require having two particles close enough that the interaction potential is nonzero. Roughly speaking, the probability of having a single particle in a volume v is proportional to vn, while the probability for having two particles goes as $(vn)^2$. In Chapter 4 we learn how to make these ideas quantitative. Here we assume that the parameters $\bar{a}$ and $\bar{b}$ are known empirically.

2.9.2 Phase Diagram

Let us now check the predictions of the van der Waals equation. It is convenient to rewrite the equation in terms of dimensionless variables:

$$\bar{p} = \frac{p\bar{b}^2}{\bar{a}} \tag{762}$$

$$\bar{T} = \frac{k_B T \bar{b}}{\bar{a}} \tag{763}$$

and

$$\bar{n} = \bar{b}n, \tag{764}$$

so that

$$\bar{p} = \frac{\bar{n}\bar{T}}{1 - \bar{n}} - \bar{n}^2. \tag{765}$$

Notice that in terms of these scaled parameters, the dependence on material parameters has been eliminated. Thus this equation of state applies to all simple fluids if we rescale n and T properly [41]. We return to this point later.

In Fig. 2.16 we plot $\bar{p}$ versus $\bar{n}$ for various temperatures. We see for high temperatures that the isotherms are monotonically increasing. As $\bar{T}$ decreases, we see that there is an inflection point near $\bar{T} \approx 0.3$. As $\bar{T}$ decreases further, the $\bar{p}$ versus $\bar{n}$ curves show a dip for nonzero $\bar{n}$, and for $\bar{T} < 0.25$ the pressure becomes negative

180 PRINCIPLES OF THERMODYNAMICS

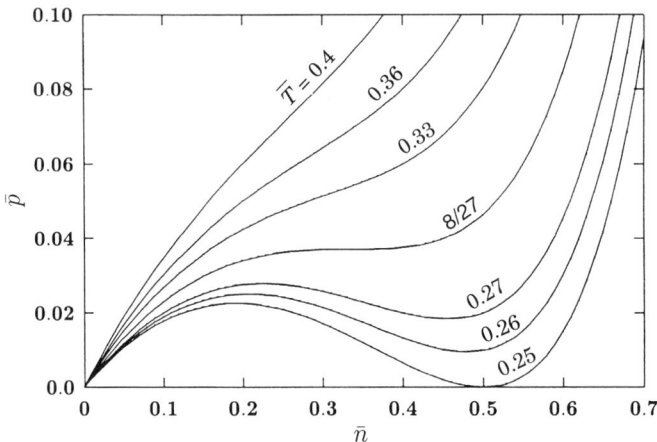

FIGURE 2.16 van der Waals equation of state in terms of scaled pressure $\bar{p}$, density $\bar{n}$, and temperature $\bar{T}$.

for $\bar{n} \approx 0.5$. Our first conclusion should be that the van der Waals equation has a disease for low temperatures: The equilibrium pressure obviously can never be negative. A related and important observation comes if we recall the stability requirement Eq. (576), which can be rewritten (for fixed N) as

$$\left(\frac{\partial p}{\partial V}\right)_T < 0. \tag{766}$$

Since

$$\left(\frac{\partial p}{\partial V}\right)_{T,N} = \left(\frac{\partial p}{\partial n}\right)_{T,N} \left(\frac{\partial n}{\partial V}\right)_{T,N}$$

$$= -\frac{N}{V^2} \left(\frac{\partial p}{\partial n}\right)_{T,N} \tag{767}$$

the stability condition can be written as

$$\left(\frac{\partial p}{\partial n}\right)_{T,N} > 0. \tag{768}$$

Looking at Fig. 2.16, we see that this inequality is satisfied for $\bar{T} > \bar{T}_c$, where $\bar{T}_c$ is the temperature corresponding to the inflection point, where

$$\left(\frac{d\bar{p}}{d\bar{n}}\right)_{\bar{T}_c, N} = 0 \tag{769}$$

and

$$\left(\frac{d^2\bar{p}}{d\bar{n}^2}\right)_{T_c,N} = 0. \tag{770}$$

For all temperatures less than $\bar{T}_c$ there will be a region ($\bar{n}_G < \bar{n} < \bar{n}_L$ in Fig. 2.16) where $d\bar{p}/d\bar{n} < 0$. Such a region is not thermodynamically stable, and van der Waals equation cannot reflect the *true* $\bar{p}$ versus $\bar{n}$ curve for a physical system. This instability is another example of the phase separation phenomena discussed above. We see from the figure that for $\bar{p} < \bar{p}_c$ (the pressure where the inflection point occurs) and for $\bar{T} < \bar{T}_c$, for a given pressure there are three possible values for the density according to the van der Waals equation of state. In the case of phase separation, we remember that for a given pressure there are two physical values of the density.

2.9.3 Maxwell Construction

From our previous discussion of the liquid–gas phase transition we know that the physical $\bar{p}$ versus $\bar{n}$ curve should look as shown in Fig. 2.17. For low densities we are in the gas phase. As we decrease the volume of the system at constant temperature, the pressure will increase until one reaches $p(T)$, which is *the* coexistence pressure for the given fixed temperature. As we decrease the volume further, the pressure remains constant and gas goes into liquid until $n = n_L$ and the system is all liquid. Further decrease in the volume will lead (as we expect for a liquid) to a rapid increase in pressure.

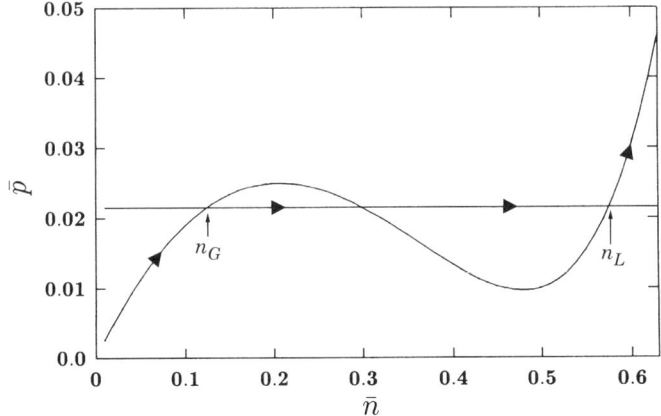

FIGURE 2.17 Isotherm, for a scaled temperature $\bar{T} = 0.26$, for the van der Waals equation of state using Maxwell's construction. The straight horizontal line gives the coexistence pressure corresponding to this temperature. Arrows indicate the equilibrium isotherm. Liquid, $\bar{n}_L$, and gas, $\bar{n}_G$, values for the density for the coexisting phases are indicated.

How can we extract from the van der Waals equation the correct coexistence behavior? Consider the following manipulation. The change in the chemical potential from one value of the density to another at constant N and T can be written

$$\mu(n_L, T) - \mu(n_G, T) = \int_{n_G}^{n_L} d\mu. \tag{771}$$

Since, however, for fixed T, from the Gibbs–Duhem relation,

$$n d\mu = dp, \tag{772}$$

we have

$$\mu(n_L, T) - \mu(n_G, T) = \int_{n_G}^{n_L} \frac{dp}{n}. \tag{773}$$

Let us focus on the integral, where we can write

$$\int_{n_G}^{n_L} \frac{dp}{n} = \int_{n_G}^{n_L} \left[d\left(\frac{p}{n}\right) - p d\left(\frac{1}{n}\right) \right]$$
$$= \frac{p(n_L, T)}{n_L} - \frac{p(n_G, T)}{n_G} + \int_{n_G}^{n_L} \frac{p(n, T)}{n^2} dn. \tag{774}$$

Inserting this result back into Eq. (773), we obtain

$$\mu(n_L, T) - \mu(n_G, T) = \frac{p(n_L, T)}{n_L} - \frac{p(n_G, T)}{n_G} + \int_{n_G}^{n_L} \frac{p(n, T)}{n^2} dn. \tag{775}$$

We can then demand that the general conditions for phase coexistence,

$$\mu(n_G, T, N) = \mu(n_L, T, N) \tag{776}$$

and

$$p(n_G) = p(n_L) \equiv p(T) \tag{777}$$

hold. Equation (775), determining n_G and n_L, reduces to

$$p(T)\left[\frac{1}{n_L} - \frac{1}{n_G}\right] + \int_{n_G}^{n_L} \frac{p(n, T)}{n^2} dn = 0. \tag{778}$$

If we write this equation in terms of the volume instead of the density, we have

$$p(T)[V_G - V_L] = \int_{V_L}^{V_G} p(V, T) dV. \tag{779}$$

This equation has a simple geometrical interpretation. It says that we can determine the appropriate $p(T)$, V_L, and V_G by choosing the hatched areas as shown in Fig. 2.22 below to have equal areas. This prescription for interpreting the van der Waals equation is known as a *Maxwell construction* [42].

Let us now apply Eq. (778) to the case of the van der Waals equation. This is fundamentally an equation relating n_G and n_L. This is because the conditions of coexistence give one $T = T(n_G, n_L)$ and $p(T) = p(n_G, n_L)$. For a given pressure on the coexistence curve,

$$p = \frac{n_G T}{1 - n_G} - n_G^2 = \frac{n_L T}{1 - n_L} - n_L^2 \tag{780}$$

which we can solve to obtain, for $T < T_c$,

$$T = (1 - n_G)(1 - n_L)(n_G + n_L). \tag{781}$$

Also along the coexistence curve,

$$p(T) = \frac{n_L}{1 - n_L} T(n_G, n_L) - n_L^2$$

$$= n_L n_G (1 - n_G - n_L). \tag{782}$$

Next we need to evaluate the integral that appears in Eq. (778):

$$\int_{n_G}^{n_L} \frac{p(n, T)}{n^2} dn = \int_{n_G}^{n_L} \frac{dn}{n^2} \left[\frac{n}{1 - n} T(n_G, n_L) - n^2 \right]$$

$$= T(n_G, n_L) \ln \left(\frac{n_L(1 - n_G)}{n_G(1 - n_L)} \right) + n_G - n_L. \tag{783}$$

Putting all of these results together in Eq. (778), we obtain the equation determining the coexistence line:

$$(n_L - n_G)(2 - n_G - n_L) = (1 - n_G)(1 - n_L)(n_G + n_L) \ln \left(\frac{n_L(1 - n_G)}{n_G(1 - n_L)} \right). \tag{784}$$

The numerical solution of this equation, giving n_G versus n_L, is given in Fig. 2.18. One can then use Eq. (781) to plot the coexistence curve versus T as shown in Fig. 2.19. The coexistence curve versus p follows from Eq. (782) and is shown in Fig. 2.20. The coexistence line giving p versus T is shown in Fig. 2.21. In the geometrical interpretation of the Maxwell construction we need to plot, for a given temperature, p versus $1/n$, as shown in Fig. 2.22.

184 PRINCIPLES OF THERMODYNAMICS

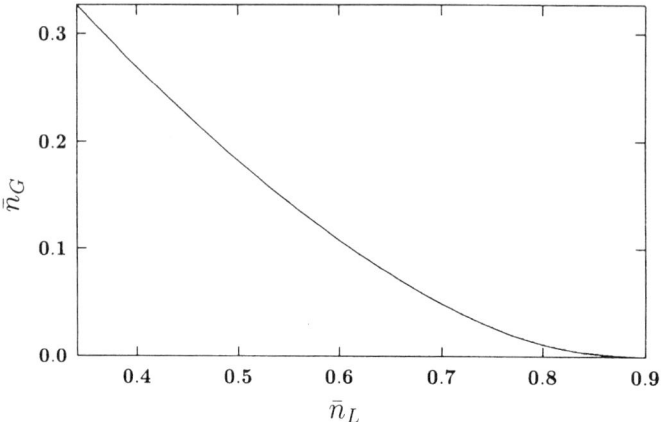

FIGURE 2.18 Values of coexisting liquid, $\bar{n}_L$, and gas, $\bar{n}_G$, densities for the van der Waals equation of state.

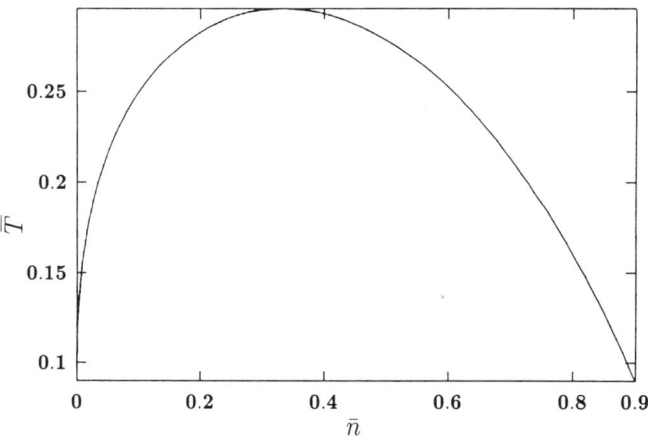

FIGURE 2.19 Coexistence curve for the van der Waals equation of state. On the left are the values of the gas-phase densities, $\bar{n}_G$, on the right the values of the liquid-phase densities, $\bar{n}_L$.

2.9.4 Critical Point

It is clear that the inflection point we discussed above is very special. Mathematically, it corresponds to that point where

$$\left(\frac{\partial p}{\partial n}\right)_T = 0 \tag{785}$$

$$\left(\frac{\partial^2 p}{\partial n^2}\right)_T = 0. \tag{786}$$

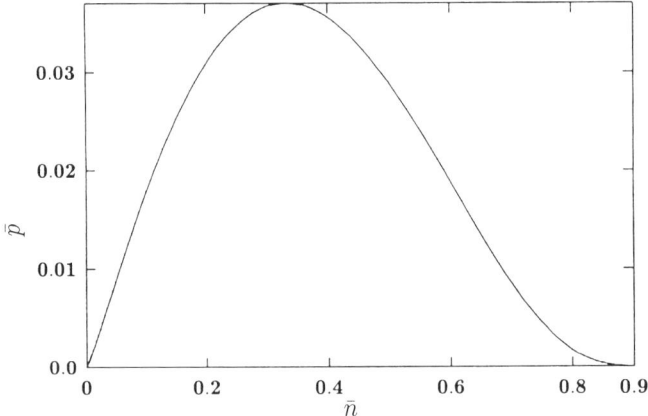

FIGURE 2.20 Coexistence curve for the van der Waals equation of state as a function of scaled pressure. On the left are the values of the gas-phase densities, $\bar{n}_G$, on the right the values of the liquid-phase densities, $\bar{n}_L$.

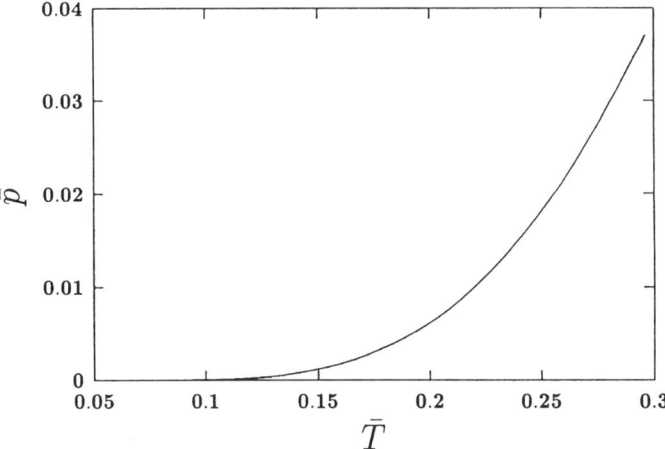

FIGURE 2.21 Gas–liquid coexistence curve for the van der Waals equation of state.

These equations determine the critical values p_c, and T_c. Physically, we see that as we increase p from below p_c, the difference $n_G - n_L$ decreases until $p = p_c$, where $n_G = n_L$. If we lower p from above p_c, the critical point is where the system must first choose between being a liquid or a gas. If we perform an experiment where we fix $p = p_c$ and vary the temperature near the critical temperature, there will be a very small latent heat since $V_G = V_L$ for $T = T_c$ and $p = p_c$. The phase transition is not first order. Nonetheless, strange things happen near the critical point. As in our discussion of the Landau theory, observables in the fluid show anamalous behavior near the critical point (discontinuities and divergences). Let us first locate the critical

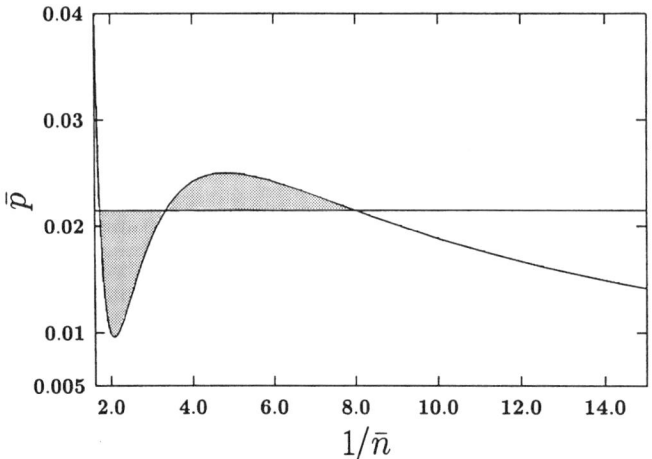

FIGURE 2.22 Maxwell's construction for the van der Waals equation of state in terms of the conventional variables; pressure versus reduced volume.

point for the van der Waals equation of state. Setting the first two derivatives of p with respect to n equal to zero:

$$\frac{\partial \bar{p}}{\partial n} = \frac{\bar{T}}{1-\bar{n}} + \frac{\bar{n}\bar{T}}{(1-\bar{n})^2} - 2\bar{n}$$

$$= \frac{\bar{T}}{(1-\bar{n})^2} - 2\bar{n} = 0 \tag{787}$$

and

$$\frac{\partial^2 \bar{p}}{\partial \bar{n}^2} = \frac{2T}{(1-\bar{n})^3} - 2 = 0. \tag{788}$$

We can easily solve Eqs. (787) and (788) to obtain the values of the critical parameters:

$$\bar{n}_c = \tfrac{1}{3} \tag{789}$$
$$\bar{T}_c = \tfrac{8}{27}. \tag{790}$$

Putting these back into the van der Waals equation of state gives

$$\bar{p}_c = \tfrac{1}{27}. \tag{791}$$

Near the critical point we can write

$$n_L = \tfrac{1}{3} + \Delta_L \tag{792}$$
$$n_G = \tfrac{1}{3} - \Delta_G \tag{793}$$

and expand Eq. (784) in powers of Δ_L and Δ_G (see Problem 2.31). To leading order in Δ_L and Δ_G, we find that

$$\Delta_L = \Delta_G. \tag{794}$$

If we then look at $T \to T_c$ along the coexistence curve, we can write

$$T - T_c = (1 - n_G)(1 - n_L)(n_G + n_L) - \tfrac{8}{27}, \tag{795}$$

and expand the right-hand side in powers of Δ_L and Δ_G to obtain

$$T - T_c = -\tfrac{2}{3}\Delta_G \Delta_L. \tag{796}$$

Using Eq. (794), we have

$$n_L - n_c = n_c - n_G = \left[\tfrac{3}{2}(T_c - T)\right]^{1/2}, \tag{797}$$

which gives the critical index $\beta = \tfrac{1}{2}$.

We turn next to the case $T = T_c$, where we let the pressure and density approach the critical point. Thus we write

$$p - p_c = \frac{n}{1-n}T_c - n^2 - p_c, \tag{798}$$

let $n = n_c + \Delta$, and expand in powers in Δ. It is shown in Problem 2.32 that

$$n - n_c = \left[\tfrac{2}{3}(p - p_c)\right]^{1/3}, \tag{799}$$

and we obtain the critical index $\delta = 3$.

Let us turn next to the *second derivatives*, such as the isothermal compressibility,

$$\kappa_T = -\frac{1}{V}\left(\frac{\partial V}{\partial p}\right)_{N,T} = \frac{1}{n}\left(\frac{\partial n}{\partial p}\right)_{N,T} \tag{800}$$

which can be measured accurately using light-scattering techniques. We easily find from the van der Waals equation of state

$$\kappa_T = \frac{\bar{b}^2}{\bar{a}}\frac{1}{\bar{n}}\left(\frac{\partial \bar{n}}{\partial \bar{p}}\right)_{N,T} \tag{801}$$

and

$$\left(\frac{\partial \bar{p}}{\partial \bar{n}}\right)_{N,T} = \frac{\bar{T}}{(1-\bar{n})^2} - 2\bar{n}. \tag{802}$$

Inserting Eq. (802) in Eq. (801), setting $\bar{n} = \bar{n}_c = \frac{1}{3}$, and using Eq. (790), gives

$$\kappa_T(T, \bar{n}_c) = \frac{4\bar{b}^2}{3\bar{a}} \frac{1}{\bar{T} - \bar{T}_c}. \tag{803}$$

Therefore, the compressibility diverges at the critical point. It is not a coincidence that the divergence of $\kappa_T \approx (T - T_c)^{-1}$ for a gas and of the magnetic susceptibility $\chi \approx (T - T_c)^{-1}$ from the Landau theory have the same exponent. In both cases we have implicitly made the assumption that equations of state are analytic near T_c in the appropriate variables. We discuss this point in greater detail in Chapter 5.

Finally, let us return to the physical variables, where we have

$$p_c = \frac{\bar{a}}{\bar{b}^2} \frac{1}{27} \tag{804}$$

$$k_B T_c = \frac{\bar{a}}{\bar{b}} \frac{8}{27} \tag{805}$$

and

$$n_c = \frac{1}{\bar{b}} \frac{1}{3}. \tag{806}$$

Given p_c, $k_B T_c$, and n_c, we can determine $\bar{a}$ and $\bar{b}$. In fact, they are overdetermined (three equations and two unknowns). This implies that p_c, $k_B T_c$, and n_c are not independent but are related for all systems by

$$\frac{p_c}{n_c k_B T_c} = \frac{3}{8} = 0.375. \tag{807}$$

This ratio for a selection of systems is shown in Table 2.3. We find that while the $\frac{3}{8}$ is a bit large, this ratio is relatively constant (about 0.291) over a wide range of systems.

This brings us back to our earlier statement that when written in terms of scaled variables, the equation of state is a *universal function*. This is known as the *principle of corresponding states* and is discussed further in Chapter 4.

TABLE 2.3 Critical Point Ratio from Eq. (807)

System	Ratio	System	Ratio
Ne	0.305	N_2	0.292
Ar	0.292	O_2	0.292
Kr	0.290	CO	0.294
Xe	0.278	CH_4	0.289

Source: Data from Ref. 43.

REFERENCES

[1] Historically, this entropy principle, the second law of thermodynamics, was developed in the very practical context of looking at cyclic processes associated with *heat engines*. For example, R. Becker, *Theory of Heat*, Springer-Verlag, New York, 1967, states on page 17: "It is impossible to construct an engine that works periodically such that after one period the only changes in its surroundings are: a gain in mechanical or equivalent (e.g., electric) work and a corresponding loss of heat from one heat bath." The second law precludes perpetual motion machines.

[2] Because of the history of thermodynamics, where it was formulated on an axiomatic basis independent of controversial microscopic models, a prejudice evolved that it is more *fundamental* than statistical mechanics. From our current perspective this view is archaic. Thermodynamics is simply the statistical mechanical treatment of the longest-length scales in systems that allow a separation between microscopic and macroscopic length and time scales. Indeed, there are places, such as at a critical point, where such a separation is not possible and where thermodynamics is mute. At the most basic level, the second law, as formulated in Chapter 1, is fundamental to both statistical mechanics and thermodynamics.

[3] As discussed in Chapter 1, and as will be discussed further in Volume 2 , statistical mechanics identifies, via symmetry and constraint conditions, the proper macrovariables. This is particularly true in the cases of broken symmetry states (e.g., solids, superfluids, liquid crystals, etc.), where we must associate a thermodynamic variables with Nambu–Goldstone modes.

[4] It is, of course, the role of statistical mechanics, among other things, to determine these equations of state.

[5] The development of thermodynamics from the point of view of cyclic engines is a beautiful exercise in logical development. See, for example, M. Planck, *Thermodynamics*. I was treated to this material in a set of lectures by Professor Felix Bloch, Stanford University, Winter 1967, unpublished.

[6] Our development here follows that of H. B. Callen, *Thermodynamics*, Wiley, New York, 1960.

[7] Notice that this development is simply giving a physical context for the discussion given in Section 1.10.3.

[8] The identification of F_i with forces is in analogy with mechanics where the force is the coefficient relating work (change in the energy) and displacement.

[9] The potentials which are Legendre transformations on the entropy were invented by M. Massieu, *C. R. Trav.* **69**, 858, 1057 (1869) and are called generalized *Massieu functions* These predated the energy transforms introduced by J. W. Gibbs, "On the equilibrium of heterogeneous substances", *Trans. Conn. Acad.* **3**, 108 (1876); 343 (1878); reprinted in J. W. Gibbs, *The Collected Works*, Longmans, New York, 1931.

[10] It was Clausius who recognized that in an isothermal process the change in heat ΔQ is related to the entropy by $\Delta Q - T\, dS$.

[11] S. Carnot, 1824; reprinted by Hermann, Paris, 1912; J. P. Joule, *Philos. Mag.* **27**, 205 (1845); H. von Helmholtz, "Über die Erhaltung der Kraft", *Phys. Ges. Ber.* (1847); R. Clausius, *Ann. Phys. (Leipzig)* **79**, 368, 500 (1850).

[12] We should remember that in statistical mechanics we can investigate nonthermodynamic properties such as equilibrium averaged space and time correlation functions. Such quantities are measured in scattering and response experiments.

[13] See H. Goldstein, *Classical Mechanics*, 2nd ed., Addison-Wesley, Reading Mass., 1980, p. 62, for discussion.

[14] J. W. Gibbs, "On the equilibrium of heterogeneous substances, "*Trans. Conn. Acad.* **3**, 108 (1876); 343 (1878); reprinted in J. W. Gibbs, *The Collected Works*, Longmans, New York, 1931; P. Duhem, *Le Potentiel thermodynamique et ses applications*, Hermann, Paris, 1886.

[15] We distinguish between thermodynamic information and fluctuation phenomena. As indicated in Chapter 1, when we look at fluctuation effects we see differences in the ensembles.

[16] The geometrical interpretation of Legendre transformations is discussed by R. Courant and D. Hilbert, *Methods of Mathematical Physics*, Vol. 2 Interscience, New York, 1962, p. 32. See also the discussion in H. Goldstein, *Classical Mechanics*, 2nd ed., Addison-Wesley, Reading, Mass., 1980.

[17] In Legendre-transformed potentials we explicitly list only the intensive quantities that appear among the independent variables.

[18] See the discussion in L. D. Landau and E. M. Lifshitz, *Statistical Physics*, Pergamon, Press, Oxford, 1980.

[19] These might also depend explicitly on V and T through the factors of these quantities appearing in Eqs. (500), (501), and (502).

[20] J. C. Maxwell, *Theory of Heat*, Longmans, New York, 1885, p. 165.

[21] By thermodynamic ensemble we mean a particular set of independent variables and the associated thermodynamic potential. This is the same as choosing an ensemble in the statistical mechanics context.

[22] N. Shaw, *Philos Trans. R. Soc. London* A **234**, 299 (1935); D. E. Christie, *Am. J. Phys.* **25**, 486 (1957).

[23] First proposed by Kamerlingh Onnes; see A. W. Porter, *Faraday Soc.* **18**, 140 (1922).

[24] E. Clapeyron, *J. Ecole Polytech. (Paris)* **14** (23), 153 (1834). R. Clausius (1850) gave a more modern derivation.

[25] G. A. Cook, ed., *Data from Argon, Helium and the Rare Gases*, Vol. 1, Interscience, New York, 1961.

[26] "The Ehrenfest classification of phase transitions" was presented by P. Ehrenfest, *Commum. Kamerlingh Onnes Lab. Univ. Leiden Suppl.* **75b** (1933). See also A. Münster, *Statistical Thermodynamics*, Springer-Verlag, Berlin, 1969, p. 262.

[27] See, for example, the discussion in L. D. Landau and E. M. Lifshitz, *Statistical Physics*, Pergamon, Press, Oxford, 1980 on multicomponent phase diagrams.

[28] J. W. Gibbs (*Elementary Principles in Statistical Mechanics*, Ox Bow Press, Woodbridge, Conn., 1981) understood the need to treat generic phases where these is a permutation symmetry under exchange of identical particles as essential not paradoxical. See the discussion in Chapter 1.

[29] J. H. van't Hoff, *Z. Phys. Chem.* **1**, 481 (1887).

[30] P. Waage and C. M. Guldberg, *Studies Concerning Affinity*, Forhandlinger Videnskabs-Selskabet, Christiana (Olso), Norway, 35 1864.

[31] L. D. Landau, "On the theory of phase transitions", *Phys. Z. Sov. Union* **11**, 26, (1937); reprinted in *Collected Papers of L. D. Landau*, Pergamon Press, Elmsford, N. Y., 1965.

[32] A thorough discussion of the development of the statistical mechanics of magnetic systems is given in Chapter 5.

[33] J. Bardeen, L. N. Cooper, and J. R. Schrieffer, *Phys. Rev.* **106**, 162 (1957): **108**, 1175 (1957). For this work they were awarded a Nobel Prize in 1972.

[34] The microscopic details associated with magnetic models, including antiferromagnets, are developed in Chapter 5.

[35] In many cases the choice of the order parameter is aided by symmetry. However in the liquid–gas case, no symmetry breaking is associated with the transition. We can still define an order parameter with the desired properties. If $n_0 = N/V$ is the density of the system in the high-temperature regime where we keep the volume and number of particles fixed and vary temperature and pressure, the quantity $n - n_0$ has zero average in the high-temperature regime. However, this quantity is nonzero in the phase coexistence regime, where it has either the value $n_G - n_0$ or $n_L - n_0$.

[36] More discussion of critical phenomena is given in Chapter 5, with a much more detailed treatment in Volume 2.

[37] This is an approximation that will be seen to be inadequate. An improved approximation requires some development and is discussed in Volume 2.

[38] Recall that like the energy, the free energy must be a minimum with respect to those variables that are not held under control. The requires that the equilibrium free energy be convex. This is, of course, related to the fact that the entropy must be a concave function, as shown in Fig. 2.2.

[39] This result is based on the approximate notion that the entropy is only weakly affected by the interactions. Alternatively, these corrections are of the same form as E and can be absorbed into an effective set of interaction parameters U.

[40] J. D. van der Waals, *Over der continuiteit van den gas-en vloeisoftoestand*. (*The Continuity of the Liquid and Gaseous States of Matter*), Sijthoff Leiden; The Netherlands, 1873 in *Physical Memoirs*, Vol. 1, Part 3, Taylor & Francis, for the Physical Society, London, 1890.

[41] This *scaling behavior* is more general than implied by the discussion here. The more general development, called the *principle of corresponding states*, is discussed in Chapter 4.

[42] J. C. Maxwell, "On the dynamical evidence of the molecular constitution of bodies", *Nature* **11**, 357 (1875).

[43] E. A. Guggenheim, *J. Chem. Phys.* **13**, 253 (1945).

PROBLEMS

2.1. Given the two equations of state

$$\frac{1}{k_B T} = \frac{3N}{2E}$$
$$\frac{p}{T} = \frac{Nk_B}{V},$$

use the Gibbs–Duhem relation $\sum_{i=0}^{t} X_i dF_i = 0$ to construct the third equation of state, $\mu = \mu(N, E, V)$. From the three equations of state, construct the entropy $S = S(N, E, V)$.

2.2. The *third law of thermodynamics* states: The entropy of a system at absolute zero is a universal constant, which may be taken to be zero. This means that $S = 0$ at $T = 0$ for any system regardless of the values of any other parameters of which S may be a function.

(a) Show that the third law implies than any heat capacity of a system must vanish at absolute zero.

(b) Show that at absolute zero the coefficient of thermal expansion

$$\alpha \equiv \frac{1}{V}\left(\frac{\partial V}{\partial T}\right)_p$$

vanishes.

2.3. Go to the literature and find the experimentally determined behavior of C_p, α, and κ_T as a function of temperature at atmospheric pressure for argon.

2.4. Work out the quantities C_p, α, and κ_T for an ideal gas.

2.5. A piezoelectric substance is one in which a mechanical deformation produces an electric polarization, and vice versa. Consider one-dimensional deformation of such a material. The force per unit area (stress) in the x-direction σ is found from experiments to produce polarization P proportional to σ,

$$\alpha \equiv \left(\frac{\partial P}{\partial \sigma}\right)_{\epsilon, T}$$

where α is a constant and ϵ is the electric field. Starting from the expression for the energy change

$$dE = T\, dS - \sigma A\, dL + \epsilon\, dP$$

where A is the area and L is the width of the sample, derive the coefficient $(\partial L/\partial \epsilon)_{\sigma, T}$ in terms of α and whatever other quantities are needed.

2.6. The adiabatic speed of sound, C, is the physical speed of sound and is given by the thermodynamic derivative

$$C^2 = \left(\frac{\partial p}{\partial (mn)}\right)_{S,N}$$

where m is the mass of a particle. Express C in terms of C_p, α, and κ_T.

2.7. Starting with the derivative relations in the grand canonical ensemble,

$$E = \frac{\partial(\beta\Omega)}{\partial\beta}, \qquad N = \frac{\partial(\beta\Omega)}{\partial\alpha}, \qquad p = -\frac{\partial\Omega}{\partial V},$$

show that

$$S = -\frac{\partial \Omega}{\partial T}, \quad N = -\frac{\partial \Omega}{\partial \mu}, \quad p = -\frac{\partial \Omega}{\partial V}$$

which follows from working in the energy representation and choosing the independent variables T, μ, and V.

2.8. Express the Joule–Kelvin coefficient $(\partial T/\partial p)_{H,N}$, where $H = E + pV$, in terms of C_p, α, and κ_T. Show that this derivative vanishes for an ideal gas.

2.9. The equation of state for a large class of ideal gases (classical or quantum) can be written in the form

$$pV = gE$$

where g is an appropriate constant ($g = \frac{2}{3}$ for a monatomic classical gas and for electrons, $g = \frac{1}{3}$ for photons and phonons). The purpose of this problem is to investigate the restrictions imposed on the energy E by the laws of thermodynamics.

(a) Prove that E must satisfy

$$E = T\left(\frac{\partial E}{\partial T}\right)_V - \frac{V}{g}\left(\frac{\partial E}{\partial V}\right)_T. \tag{808}$$

Show that in order to satisfy this equation, E must be of the form

$$E = V^{-g} f(TV^g)$$

where f is an undetermined function of the variable $x = TV^g$.

(b) Solve Eq. (808) in the case where $E = V\epsilon(T)$ and $g = \frac{1}{3}$ (photons or phonons). In other words, determine the temperature dependence of the energy density $\epsilon(T)$.

2.10. Verify the thermodynamic relations

$$\left(\frac{\partial T}{\partial V}\right)_E = \left[p - T\left(\frac{\partial p}{\partial T}\right)_V\right]\bigg/ C_V$$

and

$$\left(\frac{\partial C_V}{\partial V}\right)_T = T\left(\frac{\partial^2 p}{\partial T^2}\right)_V.$$

2.11. Suppose that we have carried out measurements on a magnetic system with independent variables the volume V, temperature T, and applied field h. It is assumed that we have measured the specific heat,

$$C_V = \left(\frac{\partial E}{\partial T}\right)_{h,V},$$

the magnetic susceptibility χ, and the change in the magnetization with temperature

$$\kappa_M = \frac{1}{V}\left(\frac{\partial M}{\partial T}\right)_{h,V}.$$

Suppose now that we are interested in the *adiabatic* susceptibility

$$\chi_S = \left(\frac{\partial M}{\partial h}\right)_{S,V}.$$

Express the unknown χ_S in terms of the known quantities C_V, χ, and κ_M.

2.12. In our analysis of the second derivatives in the Gibbs ensemble, we held the average particle number fixed. Show that the derivatives

$$\frac{\partial^2 G}{\partial N^2}, \quad \frac{\partial^2 G}{\partial T \partial N}, \quad \frac{\partial^2 G}{\partial p \partial N}$$

can be expressed in terms of the first and second derivatives $\partial G/\partial T$, $\partial G/\partial p$, $\partial^2 G/\partial T^2$, $\partial^2 G/\partial T \partial p$, and $\partial^2 G/\partial p^2$.

2.13. The equation of state of a gas is assumed to be of the form

$$pV = Nk_BT - \frac{aNp}{T}$$

where a is a constant independent of $p, V, N,$ and T. The heat capacity at constant pressure is of the form

$$C_p = \tfrac{5}{2}Nk_B + 2aNpS(T)$$

where $S(T)$ is a function of temperature. Find $S(T)$.

2.14. Suppose that we have a dynamical system in contact with a *heat bath*, but now, in addition to being able to exchange energy with the heat bath, suppose that the system can *exchange volume* with the bath (i.e., the wall separating the systems is mobile). (Physically, this corresponds to *holding our system at fixed pressure*.) Show that the probability that our system is in a given macroscopic state is proportional to e^{-G/k_BT}, with $G \equiv E - T_0 S + p_0 V$, where T_0 and p_0 are the temperature and pressure of the heat bath, E and V are the energy and volume of our system, and S is its entropy (for the given value of E and V).

2.15. The *Joule–Thomson effect*: Consider a system consisting of two chambers filled with the same type of gas each held at constant pressures p_1 and p_2. By compressing the gas at pressure p_1 from volume V_1^I to V_1^F, some of the gas is forced through an *expansion valve* into the system at pressure p_2, which expands that chamber from volume V_2^I to V_2^F. There is no heat loss during this process. Show that this process is carried out at constant enthalpy, defined by $H = E + pV$. Assume that the gas is governed by the van der Waals equation of state. For small $p_2 - p_1$, find the change in the temperature in chamber 2 during this process. Discuss when this process could be used to cool a system. Assume that the energy is given by

$$E = \frac{3}{2}Nk_BT - N^2\frac{a}{V}$$

where a is a constant.

2.16. Consider a mixture of two ideal classical gases. Assume that we know the total average energy is given by E_T. The average number of particles of type 1 is $\bar{N}_1$ and the average number of particles of type 2 is $\bar{N}_2$. Particles of type 1 have mass M_1, while particles of type 2 have mass M_2. The Hamiltonian describing the total system is

$$H_T = \sum_{i=1}^{N_1} \frac{\mathbf{p}_i^2}{2M_1} + \sum_{j=1}^{N_2} \frac{\mathbf{p}_j^2}{2M_2}.$$

Determine the pressure of this system in terms of the known quantities.

2.17. Consider an isolated system made up of two subsystems A and B. System A is in thermal contact with B, which is at temperature T. Using the extremum principle for the entropy, which applies to the combined system, show that for system A alone one has the extremum principle that $F = E - TS$ (for A) is a minimum for stable equilibrium. Explain how this extremum principle follows directly from our information-theoretic approach. Find the corresponding extremum principle for the general thermodynamic potential $S[F_0, F_1, \ldots, F_i]$.

2.18. Consider a system consisting of n different substances which are coexisting in r phases in contact (each phase containing, in general, all of the substances). Using the equilibrium conditions derived in the chapter to find the maximum number r for a given n. The correct result is known as *Gibbs phase rule*.

2.19. A monatomic ideal gas consisting of N particles is initially at temperature T_0 and confined to a volume V_0 by a piston of cross-sectional area A. Initially, the piston is rigidly held in place. The walls of the container are composed of a thermally insulating material. A spring, of spring constant K, is now attached to the piston so that it is initially unstretched (i.e., exerting no force). The supports holding the piston are now removed and the gas is allowed to evolve to a new equilibrium configuration. What is the final temperature T of the gas?

2.20. A box of total volume V is divided into two equal halves by a partition. In the first volume is a classical ideal (monatomic) gas consisting of N_1 particles of species 1, at temperature T_1. In the second volume are N_2 particles of species 2, at temperature T_2. The partition is now removed and the total system is allowed to come to thermal equilibrium. Calculate the change in entropy occurring in this process when **(a)** species 1 and 2 are distinguishable and **(b)** species 1 and 2 are indistinguishable.

2.21. Consider an adsorbent surface having N sites. Each site can adsorb one gas molecule. Assume that this surface is in contact with a classical monatomic ideal gas with chemical potential μ (determined by the pressure p and temperature T), and that an adsorbed molecule has energy $-\epsilon_0$ compared to one in the free state. Assume that there is no interaction between the particles on the surface. Find the average fraction of sites occupied as a function of p and T.

2.22. A dilute, classical gas is in equilibrium with a solid surface, all at temperature T. The surface may be regarded as a two-dimensional lattice of sites, each of which may or may not be occupied by a gas molecule. We define the partition function of an empty site as 1; that of an occupied site is $q(T) \gg 1$.
  (a) Find the partition function of a site in equilibrium with the gas at chemical potential μ.
  (b) Find the fraction of sites occupied as a function of the gas pressure p for a given $q(T)$. This is called the *Langmuir isotherm*.

2.23. Consider a dynamical system contains three species of particles, labeled A, B, and C. Suppose that the chemical reaction $A + B \leftrightarrow C$ can occur. Suppose that each of the particle species A, B, C can be treated as a monatomic ideal gas. (For example, A could represent free electrons, B could represent free protons, and C could represent neutral hydrogen atoms.) Suppose further that the total energy is

$$E = E_A + E_B + E_C - \epsilon N_C.$$

Here E_A, E_B, and E_C are the usual kinetic energies of the three gases, and ϵ is the binding energy of the C particle. Assume, for simplicity, that $N_A = N_B$. The system is in thermal equilibrium at temperature T. Calculate the ratio N_A/N_C as a function of T, ϵ, and the masses of the particles.

2.24. Consider a system that is initially partitioned into two equal volumes V_0 which are not in contact. Compartment 2 contains N_A^0 particles of type A with mass m_A, and compartment 1 contains N_B^0 particles of type B with mass m_B. Both compartments are in equilibrium with a surrounding thermal bath at temperature T. At some time the partition dividing the systems is removed to reveal a membrane through which A particles can pass, but not B particles. We also know that, when allowed to come together, particles A and B undergo a chemical reaction to form a molecule C with mass m_C:

$$A + B \rightleftharpoons C.$$

The particles C can pass through the membrane. Find the equilbrium densities of A, B, and C assuming that each can be treated as a classical ideal gas. Assume that $N_A^0 = N_B^0$.

2.25. Consider the situation where we have two ideal gases at the same temperature but separated by a partition. Assume initially that the gases have different masses m_1 and m_2. Compute the total entropy of the system S_T^I. Remove the partition and allow the gases to mix. Compute the total entropy after mixing S_T^F. The change in the value of S_T is called the *entropy of mixing*:

$$\Delta S = S_T^F - S_T^I.$$

Can we guess the sign of ΔS? Suppose that the densities of the two gases are initially the same; what is ΔS? Suppose that $m_1 = m_2$ (the particles in 1 and 2 are the same) and the densities are equal. Compute ΔS for this case. What does one expect from physical arguments for this case? Explain.

2.26. Consider the Landau free energy

$$\frac{F}{V} = a_o + \frac{1}{2}a_2\psi^2 + \frac{1}{4!}a_4\psi^4 + \frac{1}{6!}a_6\psi^6$$

where ψ is the magnetization density, $a_2 = a_2^0(T - T_c)$, $a_4 = a_4^0$, and $a_6 = a_6^0$ for temperatures T near the transition temperature T_c. a_2^0, a_4^0, and a_6^0 are positive constants independent of T. Find the temperature dependence of the average magnetization and the magnetic susceptibility for T near T_c.

2.27. In the chapter we discussed the Landau theory for a simple scalar order parameter. Consider now a system with a complex order parameter, ψ, which enters into the description of model systems known as the *p-state clock models*. The associated Landau free energy is given by

$$\frac{F}{V} = r|\psi|^2 + \frac{u}{2}|\psi|^4 - \frac{v}{2p}[\psi^p + (\psi^*)^p]$$

where p is an integer, $r = r_0(T - T_c)$, $u > 0$, and the range of values for v that leads to ordering depends on u and p. ψ^* is the complex conjugate of ψ. Determine the form of ψ in the ordered phase ($T < T_c$) as functions of r, u, v, and p. Assume that $\psi = fe^{is}$, where f and s are real. Find s and express f in terms of a transcendental equation for general p. Solve explicitly for f for the case $p = 4$. What is the condition on v to obtain ordering in this case?

2.28. Consider a system where the free-energy density depends on the vector order parameter $\boldsymbol{\psi}$ such that

$$f = \frac{r}{2}\boldsymbol{\psi}^2 + \frac{u}{4}(\boldsymbol{\psi}^2)^2 + \frac{1}{2}(\mathbf{h} \cdot \boldsymbol{\psi})^2$$

where $r = r_0(T/T_0 - 1)$ and r_0, T_0, and u are positive and temperature independent. This represents an over simplified model for a nematic liquid crystal in the presence of an applied uniform magnetic field $\mathbf{h}$. At high temperatures one has the $\boldsymbol{\psi} \to -\boldsymbol{\psi}$ symmetry and $\boldsymbol{\psi} = 0$ minimizes the free energy. As one lowers the temperature, one finds a phase transition and symmetry breaking. Identify the nature of the ordering and find the value of the transition temperature.

2.29. Assume the validity of van der Waals equation of state for a substance at an absolute temperature T equal to or slightly below the critical temperature T_c.

198   PRINCIPLES OF THERMODYNAMICS

In terms of the critical values p_c, n_c, T_c, and Boltzmann's constant, find the density ratio of liquid to gas, the vapor pressure, and the latent heat of vaporization, keeping in each of these quantities only the lowest power of $T_c - T$.

2.30. Show that the critical properties of the van der Waals equation of state are shared by the more general equation of state

$$p = -At(\delta n) - B(\delta n)^3 + f(t) \qquad (809)$$

where $\delta n = n - n_c$, $t = T - T_c$, and $f(t)$ is an unimportant function of t. Equation (809) is equivalent to

$$\left(\frac{\partial p}{\partial n}\right)_T = -At - 3B(\delta n)^2.$$

Explain why this is a very reasonable series expansion for $(\partial p/\partial n)_T$.

2.31. Let us consider the van der Waals equation along the coexistence line as given by Eq. (784). Near the critical point we can write

$$n_L = \tfrac{1}{3} + \Delta_L$$
$$n_G = \tfrac{1}{3} - \Delta_G$$

and expand in powers of Δ_L and Δ_G. Working to leading order in Δ_L and Δ_G, show that

$$\Delta_L = \Delta_G.$$

2.32. Consider the van der Waals equation near the critical point from which we can subtract the equation at the critical point to obtain

$$p - p_c = \frac{n}{1-n} T_c - n^2 - p_c.$$

If we write $n = n_c + \Delta$ and expand in powers in Δ, show that

$$n - n_c = \left[\tfrac{2}{3}(p - p_c)\right]^{1/3}.$$

Extract from this the critical index δ.

3 Quantum Statistical Mechanics

3.1 GENERAL PRINCIPLES

The general foundations of statistical mechanics developed in Chapter 1 also hold true for quantum systems. Indeed, in quantum statistical mechanics [1], we follow essentially the same steps as in the classical case:

1. Enumerate the configurations (allowing for the fact that observables are now associated with operators defined on a Hilbert space).
2. Introduce the probability of being in that configuration.
3. Express the entropy or missing information in terms of the probabilities.
4. Maximize the entropy consistent with the existing constraints acting on the system.

In our treatment of classical fluids we specified the available configurations via the phase-space coordinates $\{\mathbf{r}_1, \ldots \mathbf{r}_N, \mathbf{p}_1, \ldots, \mathbf{p}_N\}$. Such a specification in the quantum mechanical case is prohibited by the Heisenberg uncertainty principle. Thus we must develop an entirely different approach for the specification of configurations. This question is, of course, central to the development of quantum mechanics. We state here the basic postulates of quantum mechanics relevant to our development:

1. The state of a system can be specified completely by a state vector $|\psi>$ that is defined on a Hilbert space. (For an introduction to abstract vector spaces and Dirac notation, see Appendix J.)
2. The collection of such physically realizable states (the possible configurations) are complete. They *span* the Hilbert space. This completeness can be expressed in the form

$$\sum_{\psi} |\psi\rangle\langle\psi| = 1, \tag{810}$$

   where the sum is over all members of the set.
3. These states can be constructed to be orthogonal. Thus we can write

$$\langle\psi|\psi'\rangle = \delta_{\psi,\psi'}. \tag{811}$$

4. If an operator $\hat{A}$ is associated with an observable A, the average value of A, when the system is in state $|\psi\rangle$, is

$$A_\psi = \langle\psi|\hat{A}|\psi\rangle. \tag{812}$$

At this stage we will not worry about the construction of the state vectors $|\psi\rangle$ (which may be extremely difficult in practice). The important point is that they label the possible physical configurations.

We next introduce the probability $P[\psi]$ that the system is in the state $|\psi\rangle$. It has the same interpretation as $P[q]$ in Chapter 1 and includes the uncertainty about the preparation of the system. P_ψ must be normalized such that

$$\sum_\psi P[\psi] = 1. \tag{813}$$

This probability should not be confused with the probability $|\langle i|\psi\rangle|^2$ that in a measurement on a system in state $|\psi\rangle$, we will obtain a value A_i for an observable A, where $|i\rangle$ is an eigenstate of the operator $\hat{A}$:

$$\hat{A}|i\rangle = A_i|i\rangle. \tag{814}$$

This is an entirely different issue, which we will return to in the next section. The thermodynamic average is given, much as in Chapter 1, by

$$\langle A \rangle = \sum_\psi P[\psi] A_\psi, \tag{815}$$

and the entropy is defined, as before, as

$$S = -k_B \sum_\psi P[\psi] \ln P[\psi]. \tag{816}$$

Let us look at the quantum version of the physical system emphasized in Chapter 1—a set of N particles governed by a Hamiltonian $\hat{H}$. In this case we have a fixed number of particles in a physically realizable situation, so the state vectors are labeled by the number of particles, $|\psi\rangle_N$. In quantum mechanics we *cannot* distinguish among identical particles. It is mandatory therefore, just as in the classical case, to divide by $N!$ if the sum over all the states includes all possible exchanges. Working in the grand canonical ensemble we have the normalization condition

$$\sum_{N=0}^\infty \frac{1}{N!} \sum_{\psi_N} P_N[\psi] = 1, \tag{817}$$

and the entropy

$$S = -k_B \sum_{N=0}^\infty \frac{1}{N!} \sum_{\psi_N} P_N[\psi] \ln P_N[\psi]. \tag{818}$$

GENERAL PRINCIPLES

In equilibrium the entropy is maximized while holding

$$E = \langle \hat{H} \rangle \tag{819}$$

and

$$\bar{N} = \langle N \rangle \tag{820}$$

fixed, where averages are given by

$$\langle A \rangle = \sum_{N=0}^{\infty} \frac{1}{N!} \sum_{\psi_N} P_N[\psi]_N \langle \psi | \hat{A} | \psi \rangle_N. \tag{821}$$

The problem we encounter here of maximizing the entropy to determine the equilibrium probability distribution is formally equivalent to the classical problem we solved in Chapter 1. Again we can introduce Lagrange multipliers and follow closely the analysis used in Chapter 1. This leads to the equilibrium probability distribution in the grand canonical ensemble given by

$$P_N[\psi] = \frac{e^{-\beta(H_\psi - \mu N)}}{Z} \tag{822}$$

where the grand partition function is given by

$$Z = \sum_{N=0}^{\infty} \frac{1}{N!} \sum_{\psi_N} e^{-\beta(H_\psi - \mu N)} \tag{823}$$

and

$$H_\psi = {}_N\langle \psi | \hat{H} | \psi \rangle_N. \tag{824}$$

Clearly, β and μ retain the interpretation as the inverse temperature and chemical potential.

The equilibrium average of some observable is given by inserting this expression for $P_N[\psi]$ into Eq. (821). From the point of view of statistical mechanics, this is all there is to the extension to quantum mechanics. All the derivative relations and thermodynamic arguments go through exactly as before. We have, for example, that the average energy is given by

$$E = \langle \hat{H} \rangle = \frac{\partial}{\partial \beta}(\beta \Omega) \tag{825}$$

where, as before, the grand potential is defined by

$$\Omega = -\beta^{-1} \ln Z. \tag{826}$$

3.2 EQUILIBRIUM PROBABILITY OPERATOR

Although our statistical mechanical description in the quantum mechanical case is well posed, it is not very convenient. The reason is that unlike the classical case, the nature of the sum over configurations is not known until we have solved the full dynamical problem. Thus, in general, construction of the physical states $|\psi\rangle_N$ for an N-body system is extremely difficult. Fortunately, this is not as much of a limitation as it may appear. The way we have set up our problem, as in the classical case, the physically realizable states for an isolated system have constant energy. This means that $|\psi\rangle_N$ is an energy eigenstate,

$$\hat{H}|\psi\rangle_N = E_\psi^N |\psi\rangle_N, \tag{827}$$

where E_ψ^N is the energy quantum number or eigenvalue. Consider the partition function given by Eq. (823). Since

$$H_\psi = {}_N\langle\psi|\hat{H}|\psi\rangle_N = E_\psi^N, \tag{828}$$

we can write

$$Z = \sum_{N=0}^{\infty} \frac{1}{N!} \sum_{\psi_N} e^{-\beta(E_\psi^N - \mu N)}. \tag{829}$$

Since ${}_N\langle\psi|\psi\rangle_N = 1$, Eq. (829) can be rewritten as

$$Z = \sum_{N=0}^{\infty} \frac{1}{N!} \sum_{\psi_N} e^{-\beta(E_\psi^N - \mu N)} {}_N\langle\psi|\psi\rangle_N$$

$$= \sum_{N=0}^{\infty} \frac{1}{N!} \sum_{\psi_N} {}_N\langle\psi|e^{-\beta(E_\psi^N - \mu N)}|\psi\rangle_N. \tag{830}$$

Since $|\psi\rangle_N$ is an eigenstate of $\hat{H}$, we can replace E_ψ^N by $\hat{H}$ inside the brackets and obtain

$$Z = \sum_{N=0}^{\infty} \frac{1}{N!} \sum_{\psi_N} {}_N\langle\psi|e^{-\beta(\hat{H} - \mu \hat{N})}|\psi\rangle_N. \tag{831}$$

Notice that we have also introduced the number operator $\hat{N}|\psi\rangle_N = N|\psi\rangle_N$. The partition function is then in the compact form

$$Z = \mathrm{Tr}\, e^{-\beta(\hat{H}-\mu\hat{N})} \qquad (832)$$

where Tr is the trace over the set of states spanning the Hilbert space. The important point is that the trace is independent of the representation used to evaluate it. Consider the quantity

$$\mathrm{Tr}\,\hat{A} = \sum_i \langle i|\hat{A}|i\rangle \qquad (833)$$

where the states $|i\rangle$ are complete and orthonormal. Suppose that the states $|\psi\rangle$ are also complete and orthonormal. Using the completeness of the $|\psi\rangle$, we have

$$\mathrm{Tr}\,\hat{A} = \sum_i \langle i|\hat{A} \sum_\psi |\psi\rangle\langle\psi|i\rangle = \sum_i \sum_\psi \langle\psi|i\rangle\langle i|\hat{A}|\psi\rangle$$
$$= \sum_\psi \langle\psi| \sum_i |i\rangle\langle i|\hat{A}|\psi\rangle = \sum_\psi \langle\psi|\hat{A}|\psi\rangle. \qquad (834)$$

We can use either set of states, $|i\rangle$ or $|\psi\rangle$, to evaluate $\mathrm{Tr}\,\hat{A}$. Thus we reach the conclusion that the partition function is a trace over the operator $e^{-\beta(\hat{H}-\mu N)}$ and is not dependent on the original representation $|\psi>$. We can evaluate Z using *any* complete and orthonormal representation; they need not be energy eigenstates. We exploit this fact in the next section.

The analysis above can be extended to the calculation of any thermodynamic average. We have, from Eq. (821),

$$\langle A \rangle = \sum_{N=0}^{\infty} \frac{1}{N!} \sum_{\psi_N} \frac{e^{-\beta(E_\psi^N - \mu N)}}{Z}\, {}_N\langle\psi|\hat{A}|\psi\rangle_N$$
$$= \sum_{N=0}^{\infty} \frac{1}{N!} \sum_{\psi_N} {}_N\langle\psi| \frac{e^{-\beta(\hat{H}-\mu\hat{N})}}{Z} \hat{A}|\psi\rangle_N$$
$$= \mathrm{Tr}\left(\frac{e^{-\beta(\hat{H}-\mu\hat{N})}}{Z} \hat{A}\right). \qquad (835)$$

We can define the equilibrium *probability operator*

$$\hat{P}_{eq} = \frac{e^{-\beta(\hat{H}-\mu\hat{N})}}{Z} \qquad (836)$$

and averages can be expressed in the simple form

$$\langle A \rangle = \text{Tr}\, \hat{P}_{\text{eq}} \hat{A}. \tag{837}$$

There are situations, not necessarily in equilibrium, where we are interested in a particular observable $\hat{B}$ (energy, position of the center of mass, angular momentum, etc.) which has an eigenvalue spectrum resulting from the eigenvalue problem:

$$\hat{B}|B_i\rangle = B_i|B_i\rangle. \tag{838}$$

Thus a measurement of $\hat{B}$ will always give a value corresponding to one of the eigenvalues B_i. Quite generally, the average measured value of $\hat{B}$ is given by

$$\langle B \rangle = \sum_\psi P[\psi]\langle\psi|\hat{B}|\psi\rangle \tag{839}$$

where the $|\psi\rangle$ correspond to the complete set of the physically realizable states. As discussed above, in many situations we do not know the $|\psi\rangle$. It is then convenient to rewrite the expression above in terms of the eigenstates of $\hat{B}$. Using the completeness of the $|B_i\rangle$, we have

$$\langle B \rangle = \sum_\psi P[\psi]\langle\psi| \sum_{B_i} |B_i\rangle\langle B_i|\hat{B}|\psi\rangle$$
$$= \sum_{B_i} B_i \sum_\psi \langle B_i|\psi\rangle P[\psi]\langle\psi|B_i\rangle. \tag{840}$$

If we introduce the probability operator

$$\hat{P} = \sum_\psi |\psi\rangle P[\psi]\langle\psi|, \tag{841}$$

the probability of obtaining the value B_i is a measurement given by

$$P[B_i] = \langle B_i|\hat{P}|B_i\rangle. \tag{842}$$

Clearly, full specification of the operator $\hat{P}$ requires evaluation of the matrix

$$P_{ij} = \langle B_i|\hat{P}|B_j\rangle \tag{843}$$

where P_{ij} is known as the *density matrix* [2].

The operator $\hat{P}$ tells us *how much*, on average, the system is in a given configuration. If we know that the system is in a state $|\psi_0\rangle$, then

$$P[\psi] = \delta_{\psi,\psi_0} \tag{844}$$

and

$$\hat{P}_0 = |\psi_0\rangle\langle\psi_0|. \tag{845}$$

In this case the system in a *pure state*. If the system is in a pure state, then the density operator is simply a projection operator. The definition of a projection operator $\hat{P}$ is

$$\hat{P}^2 = \hat{P}. \tag{846}$$

Clearly, $\hat{P}_0$ is a projection operator. If a system is characterized by a density operator that is *not* a projection operator the system is said to be in a *mixed state*. In this case

$$\hat{P}^2 \neq \hat{P}. \tag{847}$$

Note that, in general,

$$\hat{P}^2 = \sum_\psi \sum_{\psi'} |\psi\rangle P[\psi]\langle\psi|\psi'\rangle P[\psi']\langle\psi'| \tag{848}$$

$$= \sum_\psi \sum_{\psi'} |\psi\rangle P[\psi]\delta_{\psi,\psi'} P[\psi']\langle\psi'| = \sum_\psi |\psi\rangle P^2[\psi]\langle\psi| \tag{849}$$

which equals $\hat{P}$ only if $P^2[\psi] = P[\psi]$. It is clear that the thermal equilibrium state is a mixed state. In nonthermal situations we may be interested in whether $\hat{P}$ is in a pure state with respect to a given set of states. Is the system in a definite energy state?

3.3 STATISTICAL MECHANICS IN SECOND-QUANTIZED LANGUAGE

Because the trace operation occurring in the averages for systems in equilibrium is independent of the basis used, we can choose the most convenient complete basis for the evaluation independent of whether the set gives a good representation of the actual physical states. A very convenient choice is that of the *occupation* basis, where we assume that the N-particle Hilbert space is spanned by a properly symmetrized product of single-particle states. Although such states may individually give quite a poor description for a strongly interacting system, the physical state can be represented accurately by a linear combination of these product states.

We introduce these states because they can be constructed in a straightforward manner that incorporates the quantum statistics in the problem in a natural way. Mathematically, the analysis is very similar to the construction of angular momentum states $|l, m\rangle$ using raising and lowering operators. The details of the development of the occupation basis and the *language* of second quantization [3] are discussed in Appendix K. Here we point out the main aspects.

Suppose that in a problem of interest such as that for a quantum fluid, there is some natural single-particle problem described by a Hamiltonian $\hat{h}_0^\alpha$ where α labels the αth particle in the system and

$$\hat{H}_0 = \sum_{\alpha=1}^{N} \hat{h}_0^\alpha \qquad (850)$$

gives the Hamiltonian for the set of noninteracting single particles. We then assume that we can solve the single-particle Schrödinger equation

$$\hat{h}_0^\alpha |i\rangle = \epsilon_i |i\rangle, \qquad (851)$$

to obtain the complete and orthonormal set of single-particle states $|i\rangle$ and the associated single-particle energies ϵ_i. In the coordinate representation we have the single-particle wavefunctions

$$\phi_i(\mathbf{r}) = \langle \mathbf{r} | i \rangle. \qquad (852)$$

In constructing the second-quantized description we introduce the creation operator a_i^+ and the annihilation operator a_i. a_i^+ has the physical interpretation of creating a particle in the *single-particle eigenstate* labeled by i. a_i removes a particle from the state i. a_i is mathematically the Hermitian conjugate of a_i^+,

$$a_i^+ = (a_i)^+. \qquad (853)$$

The fundamental properties satisfied by the creation and annihilation operators for bosons (particles with integral spin [4]) are the canonical commutation relations,

$$[a_i, a_j^+] \equiv a_i a_j^+ - a_j^+ a_i = \delta_{ij} \qquad (854)$$
$$[a_i, a_j] = [a_i^+, a_j^+] = 0. \qquad (855)$$

We can also introduce creation and annihilation operators for Fermi particles (systems with half-integral spin) with the same physical interpretation. The only difference is that they obey *anticommutation relations*

$$[a_i, a_j^+]_+ \equiv a_i a_j^+ + a_j^+ a_i = \delta_{ij} \qquad (856)$$
$$[a_i, a_j]_+ = [a_i^+, a_j^+]_+ = 0. \qquad (857)$$

We discuss in Appendix K how these operators can be used to construct a many-particle quantum state and the effect of operating on these states with a_i or a_i^+.

It is also convenient to introduce field operators $\psi^+(\mathbf{r})$ and $\psi(\mathbf{r})$ which create or destroy particles at the point $\mathbf{r}$ in space. These field operators are related to a_i^+ and a_i by

$$\psi^+(\mathbf{r}) = \sum_i \phi_i^*(\mathbf{r}) a_i^+ \tag{858}$$

and

$$\psi(\mathbf{r}) = \sum_i \phi_i(\mathbf{r}) a_i \tag{859}$$

where the ϕ_i's are the complete and orthonormal set of single-particle eigenfunctions defined above. It is straightforward to show that the field operators satisfy the commutation relations

$$[\psi(\mathbf{r}), \psi^+(\mathbf{r}')] = \delta(\mathbf{r} - \mathbf{r}') \tag{860}$$
$$[\psi(\mathbf{r}), \psi(\mathbf{r}')] = [\psi^+(\mathbf{r}), \psi^+(\mathbf{r}')] = 0 \tag{861}$$

for bosons, and the anticommutation relations

$$[\psi(\mathbf{r}), \psi^+(\mathbf{r}')]_+ = \delta(\mathbf{r} - \mathbf{r}') \tag{862}$$
$$[\psi(\mathbf{r}), \psi(\mathbf{r}')]_+ = [\psi^+(\mathbf{r}), \psi^+(\mathbf{r}')]_+ = 0. \tag{863}$$

for fermions. A key point in the development is that any quantum mechanical operator of interest can be written in second-quantized form. For example, the number of particles in the single-particle quantum state i is given by

$$\hat{n}_i = a_i^+ a_i, \tag{864}$$

so the total number of particles is the sum over all the possible single-particle states:

$$\hat{N} = \sum_i \hat{n}_i = \sum_i a_i^+ a_i. \tag{865}$$

For a set of *independent particles* with single-particle energy ϵ_i in state i, the total energy operator, the Hamiltonian, is

$$\hat{H} = \sum_i \epsilon_i a_i^+ a_i. \tag{866}$$

If we work in terms of field operators, where for the moment we suppress the spin index, the particle density at position $\mathbf{r}$ is

$$\hat{n}(\mathbf{r}) = \psi^+(\mathbf{r}) \psi(\mathbf{r}) \tag{867}$$

and the operator for the total number of particles is

$$\hat{N} = \int d^3r\, \psi^+(\mathbf{r})\psi(\mathbf{r}). \tag{868}$$

Similarly, we can show that the kinetic energy operator for a system of particles with mass m can be written as

$$\hat{K} = \int d^3r\, \psi^+(\mathbf{r}) \frac{-\hbar^2 \nabla^2}{2m} \psi(\mathbf{r}). \tag{869}$$

The quantization of a typical potential energy contribution for a system of particles interacting by two-body interactions can be written in the form

$$V = \tfrac{1}{2} \sum_{i,j} V(\mathbf{r}_i - \mathbf{r}_j) = \tfrac{1}{2} \int d^3x\, d^3x'\, n(\mathbf{x}) n(\mathbf{x}') V(\mathbf{x} - \mathbf{x}') \tag{870}$$

and follows simply by replacing the classical density $n(\mathbf{x})$ with its second-quantized equivalent, $\hat{n}(\mathbf{x}) = \psi^+(\mathbf{x})\psi(\mathbf{x})$. These results are discussed in more detail in Appendix K.

It is crucial to realize that once we know the effect of operating on a quantum state with a, ψ, or their conjugates, we know the effect of operating on these states with any operator formed from the creation and annihilation operators. Since the quantum states can be constructed using the creation and annihilation operators, we should concentrate on these operators.

The partition function can easily be expressed in second-quantized language. Working in the grand canonical ensemble, we have

$$\begin{aligned} Z &= \operatorname{Tr} e^{-\beta(\hat{H} - \mu \hat{N})} \\ &= \sum_{N=0}^{\infty} \frac{1}{N!} \int d^3r_1\, d^3r_2 \cdots d^3r_N \langle \mathbf{r}_1 \mathbf{r}_2 \cdots \mathbf{r}_N | e^{-\beta(\hat{H} - \mu \hat{N})} | \mathbf{r}_1 \mathbf{r}_2 \cdots \mathbf{r}_N \rangle \end{aligned} \tag{871}$$

where $\hat{H}$ and $\hat{N}$ are expressed in their second-quantized form and $\langle \mathbf{r}_1 \mathbf{r}_2 \cdots \mathbf{r}_N |$ is the N-body occupation basis state in the coordinate representation.

Clearly all of this is pretty abstract. We need to move on to discuss some examples.

3.4 QUANTUM HAMILTONIANS

It is worthwhile at this point to stop and consider some of the Hamiltonians describing real quantum systems. We cannot hope to be complete, but we can see what is involved in some important problems in quantum statistical mechanics.

3.4.1 Electron Gas

The most studied system in quantum statistical mechanics is the electron gas [5]. The electron gas gives, to a first approximation, the behavior of conduction electrons in a metal. The important element here is that the electrons are spin $-\frac{1}{2}$ fermions, and the associated field operators must be labeled by a spin index $\sigma = \pm 1$. Similarly, the density can be labeled by a spin index

$$\hat{n}_\sigma(\mathbf{x}) = \psi_\sigma^+(\mathbf{x})\psi_\sigma(\mathbf{x}) \tag{872}$$

and the total density is obtained by summing over σ:

$$\hat{n}(\mathbf{x}) = \sum_\sigma \hat{n}_\sigma(\mathbf{x}). \tag{873}$$

The Hamiltonian is just the sum of kinetic and potential energies,

$$\hat{H} = \sum_\sigma \int d^3x \, \psi_\sigma^+(\mathbf{x}) \left(-\frac{\hbar^2 \nabla^2}{2m}\right) \psi_\sigma(\mathbf{x}) + \frac{1}{2} \int d^3x d^3x' \, V(\mathbf{x} - \mathbf{x}') \hat{n}(\mathbf{x}) \hat{n}(\mathbf{x}'), \tag{874}$$

where the interaction potential between electrons is just the Coulomb repulsion

$$V(\mathbf{x}) = \frac{4\pi e^2}{|\mathbf{x}|} \tag{875}$$

where e is the electronic charge. In this problem one must embed the electrons in a background of positive charge if there is to be overall charge neutrality and the possibility of thermal equilibrium.

3.4.2 Helium

The Hamiltonian for the quantum fluid ^{4}He has the same form as for the electron gas except that the field operators are those for bosons (^{4}He is spinless) and the interaction potential is of the Lennard-Jones form [6], which is short-ranged with both repulsive and attractive pieces. The electron interaction is repulsive and *long ranged*. The many-body methods for handling these interactions are quite different [3]. The less abundant form of helium, ^{3}He, has the same Hamiltonian as that for ^{4}He except that ^{3}He is a system of fermions. The properties of ^{3}He and ^{4}He are *very* different for low temperatures. ^{4}He becomes a superfluid [7] at approximately 2.18 K. ^{3}He becomes a very different superfluid at 0.00265 K (this phase of ^{3}He was not discovered until 1972 [8]).

3.4.3 Electromagnetic Radiation

The free electromagnetic field is described by the Hamiltonian [9],

$$H_{EM} = \frac{1}{8\pi} \int d^3r [\mathbf{E}^2(\mathbf{r}) + \mathbf{B}^2(\mathbf{r})], \tag{876}$$

where $\mathbf{E}$ is the electric field and $\mathbf{B}$ is the magnetic field. The quantization of $\mathbf{E}$ and $\mathbf{B}$ involves the introduction of a set of boson creation and annihilation operators $\mathbf{a}_\mathbf{k}^+$ and $\mathbf{a}_\mathbf{k}$ via

$$\mathbf{E}(\mathbf{r}) = i \sum_\mathbf{k} \left(\frac{\hbar \omega_k}{V}\right)^{1/2} e^{+i\mathbf{k}\cdot\mathbf{r}} (\mathbf{a}_\mathbf{k} - \mathbf{a}_{-\mathbf{k}}^+) \tag{877}$$

and

$$\mathbf{B}(\mathbf{r}) = i \sum_\mathbf{k} \left(\frac{\hbar c^2}{\omega_k V}\right)^{1/2} e^{+i\mathbf{k}\cdot\mathbf{r}} \mathbf{k} \times (\mathbf{a}_\mathbf{k} + \mathbf{a}_{-\mathbf{k}}^+). \tag{878}$$

The $\mathbf{a}_\mathbf{k}$, and $\mathbf{a}_\mathbf{k}^+$ destroy and create *photons* with momentum $\hbar \mathbf{k}$ and energy $\hbar \omega_k = c\hbar k$, where c is the speed of light and V is the volume. The vector nature of $\mathbf{a}_\mathbf{k}$ is constrained by the condition that photons have only transverse polarizations, so

$$\mathbf{k} \cdot \mathbf{a}_\mathbf{k} = 0 \tag{879}$$

and the other two components of $\mathbf{a}_\mathbf{k}$ determine the polarization of the photon. Substituting for $\mathbf{E}$ and $\mathbf{B}$ in the Hamiltonian, we find that

$$\hat{H}_{EM} = \sum_{\mathbf{k},\alpha} \hbar \omega_k a^+_{\mathbf{k},\alpha} a_{\mathbf{k},\alpha} \tag{880}$$

where α is a two-component polarization index. This Hamiltonian is the same as for a set of uncoupled harmonic oscillators.

Remember that in general the electromagnetic field is coupled to any charged particles in the system (as in Maxwell's equations) which would generate a coupling $\hat{H}_I$ between electromagnetic field operators and particle field operators.

3.4.4 Quadratic Hamiltonians

If we now look back over these model Hamiltonians, we see that they are all of the form

$$\hat{H} = \hat{H}_0 + \hat{H}_I \tag{881}$$

where $\hat{H}_0$ is quadratic in the creation and annihilation operators, the kinetic energy for electrons and helium, and $\hat{H}_I$ gives the interaction among the particles (electrons or helium atoms). The interaction term usually involves higher-than-quadratic products of field operators. If we introduce the creation and annihilation operators $a^+_{\mathbf{k},\sigma}$ and $a_{\mathbf{k},\sigma}$ corresponding to a single-particle energy state $\epsilon_k = (\hbar k)^2/2m$ via

$$\psi_\sigma(\mathbf{x}) = \sum_{\mathbf{k}} \frac{e^{+i\mathbf{k}\cdot\mathbf{x}}}{\sqrt{V}} a_{\mathbf{k},\sigma} \tag{882}$$

(which assumes translational invariance), the kinetic energy in Eq. (874) can be put in *diagonal* form,

$$\hat{H}_0 = \sum_\sigma \sum_{\mathbf{k}} \epsilon_{\mathbf{k}} a^+_{\mathbf{k},\sigma} a_{\mathbf{k},\sigma}. \tag{883}$$

This is of the same form as for the electromagnetic radiation.

We can gain an approximate understanding of the properties of these quantum mechanical systems if we can work out the statistical mechanical properties for the *ideal* case where $\hat{H} = \hat{H}_0$. We carry out this analysis in the next section. It is more difficult to come back and include the effects of the interaction using perturbation theory (expansions in powers of the interaction $\hat{H}_I$), as discussed in Chapter 5.

3.5 PARTITION FUNCTION FOR A QUADRATIC HAMILTONIAN

We want to calculate the grand potential

$$\Omega = -\beta^{-1} \ln Z, \tag{884}$$

where the grand partition function is given by

$$Z = \text{Tr}\, e^{-\beta(\hat{H}_0 - \mu \hat{N})} \tag{885}$$

and $\hat{H}_0$ is the quadratic Hamiltonian

$$\hat{H}_0 = \sum_i \epsilon_i a^+_i a_i. \tag{886}$$

The specific nature of the single-particle states is yet to be specified. In Eq. (885) the number operator is given by

$$\hat{N} = \sum_i a^+_i a_i. \tag{887}$$

212 QUANTUM STATISTICAL MECHANICS

It will be useful in evaluating Ω to make use of the result that the average number of particles is given by

$$-\frac{\partial \Omega}{\partial \mu} = \bar{N} = \sum_i \langle a_i^+ a_i \rangle. \tag{888}$$

We note that in the limit $\mu \to -\infty$ only the $N = 0$ term survives in the trace over $e^{-\beta(\hat{H}_0 - \mu N)}$, and therefore

$$\lim_{\mu \to -\infty} \Omega = -\beta^{-1} \ln\left[1 + \mathcal{O}(e^{\beta \mu})\right] = 0. \tag{889}$$

This is just another way of saying that the pressure goes to zero with the number of particles, so we can integrate to find

$$\Omega = -\sum_i \int_{-\infty}^{\mu} d\mu' \, \bar{n}_i(\mu') \tag{890}$$

where the average occupation number for state i is defined by

$$\bar{n}_i \equiv \langle a_i^+ a_i \rangle \tag{891}$$

and we hold β fixed. We therefore need to calculate $\langle a_i^+ a_i \rangle$. It is convenient to define

$$\hat{K} = \hat{H}_0 - \mu \hat{N} = \sum_j (\epsilon_j - \mu) a_j^+ a_j \tag{892}$$

and look at the numerator in the average $\langle a_i^+ a_i \rangle$:

$$\text{Tr} \, e^{-\beta \hat{K}} a_i^+ a_i = \text{Tr} \, e^{-\beta \hat{K}} a_i^+ e^{\beta \hat{K}} e^{-\beta \hat{K}} a_i \tag{893}$$

where we have inserted $e^{\beta \hat{K}} e^{-\beta \hat{K}} = 1$ on the right-hand side of Eq. (893). Let us concentrate on the quantity

$$a_i^+(\beta) \equiv e^{-\beta \hat{K}} a_i^+ e^{\beta \hat{K}} \tag{894}$$

where

$$a_i^+(0) = a_i^+. \tag{895}$$

If we take a derivative with respect to β, we find that

$$\frac{\partial a_i^+(\beta)}{\partial \beta} = -e^{-\beta \hat{K}} [\hat{K}, a_i^+] e^{\beta \hat{K}}. \tag{896}$$

PARTITION FUNCTION FOR A QUADRATIC HAMILTONIAN

We can evaluate the commutator explicitly. For bosons *and* fermions,

$$[\hat{K}, a_i^+] = (\epsilon_i - \mu)a_i^+, \tag{897}$$

and Eq. (896) reduces to

$$\frac{\partial}{\partial \beta} a_i^+(\beta) = -e^{-\beta \hat{K}}(\epsilon_i - \mu)a_i^+ e^{\beta \hat{K}}$$
$$= -(\epsilon_i - \mu)a_i^+(\beta). \tag{898}$$

Integrating Eq. (898), we obtain

$$a_i^+(\beta) = e^{-\beta(\epsilon_i - \mu)} a_i^+, \tag{899}$$

which can be put back into Eq. (893) to obtain

$$\operatorname{Tr} e^{-\beta \hat{K}} a_i^+ a_i = e^{-\beta(\epsilon_i - \mu)} \operatorname{Tr} a_i^+ e^{-\beta \hat{K}} a_i. \tag{900}$$

Using the cyclic invariance of the trace, this can be rewritten as

$$\operatorname{Tr} e^{-\beta \hat{K}} a_i^+ a_i = e^{-\beta(\epsilon_i - \mu)} \operatorname{Tr} e^{-\beta \hat{K}} a_i a_i^+. \tag{901}$$

However, from the commutation or anticommutation relations,

$$a_i a_i^+ = 1 + \eta a_i^+ a_i \tag{902}$$

where $\eta = +1$ for bosons and -1 for fermions, and we have

$$\operatorname{Tr} e^{-\beta \hat{K}} a_i^+ a_i = e^{-\beta(\epsilon_i - \mu)} \operatorname{Tr} e^{-\beta \hat{K}}(1 + \eta a_i^+ a_i). \tag{903}$$

Dividing this equation by $Z = \operatorname{Tr} e^{-\beta \hat{K}}$, we obtain

$$\langle a_i^+ a_i \rangle = e^{-\beta(\epsilon_i - \mu)}(1 + \eta \langle a_i^+ a_i \rangle) \tag{904}$$

which can be solved to obtain the average occupation number for single-particle state i:

$$\bar{n}_i = \langle a_i^+ a_i \rangle = \frac{1}{e^{\beta(\epsilon_i - \mu)} - \eta}. \tag{905}$$

Using this result in Eq. (890), we find that the thermodynamic potential is given by

$$\Omega = -\sum_i \int_{-\infty}^{\mu} d\mu' \frac{1}{e^{\beta(\epsilon_i-\mu')} - \eta}$$

$$= \sum_i \int_{-\infty}^{\mu} d\mu'(\eta^{-1}\beta^{-1}) \frac{\partial}{\partial \mu'} \ln\left(1 - \eta e^{-\beta(\epsilon_i-\mu')}\right)$$

$$= \eta^{-1}\beta^{-1} \sum_i \ln\left(1 - \eta e^{-\beta(\epsilon_i-\mu)}\right). \tag{906}$$

This gives us the thermodynamic potential for a quantum Hamiltonian that is quadratic in the creation and annihilation operators. All we have to do to complete the solution is to specify the single-particle energies ϵ_i and the nature of the single-particle quantum states, the i's.

It should be pointed out that the method we have used for calculating $\langle a_i^+ a_i \rangle$ is more general than it might appear. The same manipulations are used in the development of many-body theory techniques [10] for treating interacting quantum systems. Notice that the analysis depended essentially only on the commutation relations satisfied by the creation and annihilation operators. It is also worth noting that the analysis that we have carried out is appropriate only to quantum Hamiltonians quadratic in operators satisfying canonical commutation relations. Consider the *magnetic* quantum Hamiltonian

$$\hat{H} = \tfrac{1}{2} \sum_{i,j} J_{ij} \mathbf{S}_i \cdot \mathbf{S}_j \tag{907}$$

where S_i^α is the αth component of a spin operator located on a discrete lattice at site i and J_{ij} is the interaction between the spins. Although this is also a quadratic Hamiltonian, these spin operators obey the *angular momentum* commutation relations

$$[S_i^\alpha, S_j^\beta] = i\hbar \sum_\gamma \epsilon_{\alpha\beta\gamma} \delta_{ij} S_i^\gamma \tag{908}$$

where $\epsilon_{\alpha\beta\gamma}$ is the antisymmetric tensor. The statistical mechanics associated with the Hamiltonian given by Eq. (907) requires a treatment (see Chapter 5) different from that given above.

3.6 SINGLE-PARTICLE DENSITY OF STATES

We must now specify how we are to treat the single-particle quantum numbers, the i's that we sum over in H_0. For a system of free particles, using periodic boundary

SINGLE-PARTICLE DENSITY OF STATES

conditions in a cube of side L, the single-particle wave functions are plane-wave states of the form

$$\psi_{\mathbf{k},\lambda}(\mathbf{r}) = \frac{1}{\sqrt{V}} e^{i\mathbf{k}\cdot\mathbf{r}} \chi_\lambda \qquad (909)$$

where the k_i are wavenumbers given by

$$k_x = \frac{2n_x\pi}{L}, \qquad k_y = \frac{2n_y\pi}{L}, \qquad k_z = \frac{2n_z\pi}{L} \qquad (910)$$

where $V = L^3$ is the volume of the system, n_x, n_y, n_z are the integers $n_i = 0, \pm 1, \pm 2, \ldots$, and λ is a spin or polarization quantum number labeling a spin eigenvector χ_λ. We expect the total number of single-particle states to be proportional to the size of the system (it is an extensive quantity). Therefore, the sum $\sum_i f_i$ should be proportional to V in the limit of large V. We therefore write $\sum_i f_i = V(1/V) \sum_i f_i$ and investigate the intensive quantity:

$$\lim_{V\to\infty} \frac{1}{V} \sum_i f_i = \lim_{V\to\infty} \frac{1}{V} \sum_{n_x,n_y,n_z} \sum_\lambda f_\lambda\left(\frac{2\pi \mathbf{n}}{L}\right). \qquad (911)$$

Using the fundamental definition of an integral, we obtain for one dimension

$$\lim_{L\to\infty} \frac{1}{L} \sum_{n=-\infty}^{+\infty} f\left(\frac{2\pi n}{L}\right) = \int_{-\infty}^{+\infty} \frac{dk}{2\pi} f(k). \qquad (912)$$

This is easily generalized from one to three dimensions, with the result, including a sum over the quantum number λ,

$$\lim_{V\to\infty} \frac{1}{V} \sum_i f_i = \sum_\lambda \int \frac{d^3k}{(2\pi)^3} f_\lambda(\mathbf{k}). \qquad (913)$$

In many cases we are interested in evaluating sums over functions that depend only on the value of the single-particle energy eigenstates,

$$S = \sum_i f(\epsilon_i) \qquad (914)$$

where $f(\epsilon)$ is a function of ϵ. In the limit of a large volume we can write

$$S = V \sum_\lambda \int \frac{d^3k}{(2\pi)^3} f(\epsilon_{\hbar\mathbf{k}}), \qquad (915)$$

where for nonrelativistic particles

$$\epsilon_{\hbar k} = \frac{(\hbar k)^2}{2m}, \tag{916}$$

while for photons and phonons (lattice vibrations in solids, see Chapter 6)

$$\epsilon_{\hbar k} = \hbar c k, \tag{917}$$

and c is the speed of light or the average speed of sound in solids. If $f(\epsilon_{\hbar k})$ does not depend on λ, we can replace the sum, $\sum_\lambda$, by g, the spin or polarization degeneracy,

$$S = Vg \int \frac{d^3k}{(2\pi)^3} f(\epsilon_{\hbar k}). \tag{918}$$

It is conventional to write this integral in terms of the momentum $\mathbf{p} = \hbar \mathbf{k}$,

$$S = gV \int \frac{d^3p}{(2\pi\hbar)^3} f(\epsilon_p). \tag{919}$$

Here we see how the dimensionless quantity $d^3pV/(2\pi\hbar)^3$, introduced in Chapter 1, enters into our analysis. Since the integrand in Eq. (919) depends only on $\epsilon_p = \epsilon(|\mathbf{p}|)$, it is convenient to make a change of variables from $|\mathbf{p}|$ to ϵ, and write

$$S = V \int_0^\infty d\epsilon\, D(\epsilon) f(\epsilon). \tag{920}$$

$D(\epsilon)$ is called the *density of energy states* or simply the *density of states* [11]. Comparing Eqs. (919) and (920), we can write

$$D(\epsilon) = g \frac{4\pi}{(2\pi\hbar)^3} p^2(\epsilon) \frac{dp(\epsilon)}{d\epsilon} \tag{921}$$

where 4π comes from the angular integration over $\hat{p}$ and $dp(\epsilon)/d\epsilon$ from the Jacobian in the coordinate transformation. Using these results, the thermodynamic potential, given by Eq. (906), can be written in the two equivalent forms

$$\Omega = \eta^{-1}\beta^{-1} gV \int \frac{d^3p}{(2\pi\hbar)^3} \ln(1 - \eta e^{-\beta(\epsilon_p - \mu)}) \tag{922}$$

$$= \eta^{-1}\beta^{-1} V \int_o^{+\infty} d\epsilon\, D(\epsilon) \ln(1 - \eta e^{-\beta(\epsilon - \mu)}). \tag{923}$$

SINGLE-PARTICLE DENSITY OF STATES

We will be particularly interested in the density of states for two cases:

1. *Nonrelativistic massive particles.* Nonrelativistic massive particles have the energy–momentum relation

$$\epsilon = \frac{p^2}{2m}, \tag{924}$$

and using Eq. (921) we obtain the density of states:

$$D(\epsilon) = \frac{g}{4\hbar^3 \pi^2} (2m)^{3/2} \sqrt{\epsilon}. \tag{925}$$

This expression can be put into a more appealing form if we realize that we can typically find a characteristic energy ϵ_* that allows us to write

$$D(\epsilon) = \frac{g}{4\hbar^3 \pi^2} (2m)^{3/2} \sqrt{\epsilon_*} \sqrt{\frac{\epsilon}{\epsilon_*}}. \tag{926}$$

For example, one can define a de Broglie wavelength associated with the energy scale ϵ_*,

$$\lambda_* = \hbar \left(\frac{2\pi}{m\epsilon_*}\right)^{1/2} \tag{927}$$

and write

$$D(\epsilon) = g \frac{1}{\lambda_*^3} \frac{2}{\sqrt{\pi}} \frac{1}{\epsilon_*} \sqrt{\frac{\epsilon}{\epsilon_*}}. \tag{928}$$

A key aspect in the development below will be to identify the natural choice for ϵ_* for both nonrelativistic bosons and fermions. As we shall see, in addition to the thermal energy $k_B T$, one can also construct a quantum energy from the kinetic energy $\approx \hbar^2/2ml^2$, where l is the typical distance between particles.

2. *Photons and phonons.* Massless bosons such as photons of light and sound waves in solids (phonons) have energy–momentum dispersion relations of the form

$$\epsilon = cp, \tag{929}$$

where c is the speed of light or sound. The associated density of states is easily obtained using Eq. (921):

$$D(\epsilon) = \frac{g}{2\pi^2} \frac{1}{(\hbar c)^3} \epsilon^2. \tag{930}$$

We see that for a number of cases of interest,

$$D(\epsilon) = \bar{D} \epsilon^s \tag{931}$$

where $\bar{D}$ is a constant.

3.7 THE CLASSICAL LIMIT

Let us now investigate how our quantum thermodynamic potential can be reduced to the classical result we found earlier. Let us focus on spinless particles (e.g., argon) where $g = 1$ and the density of states is given by Eq. (925). As we shall demonstrate self-consistently, the classical limit corresponds to

$$e^{\beta\mu} \ll 1, \tag{932}$$

or $\mu \to -\infty$. Expanding to lowest order in $e^{\beta\mu}$, we obtain from Eq. (923) that

$$\Omega_{cl} = \eta^{-1}\beta^{-1}V \int_0^\infty D(\epsilon)d\epsilon(-\eta)e^{-\beta(\epsilon-\mu)}. \tag{933}$$

Making the substitution $\epsilon = \beta^{-1}x$ in the integral and introducing the thermal wavelength $\lambda = \hbar(2\pi\beta/m)^{1/2}$, we obtain

$$\Omega_{cl} = -\beta^{-1}V \frac{e^{\beta\mu}}{\lambda^3} \frac{2}{\sqrt{\pi}} \int_0^{+\infty} \sqrt{x}\, dx\, e^{-x}. \tag{934}$$

Evaluating the elementary integral, we obtain

$$\Omega_{cl} = -\beta^{-1}V \frac{e^{\beta\mu}}{\lambda^3}. \tag{935}$$

If we check back to Chapter 1 we see that this is precisely the expression we found previously for the thermodynamic potential of an ideal classical fluid. Our method for counting states introduced in Chapter 1, which involved the introduction of $\hbar$, is justified at this point.

We consider now the self-consistency of our argument that the classical limit corresponds to

$$e^{\beta\mu} \ll 1. \tag{936}$$

Since the particle density, in the classical limit, is given by

$$n = \frac{e^{\beta\mu}}{\lambda^3}, \tag{937}$$

Eq. (936) becomes

$$n\lambda^3 \ll 1. \tag{938}$$

We can write $n = l^{-3}$, where l is the average distance between particles and the inequality becomes

$$\left(\frac{\lambda}{l}\right)^3 \ll 1. \tag{939}$$

Since λ is the de Broglie wavelength for a particle with velocity $v = (2k_B T/m)^{1/2}$, the condition for classical mechanics to be valid is that the distance between particles is large enough that their wavefunctions (with spread $\approx \lambda$) do not overlap significantly. Notice that we can achieve this inequality either by increasing l by lowering the density or by increasing the temperature and decreasing λ. The average number of particles in state i is given, for the various different statistics, by the famous occupation number expressions:

$$n_i = \begin{cases} \dfrac{1}{e^{\beta(\epsilon_i - \mu)} + 1} & \text{Fermi–Dirac} \\ \dfrac{1}{e^{\beta(\epsilon_i - \mu)} - 1} & \text{Bose–Einstein} \\ e^{-\beta(\epsilon_i - \mu)} & \text{Maxwell–Boltzmann.} \end{cases} \tag{940, 941, 942}$$

The Fermi–Dirac [12] and Bose–Einstein [13] distributions reduce to the common Maxwell–Boltzmann [14] form in the classical limit. The classical limit is therefore independent of the quantum statistics.

3.8 EQUATION OF STATE FOR IDEAL QUANTUM SYSTEMS

From our analysis of the grand canonical ensemble in Chapter 1, we know that the pressure is given by

$$pV = -\Omega. \tag{943}$$

Similarly, the average energy is given by

$$E = \frac{\partial}{\partial \beta}(\beta\Omega)\bigg|_{\alpha=-\beta\mu} = V \int_0^{+\infty} d\epsilon \, D(\epsilon) \epsilon n(\epsilon) \tag{944}$$

where

$$n(\epsilon) = \left[e^{\beta(\epsilon-\mu)} - \eta\right]^{-1}. \tag{945}$$

Finally, the average number of particles is given by

$$N = -\frac{\partial(\beta\Omega)}{\partial \alpha} = V \int_0^{\infty} d\epsilon \, D(\epsilon) n(\epsilon). \tag{946}$$

If we restrict the analysis to the case of a power law density of states,

$$D(\epsilon) = \bar{D}\epsilon^s, \tag{947}$$

where, for example, $s = \frac{1}{2}$ for nonrelativistic particles and $s = 2$ for photons and phonons, we can relate the pressure and energy. Let us integrate the expression for Ω, Eq. (923), by parts:

$$\begin{aligned}\Omega &= \eta\beta^{-1}V\bar{D}\int_0^\infty d\epsilon \ln(1 - \eta e^{-\beta(\epsilon-\mu)}) \frac{1}{s+1}\frac{d}{d\epsilon}\epsilon^{s+1} \\ &= \eta\frac{V\bar{D}\beta^{-1}}{s+1}\left[\epsilon^{s+1}\ln(1 - \eta e^{-\beta(\epsilon-\mu)})\big|_0^\infty \right. \\ &\quad \left. - \int_0^\infty d\epsilon \frac{\epsilon^{s+1}(-\eta)(-\beta)e^{-\beta(\epsilon-\mu)}}{1 - \eta e^{-\beta(\epsilon-\mu)}}\right]. \end{aligned} \tag{948}$$

Assuming that $s + 1 > 0$, then $\lim_{\epsilon\to 0} \epsilon^{s+1} = 0$, and we find that

$$pV = -\Omega = \frac{V}{s+1}\int_0^\infty d\epsilon\, D(\epsilon)\, \epsilon n(\epsilon). \tag{949}$$

Comparing Eqs. (949) and (944), we obtain

$$pV = \frac{E}{1+s}. \tag{950}$$

For particles, where $s = \frac{1}{2}$,

$$pV = \tfrac{2}{3}E \tag{951}$$

which is in agreement with the classical results $p = nk_BT$, $E/V = \tfrac{3}{2}nk_BT$. For photons and phonons, $s = 2$ and

$$pV = \frac{E}{3}. \tag{952}$$

Note that these results are valid for both Bose and Fermi statistics.

3.9 CHEMICAL POTENTIAL FOR MASSIVE NONRELATIVISTIC PARTICLES

We can now consider the determination of the chemical potential as a function of density and temperature for ideal quantum gases for the case of nonrelativistic particles. This requires inverting the equation, giving the number of particles in terms of the temperature and chemical potential in the grand canonical ensemble. We have from Eqs. (946) and (928) that the average number of particles is given by

$$N = g \frac{V}{\lambda_*^3} \frac{2}{\sqrt{\pi}} \int_0^\infty \frac{d\epsilon}{\epsilon_*} \frac{\sqrt{\epsilon/\epsilon_*}}{e^{\beta(\epsilon-\mu)} - \eta}. \tag{953}$$

If we choose $\epsilon_* = k_B T$, then $\lambda_* = \lambda_T$ is the thermal de Broglie wavelength. Changing integration variables to $x = \epsilon/k_B T$, the particle density is given by

$$n = \frac{g}{\lambda_T^3} \frac{2}{\sqrt{\pi}} \int_0^\infty dx \sqrt{x} \frac{1}{e^{x+\alpha} - \eta} \tag{954}$$

where, as usual in the grand canonical ensemble, $\alpha = -\beta\mu$. If we define a dimensionless density

$$y = n\lambda_T^3 \frac{\sqrt{\pi}}{2g} \tag{955}$$

then Eq. (954) takes the form

$$y = \int_0^\infty dx \sqrt{x} \frac{1}{e^{x+\alpha} - \eta}. \tag{956}$$

We can solve this equation numerically for $\alpha = \alpha(y)$. The results for Fermi–Dirac, Bose–Einstein, and Maxwell–Boltzmann statistics are shown in Fig. 3.1. Later we

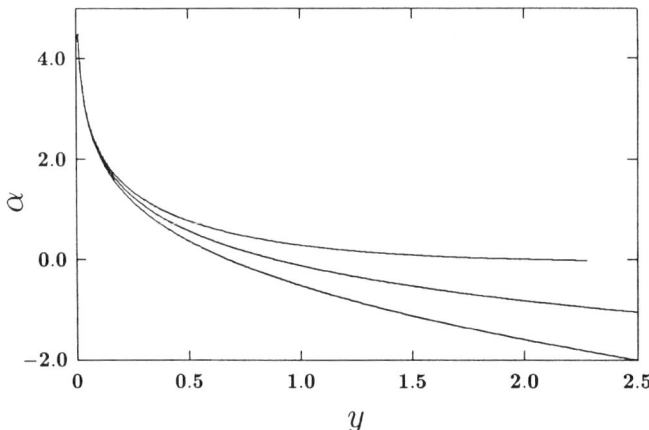

FIGURE 3.1 Scaled chemical potential $\alpha = -\mu/k_B T$ versus the dimensionless density $y = \sqrt{\pi}/2(n\lambda_T^3)$ (where λ_T is the thermal de Broglie wavelength) for ideal gases. The top curve corresponds to Bose–Einstein statistics, the lower curve to Fermi–Dirac statistics, and the middle curve to Maxwell–Boltzmann statistics.

will come back and look at these results more closely. Here we point out the interesting fact that for the Bose–Einstein case, that one loses a solution for large enough densities $[y > y_* = y(\alpha = 0)]$ where the chemical potential goes to zero.

3.10 BLACKBODY RADIATION: PHOTON GAS

In the rest of this chapter we explore our results for ideal quantum systems for a number of physical situations. We look first at the case of thermal equilibrium for electromagnetic radiation—the photon gas.

We showed above that the Hamiltonian for a photon gas can be written in second-quantized form:

$$\hat{H}_{EM} = \sum_{\mathbf{k},\alpha} \hbar\omega_k a^+_{\mathbf{k},\alpha} a_{\mathbf{k},\alpha}, \tag{957}$$

where $\omega_k = c|\mathbf{k}|$, c is the speed of light, $\hbar\mathbf{k}$ is the photon momentum, $a^+_{\mathbf{k},\alpha}$ creates a photon with momentum $\hbar\mathbf{k}$, and the two transverse polarization components of $a^+_{\mathbf{k},\alpha}$ are labeled by α. Since this is of the standard quadratic form that we treated in Section 3.5, the thermodynamic potential is given for a set of bosons, using Eq. (923), by

$$\Omega = \beta^{-1} V \int_0^{+\infty} d\epsilon D(\epsilon) \ln\left(1 - e^{-\beta(\epsilon-\mu)}\right) \tag{958}$$

where, from Eq. (930), the density of states for photons is

$$D(\epsilon) = \frac{1}{\pi^2} \frac{\epsilon^2}{(\hbar c)^3} \tag{959}$$

and we have set $g = 2$ since photons have two polarizations.

The average number of photons present in equilibrium is one of the objects of our calculation, not a given quantity. This requires that we determine μ. We can do this by noting that of all the possible values of N, the one that will characterize the thermodynamic state is that value which minimizes the free energy $F = E - TS$. (Remember that F is a minimum in a system held at constant temperature and volume.) We therefore demand that

$$\frac{\partial F[T, N, V]}{\partial N} = 0. \tag{960}$$

Since, by definition,

$$\mu = \frac{\partial F}{\partial N}, \tag{961}$$

we have
$$\mu = 0 \tag{962}$$

for a photon gas. This also matches our intuitive notion that the chemical potential is a measure of the energy we need to add a particle to a system. Since photons are massless, we can create one with essentially zero energy. The Bose–Einstein distribution for $\mu = 0$,

$$n(\epsilon) = \frac{1}{e^{\beta\epsilon} - 1} \tag{963}$$

is called the *Planck distribution* [15]. We can now compute the average number of photons,

$$N = V \int_0^\infty D(\epsilon) n(\epsilon) \, d\epsilon \tag{964}$$

and the average energy,

$$E = V \int_0^\infty D(\epsilon) n(\epsilon) d(\epsilon) \equiv \int_0^\infty E(\epsilon) \, d\epsilon \tag{965}$$

where

$$E(\epsilon) = \frac{V \epsilon^3}{\pi^2 (\hbar c)^3} \frac{1}{e^{\beta\epsilon} - 1}. \tag{966}$$

Historically, the energy density $E(\epsilon)$ played an interesting role in the development of quantum theory. If we work in terms of the frequency distribution, then

$$E(\omega) = \hbar E(\hbar\omega) = \frac{V \hbar \omega^3}{\pi^2 c^3} \frac{1}{e^{\beta\hbar\omega} - 1}. \tag{967}$$

This frequency distribution is plotted in terms of dimensionless variables in Fig. 3.2. In the low-frequency or classical limit this distribution reduces to

$$E(\omega) \rightarrow \frac{V}{\pi^2 c^3} \frac{\omega^2}{\beta}, \tag{968}$$

which is called the *Rayleigh–Jeans formula* [16]. Note, however, that when we try to calculate the total energy using the classical result, we obtain

$$E = \int_0^\infty d\omega \lim_{\hbar \to 0} \hbar E(\hbar\omega)$$
$$= \frac{V}{\beta \pi^2 c^3} \int_0^\infty \omega^2 d\omega \tag{969}$$

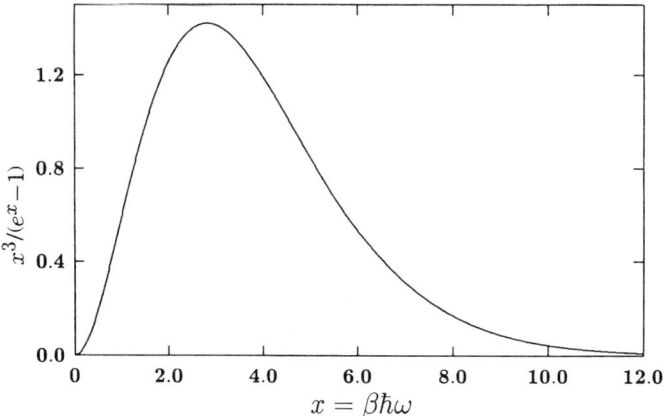

FIGURE 3.2 Scaled Planck distribution versus reduced frequency.

which is a divergent integral! This divergence was termed the *ultraviolet catastrophe*. This was a paradox in classical mechanics, and it was resolved by Planck and his *distribution*. The high-frequency limit of Eq. (966),

$$E(\epsilon) \to \frac{V\epsilon^3}{\pi^2(\hbar c)^3} e^{-\beta\epsilon}, \tag{970}$$

was obtained empirically by Wien [17], before Planck, but he did not understand its significance. If we plot $E(\epsilon)$ as a function of ϵ, we see that it starts out in agreement with the Rayleigh–Jeans law, increases to a maximum, and then falls off in agreement with the Wien law. The maximum value of the energy distribution is determined by $(x = \beta\epsilon)$

$$\frac{d}{dx}\left(\frac{x^3}{e^x - 1}\right) = 0 \tag{971}$$

or

$$\frac{x_m}{1 - e^{-x_m}} = 3. \tag{972}$$

We find numerically that $x_m = (\beta\epsilon)_{\max} = 2.823\ldots$.

The importance of this development was highlighted with the discovery of the primordial microwave background. This is radiation left over from the big bang. If one plots the intensity of the radiation versus frequency, one finds that it is well fit to the form given by Eq. (966). This is an indication that the radiation is in thermal equilibrium, and one can determine the associated temperature to be about 3 K [18].

BLACKBODY RADIATION: PHOTON GAS

Explicit calculation of the average number of photons and the average energy is straightforward. After introducing the dimensionless variable $x = \beta\epsilon$, Eqs. (964) and (965) become

$$N = \frac{V}{\pi^2(\hbar c)^3}(k_B T)^3 \int_0^\infty \frac{dx\, x^2}{e^x - 1} \tag{973}$$

and

$$E = \frac{V}{\pi^2(\hbar c)^3}(k_B T)^4 \int_0^\infty \frac{dx\, x^3}{e^x - 1}. \tag{974}$$

We evaluate integrals of the form

$$I_B(n) = \int_0^{+\infty} \frac{dx\, x^n}{e^x - 1} \tag{975}$$

in Appendix L, with the result

$$I_B(n) = \zeta(n+1)\Gamma(n+1) \tag{976}$$

where $\zeta(x)$ is the Riemann zeta function [19] and $\Gamma(x)$ is the standard Γ-function. The average number of photons is given then by

$$N = \frac{V}{\pi^2(\hbar c)^3}(k_B T)^3 \cdot 2\zeta(3) \tag{977}$$

and increases as T^3. The total energy is given by

$$E = \frac{V}{\pi^2(\hbar c)^3}(k_B T)^4 \cdot 3 \cdot 2\zeta(4) = \frac{\pi^2}{15}\frac{V}{(\hbar c)^3}(k_B T)^4, \tag{978}$$

where we have used

$$\zeta(4) = \frac{\pi^4}{90} = 1.082\ldots. \tag{979}$$

Before we knew about quantum mechanics and Planck's constant, it was conventional to write

$$E = 4\frac{\sigma}{c}VT^4 \tag{980}$$

where σ is the *Stefan–Boltzmann constant* [20], which we now know is given by

$$\sigma = \frac{\pi^2 k_B}{60 \hbar^3 c^2}. \tag{981}$$

The specific heat at constant volume can then be evaluated as

$$C_V = \left(\frac{\partial E}{\partial T}\right)_V = \frac{4\pi^2 V}{15(\hbar c)^3} k_B (k_B T)^3. \tag{982}$$

If we remember the result, Eq. (952), relating pressure and energy in an ideal quantum system of photons, we obtain

$$pV = \frac{\pi^2}{45} \frac{V}{(\hbar c)^3} (k_B T)^4. \tag{983}$$

The entropy is given in this case by the Euler equation ($\mu = 0$),

$$TS = E - \mu N + pV = E + \frac{E}{3} = \frac{4\pi^2}{45} \frac{V}{(\hbar c)^3} (k_B T)^4 \tag{984}$$

or

$$S = \frac{4\pi^2}{45} \frac{V}{(\hbar c)^3} k_B (k_B T)^3. \tag{985}$$

3.11 PROPERTIES OF IDEAL FERMI SYSTEMS AT LOW TEMPERATURES

3.11.1 Introduction

Quantum statistics are typically important in treating systems at low temperatures. In this section and Section 3.13 we investigate the low-temperature properties of collection of noninteracting (ideal) nonrelativistic particles. Properly interpreted, a gas of ideal Fermi particles at low temperatures gives a good approximation of the contribution of conduction band electrons to the thermodynamic properties of metals. The precise meaning of what we mean by low temperatures will evolve as part of our discussion.

3.11.2 $T = 0$

Let us first concentrate on the zero-temperature case, $T = 0$, where the distribution function for fermions, Eq. (490), reduces to a step function (see Fig. 3.3)

$$n(\epsilon) = \theta(\mu - \epsilon). \tag{986}$$

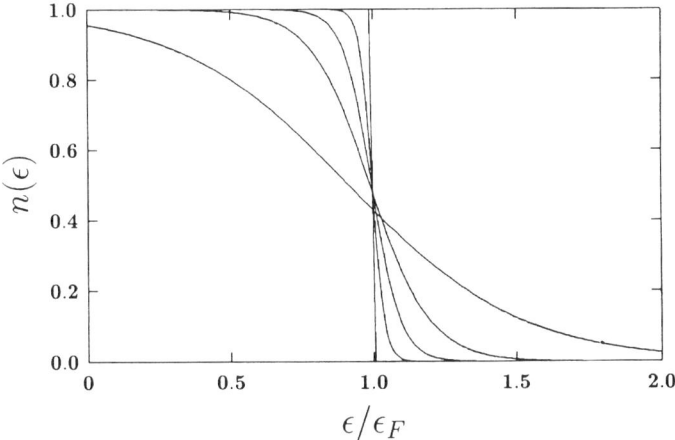

FIGURE 3.3 Fermi–Dirac average occupation number versus scaled single-particle energies for a variety of temperatures. For $\epsilon > \epsilon_F$, from bottom to top, the curves correspond to temperatures $T/T_F = 0, 0.02, 0.05, 0.1$, and 0.3.

While all of the particles try to occupy the lowest-energy states, the Pauli exclusion principle [21] keeps all but two particles, with opposite spin, from going into the lowest available state. We define the maximum occupied single-particle energy in the ground state as the Fermi energy, which we obtain from Eq. (986) as

$$\epsilon_F \equiv \mu(T=0). \tag{987}$$

We see here that the characteristic energy ϵ_*, introduced in Section 3.9, is, in this case, the Fermi energy. The thermodynamic quantities are given at $T = 0$ (remember that $pV = \frac{2}{3}E$) by

$$N(T=0) = V \int_0^{+\infty} d\epsilon\, D(\epsilon)\theta(\epsilon_F - \epsilon) = V \int_0^{\epsilon_F} d\epsilon\, D(\epsilon) \tag{988}$$

and

$$E(T=0) = V \int_0^{\infty} d\epsilon\, D(\epsilon)\epsilon\,\theta(\epsilon_F - \epsilon) = V \int_0^{\epsilon_F} d\epsilon\, \epsilon\, D(\epsilon). \tag{989}$$

Using the density of states for electrons given by Eq. (925), we can easily integrate Eqs. (988) and (989), to obtain

$$N(T=0) = \frac{gV}{6\pi^2}\left(\frac{2m\epsilon_F}{\hbar^2}\right)^{3/2} \tag{990}$$

$$E(T=0) = \frac{gV}{10\pi^2}\left(\frac{2m\epsilon_F}{\hbar^2}\right)^{3/2}\epsilon_F. \tag{991}$$

Equations (990) and (991) can be combined to obtain

$$E(T=0) = \tfrac{3}{5} N(T=0)\epsilon_F. \tag{992}$$

For fixed density, $n = N/V$, we can solve Eq. (990) for the Fermi energy,

$$\epsilon_F = \frac{\hbar^2}{2m}\left(\frac{6\pi^2 n}{g}\right)^{2/3}, \tag{993}$$

which we can estimate as follows. Let us introduce the distance between fermions, r_s, via

$$n^{-1} = \frac{4\pi}{3} r_s^3 \tag{994}$$

and set the degeneracy factor $g = 2$, to obtain

$$\epsilon_F = \frac{\hbar^2}{2mr_s^2}\left(\frac{9\pi}{4}\right)^{2/3}. \tag{995}$$

If we introduce the Bohr radius $a_0 = \hbar^2/me^2$, we can write

$$\epsilon_F = \frac{e^2}{2a_0}\left(\frac{a_0}{r_s}\right)^2\left(\frac{9\pi}{4}\right)^{2/3}. \tag{996}$$

We can identify $e^2/2a_0$ as the ionization energy for hydrogen ≈ 13.6 eV. If $r_s \approx a_0$, we can estimate $\epsilon_F \approx 10$ eV. Values of the Fermi energy for some typical systems are given in Table 3.1. At room temperature, $k_B T \approx \frac{1}{40}$ eV. Consequently, $k_B T/\epsilon_F \approx 1/400 \ll 1$. In this sense, a room-temperature solid can be treated as a *low-temperature system*.

TABLE 3.1 Parameters for an Ideal Fermi Gas[a]

Element	r_s/a_0	ϵ_F(eV)	γ/γ_0	Element	r_s/a_0	ϵ_F(eV)	γ/γ_0
Li	3.25	4.74	2.3	Ag	3.02	5.49	1.1
Na	3.93	3.24	1.3	Au	3.01	5.53	1.1
K	4.86	2.12	1.2	Be	1.87	14.3	0.42
Rb	5.20	1.85	1.3	Mg	2.66	7.08	1.3
Cs	5.62	1.59	1.5	Ca	3.27	4.69	1.8
Cu	2.67	7.00	1.3				

Source: Data from Ref. 22.

[a] r_s is defined by Eq. (994), ϵ_F is the measured Fermi energy, and γ/γ_0 is the ratio of the measured to the ideal specific heat at low temperatures.

PROPERTIES OF IDEAL FERMI SYSTEMS AT LOW TEMPERATURES   229

3.11.3 $T > 0$

For nonzero temperatures, the Fermi–Dirac distribution function smooths out as shown in the Fig. 3.3. In this regime some of the particles are excited to energies above the Fermi energy. As a crude estimate, the number of particles excited to an energy $k_B T$ above ϵ_F is proportional to $(k_B T / \epsilon_F) N$. Quantitative analytic evaluation of E and N, using Eqs. (944) and (946) for $T > 0$, requires the development of a technique, due to Sommerfeld [23], for evaluation integrals of the form

$$\int_0^{+\infty} d\epsilon \, f(\epsilon) n(\epsilon)$$

[where $f(\epsilon)$ is a smooth function of ϵ and $n(\epsilon)$ is the Fermi–Dirac distribution function] as a power series in the temperature. We present the analysis in Appendix M, the final result is

$$\int_0^{+\infty} d\epsilon \, f(\epsilon) n(\epsilon) = \int_0^{\mu} d\epsilon f(\epsilon) + \frac{\pi^2}{6}(k_B T)^2 \frac{df(\mu)}{d\mu} + \frac{7\pi^4}{360}(k_B T)^4 \frac{d^3 f(\mu)}{d\mu^3} + \mathcal{O}(k_B T)^6. \tag{997}$$

Using this result to evaluate N and E for low temperatures, we have

$$\frac{N}{V} = \int_0^{\mu} d\epsilon \, D(\epsilon) + \frac{\pi^2}{6}(k_B T)^2 D'(\mu) + \cdots \tag{998}$$

and

$$\frac{E}{V} = \int_0^{\mu} d\epsilon \, \epsilon D(\epsilon) + \frac{\pi^2}{6}(k_B T)^2 \frac{d}{d\mu}[\mu D(\mu)] + \cdots \tag{999}$$

Since $D(\epsilon) = \bar{D} \epsilon^{1/2}$ for nonrelativistic massive fermions, Eqs. (998) and (999) can be written as

$$\frac{N}{V} = \bar{D}\frac{2}{3}\mu^{3/2} + \frac{\pi^2}{6}(k_B T)^2 \frac{\bar{D}}{2\sqrt{\mu}} + \cdots \tag{1000}$$

$$\frac{E}{V} = \bar{D}\frac{2}{5}\mu^{5/2} + \frac{\pi^2}{6}(k_B T)^2 \cdot \frac{3}{2}\bar{D}\mu^{1/2} + \cdots. \tag{1001}$$

These expressions give $N = N(T, V, \mu)$ and $E = E(T, V, \mu)$. It is more convenient to treat the density of fermions, $n = N/V$, as the independent variable. Thus we need to solve for $\mu = \mu(N, T, V)$. We do this, just as in the classical case, by inverting Eq. (1000). We first rewrite Eq. (1000) in the form

$$\frac{3N}{2V\bar{D}} = \mu^{3/2}\left[1 + \frac{\pi^2}{8}\left(\frac{k_BT}{\mu}\right)^2 + \cdots\right]. \tag{1002}$$

From our previous analysis of the zero-temperature limit

$$\epsilon_F^{3/2} = \mu^{3/2}(N, 0, V) = \frac{3N}{2V\bar{D}}, \tag{1003}$$

and Eq. (1002) can be rewritten as

$$\epsilon_F^{3/2} = \mu^{3/2}\left[1 + \frac{\pi^2}{8}\left(\frac{k_BT}{\mu}\right)^2 + \cdots\right]. \tag{1004}$$

To lowest order in T, we can replace μ by ϵ_F in the term of $\mathcal{O}(T^2)$ in Eq. (1004), and solve for μ to obtain

$$\mu = \epsilon_F\left[1 - \frac{\pi^2}{12}\left(\frac{k_BT}{\epsilon_F}\right)^2 + \cdots\right]. \tag{1005}$$

The complete numerical solution giving μ/ϵ_F as a function of T/T_F is shown in Fig. 3.4. The nonzero-temperature corrections acts to lower the chemical potential.

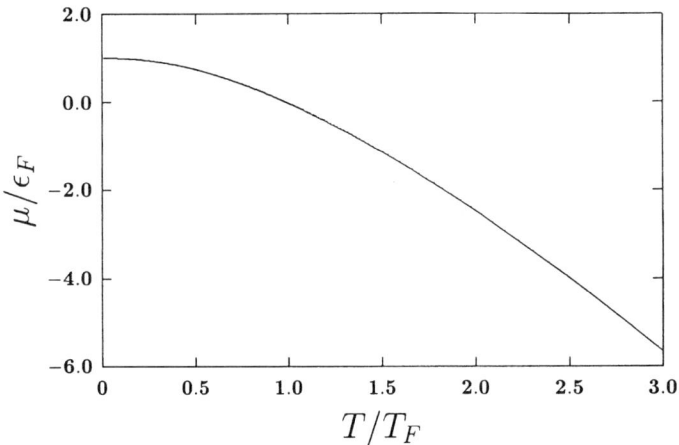

FIGURE 3.4 Scaled chemical potential versus temperature for an ideal gas obeying Fermi–Dirac statistics.

Using Eq. (1005) in Eq. (1001) and expanding in powers of T, the average energy can be written in the low-temperature form:

$$E = \frac{3N}{5}\epsilon_F \left[1 + \frac{5\pi^2}{12}\left(\frac{k_B T}{\epsilon_F}\right)^2 + \cdots\right]. \qquad (1006)$$

As expected the energy increases as the temperature increases. Since we now have $E(T,V,N)$, we can compute the specific heat,

$$C_V = \left(\frac{dE}{dT}\right)_{N,V} = \frac{3}{5}N\epsilon_F \frac{5\pi^2}{12} 2k_B \frac{k_B T}{\epsilon_F^2}$$

$$= Nk_B \frac{\pi^2}{2}\frac{k_B T}{\epsilon_F}. \qquad (1007)$$

The linear dependence of the specific heat on temperature at low temperatures is observed in metallic systems. Experimentalists plot their data in the form

$$C_V = \gamma NT. \qquad (1008)$$

Results are presented in terms of the ratio γ/γ_0, where $\gamma_0 = k_B^2\pi^2/2\epsilon_F$. The assumption of nearly free electrons is reasonable for most of the metals shown in Table 3.1.

We know that the electronic specific heat must approach $\frac{3}{2}Nk_B$ as $T \to \infty$. This is because as T increases, $\lambda = \hbar(2\pi/mk_B T)^{1/2}$ decreases, $\mu \to -\infty$, and we approach the classical limit. The quantitative behavior of the specific heat as a function of temperature is given in Fig. 3.5. The qualitative reason for the sharp decrease in the

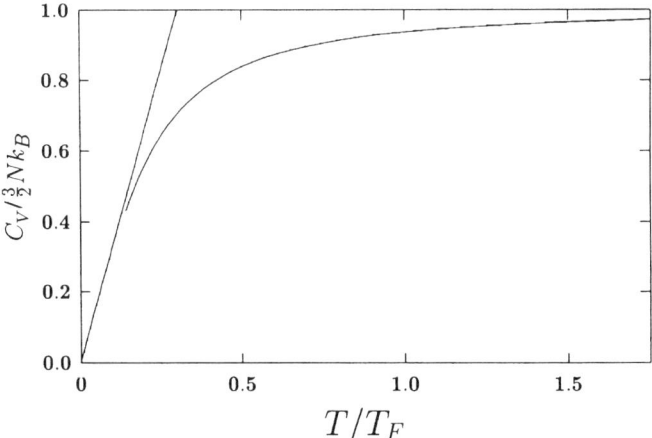

FIGURE 3.5 Specific heat for an ideal Fermi gas as a function of temperature. Also shown is the low-temperature linear term given by $(\pi^2/3)(T/T_F)$.

specific heat at low temperatures compared to the classical result is that there are only a small fraction $\approx (k_B T/\epsilon_F)N$ of the particles thermally excited compared to all N particles in the classical limit. The total excitation energy above the ground state for low temperatures is therefore approximately

$$E - E(T=0) \approx \left(\frac{k_B T}{\epsilon_F} N\right) k_B T \tag{1009}$$

from which it follows that

$$C_V \approx \frac{k_B T}{\epsilon_F} N k_B. \tag{1010}$$

3.12 PHOTON–ELECTRON–POSITRON GAS

Consider now a gas of electrons, positrons, and photons in thermal equilibrium. We know that electrons and positrons annihilate to produce photons, and a photon can lead to pair production. We must therefore treat the *chemical* reaction

$$e + e^+ \leftrightarrow n\gamma, \tag{1011}$$

where n is the number of photons in the process. Remembering our analysis of reactions in Chapter 2, we should write this reaction in the form

$$\sum_i b_i B_i = 0 \tag{1012}$$

where $B_e = e$, $B_{e^+} = e^+$, and $B_\gamma = \gamma$. We then find that

$$b_e = 1 \tag{1013}$$
$$b_{e^+} = 1 \tag{1014}$$

and

$$b_\gamma = -n. \tag{1015}$$

We then have the fundamental relation for equilibrium satisfied by the chemical potentials of each component:

$$0 = \sum_i b_i \mu_i \tag{1016}$$

or

$$\mu_e + \mu_{e^+} = n\mu_\gamma. \tag{1017}$$

We again note that the condition determining the appropriate number of photons in equilibrium is that the free energy be a minimum:

$$\mu_\gamma = \frac{\partial F}{\partial N_\gamma} = 0 \tag{1018}$$

just as we found previously. Thus we see that the precise number of photons in the chemical reaction Eq. (1016) is irrelevant for our purposes. We then have that

$$\mu_e = -\mu_{e^+}. \tag{1019}$$

We assume that the Hamiltonian for this system can be written as

$$H = H_e + H_{e^+} + H_\gamma \tag{1020}$$

where

$$H_e = \sum_{p,\sigma} \epsilon_p a^+_{p,\sigma} a_{p,\sigma} \tag{1021}$$

$$H_{e^+} = \sum_{p,\sigma} \epsilon_p b^+_{p,\sigma} b_{p,\sigma} \tag{1022}$$

and

$$H_\gamma = \sum_{p,\alpha} \hbar\omega_p d^+_{p,\alpha} d_{p,\alpha} \tag{1023}$$

where $a^+_{p,\sigma}$, $b^+_{p,\sigma}$, and $d^+_{p,\alpha}$ are the creation operators for electrons, positrons, and photons, respectively. We also have the relativistic energy–momentum relations:

$$\epsilon_p = \sqrt{m^2c^4 + p^2c^2} \tag{1024}$$

$$\hbar\omega_p = cp. \tag{1025}$$

We have neglected the interactions among electrons, positrons, and photons treated in quantum electrodynamics. The partition function of interest is

$$Z = \text{Tr}\, e^{-\beta(H - \mu_e N_e - \mu_{e^+} N_{e^+})}. \tag{1026}$$

234 QUANTUM STATISTICAL MECHANICS

Since $H_e - \mu_e N_e$, $H_{e^+} - \mu_{e^+} N_{e^+}$, and H_γ commute among themselves, the quantum mechanical trace factors into the product

$$Z = Z_e Z_{e^+} Z_\gamma \tag{1027}$$

where

$$Z_e = \text{Tr}\, e^{-\beta(H_e - \mu_e N_e)} \tag{1028}$$

$$Z_{e^+} = \text{Tr}\, e^{-\beta(H_{e^+} - \mu_{e^+} N_{e^+})} \tag{1029}$$

$$Z_\gamma = \text{Tr}\, e^{-\beta H_\gamma}. \tag{1030}$$

Let us concentrate on the average number of each species:

$$N_\gamma = 2V \int \frac{d^3p}{(2\pi\hbar)^3} \frac{1}{e^{\beta cp} - 1} \tag{1031}$$

$$N_e = 2V \int \frac{d^3p}{(2\pi\hbar)^3} \frac{1}{e^{\beta(\epsilon_p - \mu_e)} + 1} \tag{1032}$$

$$N_{e^+} = 2V \int \frac{d^3p}{(2\pi\hbar)^3} \frac{1}{e^{\beta(\epsilon_p - \mu_{e^+})} + 1}. \tag{1033}$$

The photon case is the same as we treated earlier when we found that

$$N_\gamma = \frac{V}{\pi^2 (\hbar c)^3} (k_B T)^3 \cdot 2\zeta(3). \tag{1034}$$

In treating N_e and N_{e^+}, it is convenient to define

$$\mu \equiv \mu_e = -\mu_{e^+}. \tag{1035}$$

An interesting question concerns the ground state. Let us set $T = 0$; then we can carry out the momentum integrals explicitly:

$$\begin{aligned}
N_e &= 2V \int \frac{d^3p}{(2\pi\hbar)^3} \theta(\mu - \sqrt{m^2c^4 + p^2c^2}) \\
&= \frac{2V \cdot 4\pi}{8\pi^3 \hbar^3} \theta(\mu) \int_0^{+\infty} p^2\, dp\, \theta(\mu^2 - m^2c^4 - p^2c^2) \\
&= \frac{\theta(\mu) V}{\pi^2 \hbar^3} \int_0^{\sqrt{(\mu^2 - m^2c^4)}/c^2} p^2 dp\, \theta(\mu - mc^2) \\
&= \frac{V\theta(\mu - mc^2)}{\pi^2 \hbar^3} \frac{1}{3} \left(\frac{\mu^2 - m^2c^4}{c^2}\right)^{3/2} \tag{1036}
\end{aligned}$$

and

$$N_{e^+} = \frac{V\theta(-\mu - mc^2)}{3\pi^2\hbar^3} \frac{(\mu^2 - m^2c^4)^{3/2}}{c^3}. \tag{1037}$$

We see immediately that if $\mu > mc^2$, then $N_{e^+} = 0$; if $\mu < -mc^2$, then $N_e = 0$; and if $-mc^2 < \mu < mc^2$, then N_{e^+} and N_e must both be zero at $T = 0$. In our world, N_e is nonzero and we require that $\mu > mc^2$. We can invert the equation relating N_e and μ to obtain

$$\mu = \left[m^2c^4 + \left(\frac{3\pi^2 N_e}{V}\right)^{2/3} \hbar^2 c^2\right]^{1/2}. \tag{1038}$$

In the nonrelativistic limit $c \to \infty$ we find that

$$\mu = mc^2 \left[1 + \left(\frac{3\pi^2 N_e}{V}\right)^{2/3} \frac{\hbar^2 c^2}{m^2 c^4}\right]^{1/2}$$

$$\approx mc^2 + \frac{1}{2} \frac{(3\pi^2 N_e/V)^{2/3} \hbar^2}{m} + \cdots \tag{1039}$$

We can rewrite this as

$$\mu = mc^2 + \epsilon_F + \cdots \tag{1040}$$

where ϵ_F is the Fermi energy we introduced in the preceding section. We find that the chemical potential we defined here differs from that in the nonrelativistic case by the rest-mass energy mc^2.

We have discussed the ground state corresponding to our present world—we have an excess of electrons. It is possible, however, to conceive of a world (possibly the early universe [18]) where there is complete symmetry between positrons and electrons. In this case

$$N_e = N_{e^+} \tag{1041}$$

which requires that $\mu = -\mu$ or $\mu = 0$. We can write in this case

$$N_e = 2V \int \frac{d^3p}{(2\pi\hbar)^3} \frac{1}{e^{\beta\epsilon_p} + 1}$$

$$= \frac{V}{\pi^2 \hbar^3 (\beta c)^3} \int_0^{+\infty} x^2 dx \frac{1}{\exp\{[(\beta mc^2)^2 + x^2]^{1/2}\} + 1} \tag{1042}$$

where we have set $x = \beta pc$ in the integral. In the early universe, temperatures were enormous, so $\beta mc^2 \ll 1$, and

$$N_e = \frac{V(k_B T)^3}{\pi^2 \hbar^3 c^3} \int_0^{+\infty} \frac{x^2 dx}{e^x + 1}. \tag{1043}$$

The integral has been evaluated in Appendix M:

$$\int_0^{+\infty} \frac{x^2 dx}{e^x + 1} = \frac{1}{6} I_3(\infty) = \left(\frac{1}{6}\right) 2(3!)(1 - 2^{1-3})\zeta(3)$$

$$= \frac{3\zeta(3)}{2} \tag{1044}$$

so

$$N_e = N_{e^+} = \frac{V}{\pi^2 (\hbar c)^3} (k_B T)^3 \left(\frac{3}{2}\right) \zeta(3). \tag{1045}$$

We see that the number of electrons and positrons is $\frac{3}{4}$ the number of photons. As this gas cooled, there must have been spontaneous symmetry breaking which destroyed the symmetry between electrons and positrons.

3.13 BOSE–EINSTEIN CONDENSATION

3.13.1 Uniform Systems

Let us now turn our attention to the statistical mechanics of massive non-interacting Bose particles at low temperatures. We have immediately from the analysis in Section 3.9 that the average number of particles and average energy are given by

$$N = \frac{gV}{4\pi^2} \left(\frac{2m}{\hbar^2}\right)^{3/2} \int_0^\infty \frac{d\epsilon \sqrt{\epsilon}}{e^{\beta(\epsilon - \mu)} - 1} \tag{1046}$$

$$E = \frac{gV}{4\pi^2} \left(\frac{2m}{\hbar^2}\right)^{3/2} \int_0^\infty \frac{d\epsilon\, \epsilon^{3/2}}{e^{\beta(\epsilon - \mu)} - 1} \tag{1047}$$

where we used the density of states for massive nonrelativistic particles. As in treating the electron gas, we want to keep the average number of particles fixed and determine $\mu = \mu(N, V, T)$. This analysis was discussed earlier from a numerical point of view and shown in Fig. 3.1. Here we want to look at the problem analytically.

Analysis of the Bose–Einstein distribution, Eq. (941), requires for all single-particle energies ϵ that

$$\epsilon - \mu \geq 0. \tag{1048}$$

Otherwise, the distribution would be less than zero for some values of ϵ. Since ϵ can vanish, the chemical potential of an ideal Bose gas must satisfy the condition $\mu \leq 0$. We now want to invert Eq. (1046), rewritten as

$$N = V\bar{D} \int_0^\infty \frac{d\epsilon\, \epsilon^{1/2}}{e^{\beta(\epsilon-\mu)} - 1}, \tag{1049}$$

to find $\mu = \mu(N, T, V)$. We know that in the high-temperature limit the chemical potential must be given by the classical value

$$\mu_c = k_B T \ln\left(\frac{N}{V}\lambda^3\right) \tag{1050}$$

which is large and negative. As we lower the temperature and β increases, $|\mu|$ must decrease such that the factor $e^{\beta|\mu|}$ does not change by orders of magnitude. From the form of Eq. (1049), it appears, as we decrease the temperature, that there will be some temperature T_0 where the chemical potential vanishes, $\mu(N, T_0, V) = 0$. The temperature T_0 is determined by the condition

$$N = V\bar{D} \int_0^\infty d\epsilon \frac{\epsilon^{1/2}}{\exp(\epsilon/k_B T_0) - 1} = V\kappa\bar{D}(k_B T_0)^{3/2}, \tag{1051}$$

where the integral

$$\int_0^\infty \frac{\sqrt{x}\,dx}{e^x - 1} = \zeta\left(\frac{3}{2}\right)\Gamma\left(\frac{3}{2}\right) \equiv \kappa \tag{1052}$$

is evaluated in Appendix L. Equation (1051) can be inverted to give

$$k_B T_0 = \left(\frac{N}{V\bar{D}\kappa}\right)^{2/3} = \left(\frac{4\pi^2}{g\kappa}\right)^{2/3} \frac{\hbar^2}{2ma^2} \tag{1053}$$

where $V/N \equiv a^3$ gives the interparticle separation. The energy $\hbar^2/2ma^2$ is essentially the zero-point energy associated with localizing a particle of mass m in a volume a^3. The density of liquid ^{4}He at low temperatures is $\approx 0.145\,\mathrm{g/cm^3}$. We can use this to obtain $T_0 = 3.14\,\mathrm{K}$.

238   QUANTUM STATISTICAL MECHANICS

Let us next investigate the manner in which $\mu \to 0$ as $T \to T_0$. As a first step in this analysis it is convenient to introduce the quantity

$$\bar{N}_0(T) = V\bar{D} \int_0^\infty d\epsilon \frac{\epsilon^{1/2}}{e^{\beta\epsilon} - 1} \tag{1054}$$

which is the fictitious number of particles computed for $\mu = 0$ and $T > T_0$ and reduces to the physical particle number for $T = T_0$:

$$\bar{N}_0(T_0) = N. \tag{1055}$$

We easily evaluate

$$\bar{N}_0(T) = \kappa V\bar{D}(k_BT)^{3/2} \tag{1056}$$

and obtain

$$\frac{\bar{N}_0(T)}{N} = \frac{\bar{N}_0(T)}{\bar{N}_0(T_0)} = \left(\frac{T}{T_0}\right)^{3/2}. \tag{1057}$$

It is then convenient to write

$$N - \bar{N}_0(T) = V\bar{D} \int_0^{+\infty} d\epsilon \sqrt{\epsilon} \left(\frac{1}{e^{\beta(\epsilon-\mu)} - 1} - \frac{1}{e^{\beta\epsilon} - 1}\right). \tag{1058}$$

If we set $x = \beta\epsilon$ in the integral and define $\bar{\mu} = \beta|\mu|$, we can write

$$N - \bar{N}_0 = V\bar{D}(k_BT)^{3/2} I(\bar{\mu}) \tag{1059}$$

where

$$I(\bar{\mu}) = \int_0^{+\infty} dx \sqrt{x} \left(\frac{1}{e^{x+\bar{\mu}} - 1} - \frac{1}{e^x - 1}\right). \tag{1060}$$

By construction, the left-hand side of Eq. (1058) vanishes as $T \to T_0$, and $\mu \to 0$. We want to evaluate $I(\bar{\mu})$ for small $\bar{\mu}$, then invert Eq. (1058) to obtain $\bar{\mu}$ as $T \to T_0$. We show in Problem 3.31 that

$$I(\bar{\mu}) = -\pi\bar{\mu}^{1/2} + \mathcal{O}(\bar{\mu}), \tag{1061}$$

so that

$$N - \bar{N}_0(T) = -V\bar{D}\pi(k_BT)^{3/2}\bar{\mu}^{1/2} + \mathcal{O}(\bar{\mu}). \tag{1062}$$

If we divide both sides of Eq. (1062) by $\bar{N}_0(T)$, we obtain

$$\frac{N}{\bar{N}_0(T)} - 1 = -\frac{\pi}{\kappa}\left(\frac{-\mu}{k_B T}\right)^{1/2} + \cdots \qquad (1063)$$

Using $\bar{N}_0(T)/N = (T/T_0)^{3/2}$, we can invert Eq. (1063) to obtain the chemical potential:

$$\mu = -k_B T \left(\frac{\kappa}{\pi}\right)^2 \left[1 - \left(\frac{T_0}{T}\right)^{3/2}\right]^2. \qquad (1064)$$

For T near T_0, defining

$$\epsilon = \frac{T - T_0}{T_0}, \qquad (1065)$$

and expanding for small ϵ, we obtain to lowest order in ϵ,

$$\mu = -k_B T_0 \left(\frac{\kappa}{\pi}\right)^2 \frac{9}{4}\epsilon^2. \qquad (1066)$$

We see that μ vanishes quadratically as $T \to T_0$.

Let us now return to our expression for the number of particles, Eq. (1049). Thus far we have established that $|\mu|$ decreases to zero as $T \to T_0$. What happens for $T < T_0$? Since $N = \bar{N}(T_0)$, we can express N in terms of an integral over the Bose–Einstein distribution at temperature T_0, and after canceling a common factor of $\bar{D}$, we can rewrite Eq. (1049) as

$$\int_0^{+\infty} \frac{d\epsilon\, \epsilon^{1/2}}{e^{\beta_0 \epsilon} - 1} = \int_0^{+\infty} \frac{d\epsilon\, \epsilon^{1/2}}{e^{\beta(\epsilon - \mu)} - 1} \qquad (1067)$$

where $\beta_0 = 1/k_B T_0$. After simple manipulations, we can rewrite Eq. (1067) as

$$\int_0^{+\infty} \frac{\epsilon^{1/2} d\epsilon}{(e^{\beta_0 \epsilon} - 1)(e^{\beta(\epsilon - \mu)} - 1)} (e^{\beta(\epsilon - \mu)} - e^{\beta_0 \epsilon}) = 0. \qquad (1068)$$

If $T < T_0$ ($\beta > \beta_0$), then since $\mu \leq 0$,

$$e^{\beta(\epsilon + |\mu|)} > e^{\beta_0 \epsilon} \qquad (1069)$$

and the integrand on the left-hand side of Eq. (1068) is positive definite. This leads to the conclusion that we cannot find a μ such that Eq. (1049) is satisfied for $T < T_0$.

240 QUANTUM STATISTICAL MECHANICS

What is the problem? It is clear physically that as we lower the temperature, many particles will begin to occupy the lowest available single-particle state—the ground state $\epsilon = 0$. For bosons there is no restriction on the number of particles we can put into a single state. It is clear that the state $\epsilon = 0$ is special and must be treated with some care as the temperature becomes low. We were not careful enough when we replaced the sum

$$N = \sum_i \frac{1}{e^{\beta(\epsilon_i - \mu)} - 1} \tag{1070}$$

with the integral given by Eq. (964). All of the states but the ground state can be approximated by our continuum approximation; these terms give the integral we have evaluated. In contrast, the ground-state term has been lost in passing to the integral because the $\sqrt{\epsilon}$ in the density of states vanishes at $\epsilon = 0$. We write more accurately

$$N = \frac{1}{e^{\beta|\mu|} - 1} + V\bar{D} \int_0^{+\infty} \frac{d\epsilon \sqrt{\epsilon}}{e^{\beta(\epsilon - \mu)} - 1}. \tag{1071}$$

Normally, in the thermodynamic limit,

$$\lim_{V \to \infty} \frac{1}{V} \frac{1}{e^{\beta|\mu|} - 1} \to 0 \tag{1072}$$

and our previous analysis holds. If $\mu \to 0$, however, the occupation of the ground state,

$$N_0 = \frac{1}{e^{\beta|\mu|} - 1}, \tag{1073}$$

may become macroscopically large and

$$\lim_{V \to \infty} \frac{N_0}{V} \equiv n_0 \tag{1074}$$

may become finite. This clearly requires that

$$N_0 \approx \frac{1}{\beta|\mu|} + \mathcal{O}(1) \tag{1075}$$

or

$$|\mu| = \frac{k_B T}{V n_0}. \tag{1076}$$

We see that for $T > T_0$ there is no macroscopic occupation of the ground state and μ is a nonzero number in the thermodynamic limit. At $T = T_0, \mu \to 0$ in the

BOSE–EINSTEIN CONDENSATION   241

thermodynamic limit and will remain zero for $T < T_0$. We can write for $T < T_0$,

$$N = N_0 + V\bar{D} \int_0^{+\infty} \frac{d\epsilon\sqrt{\epsilon}}{e^{\beta\epsilon} - 1}$$
$$= N_0 + N\left(\frac{T}{T_0}\right)^{3/2} \tag{1077}$$

or

$$N_0 = N\left[1 - \left(\frac{T}{T_0}\right)^{3/2}\right]. \tag{1078}$$

As $T \to 0$, all of the particles go into the ground state,

$$n_0(T=0) = n. \tag{1079}$$

The phenomenon of a large number of particles collecting in the ground state of an ideal Bose system is known as *Bose–Einstein condensation*. We plot the temperature dependence of n_0/n in Fig. 3.6. The chemical potential, near the transition, can be written as

$$\mu = \begin{cases} -k_B T_0 \left(\frac{3\kappa}{2\pi}\right)^2 \left(\frac{T-T_0}{T_0}\right)^2 & T \geq T_0 \\ 0 & T < T_0. \end{cases} \tag{1080}$$

We see that the second derivative of μ is discontinuous at $T = T_0$:

$$\lim_{T \to T_0^+} \frac{d^2\mu}{dT^2} = -\frac{k_B}{T_0}\left(\frac{9}{2}\right)\left(\frac{\kappa}{\pi}\right)^2 \tag{1081}$$

$$\lim_{T \to T_0^-} \frac{d^2\mu}{dT^2} = 0. \tag{1082}$$

Since we now know $\mu = \mu(T,n)$ we can go back and compute the energy and the pressure ($= 2E/3V$). We do not need to worry about condensation effects in this case since the particles in the ground state have zero energy and zero momentum. This means that the pressure vanishes at zero temperature since particles with zero momentum can exert no force on the walls of the container.

Let us treat first the case $T \geq T_0$. We have

$$E = V\bar{D} \int_0^{+\infty} \frac{d\epsilon\, \epsilon^{3/2}}{e^{\beta(\epsilon-\mu)} - 1}$$
$$= V\bar{D}(k_B T)^{5/2} \int_0^{+\infty} \frac{dx\, x^{3/2}}{e^{x+\bar{\mu}} - 1}. \tag{1083}$$

242 QUANTUM STATISTICAL MECHANICS

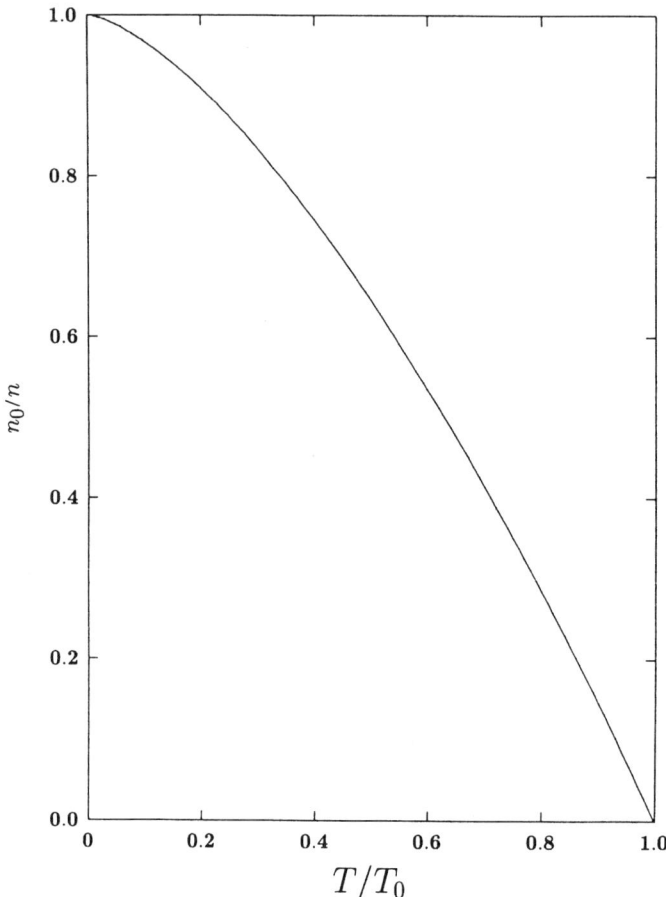

FIGURE 3.6 Condensate fraction versus temperature for an ideal Bose gas undergoing Bose–Einstein condensation.

We will be interested in the region slightly above but near T_0 where $\bar{\mu}$ is small. We can therefore expand the integral in a power series in $\bar{\mu}$. This is direct in this case because we have $x^{3/2}$ in the integrand of Eq. (1083), not $x^{1/2}$ as in the integral for N. We have immediately that

$$E = E(\mu = 0) + V\bar{D}(k_B T)^{5/2} \bar{\mu} \int_0^{+\infty} dx\, x^{3/2} \frac{d}{dx}\left(\frac{1}{e^x - 1}\right) + \mathcal{O}(\bar{\mu}^{3/2}) \quad (1084)$$

where we have used

$$\left(\frac{d}{d\bar{\mu}} f(x + \bar{\mu})\right)_{\bar{\mu}=0} = \frac{df(x)}{dx}. \quad (1085)$$

On integrating by parts, we obtain

$$E = E(\mu = 0) + V\bar{D}(k_B T)^{5/2} \bar{\mu}\left(-\frac{3}{2}\right) \int_0^{+\infty} \frac{dx\, x^{1/2}}{e^x - 1} + \mathcal{O}(\bar{\mu}^{3/2})$$

$$= E(\mu = 0) + \frac{3}{2} V\bar{D}(k_B T)^{3/2} \mu \kappa + \mathcal{O}(\mu^{3/2}) \tag{1086}$$

for $T \geq T_0$. For $T < T_0$, $\mu = 0$, so

$$E = V\bar{D}(k_B T)^{5/2} \int_0^{\infty} \frac{dx\, x^{3/2}}{e^x - 1}$$

$$= V\bar{D}(k_B T)^{5/2} \kappa_1 \tag{1087}$$

where $\kappa_1 = \zeta(\frac{5}{2})\Gamma(\frac{5}{2})$. This also gives $E(\mu = 0)$ for $T > T_0$. In summary, we have

$$E = \begin{cases} \dfrac{\kappa_1}{\kappa}\left(\dfrac{k_B T}{k_B T_0}\right)^{3/2} k_B T N & T < T_0 \\[1em] \dfrac{\kappa_1}{\kappa}\left(\dfrac{k_B T}{k_B T_0}\right)^{3/2} k_B T N + \dfrac{3}{2} N \left(\dfrac{k_B T}{k_B T_0}\right)^{3/2} \mu & T > T_0 \end{cases} \tag{1088}$$

where we used

$$N = V\bar{D}(k_B T_0)^{3/2} \kappa. \tag{1089}$$

For low temperatures ($T < T_0$) the heat capacity is given by [using $\Gamma(\frac{5}{2}) = \frac{3}{2}\Gamma(\frac{3}{2})$]

$$C_V = \frac{15}{4} \frac{\zeta(5/2)}{\zeta(3/2)} N k_B \left(\frac{T}{T_0}\right)^{3/2} \tag{1090}$$

which vanishes as $T^{3/2}$ for low temperatures and has the value $1.925 N k_B$ at $T = T_0$. An interesting feature of the specific heat is that it has a discontinuous derivative at $T = T_0$. This comes from the discontinuity in the second derivative of μ at $T = T_0$. We easily find that

$$\lim_{\epsilon \to 0}\left[\left(\frac{dC_V}{dT}\right)_{T=T_0+\epsilon} - \left(\frac{dC_V}{dT}\right)_{T=T_0-\epsilon}\right] = \frac{27}{4}\left[\frac{\kappa}{\pi}\right]^2 \frac{N k_B}{T_0}. \tag{1091}$$

The specific heat as a function of temperature has a cusp at T_0, which implies that an ideal Bose gas exhibits a type of phase transition at the temperature T_0.

3.13.2 Bose–Einstein Condensation in a Harmonic Trap

An interesting realization [24–27] of Bose–Einstein condensation occurs for a system of massive bosons in a harmonic potential trap. Such systems are now available in the laboratory. Because of the nonuniformity in space imposed by the trapping potential, these systems fall outside the typical set of systems we have treated. Thus our previous development concerning the thermodynamic limit and extensive variables does not apply directly. It is still true that we can use our development for systems in thermal equilibrium.

Although a fully realistic treatment of this problem requires inclusion of the interaction between the bosons [28], we can gain some feeling for these systems by treating the noninteracting case. Physically, one needs to treat anisotropic trapping potentials of the form

$$V(\mathbf{r}) = \frac{m}{2} \sum_{i=1}^{3} \omega_i^2 r_i^2 \tag{1092}$$

where m is the mass of the particles, the center of the trap is at the origin, and $\omega_i/2\pi$ has a typical value of 60 Hz.

The single-particle levels for this potential are given as usual for a harmonic potential by

$$E(\mathbf{n}) = \sum_{i=1}^{3} (n_i + \tfrac{1}{2})\hbar\omega_i, \tag{1093}$$

where $\mathbf{n} = (n_1, n_2, n_3)$ and $n_i = 0, 1, 2, \ldots$. Also, as usual, the thermodynamic potential is given by

$$\Omega = -\beta^{-1} \sum_{\mathbf{n}} \ln(1 - e^{-\beta(\epsilon(\mathbf{n}) - \mu)}). \tag{1094}$$

Let us restrict ourselves to the isotropic limit where $\omega_0 = \omega_1 = \omega_2 = \omega_3$. In this case the sum over $\mathbf{n}$ can be converted to a single sum (this amounts to evaluating the degeneracy of each energy state as discussed in Problem 3.35) and we have

$$\Omega = -\beta^{-1} \sum_{l=0}^{\infty} \tfrac{1}{2}(l+1)(l+2) \ln\left(1 - e^{-(l+\epsilon)/t}\right) \tag{1095}$$

where we have introduced the dimensionless temperature

$$t = \frac{k_B T}{\hbar\omega_0} \tag{1096}$$

and the dimensionless chemical potential

$$\epsilon = \frac{3}{2} - \frac{\mu}{\hbar\omega_0}. \tag{1097}$$

Taking the derivative of the thermodynamic potential with respect to $\alpha = -\beta\mu$, we obtain the expression for the average number of particles:

$$N = \frac{1}{2}\sum_{l=0}^{\infty} \frac{(l+1)(l+2)}{e^{(l+\epsilon)/t} - 1}. \tag{1098}$$

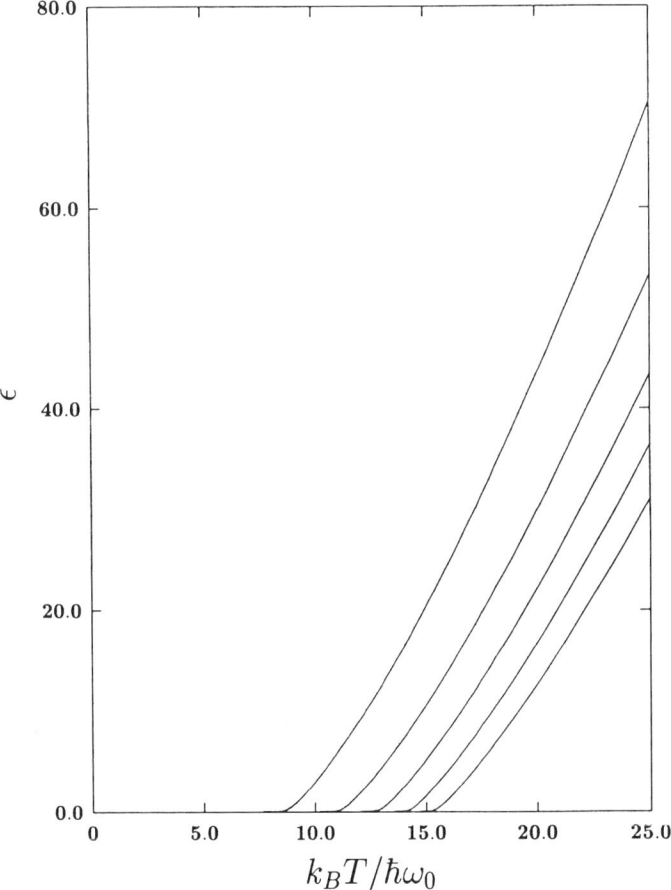

FIGURE 3.7 Dimensionless chemical potential versus reduced temperature for various numbers of trapped particles. ϵ going to zero is associated with the onset of Bose–Einstein condensation in trapped systems. From left to right the curves represent $N = 1000$, 2000, 3000, 4000, and 5000.

The number of particles in the ground state is given by

$$N_0 = \frac{1}{e^{\epsilon/t} - 1} \tag{1099}$$

since

$$\beta(E_0 - \mu) = \frac{\epsilon}{t}. \tag{1100}$$

The procedure, then, for a fixed number of trapped particles N is to solve for $\epsilon = \epsilon(T, N)$ and then obtain $N_0 = N_0(T, N)$.

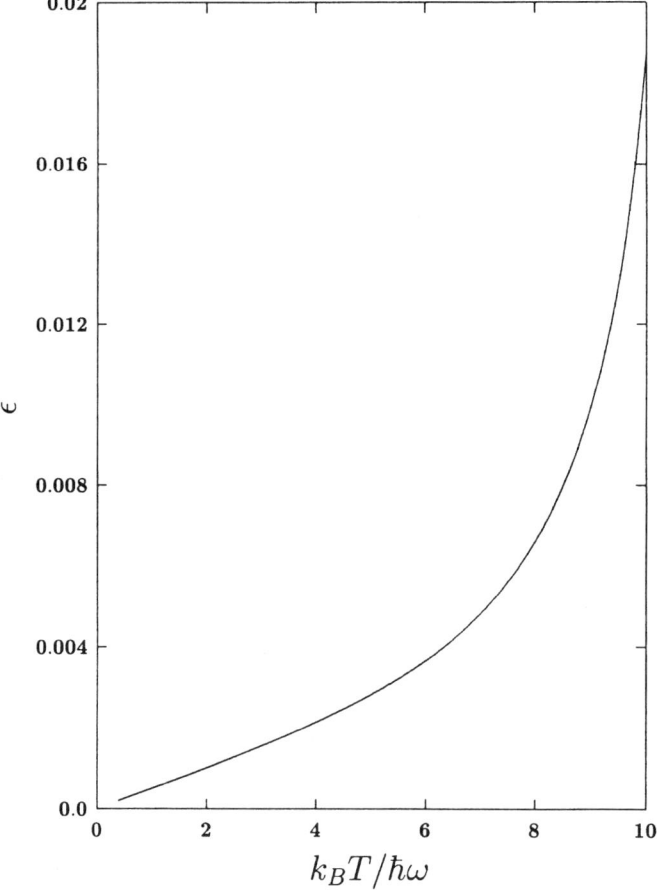

FIGURE 3.8 Restricted scale plot of Figure 3.7 for $N = 2000$. This shows that ϵ never actually hits zero.

We can work out some analytic results in the limit of large N. In this limit many terms contribute to the sum in Eq. (1098), ($l \gg \epsilon$), and we can replace the sum by an integral and set $\epsilon = 0$ to obtain

$$N = \frac{1}{2}\int_0^\infty dl\, l^2 \frac{1}{e^{l/t} - 1} = \zeta(3)t^3 \qquad (1101)$$

where we have used the result for Bose integrals from Appendix L. It is clear that we can associate the transition temperature with the temperature defined by Eq. (1101):

$$\frac{k_B T_{\text{BEC}}}{\hbar \omega_0} = \left(\frac{N}{\zeta(3)}\right)^{1/3} \approx 0.941\ldots N^{1/3}. \qquad (1102)$$

It is not difficult for a given N to numerically solve Eq. (1098) to obtain $\epsilon = \epsilon(x, N)$ and then obtain the fractional occupation of the ground state $N_0(x,N)/N$. In Fig. 3.7 we plot ϵ versus $t = k_B T/\hbar\omega$ over a broad range of temperature for the

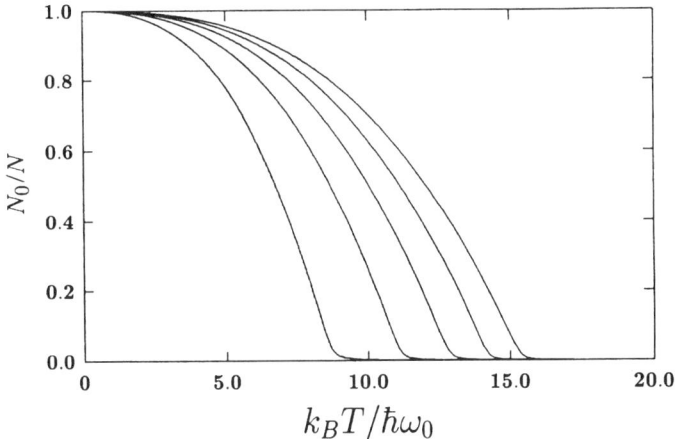

FIGURE 3.9 Fraction of trapped particles in the ground state versus reduced temperature for various numbers of trapped particles. From left to right the curves represent $N = 1000$, 2000, 3000, 4000, and 5000.

TABLE 3.2 Variation of Condensation Temperature with the Number of Trapped Particles

N	$N^{1/3}$	$k_B T_{\text{BEC}}/\hbar\omega_0$
1000	10	9.410
2000	12.599	11.855
3000	14.423	13.572
4000	15.874	14.937
5000	17.100	16.091

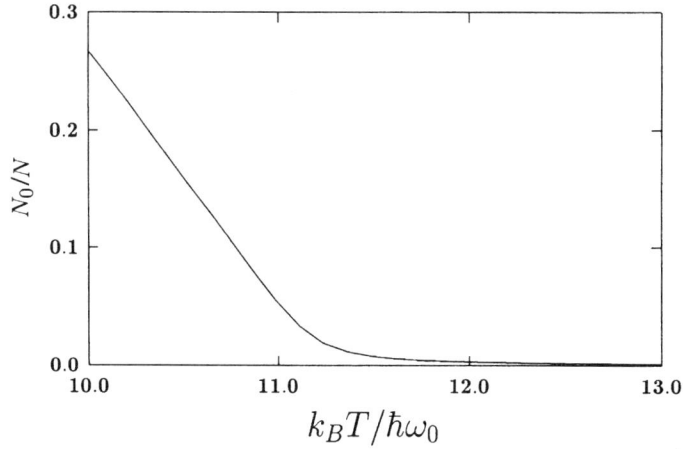

FIGURE 3.10 Figure 3.9 shown over a restricted temperature range for $N = 2000$. This shows that the number of particles in the ground state remains nonzero for all temperatures.

cases $N = 1000, 2000, 3000, 4000$, and 5000. Note that on the scale shown, the *chemical potential* goes to zero at a definite value of the temperature. On a finer scale, shown in Fig. 3.8 (for $N = 2000$), we see that ϵ has not gone to zero but has become small. For the same set of N we show the occupation of the ground state as a function of temperature in Fig. 3.9. Clearly, there is a rather well defined temperature where Bose–Einstein condensation occurs. We also give the value of T_{BEC} given by Eq. (1102) in Table 3.2. If we look more closely, as shown in Fig. 3.10 for $N = 2000$, we see that in fact the occupation of the ground state does not go to zero for $T > T_{\text{BEC}}$.

Notice that the thermodynamic limit does not exist in this system. The most obvious example is that T_{BEC} does not settle down to a *bulk* value as the number of trapped particles increases. One can go on and study the thermal properties of these systems and also investigate the role of interactions.

REFERENCES

[1] Texts that give an introduction to quantum mechanics at the level needed here are: G. Baym, *Lectures in Quantum Mechanics*, W. A. Benjamin, New York, 1969; K. Gottfried, *Quantum Mechanics*, Vol. I, W. A. Benjamin, New York, 1966; and because it is so interesting, the classic text: P. A. M. Dirac, *The Principles of Quantum Mechanics*, 4th ed., Oxford University Press, London, 1958.

[2] U. Fano, *Rev. Mod. Phys.* **29**, 74 (1957).

[3] For a good reference on second quantization, see A. Fetter and J. D. Walecka, *Quantum Theory of Many-Particle Systems*, Mc Graw-Hill, New York, 1971, Chap. 1.

[4] The spin of the electron was discovered by G. E. Uhlenbeck and S. Goudsmit [*Naturwiess* **13**, 953 (1925); *Nature* **117**, 264 (1926)]. See also Refs. 12 and 13.

REFERENCES 249

[5] See G. Baym, *Lectures in Quantum Mechanics*, W. A. Benjamin, New York, 1969, p. 434. The original work is by M. Gell-Mann and K. Brueckner, *Phys. Rev.* **106**, 364 (1957).

[6] The Lennard-Jones pair potential is discussed in some detail in Chapter 4.

[7] For a discussion on the discovery of λ-transition in ^{4}He, see S. G. Brush, *Statistical Physics and the Atomic Theory of Matter: From Boyle and Newton to Landau and Onsager*, Princeton University Press, Princeton, N.J., 1983, p. 172.

[8] D. D. Osheroff, R. C. Richardson, and D. M. Lee, *Phys. Rev. Lett.* **28**, 885 (1972), received the Nobel Prize in Physics in 1996 for the discovery of superfluid ^{3}He.

[9] K. Gottfried, *Quantum Mechanics*, Vol. I, W. A. Benjamin, New York, 1966, Chap. VIII.

[10] This equation of motion approach was developed in L. P. Kadanoff and G. Baym, *Quantum Statistical Mechanics*, W. A. Benjamin, New York, 1962 which built on the earlier work of P. C. Martin and J. Schwinger, *Phys. Rev.* **115**, 1342 (1959).

[11] The determination of the density of states for systems with more complicated symmetries may require sophisticated calculational techniques.

[12] E. Fermi, "Über die Wahrscheinlich keit der Quantenzustände," *Z. Phys.* **26**, 54 (1926); P. A. M. Dirac, "On the theory of quantum mechanics," *Proc. R. Soc. London A* **112**, 661 (1926).

[13] S. N. Bose, "Planck's Gesetz and Lichtquantenhypothese," *Z. Phys.* **26**, 178 (1924); A. Einstein, "Sitzungsberichte," *Preussische Akad. Wissenshaften* **1**, 3 (1925).

[14] J. C. Maxwell, "Illustrations of the dynamical theory of gases," *Philos. Mag. Ser.* 4, 19 (1860); "On the dynamical theory of gases," *Philos. Trans. R. Soc. London*, **157**, 49 (1867); L. Boltzmann, "Studien über das Gleichgewicht der legendige Kraft zwischen bewegten materiellen Punkten," *Sitzungsberichte, K. Akademie der Wissenschaften, Wien, Math. Naturwiss. Kl*, **58**, 517 (1868).

[15] Max Planck, "Zur theorie des Gesetzes der Energieverteilung im Normalspectrum," *Verh. Dtsch. Phys. Ges.* **2**, 237 (1900).

[16] Lord Rayleigh, "Remarks upon the laws of complete radiation," *Philos. Mag. Ser. 5* **49**, 539 (1900).

[17] W. Wien, "Eine neue Beziehung der Strahlung schwarzer Körper zum zweiten Hauptsatz der Wärmetheorie," *Sitzungsberichte, K. Preussischen Akademie der Wissenschaften, Wien, Phys. Math. Kl.*, 55 (1893).

[18] S. Weinberg, *The First Three Minutes*, Bantam Books, New York, 1977.

[19] M. Abramowitz and I. Stegun, *Handbook of Mathematical Functions*, National Bureau of Standards, Washington, D.C. 1964, p. 807.

[20] The T^4 behavior was experimentally established by Stefan in 1879, and Boltzmann derived it using thermodynamics in 1884. J. Stefan, "Über die Beziehung zwischen der Wärmestrahlung und der Temperatur," *Sitzungsberichte, K. Akademie der Wissenschaften, Wien, Math. Naturwiss. Kl.* **79**, 55 (1879); *Wien. Ber.* **79**, 391 (1879); L. Boltzmann, *Ann. Phys.* **22**, 291 (1884); Über die Eigenschaften monocyclischer und anderer damit verwandter Systeme," *Sitzungsberichte, K. Akademie der Wissenschaften*, Wien, *Math. Naturwiss. Kl.* **90**, 231 (1884).

[21] Wolfgang Pauli, *Z. Phys.* **31**, 765 (1925).

[22] N. Ashcroft and N. D. Mermin, *Solid State Physics*, Holt, Rinehart and Winston, New York, 1976.

[23] A. Sommerfeld, *Z. Phys.* **47**, 1 (1928).

[24] Helium is not a physically faithful candidate for ideal gas Bose–Einstein condensation because it is a strongly interacting system at densities compatible with Bose–Einstein condensation.
[25] M. H. Anderson, J. R. Ensher, M. R. Matthews, C. E. Wieman, and E. A. Cornell, *Science* **269**, 198 (1995).
[26] C. C. Bradley, C. A. Sackett, J. J. Tollett, and R. G. Hulet, *Phys. Rev. Lett.* **75**, 1687 (1995).
[27] K. B. Davis, M.-O. Mewes, M. R. Kurn, and W. Ketterle, *Phys. Rev. Lett.* **75**, 3969 (1995).
[28] A. L. Fetter, *Theory of a Dilute Low-Temperature Trapped Bose Condensate*, cond-mat/9811366.

PROBLEMS

3.1. Given the canonical commutation (anticommutations) relations satsified by the creation and annihilation operators, and introducing the field operators via Eqs. (858) and (859), prove that Eqs. (860) and (862) follow.

3.2. Starting with the second-quantized form for the density

$$\hat{n}(\mathbf{x}) = \psi^+(\mathbf{x})\psi(\mathbf{x})$$

for a system described by a Hamiltonian

$$H = \int d^3x\, \psi^+(\mathbf{x}) \left(-\frac{\hbar^2 \nabla^2}{2m} \right) \psi(\mathbf{x}) + \frac{1}{2} \int d^3x \int d^3x'\, V(\mathbf{x}-\mathbf{x}')\hat{n}(\mathbf{x})\hat{n}(\mathbf{x}')$$

show that one obtains a continuity equation of the form

$$\frac{\partial \hat{n}(\mathbf{x})}{\partial t} + \nabla \cdot \mathbf{J}(\mathbf{x}) = 0$$

and identify the particle current $\mathbf{J}$.

3.3. Calculate the first quantum correction to the thermodynamic potential for ideal Bose and Fermi systems.

3.4. Compute the fluctuation $\langle (\delta n_i)^2 \rangle$ in the occupation number $n_i = a_i^+ a_i$ for ideal Fermi and Bose systems where $\delta n_i = n_i - \langle n_i \rangle$. For photons and massive fermions at low temperatures, find the energy value for which fluctuations are largest.

3.5. The method we developed in the text for evaluating quantum partition functions only works for systems where the fundamental fields satisfy the canonical commutation (or anticommutations) relations,

$$[a_i, a_j^+] = \delta_{ij}.$$

In a quantum spin system, with a spin operator $\mathbf{S}_i$ defined on a lattice site i, the operators satisfy the commutation relations

$$[S_i^\alpha, S_j^\beta] = i\hbar \delta_{ij} \sum_\gamma \epsilon_{\alpha\beta\gamma} S_i^\gamma$$

where α, β, and γ label the vector components and $\epsilon_{\alpha\beta\gamma}$ is the completely antisymmetric tensor. If we have a Hamiltonian of the form

$$H = \sum_i H_i$$

where

$$H_i = -JS_i^2 - HS_i^z,$$

then the eigenstates of the system are of the form

$$|\psi\rangle = \prod_{i=1}^{N} |m_i^z, m\rangle$$

where

$$S_i^z |m_i^z, m\rangle = \hbar m_i^z |m_i^z, m\rangle$$

and

$$S_i^2 |m_i^z, m\rangle = \hbar^2 m(m+1) |m_i^z, m\rangle.$$

In these equations m is the magnitude of the spin and the m_i^z are the projections of the spin onto the z-axis and can take on values $m_i^z = [-m, -(m-1), \ldots, m-1, m]$. Calculate, using the canonical ensemble, the free energy, the energy per spin, the specific heat at constant volume, the average magnetization, and the magnetic susceptibility in zero external field. Discuss your results as a function of temperature.

3.6. Two noninteracting, identical spin 1 (and hence Bose) particles are placed in the same (spatial) state. They each have magnetic moment $\mu = \mu_0 S$, and a magnetic field **B** is applied, resulting in an interaction energy of $-\mu \cdot \mathbf{B}$ for each particle. If the particles are in contact with a thermal bath at temperature T, calculate the expected magnetization $\langle M \rangle = \langle \mu_1 + \mu_2 \rangle$.

3.7. (a) Consider a classical spinning particle with magnetic moment $\mu = \mu_0 S$. A magnetic field **B** is applied, resulting in an interaction energy $E = -\mu \cdot \mathbf{B}$. The particle is now placed in contact with a heat bath at temperature T. Find the probability distribution $P(\theta) d\theta$ for finding the angle between **S** and **B** to be within $d\theta$ of θ. If N such particles are present, what is the expected magnetization?
(b) Now consider a quantum mechanical particle of spin s, with $\mu = \mu_0 S$. Then S_z can take any of the $(2s+1)$ values, $-s, -s+1, \ldots, s-1, s$. Again a magnetic field **B** (in the z-direction) is applied and the particle is placed in contact with a heat bath at temperature T. Calculate the probability $P(S_z)$ of each of the allowed values of S_z. If N such (distinguishable) particles are present, what is the expected magnetization? Compare your answers to these questions to the classical case in the limit of large s.

252   QUANTUM STATISTICAL MECHANICS

3.8. Consider a set of N harmonic oscillators governed by a Hamiltonian

$$H = \sum_{i=1}^{N} \left(\frac{1}{2m} \mathbf{p}_i^2 + \frac{k}{2} \mathbf{x}_i^2 \right)$$

where $\mathbf{p}_i$ and $\mathbf{x}_i$ are quantum mechanical operators. Starting with the canonical commutation relations

$$[x_i^\alpha, p_j^\beta] = i\hbar \delta_{ij} \delta_{\alpha\beta},$$

where α and β label vector components, show that x_i^α and p_j^β can be written as a linear combination of creation and annihilation operators $(a_i^\alpha)^+$ and a_i^α. Reexpress the Hamiltonian in terms of these operators. Find the average energy and specific heat at constant volume as a function of temperature. Evaluate the high- and low-temperature limits as well as the classical limit $\hbar \to 0$.

3.9. What can you say about the ratio between the values of the specific heat of an ideal Fermi gas and an ideal Bose gas in the two-dimensional case? There is a very simple answer to this problem, but it is rather difficult to prove this result.

3.10. Calculate the heat capacity of a one-dimensional system of photons at low temperatures. We know that $\lim_{T \to 0} C = 0$. Find the first nonzero temperature correction to this result.

3.11. Show that the entropy for a quantum ideal gas is given by

$$S = -k_B \sum_i [n_i \ln n_i + \eta(1 - \eta n_i) \ln(1 - \eta n_i)]$$

where $\eta = 1$ for fermions and -1 for bosons and $n_i = (e^{\beta(\epsilon_i - \mu)} + \eta)^{-1}$. The index i labels the single-particle states.

3.12. Prove that

$$\sum_{k=0}^{\infty} \frac{(-1)^k}{(k+1)^n} = (1 - 2^{1-n})\zeta(n)$$

where ζ is the Riemann zeta function,

$$\zeta(n) = \sum_{s=1}^{\infty} \frac{1}{s^n}.$$

3.13. Consider a box containing a neutrino gas in thermal equilibrium at temperature T and with zero chemical potential, μ. (A box for neutrinos is, of course, unrealistic, and even if confined, because of conservation of lepton number, μ need not vanish as in the photon case. However, in the early universe the neutrinos should be in thermal equilibrium with $\mu = 0$.) Neutrinos differ from photons in that they are fermions and also in that they have only one helicity state per mode (rather than two). Calculate the partition function, Z, of this gas and use it to obtain formulas for energy, E, and entropy, S, as a function of T. [*Hint:* Use the result from Problem 3.12,

$$\sum_{k=0}^{\infty} \frac{(-1)^k}{(k+1)^n} = (1 - 2^{1-n})\zeta(n)].$$

3.14. The vibrational modes in a solid may be considered as nonconserved bosons with energy $\epsilon(k) = \hbar c|k|$, where c is the speed of sound. (This is a reasonable approximation for the phonons in typical solids below a few tens of kelvin.) Find the specific heat at constant volume C_V of a gas of such phonons in d dimensions. How does it vary with temperature?

3.15. (a) Suppose that photons had an energy $\omega_q = Dq^2$ (quadratic in q rather than linear). What would the Planck blackbody spectrum look like, and what would the temperature dependences of the total number of excited photons, the total energy, and the specific heat be? (Spin-wave excitations in a ferromagnet have this sort of ω versus q relation.)

(b) Suppose that photons were massive i.e., $[\omega_q = (q^2 + m^2)^{1/2}]$. What would the Planck blackbody relation spectrum look like in this case?

3.16. (a) Consider a subvolume V of a classical ideal gas at temperature T whose total volume is much greater than V. Suppose that on average, N particles are contained in V. Breakup the one-particle phase space of the gas in V into small cells. (1) Calculate the expected fluctuation, Δn_i, in the number of particles in the ith cell. (2) Use this result to calculate the expected fluctuation, ΔN, in the total number of particles in V.

(b) Repeat the calculation of part (a) for a photon gas at temperature T contained in a box of volume V. (1) Calculate the expected fluctuation, Δn_i, in the number of photons in the ith energy eigenstate state with energy ϵ_i. (2) Calculate the expected fluctuation, ΔN, in the total number of photons in the box.

3.17. Suppose that we can measure the energy density for cosmic radiation corresponding to two wavelengths:

λ(cm)	$E(\lambda)$
10	8.782
20	2.347

where $E(\lambda)$ is measured in arbitrary units. Determine the temperature of the associated photon gas.

3.18. Consider an ideal quantum statistical mechanical system with a density of states

$$D(\epsilon) = \frac{1}{a^3}\delta(\epsilon - \epsilon_0)$$

where a is a constant with dimensions of length and ϵ_0 is a constant with dimensions of energy.

(a) Determine the average energy per particle assuming both Fermi and Bose statistics.

(b) Determine the equation of state $p = p(n, T)$, again assuming both Fermi and Bose statistics.

3.19. Show that the low-temperature behavior for the entropy of an ideal Fermi system of massive nonrelativistic particles is of the form

$$S = -k_B N \frac{k_B T}{\epsilon_F} \gamma + \mathcal{O}\left(\frac{k_B T}{\epsilon_F}\right)^3$$

and determine the constant γ.

3.20. For a Fermi gas, we can define a temperature T_0 for which the chemical potential vanishes. Express T_0 in terms of T_F.

3.21. Determine the velocity of sound in a Fermi gas at low temperature. Express your results in terms of the Fermi velocity, which is defined by

$$\epsilon_F = \tfrac{1}{2} m v_F^2.$$

3.22. Show that in the second-order approximation at low temperatures the chemical potential of a degenerate Fermi gas is given by

$$\mu \approx \epsilon_F \left[1 - \frac{\pi^2}{12}\left(\frac{k_B T}{\epsilon_F}\right)^2 - \frac{\pi^4}{80}\left(\frac{k_B T}{\epsilon_F}\right)^4\right]$$

and that this implies that

$$E \approx \frac{3}{5} N \epsilon_F \left[1 + \frac{5\pi^2}{12}\left(\frac{k_B T}{\epsilon_F}\right)^2 - \frac{\pi^4}{16}\left(\frac{k_B T}{\epsilon_F}\right)^4\right].$$

Determine the T^3 correction to the linear specific heat and find out whether for typical simple metals it is large or small compared with the T^3 contribution due to phonons.

3.23. Consider an ideal electron gas in an external magnetic field B. The Hamiltonian, including the Zeeman term, can be written as

$$H = \sum_{p,\sigma} \epsilon_p a^+_{p,\sigma} a_{p,\sigma} - \mu B \sum_{p,\sigma} \sigma a^+_{p,\sigma} a_{p,\sigma}$$

where $\epsilon_p = p^2/2m$, $\sigma = +1$ or -1, and μ is the magnetic moment. The magnetization operator for this system is defined by

$$\hat{M} = \sum_{p,\sigma} \sigma a^+_{p,\sigma} a_{p,\sigma}.$$

Calculate the average magnetization and the magnetic susceptibility in zero external field. Discuss the temperature dependence of $\langle M \rangle$ and χ for $B = 0$.

3.24. Determine the low-temperature forms for the specific heat at constant pressure C_p, the coefficient of thermal expansion α, and the isothermal compressibility κ_T for an ideal Fermi system of massive nonrelativistic particles. Evaluate the same quantities for high temperatures. Discuss the qualitative differences in these quantities in these limits.

3.25. A system of noninteracting Fermi particles has a density of single-particle energy levels, $D(\epsilon)$, sharply peaked around energy ϵ_0, so that $D(\epsilon)$ can be idealized as

$$D(\epsilon) = D_0 \delta(\epsilon - \epsilon_0).$$

The system is now placed in contact with a heat bath or particle reservoir at temperature T and chemical potential μ.
   (a) What is the expected number of particles, $\langle N \rangle$, in the system?
   (b) What is the expected fluctuation, $(\Delta N)^2 = \langle N^2 \rangle - \langle N \rangle^2$, in the particle number?

3.26. Show that the grand canonical averages of the particle number, total energy, and pressure of a relativistic electron gas (spin $\frac{1}{2}$) are given by:

$$N = 8\pi \frac{m^3 c^3}{h^3} V \int \frac{\sinh^2 \theta \cosh \theta \, d\theta}{e^{\beta(mc^2 \cosh \theta - \mu)} + 1}$$

$$E = 8\pi \frac{m^4 c^5}{h^3} V \int \frac{\sinh^2 \theta \cosh^2 \theta \, d\theta}{e^{\beta(mc^2 \cosh \theta - \mu)} + 1}$$

$$p = \frac{8\pi}{3} \frac{m^4 c^5}{h^3} \int \frac{\sinh^4 \theta \, d\theta}{e^{\beta(mc^2 \cosh \theta - \mu)} + 1}.$$

Evaluate these quantities in the $T \to 0$ limit.

3.27. Consider a system in which the density of states $D(\epsilon)$ of the electrons is

$$D(\epsilon) = \begin{cases} D, & \epsilon > 0 \\ = 0, & \epsilon < 0. \end{cases}$$

Calculate the Fermi energy. What is condition for the system to be highly degenerate? Find the low temperature form for the specific heat.

3.28. Calculate the specific heat of a gas of neutrino–antineutrino pairs. (The total number of pairs is not constrained.)

3.29. Consider a free Bose gas in which the particle spectrum is given by

$$\epsilon = ap^s$$

where s is a positive number. Determine for which values of s the system undergoes Bose–Einstein condensation in three dimensions, two dimensions, and four dimensions. Also show that one has for this system:

$$p = \frac{s}{d}\frac{E}{V}$$

for dimensionality d.

3.30. For the cases in Problem 3.29 where Bose–Einstein condensation does occur, find out whether the specific heat C_V (not its derivative) is continuous at the transition.

3.31. Show, by breaking the integral up into small and large x-regimes, that

$$I(\bar{\mu}) = \int_0^{+\infty} dx\, x^{1/2} \left(\frac{1}{e^{x+\bar{\mu}} - 1} - \frac{1}{e^x - 1} \right).$$

$$= -\pi \bar{\mu}^{1/2} + \mathcal{O}(\bar{\mu}).$$

3.32. Consider an ideal Bose gas composed of particles with internal degrees of freedom. Assume that these are just two internal energy levels, $\epsilon_0 = 0$ and ϵ_1. Determine the Bose–Einstein condensation temperature for the case

$$\frac{\epsilon_1}{k_B T_0} \gg 1.$$

3.33. (a) Calculate the grand potential Ω for a two-dimensional ideal Bose gas ($\epsilon = p^2/2m$). In two dimensions the *volume* is replaced by the area, $V = L^2$.

(b) Find the average number of particles per unit area as a function of T and μ.

(c) Show that there is no Bose–Einstein condensation at nonzero temperatures for a two-dimensional ideal Bose gas.

(d) Discuss why dimensionality makes a difference in the occurrence of this phase transition.

3.34. Find the value of the Bose–Einstein condensation transition temperature, T_{BEC}, for a set of nonrelativistic particles (mass $= m$) in four dimensions. Show that T_{BEC} can be expressed simply in terms of the *zero-point energy*,

$$\epsilon_0 = \frac{\hbar^2}{2ml^2}$$

where the length l is related to the particle density by

$$nl^4 = 1.$$

3.35. Consider the expresssion for the thermodynamic potential for a system of trapped particles given by Eq. (1094). In the isotropic limit show that the sum over **n** can be converted to a single sum (this amounts to evaluating the degeneracy of each energy state) as given by Eq. (1095).

PROBLEMS 257

3.36. Show, starting with Eq. (1098), giving the number of particles in a harmonic trap in terms of the chemical potential and temperature, that one can find the *chemical potential* ϵ explicitly in terms of temperature and the number of trapped particles in the high-temperature limit.

3.37. Show that the boson multiparticle states,

$$|n\rangle = \frac{1}{\sqrt{n!}}(a^+)^n|0\rangle,$$

are indeed orthonormal:

$$\langle m|n\rangle = \delta_{m,n}.$$

3.38. If we define a set of two-particle states as in Eq. (2738), show that

$$\langle \mathbf{r}_1 \mathbf{r}_2 | 1_i 1_l \rangle = \phi_l(\mathbf{r}_1)\phi_i(\mathbf{r}_2) + \eta\phi_i(\mathbf{r}_1)\phi_l(\mathbf{r}_2)$$

where $\eta = 1$ for bosons and $\eta = -1$ for fermions.

4 Statistical Mechanics of Fluids

4.1 IDEAL GASES WITH INTERNAL DEGREES OF FREEDOM

In Chapters 1 and 3 we discussed the statistical mechanics of noninteracting (ideal) point particles. In this chapter we extend the discussion to interacting particles as well as particles with structure. In both cases we must look more closely at the forces acting in the system.

4.1.1 Internal Partition Function

We begin with a general discussion of noninteracting molecules or atoms. This is a large and well-studied topic and we will be satisfied with studying the simplest models. Let us first recall the analysis in Chapter 1 for ideal point particles. The Hamiltonian for the ideal gas of particles with mass m is simply the kinetic energy:

$$H = \sum_{i=1}^{N} \frac{\mathbf{p}_i^2}{2m}. \tag{1103}$$

Since the Hamiltonian for an ideal gas has the special property that it is a sum of single-particle Hamiltonians, we can directly evaluate the grand partition function in the form

$$Z_G = \exp\left(\frac{V e^{\beta \mu}}{\lambda^3}\right) \tag{1104}$$

where $\lambda = \hbar(2\pi\beta/m)^{1/2}$ is the thermal de Broglie wavelength. In the more general situation of a system of noninteracting molecules or atoms, the Hamiltonian can be written

$$H = \sum_{i=1}^{N} H_i^m, \tag{1105}$$

where H_i^m is the Hamiltonian describing the ith molecule or atom in the system. More generally, an atom is built up from the protons, neutrons, and electrons and a molecule will be comprised of a number of atoms. Thus we must include in the associated Hamiltonian the kinetic energies of all the particles and their mutual interactions. Then, for example, there must be attractive interactions between the atoms forming a molecule or it would dissociate (break up). The grand partition function can be written as

$$Z_G = \sum_{N=0}^{\infty} \frac{1}{N!} \prod_{i=1}^{N} \text{Tr}_i^m e^{-\beta(H-\mu N)} \tag{1106}$$

where Tr_i^m is the sum over all the degrees of freedom for the ith molecule or atom. Inserting the additive form for the Hamiltonian given by Eq. (1105) into Eq. (1106), we have

$$\begin{aligned} Z_G &= \sum_{N=0}^{\infty} \frac{1}{N!} \prod_{i=1}^{N} \text{Tr}_i^m e^{-\beta\left(\sum_{j=1}^{N} H_j^m - \mu N\right)} \\ &= \sum_{N=0}^{\infty} \frac{e^{\beta\mu N}}{N!} (\text{Tr}^m e^{-\beta H^m})^N = \exp(e^{-\alpha} Z_1^m) \end{aligned} \tag{1107}$$

where $\alpha = -\beta\mu$ and

$$Z_1^m = \text{Tr}^m e^{-\beta H^m} \tag{1108}$$

is the one-molecule or one-atom partition function. The grand potential is given in this case by

$$\Omega = -\beta^{-1} \ln Z_G = -\beta^{-1} e^{-\alpha} Z_1^m. \tag{1109}$$

The resulting average number of molecules is

$$\bar{N} = \frac{\partial}{\partial\alpha} (\beta\Omega) = e^{-\alpha} Z_1^m, \tag{1110}$$

while the average energy is

$$E = \frac{\partial}{\partial\beta} (\beta\Omega)\bigg|_{\alpha} = -e^{-\alpha} \frac{\partial}{\partial\beta} Z_1^m \tag{1111}$$

or, eliminating $e^{-\alpha}$ using Eq. (1110),

$$E = -\bar{N} \frac{\partial}{\partial\beta} \ln Z_1^m. \tag{1112}$$

We have reduced the problem to an evaluation of the partition function associated with the internal degrees of freedom of the particles of interest. This can be a

4.1.2 Equipartition Theorem

The simplest classical model for a molecule is to assume a *dumbbell* model of s point masses connected by springs. In this case the Hamiltonian can be written in the general form

$$H_M = \sum_{i=1}^{s} \frac{\mathbf{p}_i^2}{2m_i} + \sum_{i=1}^{r} \frac{K_i}{2} q_i^2 \tag{1113}$$

where m_i is the mass of atom i and K_i the effective spring constant associated with the ith normal-mode coordinate q_i. The value of r, the number of internal normal modes, follows from the fact that there are a total of $3s$ position coordinates, but three of these degrees of freedom must be associated with the translational motion of the center of mass. Since, due to translational invariance, H_m does not depend on the center-of-mass coordinates, $r \leq 3s - 3$. The internal partition function is given by

$$Z_1^m = \int \frac{d^3 p_1}{(2\pi\hbar)^3} \cdots \frac{d^3 p_s}{(2\pi\hbar^3)} \, dq_1 \cdots dq_r \, d^3 R \, e^{-\beta H_M} \tag{1114}$$

where $\mathbf{R}$ represents the center-of-mass coordinate. We can easily evaluate the integrals in Eq. (1114) to obtain

$$Z_1^m = \lambda^{-3s} V \prod_{i=1}^{r} \left(\frac{2\pi}{\beta K_i} \right)^{1/2} = \beta^{-(r+3s)/2} C \tag{1115}$$

where C is independent of temperature. We obtain immediately for the average energy

$$E = -\bar{N} \frac{\partial}{\partial \beta} \ln \beta^{-(r+3s)/2} = \frac{\bar{N}(r + 3s)}{2} k_B T. \tag{1116}$$

The specific heat is given by

$$C_V = \left(\frac{\partial E}{\partial T} \right)_{N, V} = \frac{r + 3s}{2} k_B \bar{N}. \tag{1117}$$

For every degree of freedom that is quadratic in the Hamiltonian, we see that we pick up a contribution $\bar{N}(kT)/2$ to the energy and $\bar{N}(k_B/2)$ to the specific heat. This result is known as the equipartition theorem [1]. We should remember that this is a classical physics result that is correct in the classical or high-temperature regime.

4.1.3 Quantum Mechanical Rotation and Vibration

A more realistic model for diatomic molecules is that of a quantum mechanical vibrating rigid rotor. At a physical level this model for a diatomic molecule seems more or less obvious. However, as discussed in Section 4.9.4, justification of this model at the atomic level of two nuclei surrounded by sets of electrons is not trivial. This is the first of a number of examples where we replace a microscopic model (two nuclei and their surrounding electrons) with a more macroscopic model (vibrating–rotating dumbbell). This general process, called coarse graining, is discussed in a more organized way below.

This model assumes that the molecule is in its electronic ground state and that vibrational and rotational motions of the nuclei are uncoupled. The quantized single-particle energy states associated with rigid-body rotations is given by

$$\epsilon_R = \frac{l(l+1)\hbar^2}{2I_0} \tag{1118}$$

where $l = 0, 1, 2, \ldots$ is the rotational quantum number, I_0 is the nuclear moment of inertia, and the rotational states have degeneracy $2l + 1$. The vibrational energy levels are quantized and given by

$$\epsilon_v = \left(\frac{1}{2} + n\right)\hbar\omega_0 \tag{1119}$$

where $n = 0, 1, 2, \ldots$ is the vibrational quantum number, ω_0 is the vibrational frequency, and the levels are nondegenerate. Typical values of $k_B\Theta_v \equiv \hbar\omega_0$ and $k_B\Theta_r = \hbar^2/2I_0$ are listed in Table 4.1, where $\Theta_v \gg \Theta_r$.

In the case of homonuclear diatomic molecules, we must treat the quantum statistics associated with the indistinguishability of like nuclei. We discuss the ramifications of these additional constraints below. For the case of unlike nuclei, the molecular partition function can be written in the product form

$$Z_1^M = Z_T Z_n Z_v Z_r, \tag{1120}$$

where Z_T is the usual contribution from the center-of-mass translational degrees of freedom

$$Z_T = V\lambda^{-3}, \tag{1121}$$

where the thermal de Broglie wavelength is given by

$$\lambda = \hbar\left(\frac{2\pi}{k_B T \mu}\right)^{1/2} \tag{1122}$$

TABLE 4.1 Characteristic Temperatures for Diatomic Molecules Θ_r and Θ_v and Dissociation Temperatures Θ_d

Molecule	Θ_r (K)	Θ_v (K)	Θ_d (K)
H_2	85.3	6215	52,000
D_2	42.7	4394	—
Cl_2	0.351	808	29,000
Br_2	0.116	463	—
I_2	0.0537	308	—
O_2	2.07	2256	59,000
N_2	2.88	3374	113,000
CO	2.77	3103	98,000
NO	2.45	2719	61,000
HCl	15.02	4227	—
HBr	12.02	3787	—
HI	9.06	3266	—
Na_2	0.221	229	—
K_2	0.081	133	—

Source: Θ_r, Θ_v adapted from Ref. 2, p. 95; Θ_d from Ref. 3.

with μ is the reduced mass of the molecule. Z_n is the nuclear contribution, which, for unlike nuclei, can be taken to be a constant. Z_v is the contribution to the partition function from the vibrational degrees of freedom:

$$Z_v = \sum_{m=0}^{\infty} e^{-(1/2+m)\beta\hbar\omega_0} = \left[\sinh\left(\frac{\beta\hbar\omega_0}{2}\right)\right]^{-1}. \qquad (1123)$$

Finally, we have the rotational contribution,

$$Z_r = \sum_{l=0}^{\infty} (2l+1) \exp\left[\frac{-l(l+1)\beta\hbar}{2I_0}\right]. \qquad (1124)$$

In general, we cannot write down a closed expression for the sum. If the energy-level spacing is small compared to $k_B T$, we can replace the sum by an integral,

$$Z_r = \int_0^{+\infty} dl(2l+1) e^{-l(l+1)\hbar^2/2I_0} = \frac{T}{\Theta_r} \qquad (1125)$$

where $\Theta_r = \hbar^2/2I_0 k_B$ is the rotational temperature. In Problem 4.6 we show how we can systematically expand Z_r given by Eq. (1124) in powers of Θ_r/T.

The product of the results from Eqs. (1121), (1123), and (1124) give the molecular partition function, which goes into Eq. (1112) to produce the total energy,

$$E = E_T + E_v + E_r \qquad (1126)$$

where E_T is the translational contribution found previously,

$$E_v = \bar{N} \frac{\partial}{\partial \beta} \ln\left[\sinh\left(\frac{\beta \hbar \omega_0}{2}\right)\right] \qquad (1127)$$

is the vibrational contribution, and

$$E_r = -\bar{N} \frac{\partial}{\partial \beta} \ln\left[\sum_{l=0}^{\infty} (2l+1) e^{-l(l+1)\beta\hbar^2/2I_0}\right] \qquad (1128)$$

is the rotational contribution to the energy.

From these results we can work out the vibrational and rotational contributions to the heat capacity,

$$C_V^{(v)} = \left(\frac{\partial E_v}{\partial T}\right)_V \qquad (1129)$$

and

$$C_V^{(r)} = \left(\frac{\partial E_r}{\partial T}\right)_V. \qquad (1130)$$

We easily find for the vibrational contribution,

$$\frac{C_V^{(v)}}{Nk_B} = f_v(t_v) = \frac{1}{t_v^2 \sinh^2(1/t_v)}, \qquad (1131)$$

where $t_v = 2k_B T/\hbar\omega_0$ is the dimensionless temperature appropriate for this quantity. $f_v(t_v)$ is plotted in Fig. 4.1. The rotational contribution is more complicated but has the form $C_V^{(r)}/Nk_B = f_r(t_r)$, where the relevant dimensionless rotational temperature is $t_r = T/\Theta_r$. $f_r(t_r)$ is also plotted in Fig. 4.1. Notice that the rotational contribution approaches its high-temperature limit from above. This is observed experimentally. The high- and low-temperature limits of these quantities can be checked analytically. In the low-temperature regime we have the analytical results

$$\frac{C_V^{(v)}}{Nk_B} = \frac{4}{t_v^2} e^{-2/t_v} \qquad (1132)$$

while

$$\frac{C_V^{(r)}}{Nk_B} = \frac{12}{t_r^2} e^{-2/t_r} \qquad (1133)$$

IDEAL GASES WITH INTERNAL DEGREES OF FREEDOM   265

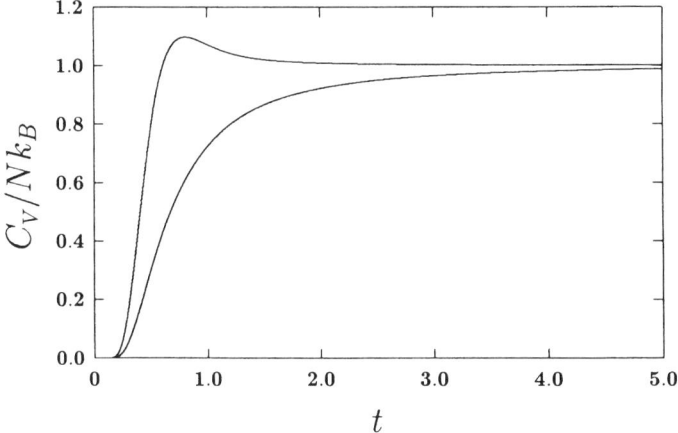

FIGURE 4.1 Vibrational (bottom curve) and rotational (top curve) contributions to the specific heat versus the appropriate scaled temperature (see the text) for diatomic molecules.

and the specific heats go exponentially to zero for low temperatures, as shown in the figure.

The full specific heat is the sum of translational, rotational, and vibrational components:

$$\frac{C_V}{Nk_B} = f_T\left(\frac{T}{T_{\text{BE}}}\right) + f_r\left(\frac{T}{T_r}\right) + f_v\left(\frac{T}{T_v}\right) \tag{1134}$$

where the translational contribution is discussed in Section 3.13.1 and the other contributions are shown in Fig. 4.1. The various characteristic temperatures, including the dissociation temperature, are shown in Table 4.1. Clearly, the various contributions turn on at very well separated temperatures. In Table 4.2 we show the transition from $C_V/k_B \approx 2.5$ to 3.5 in H_2 as a function of temperature.

At sufficiently high temperatures we see that vibrational and rotational degrees of freedom contribute equally to E, and the results are in agreement with the classical equipartition theorem. Each contribution to specific heat in the classical high-temperature limit can be treated as coming from a quadratic term in the Hamiltonian.

TABLE 4.2 Crossover for Specific Heat from Translational and Rotational Regime to Include Vibrational Contribution for H_2

$T(K)$	C_V/k_B observed	$T(K)$	C_V/k_B observed
600	2.56	1600	2.91
800	2.63	1800	2.98
1000	2.70	2000	3.05
1200	2.77	2500	3.23

Source: Data from Ref. 4.

The six $(s = 2)$ kinetic energy terms are distributed among three center-of-mass terms and two rotational and one vibrational kinetic energy term. There is a single potential energy contribution $(r = 1)$ associated with vibrations. The sum of contributions gives the specific heat

$$\frac{C}{N} = \frac{7}{2} k_B \qquad (1135)$$

in agreement with the equipartition theorem and the high-temperature limit of our quantum mechanical result.

There has been extensive comparison with experiment of this theory. Figure 4.2 shows a comparison of experiment with the theoretical predictions for f_v for a variety of systems. We see that the theory works well.

Let us turn to the interesting case of homonuclear molecules. The key point here is that due to quantum statistics, the *total* wavefunction for the molecule must be either symmetric or antisymmetric under the interchange of the two identical nuclei. It must be symmetric if the nuclei are bosons or antisymmetric if the nuclei are fermions. If the nucleus has spin I, then I is integral for bosons and half-integral for fermions. Part of the total wavefunction can be written as a product of translational, vibrational, and electronic degrees of freedom, which are typically symmetric under nuclear exchange. This is true, for example, if the electronic ground state is symmetric. However, it is easy to see that the rotational contribution is symmetric for even orbital quantum number l and antisymmetric for odd l. One must then combine the nuclear and rotational degrees of freedom to obtain a contribution to the wavefunction that gives the correct symmetry for a given pair l and I.

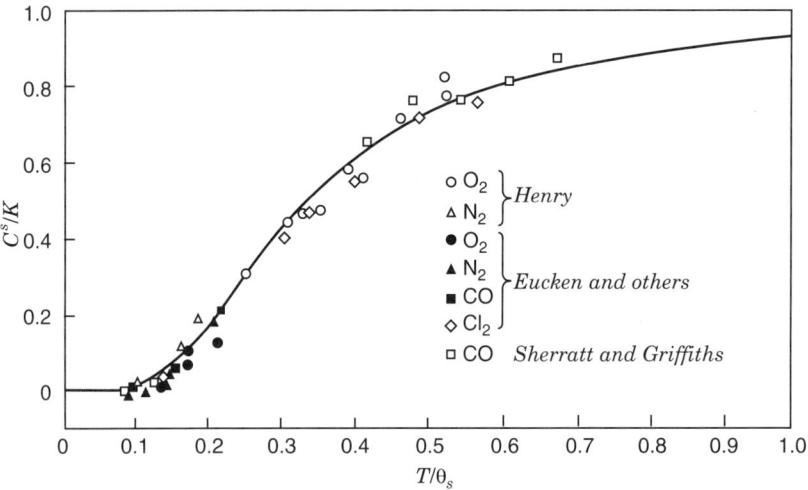

FIGURE 4.2 Vibrational heat capacity of idatomic molecules. (From R.H. Fowler and E.A. Guggenheim, *Statistical Thermodynamics*, Cambridge University Press, Cambridge, 1949.)

For the simplest case of $I = \frac{1}{2}$, we know that there are three symmetric nuclear wavefunctions describing the pair of nuclei that we must pair with odd values of l to obtain an overall antisymmetric wavefunction. There is one antisymmetric nuclear wavefunction that we pair with even l. More generally, one can show [5] that for integral I, $I(2I+1)$ antisymmetric nuclear spin functions couple to odd l, and $(I+1)(2I+1)$ symmetric nuclear spin function couple to even values of l. For half-integral I, $I(2I+1)$ antisymmetric nuclear spin functions couple with even l and $(I+1)(2I+1)$ symmetric nuclear spin functions couple with odd l.

The key point coming out of this analysis is that one cannot generally decouple the rotational and nuclear degrees of freedom. Thus we must replace $Z_r Z_n \to Z_{r,n}$, where for nuclei with half-integer spin

$$Z_{r,n} = I(2I+1) \sum_{l \text{ even}} (2l+1) \exp\left[\frac{-\Theta_r l(l+1)}{T}\right]$$
$$+ (I+1)(2I+1) \sum_{l \text{ odd}} (2l+1) \exp\left[\frac{-\Theta_v l(l+1)}{T}\right], \quad (1136)$$

and for integer spin

$$Z_{r,n} = (I+1)(2I+1) \sum_{l \text{ even}} (2l+1) \exp\left[\frac{-\Theta_r l(l+1)}{T}\right] \quad (1137)$$
$$+ I(2I+1) \sum_{l \text{ odd}} (2l+1) \exp\left[\frac{-\Theta_r l(l+1)}{T}\right]. \quad (1138)$$

Typically, at room temperature, we can use the approximate results:

$$\sum_{l \text{ even}} (2l+1) \exp\left[\frac{-\Theta_r l(l+1)}{T}\right] = \sum_{l \text{ odd}} (2l+1) \exp\left[\frac{-\Theta_r l(l+1)}{T}\right] = \frac{T}{2\Theta_r} \quad (1139)$$

which is $\frac{1}{2} Z_r$, given by Eq. (1125). Therefore, when these expressions can be replaced by their high-temperature limits, we have

$$Z_{r,n} = \frac{1}{2} Z_r Z_n \quad (1140)$$

where $Z_n = (2I+1)^2$ and the partition function does factorize. As usual, the quantum statistics comes in at low (relative to Θ_r) temperatures. Practically, these effects are most pronounced in hydrogen, where $\Theta_r = 85.3$ K. Historically, these quantum effects in H_2 were important. Initially, the quantum theory for the specific heat that follows from Eq. (1136) with $I = \frac{1}{2}$ was not observed. The reason for this

268 STATISTICAL MECHANICS OF FLUIDS

disagreement was that there is a long-lived metastable persistence of the high-temperature equilibrium ratio (3:1) of *ortho*-hydrogen (odd nuclear wavefunction) and *para*-hydrogen (even nuclear wavefunction). After ensuring equilibration, Bonhoeffer and Harteck [6] confirmed the quantum theory experimentally.

One can go on and discuss the thermodynamics of polyatomic systems [7] and the breakdown of the rotor–vibration model at high temperatures. We return to this question in Section 4.10. For a discussion of the long history of the development of the theory of specific heats of gases, see Brush [8].

4.2 AVERAGES OVER QUANTITIES DEPENDING ONLY ON MOMENTUM

Let us consider a classical system described by the Hamiltonian

$$H = \sum_{i=1}^{N} \frac{\mathbf{p}_i^2}{2m} + V_T \tag{1141}$$

where V_T is the interaction between the particles and depends only on the positions of the particles. Suppose that we are interested in the average of an observable that depends only on the momenta of the particles:

$$\langle A \rangle = \text{Tr}_{\text{cl}} P_G A \tag{1142}$$

where $A = A\{\mathbf{p}_i\}$. Averages of this type are simple because A does not depend on the positions of the particles. If A depends on the positions of the particles, we must then treat the very complicated interactions occurring via V_T as discussed below. Let us assume for simplicity that A is a sum of momentum-dependent quantities from each particle:

$$A = \sum_{i=1}^{N} \sigma(\mathbf{p}_i). \tag{1143}$$

A simple example would be the kinetic energy, where

$$\sigma(p) = \frac{p^2}{2m}. \tag{1144}$$

It turns out to be very convenient to rewrite A as

$$A = \int d^3p\, \sigma(\mathbf{p}) f(\mathbf{p}) \tag{1145}$$

where

$$f(\mathbf{p}) \equiv \sum_{i=1}^{N} \delta(\mathbf{p} - \mathbf{p}_i) \tag{1146}$$

is a *momentum density*. Then, if we can calculate the average of $f(\mathbf{p})$, we can determine the average of any quantity of the form given by Eq. (1143) simply by carrying out a three-dimensional integral over $\mathbf{p}$. Let us analyze the grand canonical ensemble average of $f(\mathbf{p})$:

$$\bar{f}(\mathbf{p}) \equiv \langle f(\mathbf{p}) \rangle = \frac{\sum_{N=0}^{\infty}(1/N!h^{3N}) \int dq_N e^{-\beta(H-\mu N)} f(\mathbf{p})}{\sum_{N=0}^{\infty}(1/N!h^{3N}) \int dq_N e^{-\beta(H-\mu N)}}. \tag{1147}$$

If we introduce the canonical partition function for N particles,

$$Z_N = \frac{1}{\lambda^{3N} N!} \int d^3 r_1 \cdots d^3 r_N e^{-\beta V_T} \tag{1148}$$

we can rewrite Eq. (1147) in the form

$$\bar{f}(p) = \frac{\sum_{N=0}^{\infty}(\lambda^3 e^{\beta\mu}/h^3)^N Z_N \int d^3 p_1 \cdots d^3 p_N e^{-\beta \sum_{j=1}^{N} p_j^2/2m} \sum_{i=1}^{N} \delta(\mathbf{p} - \mathbf{p}_i)}{\sum_{N=0}^{\infty}(\lambda^3 e^{\beta\mu}/h^3)^N Z_N \int d^3 p_1 \cdots d^3 p_N e^{-\beta \sum_{j=1}^{N} p_j^2/2m}}. \tag{1149}$$

If we carry out all of the momentum integrations in the denominator and all but the ith integral in the numerator, using

$$\frac{1}{h^3} \int d^3 p_i e^{-\beta p_i^2/2m} = \frac{1}{h^3} (2\pi m \beta^{-1})^{3/2} = \lambda^{-3}, \tag{1150}$$

we obtain

$$\bar{f}(p) = \frac{\sum_{N=0}^{\infty} e^{\beta\mu N} Z_N (\lambda^3/h^3) \sum_{i=1}^{N} \int d^3 p_i e^{-\beta p_i^2/2m} \delta(\mathbf{p} - \mathbf{p}_i)}{\sum_{N=0}^{\infty} e^{\beta\mu N} Z_N}. \tag{1151}$$

We see that the $\mathbf{p}_i$ integral gives N terms proportional to $e^{-\beta p^2/2m}$, so

$$\bar{f}(p) = e^{-\beta p^2/2m} \left(\frac{\lambda}{h}\right)^3 \frac{\sum_{N=0}^{\infty} e^{\beta\mu N} Z_N N}{\sum_{N=0}^{\infty} e^{\beta\mu N} Z_N}. \tag{1152}$$

It is easy to see that the average particle number for this system (we can do all momentum integrals directly) is given by

$$\bar{N} = \frac{\sum_{N=0}^{\infty} e^{\beta\mu N} Z_N N}{\sum_{N=0}^{\infty} e^{\beta\mu N} Z_N}, \tag{1153}$$

so Eqs. (1152) and (1153) give

$$\bar{f}(p) = \bar{N} \frac{e^{-\beta p^2/2m}}{(2\pi m \beta^{-1})^{3/2}}. \tag{1154}$$

Then, for example, the average kinetic energy is given by

$$\langle K \rangle = \int d^3 p \frac{p^2}{2m} \bar{f}(p) = \bar{N} \frac{3}{2} k_B T \tag{1155}$$

as we expect. The average kinetic energy per particle is $3k_B T/2$ for classical systems even if the particles interact. Looking at Eq. (1154), we see that $\bar{f}(\mathbf{p})/N$ can be identified with the probability distribution of momenta for individual particles. Written in terms of particle velocities, we have

$$f(\mathbf{v}) = \left(\frac{m}{2\pi k_B T}\right)^{3/2} e^{-\beta m \mathbf{v}^2/2} \tag{1156}$$

which is the famous Maxwell velocity distribution law [9]. This holds for classical particles independent of the nature of the interactions.

4.3 MAYER CLUSTER EXPANSION

4.3.1 Partition Function for Classical Interacting Fluids

The real challenge in statistical mechanics is to treat interacting systems. This is the situation where one particle feels the presence of others. Once we allow the particles to interact, our problem becomes not only nontrivial, but in most cases, insoluble. In most physically interesting situations in statistical mechanics, we cannot solve for the partition function exactly. We have to resort to approximation techniques. These usually involve some type of perturbation theory.

Our first example of an interacting system will be a strongly interacting single-component classical gas: for example, neon. We choose neon because it is a spherical atom that interacts with short-range forces. Consequently, neon gas looks quite a bit like a collection of billiard balls. We assume that the Hamiltonian for neon

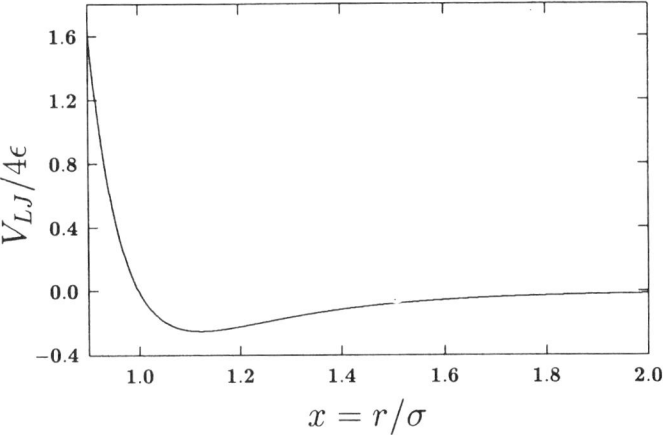

FIGURE 4.3 Scaled Lennard-Jones potential in terms of scaled particle separation.

TABLE 4.3 Lennard-Jones Potential Parameters for Noble Gases

Gas	ϵ/k_B K	σ (Å)
He	10.8	2.57
Ne	36.3	2.81
Ar	121.85	3.429
Kr	174.68	3.684
Xe	224.83	4.067

Source: Data from Ref. [10].

is of the form of Eq. (1141) with potential energy

$$V_T = \frac{1}{2} \sum_{i \neq j = 1}^{N} V(\mathbf{r}_i - \mathbf{r}_j) \tag{1157}$$

that is pairwise additive. There are two popular representations for the potential function in noble gases such as neon. The Lennard-Jones potential [11] (see Fig. 4.3) has the form

$$V_{LJ}(r) = 4\epsilon \left[\left(\frac{\sigma}{r}\right)^{12} - \left(\frac{\sigma}{r}\right)^{6} \right] \tag{1158}$$

where the parameters are given for the noble gases in Table 4.3. If we neglect the attractive part of the potential and take the billiard ball analogy seriously, we obtain the hard-sphere potential

$$V_{HS}(r) = \begin{cases} 0 & r > r_0 \\ \infty & r < r_0. \end{cases} \quad (1159)$$

The hard-sphere potential is obviously an oversimplification of the real world, but it serves as a decent first approximation to the real potential in understanding many properties of fluids.

Our goal is to compute the grand potential

$$\Omega = -\beta^{-1} \ln Z_G \quad (1160)$$

where the grand partition is given by

$$Z_G = \sum_{N=0}^{\infty} \int \frac{dq_N}{N! h^{3N}} e^{-\beta(H-\mu N)} \quad (1161)$$

and the Hamiltonian is given above. We note from our discussion in the preceding section that we can immediately perform the momentum integrals using Eq. (1150) to obtain

$$Z_G = \sum_{N=0}^{\infty} \frac{e^{\beta \mu N}}{N! \lambda^{3N}} \int d^3 r_1 \cdots d^3 r_N e^{-\beta V_T}. \quad (1162)$$

It is then convenient to define

$$Q_N = \int d^3 r_1 \cdots d^3 r_N e^{-\beta V_N} \quad (1163)$$

and the *fugacity*

$$z = \frac{e^{\beta \mu}}{\lambda^3} \quad (1164)$$

so that

$$Z_G = \sum_{N=0}^{\infty} \frac{z^N}{N!} Q_N. \quad (1165)$$

A little reflection on the structure of the Q_N, for large N, will convince us that we cannot evaluate them explicitly. In Fig. 4.4 circles are drawn representing the interaction range for that particle. If the circles representing two particles overlap,

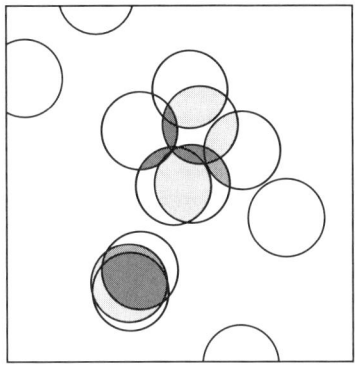

FIGURE 4.4 Internested configuration constructed by sequentially randomly depositing hard disks on a surface. Gray areas correspond to overlap of two disks. Black regions correspond to overlap of three disks.

they are interacting. The overlapping regions are hatched. In this configuration we see that many of the particles are communicating with each other via their neighbors, which interact with their neighbors, and so on. The point is that Q_N will be an enormously complicated internested multiple integral. In the noninteracting case, none of the particles *overlap*, the multiple integrals uncouple, and each integral in Eq. (1163) can be performed for each particle, giving a factor of the volume

$$Q_N = V^N \tag{1166}$$

and

$$Z_G = \exp(zV). \tag{1167}$$

Thus we regain the ideal gas result

$$n = \frac{e^{\beta\mu}}{\lambda^3} = z. \tag{1168}$$

When we turn on interactions, $n \neq z$. Consider the situation where we have interactions but the density of particles is low. Quantitatively, low densities correspond to

$$\frac{\sigma}{l} \ll 1 \tag{1169}$$

where σ is the range of the interaction between particles (on the order of 5 Å) and $l = n^{-1/3}$ is the average distance between particles. Physically we expect, for $\sigma/l \ll 1$, that the system will look very much like an ideal gas since very few of the particles, on average, are interacting with other particles. Thus, to a reasonably good approximation, $n \approx z$, and z can be taken to be small. This suggests that for low-density systems we can compute corrections to the ideal gas law by expanding in powers of z. In other words, we expect that the density and energy density can be expanded in a power series in z with the leading terms in the power series giving the ideal gas results:

$$n = z + n^{(2)}z^2 + \mathcal{O}(z^3) \tag{1170}$$

$$\epsilon = \frac{E}{V} = \frac{3}{2}k_B T z + \epsilon^{(2)}z^2 + \mathcal{O}(z^3). \tag{1171}$$

The coefficients $n^{(2)}$ and $\epsilon^{(2)}$ give the corrections to the ideal gas law.

4.3.2 Linked Cluster Expansions

Although it may be clear that we should *expand in powers of z*, it is not so clear what we should expand in z. We note immediately that the grand partition function is already expressed as a power series in z given by Eq. (1165). We see, however, that in the case of an ideal gas, the coefficient of z^N in this expansion is $V^N/N!$ and the series itself is of the form

$$Z_G = 1 + zV + \frac{z^2 V^2}{2} + \cdots \tag{1172}$$

If this expansion is to make sense, the terms should become progressively smaller. Thus we require, comparing the second and third terms, that

$$\frac{z^2 V^2}{2zV} = \frac{1}{2}zV \ll 1. \tag{1173}$$

We see, however, that we cannot satisfy this inequality in the thermodynamic limit (V large). The key point here is that in general Z_G has a rather complicated volume dependence. We know, however, the volume dependence of the thermodynamic potential, Ω: It is an extensive variable,

$$\Omega = f(\beta, \mu)V, \tag{1174}$$

in the thermodynamic limit. It then makes sense to expand Ω in powers of z, being careful to keep track of the single power of the volume. We assume that we can write Ω as a Taylor series,

$$\Omega = -\beta^{-1} \sum_{l=0}^{\infty} \frac{z^l}{l!} B_l \tag{1175}$$

so that the coefficients are given by

$$B_l = \lim_{z \to 0} \frac{\partial^l}{\partial z^l}(-\beta\Omega). \tag{1176}$$

From our discussion above we expect that each contribution B_l is proportional to the volume. If we remember, using Eq. (1165), that

$$-\beta\Omega = \ln \sum_{N=0}^{\infty} \frac{z^N}{N!} Q_N, \qquad (1177)$$

then

$$B_l = \lim_{z \to 0} \frac{\partial^l}{\partial z^l} \ln \sum_{N=0}^{\infty} \frac{z^N}{N!} Q_N. \qquad (1178)$$

Let us look at a few of these coefficients. Since $Q_0 = 1$,

$$B_0 = \ln Q_0 = \ln 1 = 0. \qquad (1179)$$

Next we have

$$\begin{aligned} B_1 &= \lim_{z \to 0} \frac{\partial}{\partial z} \ln \sum_{N=0}^{\infty} \frac{z^N}{N!} Q_N \\ &= \lim_{z \to 0} \frac{\sum_{N=1}^{\infty} (N/N!) z^{(N-1)} Q_N}{\sum_{N=1}^{\infty} (z^N/N!) Q_N} = Q_1, \end{aligned} \qquad (1180)$$

and

$$\begin{aligned} B_2 &= \lim_{z \to 0} \frac{\partial^2}{\partial z^2} \ln \sum_{N=1}^{\infty} \frac{z^N}{N!} Q_N \\ &= \lim_{z \to 0} \frac{\partial}{\partial z} \left(\frac{1}{\sum_{N=0}^{\infty} (z^N Q_N/N!)} \sum_{N=0}^{\infty} \frac{N z^{N-1} Q_N}{N!} \right) \\ &= \lim_{z \to 0} \left[\frac{\sum_{N=2}^{\infty} [N(N-1)/N!] z^{N-2} Q_N}{\sum_{N=0}^{\infty} (z^N Q_N/N!)} - \left(\sum_{N=1}^{\infty} \frac{N}{N!} z^{N-1} Q_N \bigg/ \sum_{N=0}^{\infty} \frac{z^N Q_N}{N!} \right)^2 \right] \\ &= Q_2 - Q_1^2. \end{aligned} \qquad (1181)$$

Before looking at the higher-order terms, let us look more closely at these first few terms. We remember from Eq. (1163) that the Q's determining B_1 and B_2 are

$$Q_1 = \int d^3 r_1 e^{-\beta V_1} \qquad (1182)$$

276 STATISTICAL MECHANICS OF FLUIDS

and
$$Q_2 = \int d^3r_1 d^3r_2 e^{-\beta V_2}. \tag{1183}$$

Clearly, $V_1 = 0$, since a particle does not interact with itself, and
$$V_2 = V(\mathbf{r}_1 - \mathbf{r}_2). \tag{1184}$$

We have immediately then that
$$Q_1 = \int d^3r_1 = V. \tag{1185}$$

There should be no notational confusion between the volume and the potential since the pair potential is a function of distance. Next we have
$$Q_2 = \int d^3r_1 d^3r_2 e^{-\beta V(\mathbf{r}_1 - \mathbf{r}_2)}. \tag{1186}$$

If we change to center-of-mass coordinates in the integral, we obtain
$$Q_2 = \int d^3R\, d^3r\, e^{-\beta V(r)} \tag{1187}$$
$$= V \int d^3r\, e^{-\beta V(r)}. \tag{1188}$$

A key point here is that for large r, $V(r)$ goes to zero and
$$e^{-\beta V(\mathbf{r})} \to 1. \tag{1189}$$

This means that the integral
$$\int d^3r\, e^{-\beta V(r)}, \tag{1190}$$

depends on the size of the system — we cannot simply let the volume become infinite. We can see how the volume dependence enters by writing
$$\int d^3r\, e^{-\beta V(r)} = \int d^3r(1 + e^{-\beta V(r)} - 1)$$
$$= V + \int d^3r(e^{-\beta V(r)} - 1). \tag{1191}$$

Since $e^{-\beta V(r)} \to 1$ as $r \to \infty$, the remaining integral exists in the limit $V \to \infty$ if $V(r) \to 0$ fast enough (this is quantified below). We then have that

$$Q_2 = V^2 + V \int d^3 r (e^{-\beta V(r)} - 1) \qquad (1192)$$

and we see the development of the complicated volume dependence discussed above. Let us now move over to the B's, which we expect to be better behaved (and linear in the volume). We have

$$B_1 = Q_1 = V \qquad (1193)$$

while

$$\begin{aligned} B_2 &= Q_2 - Q_1^2 \\ &= V^2 + V \int d^3 r (e^{-\beta V(r)} - 1) - V^2 \\ &= V \int d^3 r (e^{-\beta V(r)} - 1). \end{aligned} \qquad (1194)$$

We see that B_1 and B_2 are indeed linear in the volume as long as the integral in B_2 is finite in the infinite volume limit. The function we integrate over in B_2 plays an important part in our discussion and warrants its own symbol:

$$f(r) \equiv e^{-\beta V(r)} - 1. \qquad (1195)$$

We have plotted $f(r)$ in Fig. 4.5 for a Lennard-Jones potential. A convenient property of $f(r)$ is that it is well behaved even for a hard-sphere interaction:

$$f_{\text{HS}} = \begin{cases} -1 & r < r_0 \\ 0 & r > r_0. \end{cases} \qquad (1196)$$

Clearly, the integral over $f(r)$ is well defined as long as $V(r)$ falls off faster than $1/r^3$ for large r. In particular, for a hard-sphere interaction

$$\begin{aligned} \int d^3 r f(r) &= 4\pi \int_0^\infty r^2 \, dr f(r) \\ &= 4\pi \int_0^{r_0} r^2 dr (-1) = -\frac{4\pi r_0^3}{3} \end{aligned} \qquad (1197)$$

and

$$B_2 = -\frac{4\pi r_0^3}{3} V. \qquad (1198)$$

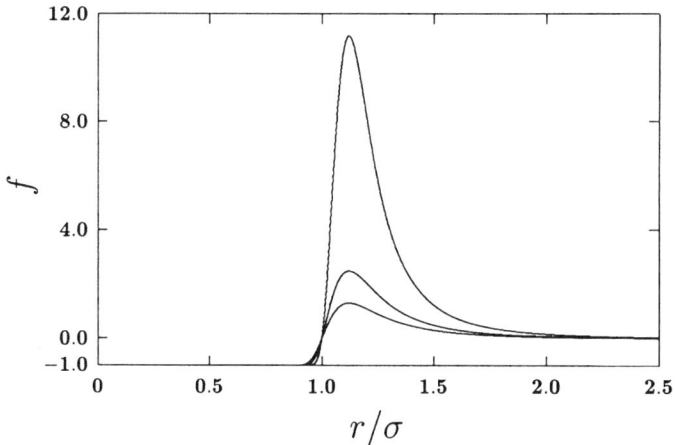

FIGURE 4.5 f function versus scaled particle separation for a Lennard-Jones potential and different scaled temperatures: $k_BT = 0.4\epsilon$ (top curve), $k_BT = 0.8\epsilon$ (middle curve), $k_BT = 1.2\epsilon$ (bottom curve).

We see then that B_2 has the appropriate volume dependence. Later we will investigate the physical consequences of B_2 (which controls the first density correction to the equation of state); for now we want to continue our analysis of the structure of the B_l's.

Let us next investigate B_3. We return to our general Eq. (1178) for the B's and obtain

$$B_3 = \lim_{z \to 0} \left[\frac{1}{Z} \frac{\partial^3 Z}{\partial z^3} - \frac{3}{Z^2} \frac{\partial Z}{\partial z} \frac{\partial^2 Z}{\partial z^2} + 2 \frac{1}{Z^3} \left(\frac{\partial Z}{\partial z} \right)^3 \right]. \tag{1199}$$

Using Eq. (1165), we easily find that

$$\lim_{z \to 0} \frac{\partial^l Z}{\partial z^l} = Q_l \tag{1200}$$

and

$$B_3 = Q_3 - 3Q_2 Q_1 + 2Q_1^3. \tag{1201}$$

If we insert our explicit expressions for the Q's, given by Eq. (1163), we can write

$$B_3 = \int d^3 r_1 d^3 r_2 d^3 r_3 \left(e^{-\beta V_3} - 3 e^{-\beta V_2} + 2 \right) \tag{1202}$$

where

$$V_3 = \frac{1}{2} \sum_{i \neq j=1}^{3} V(\mathbf{r}_i - \mathbf{r}_j)$$
$$= V(\mathbf{r}_1 - \mathbf{r}_2) + V(\mathbf{r}_1 - \mathbf{r}_3) + V(\mathbf{r}_2 - \mathbf{r}_3). \qquad (1203)$$

In writing out B_3 it is convenient to divide the three factors of $e^{-\beta V_2}$ between the three possible interactions and write

$$B_3 = \int d^3 r_1 d^3 r_2 d^3 r_3 \left[e^{-\beta [V(\mathbf{r}_1 - \mathbf{r}_2) + V(\mathbf{r}_1 - \mathbf{r}_3) + V(\mathbf{r}_2 - \mathbf{r}_3)]} - e^{-\beta V(\mathbf{r}_1 - \mathbf{r}_2)} \right.$$
$$\left. - e^{-\beta V(\mathbf{r}_1 - \mathbf{r}_3)} - e^{-\beta V(\mathbf{r}_2 - \mathbf{r}_3)} + 2 \right]. \qquad (1204)$$

It will again be convenient to introduce the quantity

$$f_{ij} = e^{-\beta V(\mathbf{r}_i - \mathbf{r}_j)} - 1 \qquad (1205)$$

or

$$e^{-\beta V(\mathbf{r}_i - \mathbf{r}_j)} = 1 + f_{ij}, \qquad (1206)$$

and express B_3 in terms of integrals over products of f's:

$$B_3 = \int d^3 r_1 d^3 r_2 d^3 r_3 (f_{12} f_{13} f_{23} + f_{12} f_{13} + f_{12} f_{23} + f_{13} f_{23}). \qquad (1207)$$

The first point we can establish is that B_3 is proportional to the volume. Consider first the term

$$I_1 = \int d^3 r_1 d^3 r_2 d^3 r_3 f_{12} f_{13} f_{23}. \qquad (1208)$$

Make the change of variables to

$$\mathbf{x}_2 = \mathbf{r}_2 - \mathbf{r}_1 \qquad (1209)$$

and

$$\mathbf{x}_3 = \mathbf{r}_3 - \mathbf{r}_1. \qquad (1210)$$

FIGURE 4.6 Basic building block for diagrams. A circle is drawn for each particle that is assigned a label. A bond is shown between particle 1 and particle 2.

Then

$$I_1 = \int d^3r_1 d^3x_2\, d^3x_3 f(\mathbf{x}_2) f(\mathbf{x}_3) f(\mathbf{x}_2 - \mathbf{x}_3)$$

$$= V \int d^3x_2 d^3x_3 f(\mathbf{x}_2) f(\mathbf{x}_3) f(\mathbf{x}_2 - \mathbf{x}_3) \qquad (1211)$$

and the remaining integrals converge as $V \to \infty$. We can show in a similar fashion that all the remaining integrals for B_3 have the same linear volume dependence.

There is a convenient graphical representation for the various contributions to the B's. We begin with B_3. Each contribution to B_3 is in the form of an integral over the positions of three particles. Let us set down the following conventions. Let us draw a circle corresponding to each of the particles and label them with a particle number. That is, we represent particle 1 by 1 inside a circle. If, in a contribution, particles are connected by an f, we draw a line or bond between the particles (see Fig. 4.6). The four contributions to B_3 can be represented graphically as shown in Fig. 4.7.

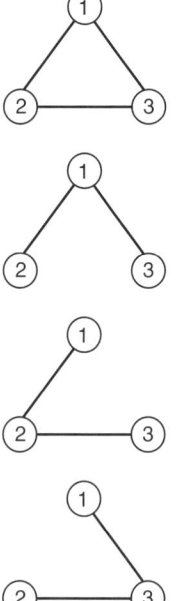

FIGURE 4.7 Connected three-particle graphs.

We should contrast these *graphs* with those that contribute to Q_3:

$$Q_3 = \int d^3r_1 \, d^3r_2 \, d^3r_3 e^{-\beta V_3}$$
$$= \int d^3r_1 d^3r_2 d^3r_3 (1+f_{12})(1+f_{13})(1+f_{23}). \quad (1212)$$

All of the terms contributing to B_3 contribute to Q_3 and we have the additional contributions shown in Fig. 4.8. The key point is that these remaining graphs contain particles that are not tied or bonded to the other particles. A graph containing a circle not connected to the other particles is called a disconnected graph. We easily see that Q_3 is the sum of all three-particle graphs with all possible bonding between the particles. We also see that B_3 is the sum of all connected three-particle graphs. We can now understand that it is the disconnected graphs that occur in the Q's that lead to the associated complicated volume-dependence problem. It is clear that each disconnected particle in a diagram will contribute a factor of the volume to Q. In B_3 we have only connected graphs. In this case we pick up only one factor of the volume corresponding to the center-of-mass coordinate for the connected *cluster* of three particles.

We can now see how things go in the N-particle case. In treating Q_N we will have N integrations over N particles, which we can label from 1 *to* N. These integrations are over $e^{-\beta V_N} = e^{-\beta \sum_{i<j} V(\mathbf{r}_i - \mathbf{r}_j)}$. If we use Eq. (1206), then

$$e^{-\beta V_N} = \prod_{i<j} (1+f_{ij}). \quad (1213)$$

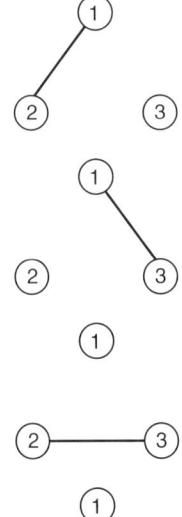

FIGURE 4.8 Disconnected three-particle graphs.

It is clear that this product generates $2^{N(N-1)/2}$ terms. We can represent each of these terms by a graph. In each graph we draw a circle for each particle and label it. We then draw lines corresponding to each f_{ij} bonding particles i and j. It is clear then that Q_N is the sum of all distinct N-particle graphs corresponding to all possible bonding configurations. The linear volume dependence of the B_N results since it includes only the connected diagrams contributing to Q_N.

As a somewhat nontrivial exercise, we can consider the four-particle case. We expect that there are a total of $2^{4(4-1)/2} = 64$ graphs contributing to Q_4. We can categorize the various graphs according to the number of bonds and whether it is connected or disconnected (see Fig. 4.9). When we count these graphs, we find a total of 64 graphs. The assertion then is that only the connected graphs contribute to B_4. This can (as discussed in Problem 4.12) be shown to follow from our general expression for the B_l's:

$$B_4 = \lim_{z \to 0} \frac{\partial^4}{\partial z^4} \ln \sum_{N=0}^{\infty} \frac{z^N}{N!} Q_N. \tag{1214}$$

We are now in a position to summarize: The grand potential can be conveniently expressed as the power series expansion

$$\Omega = -\beta^{-1} V \sum_{l=1}^{\infty} z^l b_l, \tag{1215}$$

where

$$b_l \equiv \lim_{V \to \infty} \frac{1}{l!} \frac{B_l}{V} \tag{1216}$$

and the B_l's can be calculated using

$$B_l = \lim_{z \to 0} \frac{\partial^l}{\partial z^l} \sum_{N=0}^{\infty} \frac{z^N}{N!} Q_N \tag{1217}$$

or by noting that B_l = sum of all connected l-particle graphs. (Remember that with our conventions a contribution from a graph corresponds to writing down all factors of f_{ij} corresponding to a bond between particles i and j and then integrating these factors over all l-particle coordinates in an l-particle graph.)

A number of ideas have been introduced in this section. The idea of using graphs to label contributions in a perturbation theory expansion has played a central role in quantum field theory (Feynman graphs) and lattice models (as discussed in Section 5.9). The idea that one need only evaluate linked or connected graphs is crucial and developed in this context by Joseph Mayer and collaborators [12]. The ideas of connectivity have been developed more generally beyond perturbation theory by Kubo [13] in his treatment of *cumulants*.

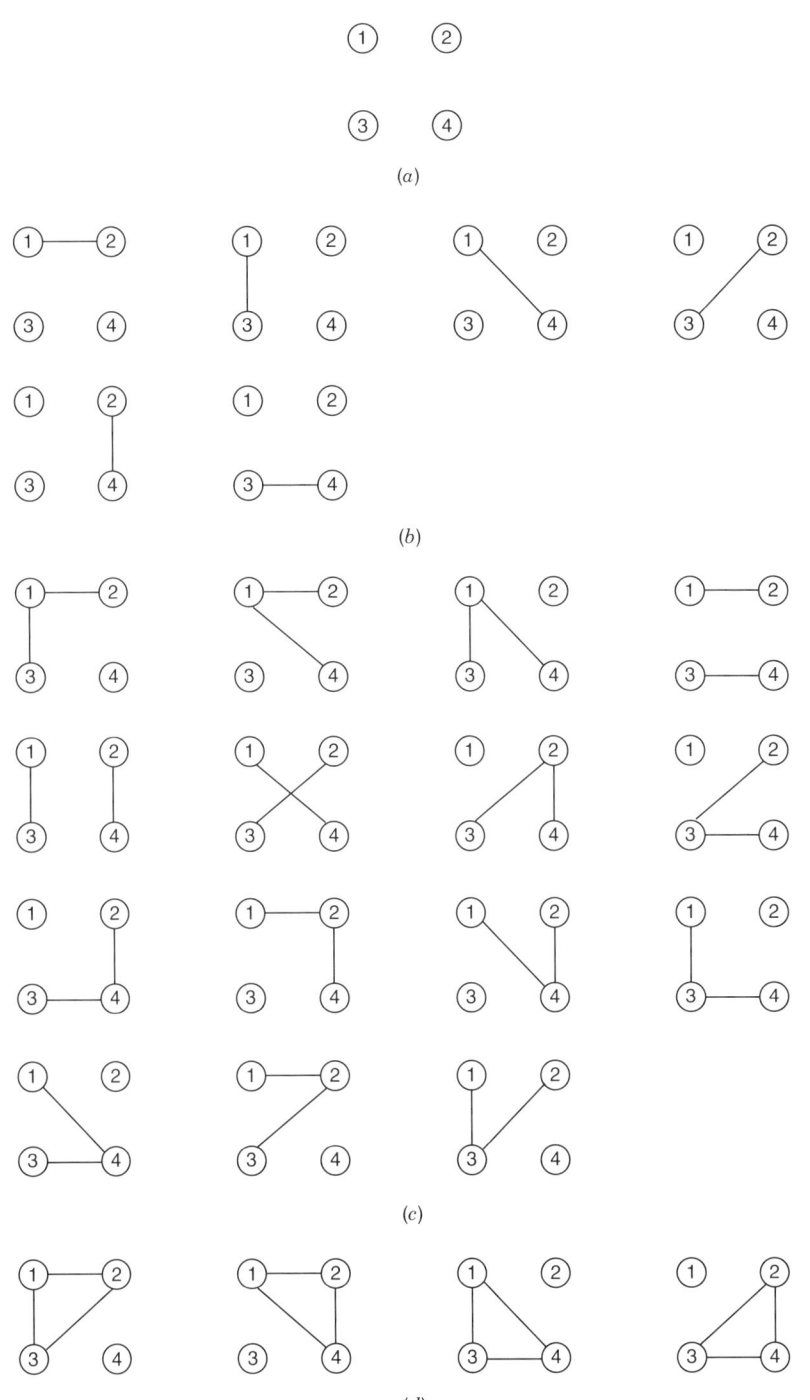

FIGURE 4.9 Full set of four-particle graphs: (*a*) no bonds, disconnected; (*b*) one bond, disconnected; (*c*) two bonds, disconnected; (*d*) three bonds, disconnected; (*e*) three bonds, connected; (*f*) four bonds, connected; (*g*) five bonds, connected; (*h*) six bonds, connected.

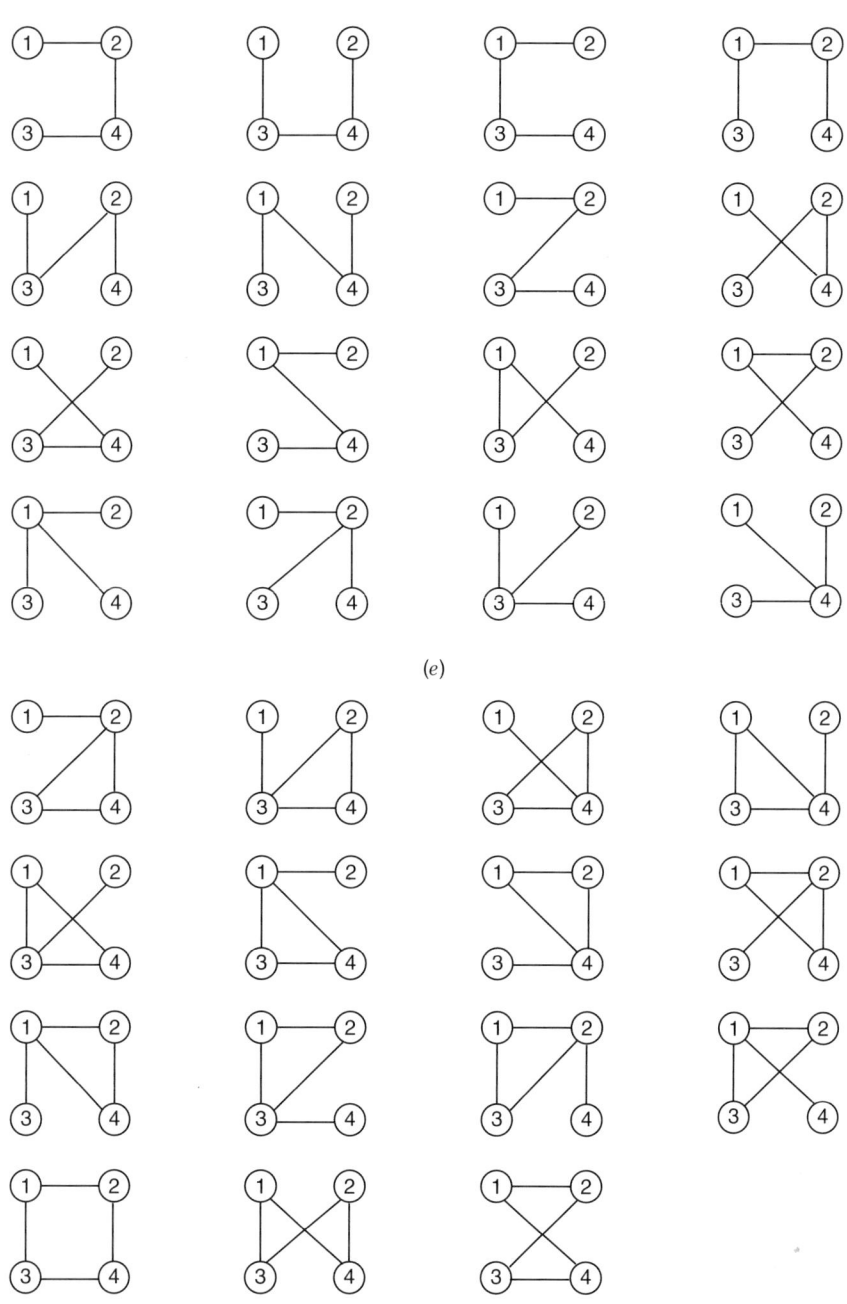

(e)

(f)

FIGURE 4.9 (*Continued*)

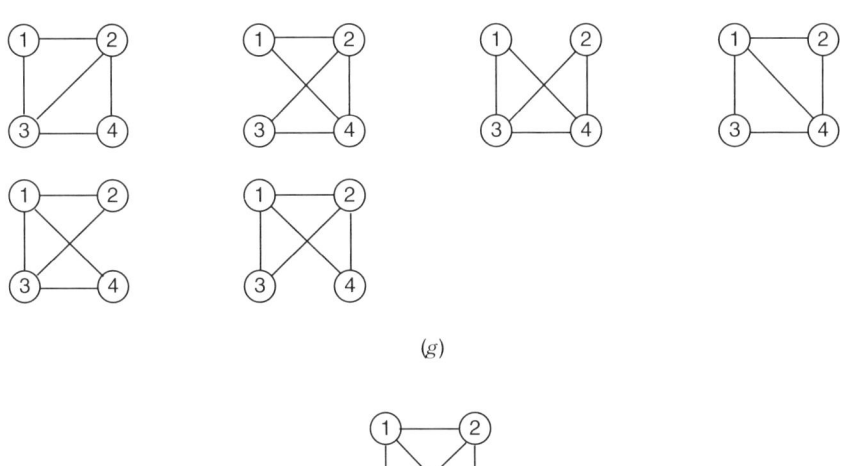

(g)

(h)

FIGURE 4.9 (*Continued*)

4.4 EQUATION OF STATE

If we can evaluate the B_l's (which are a function of temperature and the parameters characterizing the interparticle potential), we have the thermodynamic potential $\Omega(T, \mu, V)$ as a power series in z, and from this we can calculate the other thermodynamic quantities via

$$pV = -\Omega \tag{1218}$$

$$E = \frac{\partial}{\partial \beta}(\beta\Omega) \tag{1219}$$

$$\bar{N} = \frac{\partial}{\partial \alpha}(\beta\Omega) \tag{1220}$$

and

$$TS = E - \mu\bar{N} - \Omega. \tag{1221}$$

We easily obtain, remembering $b_0 = 0$, from Eq. (1215),

$$p\beta = \sum_{l=1}^{\infty} b_l z^l, \tag{1222}$$

and taking derivatives,

$$\bar{N} = V \sum_{l=1}^{\infty} l b_l z^l, \qquad (1223)$$

and, remembering $z = e^{-\alpha}/\lambda^3$,

$$E = -\frac{\partial}{\partial \beta}\left(V \sum_{l=1}^{\infty} b_l z^l\right)$$
$$= \frac{3}{2} k_B T V \sum_{l=1}^{\infty} l b_l z^l + k_B T^2 V \sum_{l=1}^{\infty} z^l \frac{db_l(T)}{dT}. \qquad (1224)$$

If we use Eq. (1223) for N, the average energy can be written as

$$E = \frac{3}{2}\bar{N} k_B T + k_B T^2 V \sum_{l=2}^{\infty} z^l \frac{db_l(T)}{dT}, \qquad (1225)$$

where we used the fact $b_1 = 1$, so $\partial b_1/\partial T = 0$. Clearly, the correction to the ideal gas contribution is given by the second term in Eq. (1225). We then have the contributions $p = p(T, \mu), n = n(T, \mu)$, and $E = E(T, \mu, V)$. It is more convenient to eliminate μ in favor of the density to obtain $p = p(T, n)$ and $E/V = \epsilon(T, n)$. We can evaluate μ, or equivalently z, in terms of n by inverting Eq. (1223), with $n = \bar{N}/V$, to obtain

$$z = \sum_{j=1}^{\infty} a_j n^j. \qquad (1226)$$

We can obtain the a_i's by substituting for z in Eq. (1223):

$$n = \sum_{l=1}^{\infty} l b_l \left(\sum_{j=1}^{\infty} a_j n^j\right)^l$$
$$= b_1(a_1 n + a_2 n^2 + a_3 n^3 + \cdots)$$
$$+ 2b_2(a_1 n + a_2 n^2 + a_3 n^3 + \cdots)^2$$
$$+ 3b_3(a_1 n + a_2 n^2 + a_3 n^3 + \cdots)^3 + \cdots, \qquad (1227)$$

and then equating the coefficients of like powers of n. The coefficients of n^1 through n^4 give the equations

$$1 = b_1 a_1 \qquad (1228)$$
$$0 = b_1 a_2 + 2b_2 a_1^2 \qquad (1229)$$
$$0 = a_3 b_1 + 4b_2 a_1 a_2 + 3b_3 a_1^3 \qquad (1230)$$
$$0 = a_4 b_1 + 2a_2^2 b_2 + 4a_3 a_1 b_2 + a_1^2 a_2 b_3 + 4a_1 b_4. \qquad (1231)$$

EQUATION OF STATE

These can easily be solved to obtain

$$a_1 = \frac{1}{b_1} = 1 \tag{1232}$$

$$a_2 = -2b_2 \tag{1233}$$

$$a_3 = -3b_3 + 8b_2^2 \tag{1234}$$

and

$$a_4 = 4b_4 - 40b_2^3 + 30b_2 b_3. \tag{1235}$$

We can then substitute Eq. (1226) into the pressure equation, Eq. (1222), to obtain

$$\begin{aligned} p\beta &= \sum_{l=1}^{\infty} b_l \left(\sum_{j=1}^{\infty} a_j n^j \right)^l \\ &= b_1(a_1 n + a_2 n^2 + a_3 n^3 + \cdots) \\ &\quad + b_2(a_1 n + a_2 n^2 + a_3 n^3 + \cdots)^2 \\ &\quad + b_3(a_1 n + a_2 n^2 + a_3 n^3 + \cdots)^3 + \cdots \end{aligned} \tag{1236}$$

Clearly, this can be rewritten in the form

$$p\beta = \sum_{l=1}^{\infty} c_l n^l, \tag{1237}$$

which, for historical reasons, is known as the *virial expansion* [14] and the c_l as the virial coefficients. Looking at Eq. (1236), we can easily identify

$$c_1 = b_1 a_1 \tag{1238}$$

$$c_2 = b_1 a_2 + b_2 a_1^2 \tag{1239}$$

$$c_3 = b_1 a_3 + 2b_2 a_1 a_2 + b_3 a_1^3. \tag{1240}$$

Inserting the expressions for the a's in terms of the b's, we obtain

$$c_1 = 1 \tag{1241}$$

$$c_2 = -b_2 \tag{1242}$$

$$c_3 = 4b_2^2 - 2b_3. \tag{1243}$$

288 STATISTICAL MECHANICS OF FLUIDS

We can then write the equation of state in the form

$$p = nk_BT + k_BT \sum_{l=2}^{\infty} c_l n^l \tag{1244}$$

and the c's for $l > 1$ give the nonideal gas corrections.

4.5 LOW-DENSITY LIMIT

The Mayer cluster expansion described in Section 4.3.2 is valid for low-density systems. In this section we investigate the first-order corrections to the ideal gas law and the range of validity of the low-density theory.

4.5.1 First-Density Correction

We can easily see from the results of Section 4.4 that first density corrections to the ideal gas expressions are given by

$$p = nk_BT + n^2 k_BT[-b_2(T)] \tag{1245}$$

$$\frac{E}{V} = \frac{3}{2} nk_BT + k_BT^2 n^2 \frac{\partial b_2(T)}{\partial T} + \mathcal{O}(n^3) \tag{1246}$$

and

$$z = n - 2b_2(T)n^2 + \mathcal{O}(n^3). \tag{1247}$$

We see immediately that the interesting information about ideal gas corrections is contained in $b_2(T)$.

The first thing we can do with these equations is to use the equations relating z and n to find the chemical potential as a function of T and n. Using Eq. (1168), we can invert Eq. (1247) to obtain

$$\mu = \beta^{-1}\ln n\lambda^3 + \beta^{-1}\ln(1 - 2nb_2). \tag{1248}$$

Since $2b_2 n \ll 1$ by hypothesis, we can use

$$\ln(1 + x) = x + \mathcal{O}(x^2), \tag{1249}$$

to obtain

$$\mu = \beta^{-1}\ln n\lambda^3 - 2\beta^{-1}b_2 n + \cdots \tag{1250}$$

The entropy can be found using the Euler relation:

$$S = \frac{1}{T}(E - \mu \bar{N} + pV)$$

$$= \bar{N}k_B \left[\frac{3}{2} + nT\frac{\partial b_2(T)}{\partial T}\right] - \bar{N}k_B(\ln n\lambda^3 - 2b_2 n) + k_B\bar{N}(1 - b_2 n)$$

$$= \bar{N}k_B \left[\frac{5}{2} - \ln n\lambda^3 + n\frac{\partial}{\partial T}(Tb_2(T))\right]. \quad (1251)$$

We see then that all the corrections to the ideal gas expressions for p, E, and S can be expressed in terms of b_2. We have discussed a number of times the set of *second derivatives* that arise in the Gibbs ensemble: C_p the specific heat at constant pressure, α the coefficient of thermal expansion, and κ_T the isothermal compressibility. The detailed evaluation of the first density corrections for C_p, α, and κ_T is discussed in Problem 4.15, where it is found that

$$C_p = Nk_B\left[\frac{5}{2} + nT^2\frac{d^2 b_2(T)}{dT^2}\right] \quad (1252)$$

$$\alpha = \frac{1}{T}\left\{1 + n\left[b_2(T) - T\frac{db_2(T)}{dT}\right]\right\} \quad (1253)$$

$$\kappa_T = \frac{1}{nk_B T}[1 + 2nb_2(T)]. \quad (1254)$$

Since we have expressed all corrections in terms of b_2, we should investigate b_2 and its temperature dependence.

4.5.2 Temperature Dependence of $b_2(T)$

To gain a feeling for the temperature dependence of $b_2(T)$ and its dependence on the form of the interaction, let us look at a particular interaction potential:

$$V(r) = \begin{cases} \infty & r < r_0 \\ -u_0\left(\frac{r_0}{r}\right)^m & r > r_0, \end{cases} \quad (1255)$$

which is known as a *generalized Sutherland potential*. This interaction is clearly an oversimplification of the potential for real fluids. However, it does contain the essential features of a harsh repulsion at short distances and a power law attraction at large distances. We now compute

$$b_2 = \lim_{V\to\infty} \frac{1}{2}\frac{1}{V} B_2 = \lim_{V\to\infty} \frac{1}{2}\frac{1}{V} \int d^3r_1 d^3r_2 f_{12}$$

$$= \frac{1}{2}\int d^3r f(r) = \frac{1}{2}(4\pi)\int_0^{+\infty} r^2 dr(e^{-\beta V(r)} - 1)$$

$$= 2\pi\left[\int_0^{r_0} r^2 dr(-1) + \int_{r_0}^\infty r^2 dr(e^{\beta u_0 (r_0/r)^m} - 1)\right]$$

$$= 2\pi\left[-\frac{r_0^3}{3} + \int_{r_0}^\infty r^2 dr(e^{\beta u_0(r_0/r)^m} - 1)\right]. \tag{1256}$$

Consider the integral involving the attractive part of the potential:

$$I = \int_{r_0}^\infty r^2 dr\left(e^{\beta u_0(r_0/r)^m} - 1\right)$$

$$= \int_{r_0}^\infty r^2 dr \sum_{N=1}^\infty \frac{1}{N!}(\beta u_0)^N \left(\frac{r_0}{r}\right)^{mN}$$

$$= \sum_{N=1}^\infty \frac{(\beta u_0)^N}{N!} r_0^3 \int_1^\infty dx\, x^{2-mN}, \tag{1257}$$

where we changed integration variables to $x = r/r_0$. For the integral to converge for $N = 1$, we require that $m > 3$ (remember that for the Lennard-Jones potential at large distances, $m = 6$), so

$$I = r_0^3 \sum_{N=1}^\infty \frac{(\beta u_0)^N}{N!(mN-3)} \tag{1258}$$

and

$$b_2(T) = 2\pi r_0^3\left[-\tfrac{1}{3} + f_m(\beta u_0)\right] \tag{1259}$$

where f_m is the dimensionless function,

$$f_m(x) = \sum_{N=1}^\infty \frac{x^N}{N!(mN-3)}. \tag{1260}$$

We see that the temperature enters b_2 only through the dimensionless parameter βu_0. We have plotted $b_2(T)/|b_2^{hs}|$ versus βu_0 in Fig. 4.10 for the physical case $m = 6$. $|b_2^{hs}| = 2\pi r_0^3/3$ is the magnitude of the hard-sphere result for b_2. If it were not for the attractive part of the interaction, this ratio would be -1.

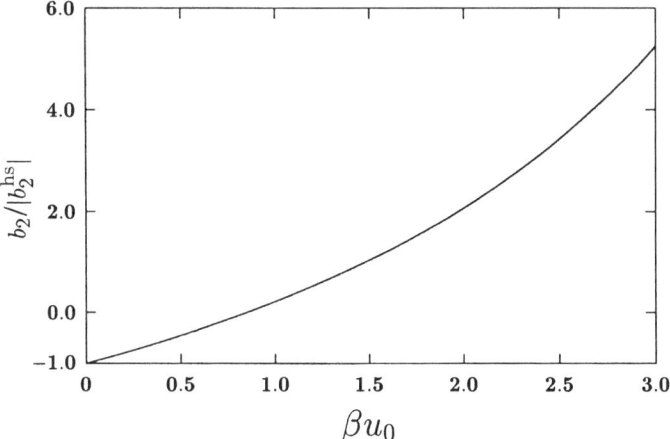

FIGURE 4.10 Second virial coefficient, scaled by the hard-sphere expression, versus the strength of the attractive part of the interaction potential measured relative to the temperature. The special potential given by Eq. (1255) with $m=6$ has been used in calculating $b_2(T)$.

The physics going into b_2 is relatively clear. The short-range repulsive forces (due to nonzero volume of a particle) tend to make b_2 negative, whereas the attractive forces [as manifested through $f_m(x)$] tend to make b_2 positive. At high temperatures $\beta u_0 \ll 1$ and the attractive part of the interaction is unimportant. Thus a hard-sphere model is appropriate for a high-temperature fluid. From a physical point of view, this says that at high temperatures the particles, on average, are moving very fast and will not sample the attractive well. They will, however, experience the harsh hard core independent of the amount of kinetic energy they carry. As we lower the temperature and $x = \beta u_0$ becomes larger, we see that $f_m(x)$ increases in magnitude rapidly with x. In Fig. 4.10 we see that at $\beta u_0 \approx 0.854$, $b_2(T) = 0$ for $m = 6$, and for lower temperatures b_2 is positive and the attractive force dominates. It is worth pointing out that the hard-sphere contribution to b_2 is temperature independent. This remains true for all the b_l's. The virial coefficients for a hard-sphere fluid are independent of temperature. Another approach toward understanding the second virial coefficient is given below in Section 4.7.4, where we discuss the principle of corresponding states.

As a simple example of the physical implications of these results for $b_2(T)$, consider the specific heat C_V given by

$$C_V = \left(\frac{\partial E}{\partial T}\right)_{N,V} = V \frac{\partial}{\partial T}\left[\frac{3}{2}nk_BT + k_BT^2 n^2 \frac{db_2(T)}{dT}\right]$$

$$= \frac{3}{2}\bar{N}k_B + \bar{N}nk_B \frac{d}{dT}\left[T^2 \frac{d}{dT}b_2(T)\right]. \qquad (1261)$$

292 STATISTICAL MECHANICS OF FLUIDS

The contribution to C_V due to interactions is given by

$$C_v^I = \bar{N} k_B n \frac{d}{dT} \left\{ T^2 \frac{d}{dT} 2\pi r_0^3 \left[-\frac{1}{3} + f_m(\beta u_0) \right] \right\}$$
$$= \bar{N} k_B (2\pi n r_0^3) \left(\frac{u_0}{k_B T} \right)^2 \frac{d^2 f_m(x)}{dx^2} \quad (1262)$$

and

$$\frac{d^2 f_m(x)}{dx^2} = \sum_{N=1}^{\infty} \frac{N(N-1) x^{N-2}}{N! (mN - 3)}. \quad (1263)$$

Let us write

$$C_V^I = \bar{N} k_B \frac{2\pi n r_0^3}{3} f_I(\beta u_0). \quad (1264)$$

We have plotted $f_I(x)$ versus x for $m = 6$ in Fig. 4.11. We see that the attractive interaction leads to an increase in C_V, and this increase is enhanced as we lower the temperature.

4.5.3 van der Waals Equation of State

If we assume we are at high temperatures where $\beta u_0 \ll 1$, then

$$f_m(\beta u_0) \approx \frac{\beta u_0}{m - 3} \quad (1265)$$

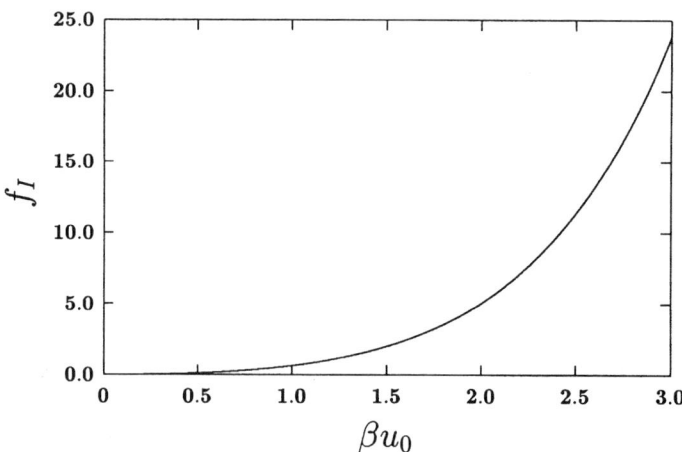

FIGURE 4.11 First density contribution to the specific heat versus inverse temperature using the same potential as in Fig. 4.10. See Eq. (1264) for the definition of f_I.

and Eq. (1259) can be written in the form

$$b_2(T) = -4\gamma + \beta\alpha, \qquad (1266)$$

where γ is the volume associated with a single particle (of radius $r_0/2$),

$$\gamma = \frac{4}{3}\pi\left(\frac{r_0}{2}\right)^3 = \frac{\pi r_0^3}{6} \qquad (1267)$$

and

$$\alpha = \frac{12\gamma u_0}{m-3} \qquad (1268)$$

is associated with the attractive part of the potential. Using these results, we can easily derive the van der Waals equation of state we studied in some detail in Chapter 2. The pressure is given by Eq. (1245) and can be rewritten as

$$p = nk_BT[1 + (4\gamma - \beta\alpha)n] \qquad (1269)$$

or as

$$p + \alpha n^2 = nk_BT(1 + 4\gamma n) \qquad (1270)$$

which for small $4\gamma n$ can be rewritten as

$$p + \alpha n^2 \approx \frac{nk_BT}{1 - 4\gamma n}. \qquad (1271)$$

We can immediately identify the parameters $\bar{a}$ and $\bar{b}$ we introduced in Chapter 2 as

$$\bar{a} = \alpha = \frac{12u_0\gamma}{m-3} \qquad (1272)$$

and

$$\bar{b} = 4\gamma = \frac{2}{3}\pi r_0^3. \qquad (1273)$$

Our derivation of the van der Waals equation of state is complete. We also see clearly the limitation on the range of validity of the phenomenological arguments used to motivate the van der Waals equation of state in Chapter 2.

4.5.4 Range of Validity of the Cluster Expansion

People have struggled very hard [15] to calculate the *virial coefficients* c_l for a hard-sphere system. The results thus far:

$$c_1 = 1 \tag{1274}$$

$$c_2 = \frac{2\pi}{3} r_0^3 \tag{1275}$$

$$c_3 = \frac{5}{8} c_2^2 \tag{1276}$$

$$c_4 = \left\{ \frac{1283}{8960} + \frac{3}{2} \left[\frac{73\sqrt{2} + 1377(\tan^{-1}\sqrt{2} - \pi/4)}{1120\pi} \right] \right\} c_2^3$$
$$= 0.28695 c_2^3 \tag{1277}$$

$$c_5 = (0.1103 \pm 0.003) c_2^4 \tag{1278}$$

$$c_6 = (0.0386 \pm 0.004) c_2^5. \tag{1279}$$

The first four contributions have been determined analytically and are due to Boltzmann [16]. The last two coefficients we calculated numerically using computer Monte Carlo techniques [17]. The increase in effort to obtain *one more term* analytically is enormous. However, even if we can obtain many terms, we expect that our cluster expansion breaks down generally when $1 \approx |b_2 n|$. For hard spheres, where $b_2 = -2\pi r_0^3/3$, the expansion breaks down when $1 \approx \frac{2}{3} \pi n r_0^3$. If we write $n = l^{-3}$, where l is the distance between particles, our condition for the expansion to break down is $1 \approx 2\pi/3 (r_0/l)^3$ or $l \approx 1.3 r_0$. As l shrinks to a size $\approx 2r_0$, we have to add up many terms to obtain convergence. This means that we have to compute b_3, $b_4, \ldots$. Each such calculation requires doing more integrals. Thus, by the time one reaches b_6, we have exhausted our calculational abilities. If $r_0 \approx 1.3 l$, all the terms in the expansion contribute with equal weight and our expansion is of no use. We no longer have a small expansion parameter. A discussion of how we proceed in the absence of a small parameter is given in Section 4.7.

4.6 STRUCTURAL INFORMATION IN FLUIDS

4.6.1 X-ray Scattering

In previous chapters [18] we limited our discussion of physical properties to thermodynamic quantities [i.e., quantities directly derivable from the thermodynamic potential $\Omega(T, \mu, V)$]. There is an entire class of important and interesting quantities that we have ignored. This information concerns the spatial or structural configurations favored in the equilibrium state. What is the average distance between neighbors? A direct and simple way of introducing the idea of structural information is through a discussion of one of the experiments used to obtain this

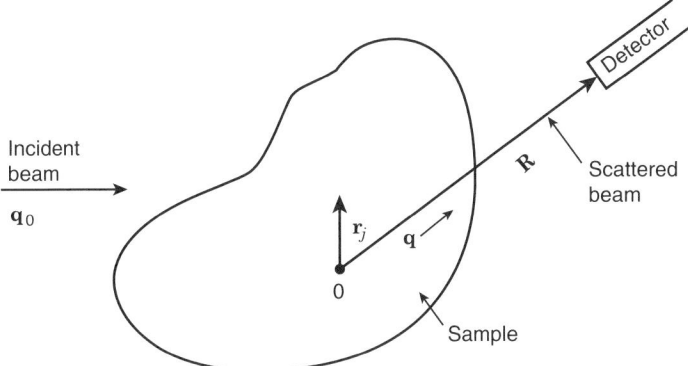

FIGURE 4.12 Schematic of scattering experiment.

information. The classic such experiment is x-ray scattering from a fluid. Our treatment will be somewhat simpleminded, but we don't want to get involved in subtleties here that are beside the main point. Assume that a plane-wave beam of x-rays is incident on a sample as shown in Fig. 4.12. The incident x-rays have wavenumber $\mathbf{q}_0$. The x-rays are scattered by the sample and the scattered beam is collected by a detector. We assume that the detector is at $\mathbf{R}$, with respect to some origin in the sample, and the scattered x-rays have momentum $\mathbf{q}$. A particle at position $\mathbf{r}_i$ in the sample will scatter the incident plane wave. The amplitude of the incident plane wave at $\mathbf{r}_i$ is proportional to $e^{i\mathbf{q}_0 \cdot \mathbf{r}_i}$ and the particle at $\mathbf{r}_i$ scatters the radiation in an outgoing spherical wave centered at $\mathbf{r}_i$. The amplitude of this contribution to the scattered beam at the detector is

$$A_i = \alpha e^{i\mathbf{q}_0 \cdot \mathbf{r}_i} \frac{e^{iq|\mathbf{R}-\mathbf{r}_i|}}{|\mathbf{R}-\mathbf{r}_i|} \tag{1280}$$

where the proportionality constant α, reflecting the interaction of the x-rays with the particles, is the same for each scatterer. Since $|\mathbf{R}| \gg |\mathbf{r}_i|$ we can write to a good approximation

$$|\mathbf{R}-\mathbf{r}_i| = R - \hat{\mathbf{R}} \cdot \mathbf{r}_i \tag{1281}$$

and since $q\hat{\mathbf{R}} = \mathbf{q}$,

$$A_i = \frac{\alpha}{R} e^{iqR} e^{i(\mathbf{q}_0-\mathbf{q}) \cdot \mathbf{r}_i}$$
$$\equiv A_0 e^{+i(\mathbf{q}_0-\mathbf{q}) \cdot \mathbf{r}_i}.$$

If we define the momentum change in the experiment as $\hbar \mathbf{k}$, then

$$\mathbf{k} = \mathbf{q} - \mathbf{q}_0. \tag{1283}$$

296 STATISTICAL MECHANICS OF FLUIDS

It is clear that the contributions to the scattering amplitude from all the particles is given by

$$A = \sum_i A_i = A_0 \sum_i e^{-i\mathbf{k}\cdot\mathbf{r}_i}. \tag{1284}$$

The intensity at the detector is just $|A|^2$. Assuming we collect data over a long time period, the measured intensity is the thermodynamic average and we can write

$$I = \langle |A|^2 \rangle = |A_0|^2 \left\langle \sum_i e^{-i\mathbf{k}\cdot\mathbf{r}_i} \sum_j e^{+i\mathbf{k}\cdot\mathbf{r}_j} \right\rangle. \tag{1285}$$

The important thing to note is that the measured intensity consists of a product of two pieces. The first piece, $|A_0|^2$, is a known factor that depends only on the external experimental parameters (e.g., R, which is a known distance from the sample to the detector). The second factor in the expression for the intensity is much more interesting. It is convenient to write

$$I = |A_0|^2 V \bar{S}(\mathbf{k}), \tag{1286}$$

where

$$\bar{S}(\mathbf{k}) = \frac{1}{V} \left\langle \sum_{i=1}^N e^{-i\mathbf{k}\cdot\mathbf{r}_i} \sum_{j=1}^N e^{+i\mathbf{k}\cdot\mathbf{r}_j} \right\rangle \tag{1287}$$

and we have separated out a factor of the volume V for convenience. $\bar{S}(\mathbf{k})$ is known as the *structure factor* for the fluid and, as we shall find below, contains interesting structural information about the system.

4.6.2 Structure Factor and Radial Distribution Function

If we remember that the particle density at point $\mathbf{x}$ in our sample is given by

$$n(\mathbf{x}) = \sum_{i=1}^N \delta(\mathbf{x} - \mathbf{r}_i), \tag{1288}$$

the Fourier transform of the density can be written as

$$n(\mathbf{k}) = \int \frac{d^3 x}{\sqrt{V}} e^{+i\mathbf{k}\cdot\mathbf{x}} n(\mathbf{x}) = \sum_{i=1}^N \frac{e^{+i\mathbf{k}\cdot\mathbf{r}_i}}{\sqrt{V}} \tag{1289}$$

and we can then easily identify

$$\bar{S}(\mathbf{k}) = \langle n(\mathbf{k})n(-\mathbf{k}) \rangle. \quad (1290)$$

We see that the structure factor is the equilibrium average of the product of two Fourier transforms of the density. It is worth emphasizing that $\bar{S}(k)$ depends only on the properties of the sample and gives us direct information about the fluid. The x-ray scattering parameters enter into $\bar{S}(k)$ only indirectly in that $\hbar\mathbf{k}$ is the momentum transferred in the scattering. We can obtain a more fundamental understanding of $\bar{S}$ if we consider the density–density correlation function

$$S(\mathbf{x}, \mathbf{x}') = \langle \delta n(\mathbf{x}) \delta n(\mathbf{x}') \rangle \quad (1291)$$

where we introduce the *fluctuation* in the density,

$$\delta n(\mathbf{x}) = n(\mathbf{x}) - \langle n(\mathbf{x}) \rangle. \quad (1292)$$

The Fourier transform of $S(\mathbf{x}, \mathbf{x}')$,

$$S(\mathbf{k}) = \int \frac{d^3x\, d^3x'}{V} e^{-i\mathbf{k}\cdot(\mathbf{x}-\mathbf{x}')} S(\mathbf{x}, \mathbf{x}'), \quad (1293)$$

is related to the structure factor by

$$S(\mathbf{k}) = \bar{S}(\mathbf{k}) - n^2 \delta(\mathbf{k}), \quad (1294)$$

where we have used the result that $\langle n(\mathbf{x}) \rangle$ is independent of $\mathbf{x}$ for translationally invariant systems. We will also refer to $S(\mathbf{k})$ as the structure factor, since in a scattering experiment the δ-function terms correspond to forward or no scattering, and outside the beam, $S = \bar{S}$.

An x-ray scattering measurement gives us information about the density fluctuations in a fluid. In particular, $S(\mathbf{x}, \mathbf{x}')$ tells us about the joint probability that if we have a particle at $\mathbf{x}'$, there will be a particle at the point $\mathbf{x}$. We *correlate* particles at point $\mathbf{x}'$ with those at $\mathbf{x}$.

Let us investigate $S(\mathbf{x}, \mathbf{x}')$ more closely. We can write

$$\begin{aligned} S(\mathbf{x}, \mathbf{x}') &= \langle n(\mathbf{x}) n(\mathbf{x}') \rangle - n^2 \\ &= \left\langle \sum_{i=1}^{N} \delta(\mathbf{x} - \mathbf{r}_i) \sum_{j=1}^{N} \delta(\mathbf{x}' - \mathbf{r}_j) \right\rangle - n^2. \end{aligned} \quad (1295)$$

It is conventional to separate out the term $i = j$ from $i \neq j$ terms inside the average to obtain

$$S(\mathbf{x}, \mathbf{x}') = \delta(\mathbf{x} - \mathbf{x}')n + n^2[g(\mathbf{x}, \mathbf{x}') - 1] \quad (1296)$$

where the pair or radial distribution function g is defined, as in Chapter 1, by

$$n^2 g(\mathbf{x}, \mathbf{x}') \equiv \left\langle \sum_{i \neq j=1}^{N} \delta(\mathbf{x} - \mathbf{r}_i) \delta(\mathbf{x}' - \mathbf{r}_j) \right\rangle. \tag{1297}$$

We multiply g by n^2, so that, to lowest order in the density, g is independent of the density. We can understand why $n^2 g(\mathbf{x}, \mathbf{x}')$ is explicitly of order n^2 quite directly. The average in Eq. (1297) is over a quantity involving *at least* two different particles. We see that the average can be written

$$\frac{\sum_{N=2}^{\infty} (z^N/N!) \int d^3 r_1 \cdots d^3 r_N \sum_{i \neq j=1}^{N} \delta(\mathbf{x} - \mathbf{r}_i) \delta(\mathbf{x}' - \mathbf{r}_j) e^{-\beta V_N}}{Z} \tag{1298}$$

where $z = e^{\beta \mu}/\lambda^3$, and it therefore begins, for small z, at $\mathcal{O}(z^2)$. Since $z \approx n$ for low densities, we obtain the desired $\mathcal{O}(n^2)$ dependence. The radial distribution function is a measure of the joint probability that if one finds a particle at $\mathbf{x}'$, one will also find a different particle at $\mathbf{x}$. It plays a fundamental role in the theory of fluids. It not only shows up in an analysis of the structure factor, but also in treatments of thermodynamic properties. We can see this by considering the general equation for the average energy,

$$E = \langle H \rangle = \frac{3}{2} \bar{N} k_B T + \frac{1}{2} \left\langle \sum_{i \neq j=1}^{N} V(\mathbf{r}_i - \mathbf{r}_j) \right\rangle \tag{1299}$$

where we have used the general result for the average kinetic energy we derived earlier. Let us now use the identity

$$\int d^3 x \, d^3 x' V(\mathbf{x} - \mathbf{x}') \delta(\mathbf{x} - \mathbf{r}_i) \delta(\mathbf{x}' - \mathbf{r}_j) = V(\mathbf{r}_i - \mathbf{r}_j) \tag{1300}$$

to rewrite Eq. (1299) in the form

$$E = \frac{3}{2} \bar{N} k_B T + \frac{1}{2} \int d^3 x \, d^3 x' V(\mathbf{x} - \mathbf{x}') \left\langle \sum_{i \neq j=1}^{N} \delta(\mathbf{x} - \mathbf{r}_i) \delta(\mathbf{x}' - \mathbf{r}_j) \right\rangle. \tag{1301}$$

We easily identify the radial distribution function in the interaction term to obtain

$$E = \frac{3}{2} \bar{N} k_B T + \frac{n^2}{2} \int d^3 x \, d^3 x' \, V(\mathbf{x} - \mathbf{x}') g(\mathbf{x}, \mathbf{x}'). \tag{1302}$$

Using the translational invariance in the system

$$g(\mathbf{x}, \mathbf{x}') = g(\mathbf{x} - \mathbf{x}') \tag{1303}$$

we can write

$$E = \frac{3}{2}\bar{N}k_BT + \frac{n^2V}{2}\int d^3r\, V(r)g(r). \tag{1304}$$

We showed in Chapter 1 and Appendix F that the equilibrium pressure can be written as

$$p = nk_BT - \frac{n^2}{6}\int d^3r\, r \frac{\partial V(r)}{\partial r} g(r) \tag{1305}$$

where we explicitly used $\epsilon_K = 3/2nk_BT$. Therefore, the ideal gas corrections are determined by $g(r)$.

4.6.3 Low-Density Limit

We can calculate $g(r)$ in the low-density limit. This involves a rather simple generalization of the ideas we developed in the Mayer cluster expansion. We first note that the fugacity dependence of the radial distribution function is of the form

$$n^2 g(\mathbf{x} - \mathbf{x}') = \frac{1}{Z} \sum_{N=2}^{\infty} \frac{z^N}{N!} \hat{g}_N(\mathbf{x} - \mathbf{x}') \tag{1306}$$

where Z is the grand partition function,

$$Z = \sum_{N=0}^{\infty} \frac{z^N}{N!} Q_N, \tag{1307}$$

and

$$\hat{g}_N(\mathbf{x} - \mathbf{x}') = \int d^3r_1 \cdots d^3r_N\, e^{-\beta V_N} \sum_{i \neq j = 1} \delta(\mathbf{x} - \mathbf{r}_j)\delta(\mathbf{x}' - \mathbf{r}_i). \tag{1308}$$

As for the thermodynamic properties, we assume an expansion of the form

$$n^2 g(\mathbf{x} - \mathbf{x}') = \sum_{l=0}^{\infty} \frac{z^l}{l!} G_l(\mathbf{x} - \mathbf{x}') \tag{1309}$$

where

$$G_l(\mathbf{x} - \mathbf{x}') = \lim_{z \to 0} \frac{\partial^l}{\partial z^l}\left[\frac{1}{Z}\sum_{N=2}^{\infty} \frac{z^N}{N!} \hat{g}_N(\mathbf{x}, \mathbf{x}')\right]. \tag{1310}$$

It is easy enough to work out the low-order results:

$$G_0 = G_1 = 0 \tag{1311}$$
$$G_2(\mathbf{x} - \mathbf{x}') = \hat{g}_2(\mathbf{x} - \mathbf{x}') \tag{1312}$$

while

$$G_3(\mathbf{x} - \mathbf{x}') = \hat{g}_3(\mathbf{x} - \mathbf{x}') - 3\hat{g}_2(\mathbf{x} - \mathbf{x}'), \tag{1313}$$

and so on. Looking at G_2 more explicitly, we easily find that

$$G_2(\mathbf{x} - \mathbf{x}') = \int d^3r_1 \, d^3r_2 \, e^{-\beta V(\mathbf{r}_1 - \mathbf{r}_2)}[\delta(\mathbf{x} - \mathbf{r}_1)\delta(\mathbf{x}' - \mathbf{r}_2) + \delta(\mathbf{x} - \mathbf{r}_2)\delta(\mathbf{x}' - \mathbf{r}_1)]$$
$$= 2e^{-\beta V(\mathbf{x} - \mathbf{x}')}. \tag{1314}$$

Therefore, to lowest order in the density,

$$g(r) = e^{-\beta V(r)}. \tag{1315}$$

As shown in Fig. 4.13, $g(r)$, in the low-density limit for a Lennard-Jones potential, goes to 1 as $r \to \infty$ and to zero for $r < r_0$ (this corresponds physically to the fact that there is zero probability that two particles will have their hard cores penetrate). The temperature dependence of $g(r)$ is also shown in Fig. 4.13. Again, the attractive part of the potential is less important at high temperatures.

A straightforward calculation gives the first-density correction to g as

$$G_3(\mathbf{x} - \mathbf{x}') = 6e^{-\beta V(\mathbf{x} - \mathbf{x}')} \int d^3r_3 (e^{-\beta V(\mathbf{x} - \mathbf{r}_3) + V(\mathbf{x}' - \mathbf{r}_3)} - 1). \tag{1316}$$

It is left to Problem 4.16 to carry out the analysis of this equation for hard spheres. The general features of $g(r)$ for dense systems are shown in Fig. 4.14 for the case of argon. As for low temperatures, $g(r)$ goes to 1 as $r \to \infty$ and is zero for r less than the hard-sphere separation. We also note that the peaks in Fig. 4.14 correspond to the location of nearest neighbors and next-nearest neighbors.

Let us now return to the density–density correlation function which we can calculate in the low-density limit since we know g. Inserting Eq. (1315) into

STRUCTURAL INFORMATION IN FLUIDS   **301**

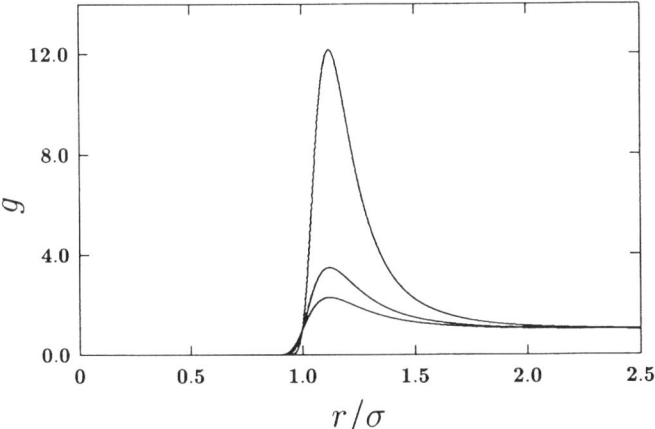

FIGURE 4.13 Radial distribution function versus scaled particle separation distance for a Lennard-Jones potential at lowest order in the density. The top curve corresponds to a temperature $k_B T = 0.4\epsilon$, the middle curve to $k_B T = 0.8\epsilon$, and the bottom curve to $k_B T = 1.2\epsilon$.

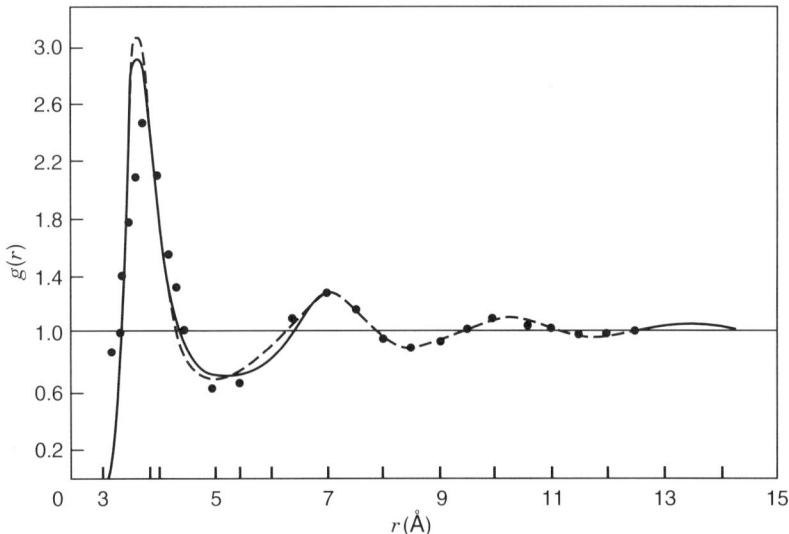

FIGURE 4.14 Calculated radial distribution function for argon at 84.4 K compared to neutron diffraction data. The solid line shows results from hypernetted chain calculation and the dashed line shows the Percus–Yevick calculation, both using the Lennard-Jones potential [from A. Kahn, *Phys. Rev. A* **134**, 367 (1964)]. The solid circles indicate the experimental points [from G.D. Henshaw, *Phys. Rev.* **105**, 976 (1957)]. (Figure from P. Egelstaff, *An Introduction to the Liquid State*, Academic Press, London, 1967.)

Eq. (1296), we obtain

$$S(\mathbf{r}) = n\delta(\mathbf{r}) + n^2(e^{-\beta V(\mathbf{r})} - 1). \tag{1317}$$

If we restrict ourselves for simplicity to hard spheres, then

$$S(\mathbf{r}) = n\delta(\mathbf{r}) + n^2[\theta|(\mathbf{r}| - r_0) - 1] \tag{1318}$$

where $\theta(x)$ is the unit-step function. Since $\theta(x) + \theta(-x) = 1$, we can write this in the form

$$S(\mathbf{r}) = n\delta(\mathbf{r}) - n^2\theta(r_0 - |\mathbf{r}|). \tag{1319}$$

The structure factor is given as the Fourier transform

$$S(\mathbf{k}) = \int d^3r\, e^{+i\mathbf{k}\cdot\mathbf{r}} S(\mathbf{r}) = n + n^2 h(\mathbf{k}) \tag{1320}$$

where

$$h(\mathbf{k}) = -\int d^3r\, e^{i\mathbf{k}\cdot\mathbf{r}} \theta(r_0 - r). \tag{1321}$$

By choosing $\mathbf{k}$ along the z-direction, this integral is easily evaluated, with the result

$$h(\mathbf{k}) = -4\pi r_0^3 \left[\frac{\sin kr_0}{(kr_0)^3} - \frac{\cos kr_0}{(kr_0)^2}\right] \tag{1322}$$

and

$$S(k) = n\left\{1 - 4\pi(nr_0^3)\left[\frac{\sin kr_0}{(kr_0)^3} - \frac{\cos kr_0}{(kr_0)^2}\right]\right\}. \tag{1323}$$

We have plotted $S(\mathbf{k})/n$ as a function of $x = kr_0$ for several values of nr_0^3 in Fig. 4.15. If there were no interaction, $S(k)/n = 1$. We see that the interactions have the largest effect near $k = 0$. We can calculate this limit explicitly to obtain

$$\frac{S(0)}{n} = 1 - \frac{4\pi}{3} nr_0^3. \tag{1324}$$

$S(\mathbf{k})/n$ raises rapidly from this value at $x = 0$ to a maximum near $x = 6$ or $k = 6/r_0$. In argon, $r_0 \approx 3.4\,\text{Å}$, so $k_{\max} = 1/1.76\,\text{Å}^{-1}$. $S(k)/n$ then undergoes damped

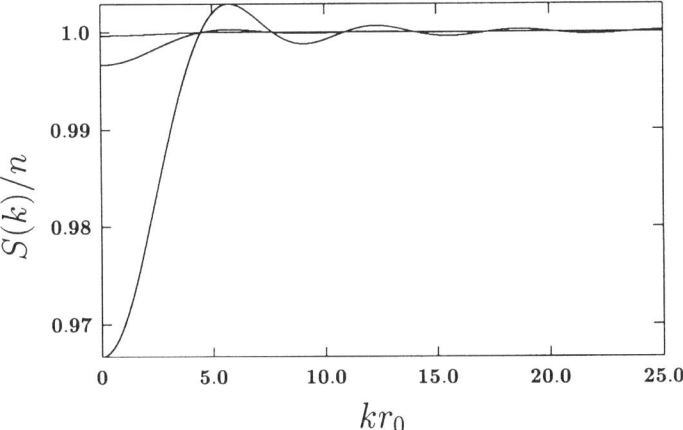

FIGURE 4.15 Low-density structure factor for a hard-sphere system versus reduced wavenumber for various values of the density. For small wavenumbers in the figure, the lowest curve corresponds to $4\pi r_0^3 n = 0.1$, the middle curve corresponds to $4\pi r_0^3 n = 0.01$, and the top curve corresponds to $4\pi r_0^3 n = 0.001$.

oscillations about $S(k)/n = 1$. This is the situation in a dilute system. In a dense system the major difference is that the peak height of the first maximum in the structure factor is much sharper and higher. Thus $S(k_{\max})/n$ may typically be 2.5.

4.6.4 Compressibility Limit

If we investigate $S(\mathbf{k})$ as a function of temperature and pressure, we find that $S(0)/n$ is small for low temperatures and high pressures, but very large near the critical point. We can understand this behavior through the following set of arguments. Remember that

$$S(\mathbf{k}) = \int \frac{d^3x\, d^3x'}{V} e^{-i\mathbf{k}\cdot(\mathbf{x}-\mathbf{x}')} \langle \delta n(\mathbf{x})\delta n(\mathbf{x}') \rangle \tag{1325}$$

so

$$S(\mathbf{0}) = \frac{1}{V}\langle (\delta N)^2 \rangle \tag{1326}$$

since

$$\int d^3x\, n(x) = N. \tag{1327}$$

We know, however, if we work in the grand canonical ensemble, that

$$\langle (\delta N)^2 \rangle = k_B T \frac{\partial}{\partial \mu} \bar{N} \tag{1328}$$

so that

$$S(\mathbf{0}) = \frac{k_B T}{V} \left(\frac{\partial \bar{N}}{\partial \mu} \right)_{T,V} \tag{1329}$$

$$= k_B T \left(\frac{\partial n}{\partial \mu} \right)_{T,V} = k_B T n^2 \kappa_T. \tag{1330}$$

Thus we see that the small-k limit of this correlation function is related to a thermodynamic derivative, the isothermal compressibility. The identity

$$\left(\frac{\partial n}{\partial \mu} \right)_{T,V} = n^2 \kappa_T \tag{1331}$$

used in the last line in Eq. (1330) is proven in Problem 4.17.

If we remember our discussion of the liquid-gas critical point, the most spectacular manifestation of the phase transition was the divergence of κ_T. We see from the result above that if we sit near $k = 0$ in an x-ray scattering experiment, the intensity will grow very rapidly near the critical point. This effect is indeed observed and is one of the most spectacular of critical phenomena. Away from the critical point and in the vicinity of the triple point we find that κ_T and $S(0)$ are very small. This is compatible with the notion that liquids are nearly incompressible.

4.7 BEYOND PERTURBATION THEORY

4.7.1 Introduction

Although the low-density expansions discussed above can be worked out in elaborate detail, there will be, as one increases the density toward the liquid state, values of the reduced density, $n\sigma^3$, which can no longer be treated as small. How does one then proceed? Development of a believable theory, in the absence of perturbation theory expansions, is a major challenge in the theory of interacting systems. Quantum electrodynamics has been successfully understood from a quantitative point of view because the fine structure constant $\alpha \approx 1/137$, and accordingly, the electron–photon interaction is small. In the case of strong interactions, where the interactions are, appropriate to the name, strong, perturbation theory does not work. Under these circumstances, where there is no natural small expansion parameter, problems such as solving analytically for the equation of state for a fluid become intractable.

In the absence of a small parameter there are a number of techniques that can work together to produce increased understanding of the problem: (1) extrapolation of theories, (2) numerical studies, and (3) scaling theories. We discuss each of these in turn.

4.7.2 Extrapolation of Theories

In the graphical analysis of the density expansion for fluids, as well as a Feynman graph development for interacting quantum systems (many-body theory), workers have developed approximations where they selectively re-sum infinite subclasses of graphs. A simple schematic example is to consider the virial expansion,

$$p\beta = \sum_{l=1}^{\infty} c_l n^l. \tag{1332}$$

Suppose that for some reason, in each order the c_l's are, except for a dimensional factor, the same: $c_l = b r_0^{3l}$. Substituting this result in Eq. (1332) yields

$$p\beta \approx b \sum_{l=1}^{\infty} (r_0^3 n)^l = b \left[\sum_{l=0}^{\infty} (r_0^3 n)^l - 1 \right] \tag{1333}$$

$$= b \left(\frac{1}{1 - nr_0^3} - 1 \right) = \frac{bnr_0^3}{1 - nr_0^3}. \tag{1334}$$

This is not a realistic approximation but an example of a *resummation* of an infinite series.

One guiding principle in such approaches is to find that quantity which is least sensitive to approximation and then work with this quantity. In the fluid case, the density correlation function is related to the radial distribution function by

$$S(\mathbf{r}) = n\delta(\mathbf{r}) + n^2(g(r) - 1). \tag{1335}$$

It is conventional, as in Eq. (1320), to define the *total correlation function*

$$h(r) = g(r) - 1 \tag{1336}$$

which we easily see is a generalization of the f-function defined by Eq. (1195). Whereas $g(r) \to 1$ as $r \to \infty$ and does not possess a well-defined Fourier transform, $h(r) \to 0$ as $r \to \infty$ and we can then Fourier transform Eq. (1335) to obtain

$$S(k) = n + n^2 h(k). \tag{1337}$$

A key building block in the theory of liquids is the *direct correlation function $c(r)$*. This quantity was introduced by Ornstein and Zernike [19] via the Ornstein–Zernike

(OZ) equation,

$$h(\mathbf{r}_1 - \mathbf{r}_2) = c(\mathbf{r}_1 - \mathbf{r}_2) + \int d^3 r_3\, h(\mathbf{r}_1 - \mathbf{r}_3) n c(\mathbf{r}_3 - \mathbf{r}_2) \qquad (1338)$$

and the direct correlation function has come to be appreciated as that quantity which is *least sensitive* to approximation for this problem. It is not difficult to relate $c(r)$ back to the density correlation function. Since the OZ equation is in the form of a convolution, it simplifies, after Fourier transformation, to

$$h(k) = c(k) + h(k) n c(k) \qquad (1339)$$

or

$$h(k) = \frac{c(k)}{1 - n c(k)}. \qquad (1340)$$

Substituting Eq. (1340) back into Eq. (1337), we easily obtain

$$S(k) = \frac{n}{1 - n c(k)} \qquad (1341)$$

and the direct correlation function is intimately related to the inverse density correlation function:

$$S^{-1}(\mathbf{r}) = \frac{1}{n} \delta(\mathbf{r}) - c(\mathbf{r}) \qquad (1342)$$

where

$$\int d^3 r'' S^{-1}(\mathbf{r} - \mathbf{r}'') S(\mathbf{r}'' - \mathbf{r}') = \delta(\mathbf{r} - \mathbf{r}'). \qquad (1343)$$

Notice that an approximation for the direct correlation function which is independent of density $c(k) \approx c_0(k)$ generates an infinite number of terms in the density expansion for $S(k)$, much as for our trivial example of a resummation discussed above.

If has been discovered by trial and error that one can find simple approximations for $c(r)$ that lead to good results for $g(r)$ and $S(k)$, in comparison with experiment and numerical results over a wide range of densities.

One very useful approximation, which arose out of a intense study of the Mayer cluster expansion, is the Percus–Yevick approximation [20], which can be written in the form

$$c(r) = e^{\beta V(r)} g(r) (e^{-\beta V(r)} - 1) \qquad (1344)$$

where $V(r)$ is the usual pair potential. It is shown in Problem 4.18 that the Percus–Yevick equation leads to the exact result for the direct correlation function at lowest order in the density.

One then has the set of equations coupling $c(r)$ and $h(r) = g(r) - 1$ given by the OZ equation and Eq. (1344). One of the key elements in choosing among the many possible resummations of the Mayer cluster expansion is *ease of handling*. Can the approximate theory be applied easily to various situations? In the case of the Percus–Yevick theory, one can make considerable analytic progress. In particular, for the case of hard spheres, Wertheim [21] found an analytic solution to this problem, with the result

$$c(r) = -\frac{\theta(\sigma - r)}{(1-\eta)^4}\left[(1+2\eta)^2 - 6\eta\left(1+\frac{\eta}{2}\right)^2\frac{r}{\sigma} + \frac{\eta}{2}(1+2\eta)^2\left(\frac{r}{\sigma}\right)^3\right] \quad (1345)$$

where $\eta = (\pi/6)\sigma^3 n$ is the reduced density and σ is the hard-sphere diameter.

Once $c(r)$ is known, $g(r)$ is known and various thermodynamic properties can be determined. Unfortunately, there are two independent ways of going from $g(r)$ to the equation of state. We can use the pressure equation given by Eq. (1305) or the *compressibility equation*, Eq. (1330), written in the form

$$S(k=0) = k_B T n \left(\frac{\partial n}{\partial p}\right)_T \quad (1346)$$

or

$$\frac{1}{k_B T}\left(\frac{\partial p}{\partial n}\right)_T = 1 - nc(k=0). \quad (1347)$$

One can work out the pressure using both Eqs. (1305) and (1347), with the results [22]

$$\frac{p}{nk_B T} = \begin{cases} \dfrac{1 + 2\eta + 3\eta^2}{(1-\eta)^2} & \text{pressure} \quad (1348) \\ \dfrac{1 + \eta + \eta^2}{(1-\eta)^3} & \text{compressibilty} \quad (1349) \end{cases}$$

Ashcroft and Lekner [23] have shown that one can work out the Fourier transforms analytically in this approximation and have compared with experimental data for Rb as shown in Fig. 4.16.

4.7.3 Numerical Work

With the development of modern computers a variety of numerical methods have evolved which can be used to determine quantities such as the radial distribution

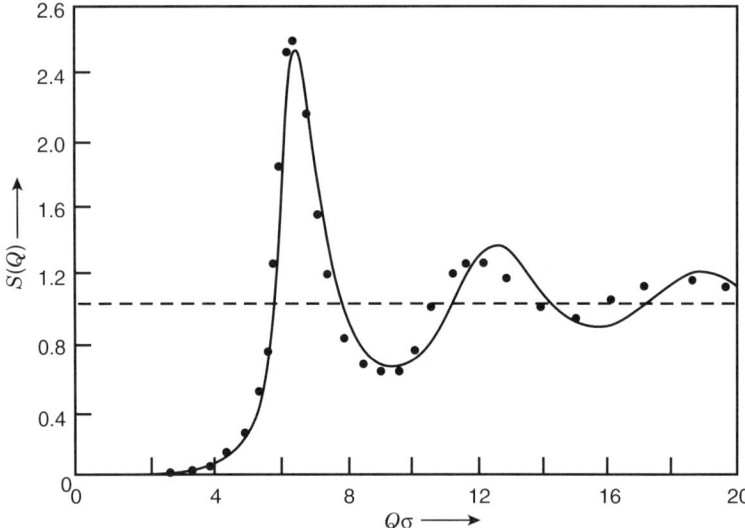

FIGURE 4.16 Comparison of hard-sphere calculation of $S(Q)$ and the experimental values for Rb at 40 °C. The solid line is the calculated curve and the solid circles are the experimental points; the value of σ was 4.30 Å and the actual liquid density was used in the calculation [from N. N. Ashcroft and J. Leckner, *Phys. Rev.* **145**, 83 (1966)]. (Figure from P. Egelstaff, *An Introduction to the Liquid State*, Academic Press, London, 1967.)

function for strongly interacting systems. Some of the earliest work was mentioned at the beginning of Chapter 1. There are two primary methods: molecular dynamics and Monte Carlo methods. Molecular dynamics involves a straight forward solution of Newton's laws. Thus one generates N-body configurations as a function of time. In the Monte Carlo method one uses statistical sampling techniques to generate configurations compaticle with the equilibrium probability distribution. The references appropriate to molecular dynamics are given at the beginning of Chapter 1 and Monte Carlo methods are discussed at the end of Chapter 5. These methods typically play a role similar to experiment except one has detailed control over the underlying system studied.

4.7.4 Principle of Corresponding States

In Chapter 2 we pointed out the scaling properties satisfied by the van der Waals equation of state,

$$\frac{p}{p_c} = \frac{8T}{T_c} \frac{n/n_c}{3 - n/n_c} - 3\left(\frac{n}{n_c}\right)^2. \tag{1350}$$

Although we know that the van der Waals equation does not give an accurate description of dense systems, the principle of corresponding states [24], in its general form,

$$\frac{p}{p_c} = u\left(\frac{T}{T_c}, \frac{n}{n_c}\right) \tag{1351}$$

is extremely useful and remarkably accurate even for very dense fluids. This is an example of a scaling theory that is not limited by constraints of perturbation theory.

As discussed by Guggenheim [24], this principle follows for systems satisfying the following basic set of assumptions:

1. Effects of quantum statistics are negligible.
2. The effect of quantization of the translational degrees of freedom is negligible.
3. The molecules are effectively spherically symmetric.
4. The intramolecular degrees of freedom are assumed to be completely independent of the particle density.
5. The potential energy is a function only of the intermolecular distances.
6. The potential energy for a pair of molecules can be written as $A\phi(R/R_0)$, where R is the intermolecular distance, A and R_0 are characteristic constants, and ϕ is a universal function.

Using these assumptions it is easy to show (see Problem 4.19) that the free energy for a system of N particles can be written in the form

$$\frac{F}{Nk_BT} = \frac{F_0}{Nk_BT} + \ln Z_0 + f\left(\frac{k_BT}{\epsilon_0}, nR_0^3\right) \tag{1352}$$

where F_0 is the free energy for an ideal gas with no internal degrees of freedom and Z_0 is the partition function for the internal degrees of freedom of a molecule, as discussed at the beginning of this chapter, and depends only on the temperature, the principle moments of inertia, and the frequencies of the normal modes. f is a complicated but universal function of $x = k_BT/\epsilon_0$ and $y = nR_0^3$, but we know that in the low-density limit, f vanishes. In the canonical ensemble the pressure is given by

$$p = -\left(\frac{\partial F}{\partial V}\right)_{T,N} = -\left(\frac{\partial F}{\partial n}\right)_{T,N}\frac{\partial n}{\partial V} = \frac{n}{V}\left(\frac{\partial F}{\partial n}\right)_{T,N}$$

$$= k_BTn - Nk_BT\frac{n}{V}\frac{\partial f(x,y)}{\partial y}R_0^3. \tag{1353}$$

If we choose $R_0^3 = n_c^{-1}$ and $\epsilon_0 = k_BT_c$, where n_c and T_c are the critical density and temperatures, respectively, then

$$p = k_BTn\left(1 - y\frac{\partial f(x,y)}{\partial y}\right) \tag{1354}$$

where $x = T/T_c$ and $y = n/n_c$. Going to the critical point, Eq. (1354) reduces to

$$p_c = k_B T_c n_c \left(1 - \left.\frac{\partial f(1,y)}{\partial y}\right|_{y=1}\right). \tag{1355}$$

Dividing Eq. (1354) by Eq. (1355) gives the fundamental expression for the principle of corresponding states:

$$\frac{p}{p_c} = \frac{T}{T_c} \frac{n}{n_c} \left(1 - y\frac{\partial f(x,y)}{\partial y}\right) \bigg/ \left(1 - y\left.\frac{\partial f(1,y)}{\partial y}\right|_{y=1}\right).$$

$$\equiv u\left(\frac{T}{T_c}, \frac{n}{n_c}\right). \tag{1356}$$

Guggenheim [24] gives many interesting applications of this principle. We review a few of these here. We start with the critical properties and the statement which follows from Eq. (1355) that $p_c/k_B T_c n_c$ is a constant. The experimental values for p_c, T_c, and n_c and the *universal ratio* are shown in Fig. 4.17. It can be seen that the values for argon, krypton, nitrogen, oxygen, carbon monoxide, and methane all lie within 1.5% of 0.292. The value of neon is larger, whereas that for xenon is a bit lower [25]. The derivation of neon from the universal value can be associated with quantum effects.

We expect that the principle of corresponding states should hold reasonably well for dilute systems. In that case the equation of state can be written in the form

$$\frac{p}{nk_B T} = 1 + B(T)n \tag{1357}$$

where $B(T) = -b_2(T)$ is the second virial coefficient. As we showed in Section 4.5.2, B is positive at high temperatures and negative at low temperatures. The temperature where B is zero is called the *Boyle temperature*, T_B. In Fig. 4.17 we give the experimentally determined T_B and the ratio T_B/T_c, which is predicted to be universal by the principle of corresponding states. By a similar argument, $B(T)n_c = B(T)/V_c$ should be a function only of T/T_c. This quantity is plotted for neon, argon, nitrogen, and oxygen in Fig. 4.18. Clearly, the principle holds rather well.

Our final example is that of the liquid–vapor coexistence lines. If n_L and n_g are the liquid and gas densities for mutual coexistence at temperature T, n_L/n_c and n_g/n_c are predicted to be universal functions of T/T_c by the principle of corresponding states. The appropriate plot is shown in Fig. 4.19 for a number of substances. All of the substances fall on a single curve except for carbon monoxide and methane. It is interesting to note that these data can be well fit by the empirical formula

$$\frac{n_L}{n_c} = 1 + \frac{3}{4}\left(1 - \frac{T}{T_c}\right) + \frac{7}{4}\left(1 - \frac{T}{T_c}\right)^{1/3} \tag{1358}$$

$$\frac{n_g}{n_c} = 1 + \frac{3}{4}\left(1 - \frac{T}{T_c}\right) - \frac{7}{4}\left(1 - \frac{T}{T_c}\right)^{1/3}. \tag{1359}$$

Formula	Ideal	Ne	A	Kr	Xe	N_2	O_2	CO	CH_4
1. M		20.18	39.94	83.7	131.3	28.02	32.00	28.00	16.03
2. $T_c/°K$		44.8	150.7	209.4	289.8	126.0	154.3	133.0	190.3
3. ρ_c/g cm^{-3}		0.484	0.5308	0.9085	1.155	0.311	0.430	0.301	0.162
4. v_c/cm^2 mole^{-1}		41.7	75.3	92.1	113.7	90.2	74.5	93.2	98.8
5. P_c/atmos.		26.9	48.0	54.1	58.2	33.5	49.7	34.5	45.7
6. $P_c v_c / RT_c$	0.292	0.305	0.292	0.290	0.278	0.292	0.292	0.294	0.289
7. $T_B/°K$		121	411.5			327		~345	491
8. T_B/T_e	2.7	2.70	2.73			2.59		2.6	2.58
9. $T_b/°K (P = 1$ atmos.)		27.2	87.3	120.9	165.1	77.3	90.1	81.6	112.5
10. $T_s/°K (P = P_c/50)$		25.2	86.9	122.0	167.9	74.1	90.1	78.9	110.5
11. T_b/T_c	0.58	0.608	0.580	0.577	0.570	0.614	0.583	0.613	0.591
12. T_s/T_c		0.563	0.577	0.582	0.580	0.588	0.583	0.593	0.581
13. $L_e/R°K$		224*	785	1086	1520	671	820	727	1023
14. L_e/RT_b	9.05	8.25	8.98	8.98	9.19	8.68	9.11	8.91	9.10
15. L_e/RT_c		8.9	9.04	8.91	9.06	9.06	9.11	9.22	9.26
16. $10^3 \alpha$			4.54			5.88	3.85	4.91	
17. αT_c	0.68		0.68			0.74	0.59	0.65	
18. $T_t/°K$		24.6	83.8	116.0	161.3	63.1	54.4	68.1	90.6
19. T_t/T_c	0.555	0.549	0.557	0.553	0.557	0.501	0.352	0.512	0.476
20. $L_f/R°K$		40.3	141.3	196.2	276	86.8	53.5	100.7	113
21. L_f/RT_t	1.69	1.64	1.69	1.69	1.71	1.37	0.98	1.48	1.25
22. P_t/atmos.		0.425	0.682	0.721	0.810				
23. $100 P_t/P_c$	1.37	1.58	1.42	1.33	1.39				
24. v_l/cm^2			28.14	34.13	42.68				
25. v_s/cm^2			24.61	29.65	37.09				
26. v_l/v_s			1.144	1.151	1.151				
27. C_P/R at T_s	5.4	5.5	5.5	5.35	5.35	6.75	6.45	7.25	

*This is not a measured value, but an approximate value based on vapor pressure measurements.

FIGURE 4.17 Experimental data presented by Guggenheim in his original discussion of the principle of corresponding states. [From E. A. Guggenheim, *J. Chem. Phys.* **13**, 253 (1945).]

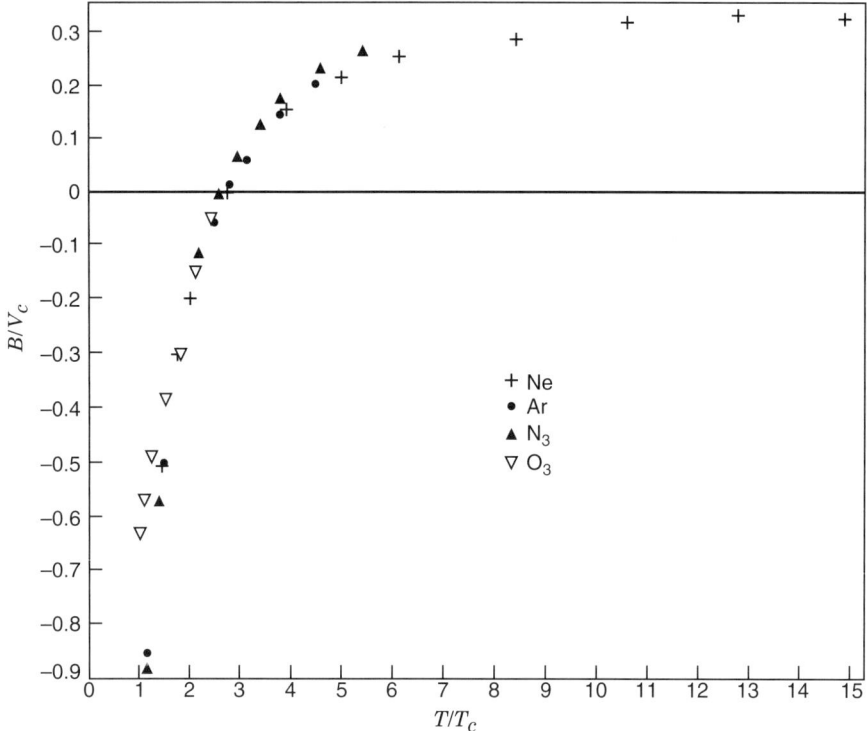

FIGURE 4.18 Plot of reduced second virial coefficient versus reduced temperature for four substances as a test of the principle of corresponding states. [From E. A. Guggenheim, *J. Chem. Phys.* **13**, 253 (1945).]

Adding these results, one obtains

$$\frac{n_L + n_g}{2n_c} = 1 + \frac{3}{4}\left(1 - \frac{T}{T_c}\right) \quad (1360)$$

which is known as the *law of rectilinear diameters*. On subtracting the two equations, we obtain

$$\frac{n_L - n_g}{n_c} = \frac{7}{2}\left(1 - \frac{T}{T_c}\right)^{1/3} \quad (1361)$$

is easily seen to be the mean-field theory result found from Landau theory and the van der Waals equation of state.

It appears that the principle of corresponding states holds well for a variety of simple fluids. One does not expect it to hold for highly polar molecules, metals, and molecules capable of forming hydrogen bonds.

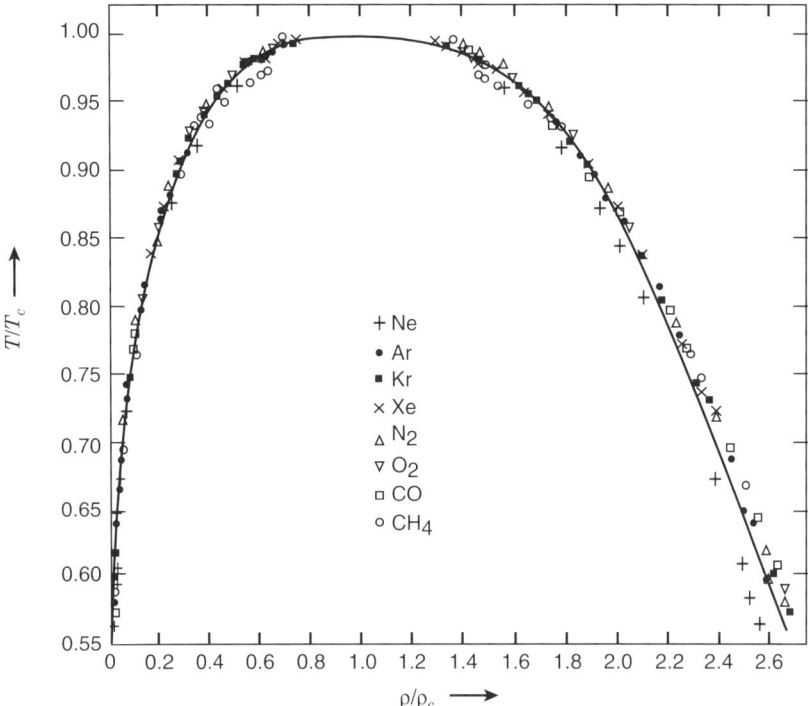

FIGURE 4.19 Coexistence curve plotted for a variety of substances in terms of reduced variables as as a test of the principle of corresponding states. [From E. A. Guggenheim, *J. Chem. Phys.* **13**, 253 (1945).]

Scaling theories such as the principle of corresponding states can be very useful in interpreting data. The basic idea is that if we can establish that there is a single system-dependent dominant length and energy, then, after rescaling all lengths and energies by these quantities, the results will be universal, independent of the particular physical system.

4.8 CHARGED SYSTEMS

4.8.1 Introduction

Thus far in this chapter we have been concerned with fluids that interact via short-range interactions which fall off at large separations at least as fast as r^{-3}. Clearly, a very important exception to such systems is the case of charged particles which interact with a r^{-1} Coulomb potential. Although there are a variety of ionized systems we could treat, we concentrate here on the one-component plasma. Although this system is slightly special, the basic ideas we develop are useful quite generally.

4.8.2 One-Component Plasma

Consider a classical system of N-point particles with charge q immersed into a uniform and inert neutralizing background enclosed in a volume V. The Hamiltonian is still of the standard form, with potential energy

$$V_N = \frac{1}{2} \sum_{i \neq j=1}^{N} \frac{q^2}{|\mathbf{r}_i - \mathbf{r}_j|} - \sum_{i=1}^{N} n \int \frac{d^3 r\, q^2}{|\mathbf{r}_i - \mathbf{r}|} + \frac{n^2}{2} \int q^2 \frac{d^3 r\, d^3 r'}{|\mathbf{r} - \mathbf{r}'|} \tag{1362}$$

where the second term on the right is due to the neutralizing background charge, $n = N/V$, and the last term on the right subtracts the constant self-energy. This problem is somewhat simpler than the problem of an ionized plasma of oppositely charged particles. In that case we must worry about the collapse of the system due to particle attraction of oppositely charged particles. The stability of such systems requires a full quantum statistical mechanical treatment as discussed by Lenard and Dyson [26]. In the present case, Lieb and Narnhofer [27] have been able to show that the one-component plasma is thermodynamically stable. This means that the total potential energy is bounded from below. In particular, they have shown that

$$\frac{V_N}{V} \geq -\frac{9}{10} \frac{q^2}{a} \tag{1363}$$

where $n^{-1} = \frac{4}{3}\pi a^3$.

Given the stability of our system with potential V_N, we can proceed to develop the associated statistical mechanics. It will be convenient to work in the canonical ensemble where the partition function, after integrating over the momenta, is given by

$$Z_N = \frac{V^N}{N! \lambda^{3N}} \mathcal{Q}_N \tag{1364}$$

where $\lambda = (2\pi m k_B T)^{-1/2}$ is the de Broglie thermal wavelength and

$$\mathcal{Q}_N = \int \frac{d^3 r_1}{V} \cdots \frac{d^3 r_N}{V} e^{-\beta V_N}. \tag{1365}$$

We have defined $\mathcal{Q}_N$ such that it is dimensionless. A bit of reflection indicates that the quantity

$$f = \lim_{N,V \to \infty} -\frac{1}{N} \ln \mathcal{Q}_N \big|_{N/V=n} \tag{1366}$$

is a function only of the dimensionless parameter

$$\Gamma = \frac{\beta q^2}{a} \tag{1367}$$

or
$$f = f(\Gamma). \tag{1368}$$

Thus the Helmboltz free energy is of the form

$$F = -\beta^{-1}\ln Z_N = F_0 + Nf(\Gamma) \tag{1369}$$

where

$$F_0 = -\beta^{-1}\ln\left(\frac{V^N}{N!\lambda^{3N}}\right) \tag{1370}$$

is the free energy for the noninteracting system. We obtain immediately for the pressure and average energy,

$$p = -\left(\frac{\partial F}{\partial V}\right)_{N,T} = k_B T n\left(1 + n\frac{\partial f}{\partial n}\right) \tag{1371}$$

and

$$E = \frac{\partial}{\partial \beta}(\beta F) = \frac{3}{2} N k_B T\left(1 + \frac{2}{3}\beta\frac{\partial f}{\partial \beta}\right). \tag{1372}$$

However, since

$$3n\frac{\partial f}{\partial n} = \beta\frac{\partial f}{\partial \beta} = \Gamma\frac{\partial f}{\partial \Gamma} \tag{1373}$$

we have

$$p = k_B T n\left(1 + \frac{\Gamma}{3}\frac{\partial f}{\partial \Gamma}\right) \tag{1374}$$

and

$$\frac{E}{N} = \frac{3}{2} k_B T\left(1 + \frac{2}{3}\Gamma\frac{\partial f}{\partial \Gamma}\right). \tag{1375}$$

Comparing these last two expressions, it is clear that we can eliminate $\Gamma \partial f/\partial \Gamma$ between them and obtain an equation relating the pressure and the energy per unit volume:

$$p = \frac{1}{2} n k_B T + \frac{1}{3}\frac{E}{V}. \tag{1376}$$

This relation is a result of the special r^{-1} behavior of the interaction potential, as can be seen directly from Eq. (1305), as discussed in Problem 4.21.

4.8.3 Spatial Correlations

Taking the direct equilibrium average of the potential energy given by Eq. (1362), we obtain

$$\frac{\langle V_N \rangle}{N} = \frac{n}{2} \int d^3 r \, V(\mathbf{r})(g(r) - 1) \tag{1377}$$

where

$$V(r) = \frac{q^2}{r} \tag{1378}$$

is the Coulomb potential and $g(r)$ the radial distributions function. The quantity $qn(g(r) - 1)$ is the average charge located at a distance r from a charge at the origin. The uniform contribution $-qn$ is due to the background charge.

If we introduce Fourier transforms, we see that the total correlation function enters, as defined by Eq. (1336), and can be expressed back in terms of the structure factor:

$$\frac{\langle V_N \rangle}{N} = \frac{1}{2} \int \frac{d^3 k}{(2\pi)^3} V(\mathbf{k})(S(\mathbf{k}) - n)/n \tag{1379}$$

where

$$V(\mathbf{k}) = \frac{4\pi q^2}{k^2} \tag{1380}$$

is the Fourier transform of the Coulomb potential.

4.8.4 Debye–Hückel Theory

In the case of short-ranged interactions, we developed a density expansion to evaluate the pressure, energy, radial distribution function, and so on. We cannot apply these methods to a plasma directly because of the long-ranged nature of the Coulomb interaction. In particular, the integral in the second virial coefficient,

$$\int d^3 r (e^{-\beta V(r)} - 1), \tag{1381}$$

does not converge for large r. Special methods must be developed for treating the long-range interaction.

The important consideration in dealing with the Coulomb interaction in fluids is that two charges do not interact with each other in a vacuum. The range of interaction is sufficiently long that any two charges will feel the influence of many other charges and the positive background. The key point is that over a sufficiently large volume, the fluid is, on average, neutral. We can quantify this idea following the pioneering work of Debye and Hückel [28]. The approach is somewhat heuristic, but it is very physical and the result is easy to appreciate.

Consider Poisson's equation for the electrostatic potential $\phi(\mathbf{r})$ at point $\mathbf{r}$ due to a charge distribution $q\rho(\mathbf{r})$,

$$-\nabla^2 \phi(\mathbf{r}) = 4\pi q \rho(\mathbf{r}). \tag{1382}$$

We assume that there is a charge at the origin, so

$$\rho(\mathbf{r}) = \delta(\mathbf{r}) + n[g(\mathbf{r}) - 1] \tag{1383}$$

where $ng(\mathbf{r})$ is the average number of charges at position $\mathbf{r}$ given that there is a charge at $\mathbf{r} = 0$, and n is the average *number* of background charges at position $\mathbf{r}$. The key assumption is that the radial distribution function is a simple Boltzmann distribution

$$g(\mathbf{r}) = e^{-\beta q \phi(\mathbf{r})}. \tag{1384}$$

Combining Eqs. (1382), (1383), and (1384), we have a self-consistent nonlinear Debye–Hückel equation determining the electric potential $\phi(\mathbf{r})$:

$$-\nabla^2 \phi(r) = 4\pi q \delta(\mathbf{r}) + 4\pi q n (e^{-\beta q \phi(\mathbf{r})} - 1). \tag{1385}$$

The general solution of this equation requires numerical analysis. We will restrict ourselves to the linear regime, where we can expand the exponential to obtain

$$-\nabla^2 \phi(\mathbf{r}) = 4\pi q \delta(\mathbf{r}) - 4\pi \beta q^2 n \phi(\mathbf{r}). \tag{1386}$$

Defining the length λ_D via

$$\lambda_D^{-2} = 4\pi \beta q^2 n, \tag{1387}$$

Eq. (1386) takes the form

$$(\lambda_D^{-2} - \nabla^2)\phi(\mathbf{r}) = 4\pi q \delta(\mathbf{r}). \tag{1388}$$

Fourier transforming over space gives

$$\phi(\mathbf{k}) = \frac{4\pi q}{k^2 + \lambda_D^{-2}}. \tag{1389}$$

It is easy to invert the Fourier transform to obtain

$$\phi(\mathbf{r}) = \int \frac{d^3k}{(2\pi)^3} \frac{e^{-i\mathbf{k}\cdot\mathbf{r}} 4\pi q}{k^2 + \lambda_D^{-2}} \tag{1390}$$

$$= \frac{q}{r} e^{-r/\lambda_D}. \tag{1391}$$

The interpretation of these results is direct and very important. The effective interaction between two charges in a medium of charges with overall neutrality is given by

$$V_E(\mathbf{r}) = q\phi(\mathbf{r}) \tag{1392}$$

$$= \frac{q^2}{r} e^{-r/\lambda_D} \tag{1393}$$

and falls off exponentially with distance due to the *screening effect* of the other charges. The screening length λ_D is also known as the *Debye screening length* and represents the average distance it takes for the surrounding charge to cancel the charge sitting at the origin. The Fourier transform of the effective potential

$$V_E(\mathbf{k}) = \frac{4\pi q^2}{k^2 + \lambda_D^{-2}} \tag{1394}$$

is now rendered integrable:

$$V_E(\mathbf{k}=0) = \int d^3 r\, V_E(\mathbf{r}) = 4\pi q^2 \lambda_D^2. \tag{1395}$$

Note that $V_E(\mathbf{k}=0)$ is now singular ($\approx 1/n$) in the low-density limit. The property of screening is a general property of charged systems with overall neutrality.

Given $\phi(\mathbf{r})$, we can determine the radial distribution function and all of the thermodynamic properties of the system. Equation (1384) in the linear regime gives

$$g(\mathbf{r}) = 1 - \beta q\phi(r) = 1 - \beta \frac{q^2}{r} e^{-r/\lambda_D} \tag{1396}$$

and the structure factor is given by

$$\frac{S(\mathbf{k})}{n} = 1 + n \int d^3 r\, e^{+i\mathbf{k}\cdot\mathbf{r}}(g(\mathbf{r}) - 1)$$

$$= 1 - \frac{\beta q^2 4\pi n}{k^2 + \lambda_D^{-2}} = 1 - \lambda_D^{-2}(k^2 + \lambda_D^{-2})^{-1}$$

$$= \frac{k^2}{k^2 + \lambda_D^{-2}}. \tag{1397}$$

This gives the result, very different from short-range forces, that the structure factor vanishes as $k \to 0$. Comparing Eqs. (1341) and (1397), we can identify the direct correlation function in the Debye–Hückel theory as

$$1 - nc_{DH}(k) = \frac{1}{k^2}(k^2 + \lambda_D^{-2}) \tag{1398}$$

or

$$c_{DH}(k) = -\beta V(k). \tag{1399}$$

Thus the linear Debye–Hückel theory is equivalent to replacing the direct correlation function by $-\beta$ times the Fourier transform of the bare Coulomb potential. This is just the result we would obtain from a brute force expansion of $c(\mathbf{r})$ in powers of $V(\mathbf{r})$ at lowest order. A systematic theory can be developed by writing

$$c(\mathbf{r}) = c_{DH}(\mathbf{r}) + c_R(\mathbf{r}) \tag{1400}$$

where $c_R(\mathbf{r})$ is the part of the direct correlation function rendered regular by the screening behavior treated by the Debye–Hückel approximation.

Finally, we can determine the thermodynamic quantities in the Debye–Hückel approximation. Using Eq. (1396) in Eq. (1377), we obtain

$$\begin{aligned}\frac{\langle V_N \rangle}{N} &= \frac{n}{2}\int d^3r \frac{q^2}{r}\left(-\frac{\beta q^2}{r}e^{-r/\lambda_D}\right) \\ &= -\beta nq^4 2\pi\lambda_D = -\frac{1}{2}\frac{q^2}{\lambda_D} \\ &= -\beta^{-1}\frac{\epsilon}{2}\end{aligned} \tag{1401}$$

where ϵ is the dimensionless parameter

$$\epsilon = \frac{\beta q^2}{\lambda_D}. \tag{1402}$$

It is left as a problem (4.22) to show that $\epsilon = \sqrt{3}\,\Gamma^{3/2}$. Then the energy per particle is

$$\frac{E}{N} = k_B T\left(\frac{3}{2} - \frac{\epsilon}{2}\right) \tag{1403}$$

and using Eq. (1376), the pressure is

$$p = nk_B T\left(1 - \frac{\epsilon}{6}\right). \tag{1404}$$

4.9 COARSE GRAINING AND EFFECTIVE HAMILTONIANS

4.9.1 Introduction

Earlier in this chapter we introduced several models for interactions in atomic and molecular systems. In the case of diatomic molecules we assumed that the internal motions of a diatomic molecule can be represented as a sum of rotational and vibrational contributions to the Hamiltonian. Later we assumed that noble gas atoms interact with a Lennard-Jones potential. In Chapter 5 we introduce models for *permanent* electric and magnetic dipole moments for atoms and molecules and the Heisenberg model govening the interaction between the magnetic moments. All of these models are written in terms of *average* properties of complicated atomic or molecular systems. How realistic are these models, and how can they be justified from first principles?

This type of question arises again and again in physics as we change the level of description. Let us be more specific. In physics we can typically identify a hierarchy of well-separated energy (or length) scales. At the highest-energy scales (10^{21} GeV) we have some supergrand theory described by something like superstrings. At *lower* energies (10^{15} GeV) we no longer see strings but instead a set of composite particles described by grand unified theories. At yet lower energies (10^3 GeV) we have the physics described by the standard model of particle physics with families of quarks and leptons. At yet lower energies (10 MeV) we have a theory with no direct evidence of quarks but with the composites associated with hadrons and nuclear physics. We are, of course, not finished yet (see Table 4.4). Hadrons, via the nuclear force, form nuclei that appear as well-behaved *particles* on the atomic energy scales of 10 eV. At the atomic scale, nuclei and electrons interact to form atoms and molecules. Finally, we are at the scale discussed earlier in this chapter. There are additional lower-energy scales where this process continues. In particular, in dealing with *complex* systems such as polymers, liquid crystals, membranes, and various biological systems, we again find a new level of description in which the underlying atomic level does not enter directly. This range is called the *mesoscopic scale*. There are, of course, at least two additional scales of interest: the *macroscopic scale*, where thermodynamics is appropriate, and *astrophysical scales*, where

Table 4.4 Energy and Length Scales in Physics

Energy	Length	Phenomenon
10^{21} GeV	—	Big bang
10^{15} GeV	—	GUTS
10^3 GeV	—	Standard model
1 MeV	—	Nuclear physics
10 eV	1 Å	Atomic physics
—	10^3 Å	Mesoscopic physics
—	1 cm	Thermodynamics
—	1 ASU	Gravitational physics

gravity becomes important. One can go on, since there are ranges of length scales in astrophysics where this coarse graining continues (solar systems to galaxies, to clusters, etc.).

At a given scale we have knowledge of the previous *more microscopic scale* only through interaction constants or potentials and masses that enter the Hamiltonian (or Lagrangian) governing the physics at the more macroscopic scale. There are many examples. The tensor nuclear force governing the interactions between nucleons is a function of a few parameters representing the low-energy effect of the fundamental meson exchange between nucleons. Similarly, the static Coulomb interaction represents the low-energy effect of photon exchange between charged particles. The couplings and masses in quantum field theories result from some more *fundamental* string(?) theory. In statistical mechanics we have a wealth of examples of this coarse-graining procedure. More specifically, by *coarse graining* we mean a procedure where degrees of freedom active on more microscopic scales are replaced by a few *average* degrees of freedom governing the physics at a more macroscopic length scale. There are numerous examples of this procedure. Below we discuss the effective interaction between spherical atoms, where we average or coarse grain over all the internal degrees of freedom of the atoms. Similarly, by averaging over the electronic degrees of freedom in a molecule, we arrive at the effective rotational and vibrational energies discussed above. In Chapter 5 we discuss how the magnetic and dielectric properties of various materials result from treating atoms and molecules as *effective* electric and magnetic dipoles. In Volume 2 we introduce the idea of coarse-graining Hamiltonians over a continuous range of length scales. This curious idea, basic to the renormalization group approach, has turned out to be fundamental for those problems, such as that of critical phenomena, where a wide variety of length scales are important and must be considered simultaneously. It is now conventional to work with effective Hamiltonians of the Ginsburg–Landau type, which are essential in treating systems (e.g., liquid crystals) at the mesoscopic length scale.

4.9.2 Formal Development

Although the technical details of coarse graining differ dramatically in going from one length scale to the next, the basic idea can be stated rather generally. Suppose that at a microscopic length scale there are a set of degrees of freedom, labeled by ϕ_i, which are coupled to another set of degrees of freedom ψ_i, which we expect to be important at a more macroscopic level. The ψ_i may be the conserved variables that are used in our development in Chapter 1 of the basics of statistical mechanics. Generally, the Hamiltonian describing the system at the microscopic level can be written in the form

$$H = H_S(\psi) + H_F(\phi) + \lambda H_I(\phi, \psi) \qquad (1405)$$

where the subscript S indicates slow and F fast. Since we do not believe that the variables ϕ are relevant at the macroscopic level, we want to average H over ϕ to obtain an effective Hamiltonian $H_E(\psi)$. Within the context of equilibrium statistical mechanics, this is, in principle, straightforward. The details are given in Appendix N. We emphasize the basic idea here.

The equilibrium probability distribution for our microscopic problem is given by

$$P(\psi, \phi) = \frac{e^{-\beta H(\psi, \phi)}}{Z} \tag{1406}$$

where Z is the normalizing partition function. We assume, for simplicity, that the fields ψ and ϕ can be treated as independent,

$$[H_S(\psi), H_F(\phi)] = 0, \tag{1407}$$

and the set of eigenstates associated with $H_s(\psi)$ and $H_F(\phi)$ form a product Hilbert space spanning the entire space on which $H(\psi, \phi)$ is defined. We can then write the trace over all degrees of freedom in the form

$$\text{Tr} = \text{Tr}^{\psi} \text{Tr}^{\phi} \tag{1408}$$

and

$$\text{Tr} P = 1. \tag{1409}$$

We now define the effective Hamiltonian $H_E(\psi)$ via the reduced probability distribution:

$$P_\psi = \frac{e^{-\beta H_E(\psi)}}{Z} = \frac{\text{Tr}^\phi e^{-\beta H(\psi,\phi)}}{Z} \tag{1410}$$

or

$$H_E(\psi) = -\beta^{-1} \ln \text{Tr}^\phi e^{-\beta H(\psi,\phi)}. \tag{1411}$$

In the classical limit, where all parts of H commute, we have immediately that

$$H_E(\psi) = H_s(\psi) - \beta^{-1} \ln Z_\phi^o - \beta^{-1} \ln \langle e^{-\beta \lambda H_I(\phi,\psi)} \rangle_0 \tag{1412}$$

where

$$Z_\phi^o = \text{Tr}^\phi e^{-\beta H_F(\phi)} \tag{1413}$$

and

$$\langle A \rangle_0 = \frac{1}{Z_\phi^o} \text{Tr}^\phi e^{-\beta H_F(\phi)} A. \tag{1414}$$

We see immediately that if there is no coupling ($\lambda = 0$) between the variables ϕ and ψ, the effective Hamiltonian is simply $H_s(\psi)$ plus a constant which guarantees that P_ψ is properly normalized:

$$P_\psi = \frac{e^{-\beta H_s(\psi)} e^{\ln Z_\phi^0}}{Z} \qquad (1415)$$

$$= \frac{e^{-\beta H_s(\psi)}}{Z_\psi^0}. \qquad (1416)$$

If the coupling between the fast and slow degrees of freedom is weak and we can expand in powers of λ, we obtain

$$H_E(\psi) = H_s(\psi) - \beta^{-1} \ln Z_\phi^0 + \lambda \langle H_I(\phi, \psi) \rangle_0$$
$$- \frac{\beta \lambda^2}{2} \langle (\delta H_I(\phi, \psi))^2 \rangle_0 + \mathcal{O}(\lambda^3) \qquad (1417)$$

where

$$\delta H_I(\phi, \psi) = H_I(\phi, \psi) - \langle H_I(\phi, \psi) \rangle_0. \qquad (1418)$$

The term of $\mathcal{O}(\lambda)$ in Eq. (1417) has a natural interpretation as the direct averaging over the ϕ degrees of freedom in $H(\phi, \psi)$. In certain situations the coupling is sufficiently weak that the linear term suffices, but in others the linear term vanishes due to symmetry and we must treat the second-order term in order to obtain the appropriate coupling.

One obvious point is that the effective Hamiltonian is apparently temperature dependent. However, if there is a clear separation of energy scales, then, as we discuss below, the effective Hamiltonian will be independent of temperature.

The full quantum mechanical treatment of this problem is given in Appendix N. There are two key assumptions beyond the weakness of the interaction between the two sets of degrees of freedom. First consider the quantity

$$H_I(\beta) \equiv e^{\beta(H_F^0 + H_S^0)} H_I e^{-\beta(H_F^0 + H_S^0)} \qquad (1419)$$

which occurs prominently in determining the effective Hamiltonian. The dependence on H_S^0 complicates the β dependence of the H_E considerably. However, if we restrict our consideration to matrix elements of the set of slow degrees of freedom between states $|m>_0^S$ and $|n>_0^S$, where $\beta |E_m - E_n| \ll 1$, we can safely write

$$H_I(\beta) = e^{\beta H_F^0} H_I e^{-\beta H_F^0}. \qquad (1420)$$

This is similar to assuming the slow variables are classical and $[H_I, H_S^0] = 0$. The second assumption is that the fast variables are governed primarily by the physics of its ground state. Thus the separation of E_G^F from the *fast* first- excited state is assumed large compared to $k_B T$:

$$E_1^F - E_G^F \equiv \Delta \gg k_B T. \tag{1421}$$

One can then show that one obtains the temperature-independent effective Hamiltonian govening the slow degrees of freedom:

$$H_S = H_S^0 + E_0^F + \lambda^2 \sum_{m>0} \frac{|\langle 0|H_I|m\rangle|^2}{E_0 - E_m} \tag{1422}$$

up to terms of $\mathcal{O}(\lambda^2)$ and the $|m\rangle$ refer to fast states. This result is independent of temperature because there is a clean separation of energy scales between E_F^o and the scales of interest $k_B T$.

4.9.3 van der Waals Interaction

As an example of the ideas developed in the two preceding sections, we can consider the long-distance interaction between two closed-shell noble gas atoms. One nucleus of mass M and charge Ze is at position $\mathbf{r}_1$ (see Fig. 4.20) surrounded by Z electrons at positions $\mathbf{r}_{i,1}$. A second, similar nucleus is located at $\mathbf{r}_2$ and is surrounded by Z electrons at positions $\mathbf{r}_{i,2}$. In the notation developed in the preceding section, the Hamiltonian associated with the *fast variables* is

$$H_0^F = H_{0,1}^F + H_{0,2}^F \tag{1423}$$

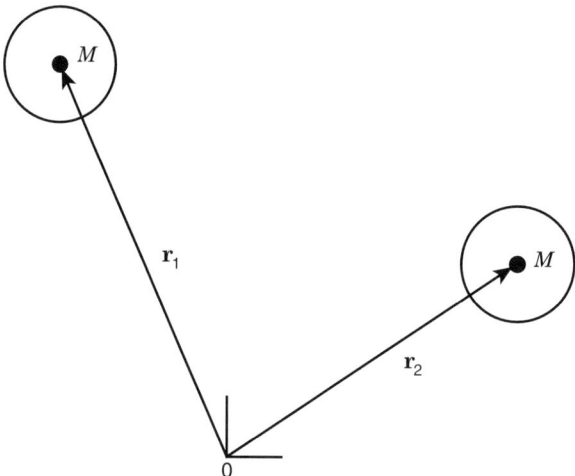

FIGURE 4.20 Geometry for the interaction of two composite particles.

where $H_{0,1}^F$ is the Hamiltonian describing the electrons associated with atom 1 with the nucleus located at position $\mathbf{r}_1$; similarly for $H_{0,2}^F$. The contribution from the slow variables to H_0^s is given by the center-of-mass contribution to the kinetic energy, which is, to a good approximation, just the nuclear kinetic energies. The interaction between the fast and slow variables is given by the Coulomb interaction:

$$V = \frac{Z^2 e^2}{|\mathbf{r}_1 - \mathbf{r}_2|} + e^2 \sum_{i,j=1}^{Z} \frac{1}{|\mathbf{r}_{i,1} - \mathbf{r}_{j,2}|} - Ze^2 \sum_{i=1}^{Z} \left(\frac{1}{|\mathbf{r}_1 - \mathbf{r}_{i,2}|} + \frac{1}{|\mathbf{r}_2 - \mathbf{r}_{i,1}|} \right).$$
(1424)

Let us write

$$\mathbf{r}_{i,1} = \mathbf{r}_1 + \boldsymbol{\xi}_{i,1} \tag{1425}$$

$$\mathbf{r}_{i,2} = \mathbf{r}_2 + \boldsymbol{\xi}_{i,2} \tag{1426}$$

and assume that the Z electrons surrounding each nucleus are localized about that nucleus. For spherical closed- shell atoms, this is correct. Next, assume that the atoms are well separated, which is equivalent to assuming that the magnitude of $\mathbf{R} = \mathbf{r}_1 - \mathbf{r}_2$ is large compared to all $|\boldsymbol{\xi}_{i,1}|$ or $|\boldsymbol{\xi}_{i,2}|$. We can then expand the interaction potential V in powers of $|\boldsymbol{\xi}_{i,1}|/R$ and $|\boldsymbol{\xi}_{i,2}|/R$ and obtain, after a bit of algebra explored in Problem 4.24,

$$V = \frac{e^2}{R^3} \sum_{i,j} (\boldsymbol{\xi}_{i,1} \cdot \boldsymbol{\xi}_{j,2} - 3\hat{R} \cdot \boldsymbol{\xi}_{i,1} \hat{R} \cdot \boldsymbol{\xi}_{j,2}). \tag{1427}$$

This is just the interaction between two electric dipoles with dipole moments $\boldsymbol{\mu}_1 = e \sum_{i=1}^{Z} \boldsymbol{\xi}_{i,1}$ and $\boldsymbol{\mu}_2 = e \sum_{i=1}^{Z} \boldsymbol{\xi}_{i,2}$. The higher-order terms give, for examples, the dipole–quadrapole (R^{-4}) interaction and the quadrapole–quadrapole interaction (R^{-5}). The effective Hamiltonian for the interacting atoms, assuming that we can treat the center-of-mass motion of the atoms classically and the atoms are in their ground states, is given by Eq. (1422), which reduces in this case to

$$H_s = \frac{P_1^2}{2M} + \frac{P_2^2}{2M} + \frac{e^2}{R^3} \langle 0 | \boldsymbol{\mu}_1 \cdot \boldsymbol{\mu}_2 - 3\hat{R} \cdot \boldsymbol{\mu}_1 \hat{R} \cdot \boldsymbol{\mu}_2 | 0 \rangle \tag{1428}$$

$$+ \frac{1}{2} \frac{1}{R^6} \sum_{m=1}^{\infty} \frac{|\langle 0 | (\boldsymbol{\mu}_1 \cdot \boldsymbol{\mu}_2 - 3\hat{R} \cdot \boldsymbol{\mu}_1 \hat{R} \cdot \boldsymbol{\mu}_2) | m \rangle|^2}{E_F^0 - E_F^m}. \tag{1429}$$

For our closed-shell system, where we have a product ground state $|0\rangle = |0\rangle_1 \times |0\rangle_2$, there is no net electric dipole moment

$$_1\langle 0 | \boldsymbol{\mu}_1 | 0 \rangle_1 = 0. \tag{1430}$$

This is just the statement that the ground-state wavefunction is isotropic in space and there is no long-range dipole–dipole interaction. See Chapter 5 for a discussion of systems with this type of interaction. Thus the interaction falls off at least as fast as R^{-6} for large separations and, since $E_F^0 - E_F^m < 0$, is *attractive*. This attractive tail was called the *dispersion force* by London [29] but is typically called a van der Waals interaction. We can gain an estimate for the size of the coefficient of $-R^{-6}$,

$$A = \frac{1}{2}\sum_{m\neq 0}\frac{|\langle 0|(\boldsymbol{\mu}_1\cdot\boldsymbol{\mu}_2 - 3\hat{R}\cdot\boldsymbol{\mu}_1\hat{R}\cdot\boldsymbol{\mu}_2)|m\rangle|^2}{|E_F^0 - E_F^m|}, \tag{1431}$$

by noting the inequality

$$\frac{1}{|E_F^0 - E_F^1|} \geq \frac{1}{|E_F^0 - E_F^m|}, \tag{1432}$$

where E_F^1 is the first excited state of the atomic system. Using Eq. (1432) in Eq. (1431), we obtain

$$A \geq \frac{1}{2|E_F^0 - E_F^1|}\sum_{m\neq 0}|\langle 0|(\boldsymbol{\mu}_1\cdot\boldsymbol{\mu}_2 - 3\hat{R}\cdot\boldsymbol{\mu}_1\hat{R}\cdot\boldsymbol{\mu}_2)|m\rangle|^2. \tag{1433}$$

However, since the $|m\rangle$ form a complete set

$$\sum_{m\neq 0}|m\rangle\langle m| = 1 - |0\rangle\langle 0|, \tag{1434}$$

we obtain, remembering Eq. (1430),

$$A \geq \frac{1}{2|E_F^0 - E_F^1|}\langle 0|(\boldsymbol{\mu}_1\cdot\boldsymbol{\mu}_2 - 3\hat{R}\cdot\boldsymbol{\mu}_1\hat{R}\cdot\boldsymbol{\mu}_2)^2|0\rangle. \tag{1435}$$

For an isotropic ground state (Problem 4.25) we have

$$\langle 0|(\boldsymbol{\mu}_1\cdot\boldsymbol{\mu}_2 - 3\hat{R}\cdot\boldsymbol{\mu}_1\hat{R}\cdot\boldsymbol{\mu}_2)^2|0\rangle = \frac{2}{3}\langle 0|\boldsymbol{\mu}^2|0\rangle^2 \tag{1436}$$

and

$$A \geq \frac{1}{3|E_F^0 - E_F^1|}(\langle 0|\boldsymbol{\mu}^2|0\rangle)^2. \tag{1437}$$

As discussed in Problem 4.26 and in Chapter 5, to the same level of approximation we can express the related atomic polarizability in the form

$$\alpha \geq \frac{2}{3|E_F^0 - E_F^1|} \langle 0|\mu^2|0\rangle. \quad (1438)$$

It is then reasonable to assume equalities in both of the last two equations, eliminate $\langle 0|\mu^2|0\rangle$ in favor of α, and obtain the estimate

$$A = \tfrac{3}{4}|E_F^0 - E_F^1|\alpha^2. \quad (1439)$$

In the theories of London [30] and Margeneau [31], they argue that this result can be generalized to pairs of different closed-shell atoms if one writes

$$A_{ij} = \frac{3}{2}\alpha_i\alpha_j \frac{\Delta E_i \Delta E_j}{\Delta E_i + \Delta E_j} \quad (1440)$$

where α_i is the polarizability of atom i and $\Delta E_i = |E_i^0 - E_i^1|$ is the separation between the ground and first excited states in atom i. Setting $i = j$ in Eq. (1440), we return to Eq. (1439). It is reasonable to identify ΔE_i with the first ionization energy I_i for atom i. Then if we can determine I_i and α_i, we can estimate the strength of the van der Waals attraction between two atoms. In Table 4.5 we list the theoretical values for the A_{ij} determined in this way for the noble gases. In Table 4.6 we give the corresponding experimental results. We see that the agreement is reasonable.

Using this coarse-graining approach, we have justified the $-V_0(\sigma/R)^6$ attractive part of the Lennard-Jones potential. What about the short-range repulsion, which we represent with the $V_0(\sigma/R)^{12}$ term in the Lennard-Jones potential? One can attempt to determine the short-range repulsive part of the interaction using some of the same methods discussed in the next section, where we consider diatomic molecules. A wide variety of models and fits to data are discussed by Reed and Gubbins [34]. Various forms fitting the short-range behavior are possible and the R^{-12} form does not stand out on its merits. Nonetheless, the R^{-12} form has become standard for historical reasons, as discussed by Brush [35].

TABLE 4.5 R^{-6} Attraction Coefficients Given by Eq. (1440) in Atomic Units[a]

Gas	I (eV)	$\alpha(\text{Å}^3)$	He	Ne	Ar	Kr	Xe
He	24.45	0.2036	1.27	2.29	8.03	11.1	16.5
Ne	21.47	0.3925	—	4.15	14.7	20.4	30.5
Ar	15.96	1.6264	—	—	53.0	74.6	112.2
Kr	13.94	2.4550	—	—	—	105.5	159.5
Xe	12.08	3.9989	—	—	—	—	242.4

Source: Data from Ref. 32.
[a] 1 a.u.= 0.95723×10^{-12} erg · Å^6

TABLE 4.6 R^{-6} **Attraction Coefficients from Spectroscopic Data in Atomic Units**[a]

Gas	He	Ne	Ar	Kr	Xe
He	1.456	3.01	9.63	13.5	18.7
Ne	—	6.31	19.7	27.4	37.7
Ar	—	—	65.4	92.3	131
Kr	—	—	—	131	186
Xe	—	—	—	—	269

Source: Data from Ref. 33.
[a] 1 a.u.= 0.95723×10^{-12} erg Å^6.

This is typical in the coarse-graining process. We find a parameterization on the slow variable scale which is derived carefully in some special cases and then modeled empirically more generally.

4.9.4 Vibrating-Rotor Model for Diatomic Molecules

Let us turn to the more complicated situation of the effective interaction between atoms forming a molecule. In Section 4.3 we noted that the vibrating-rotor model gives a good description of thermodynamic properties over a wide range of temperatures. The internal rotations and vibrations of diatomic molecules leads to the shifting of C_V from $\frac{3}{2}k_B$ at lower temperatures to $\frac{7}{2}k_B$ at high temperatures. How do this model and the associated effective Hamiltonian follow from the underlying atomic systems.

We have emphasized that we can derive effective Hamiltonians in the case where there is a separation of energy scales. Why do we expect such a separation for molecules? The key idea is that a molecule is a stable arrangement of well-separated nuclei, and the *large* nuclei move slowly relative to the *small* electrons. This is the essence of the Born–Oppenheimer [36] approximation. Thus the nuclei of the atoms have a well-defined average separation R_0 which is larger than the average atomic size a (of course, R_0 must be greater than $2a$) and the nuclear separation does not vary greatly from R_0 for the temperatures of interest. A major difference between the situation here and in the preceding section is that we cannot assume that the electrons are localized on either nucleus [37]. The full Hamiltonian is similar to that treated in the preceding section except that we must allow for different species of nuclei with masses M_j:

$$H = K_N + K_e + V \qquad (1441)$$

where K_N is the kinetic energy of the two nuclei, K_e the kinetic energy of all the electrons, and V the Coulomb interaction between all the charges, nuclear and electronic. Let us write

$$H = H_0 + K_N \qquad (1442)$$

and treat K_N as a perturbation. Our basic expansion parameter is m/M, which is typically of order 10^{-3} to 10^{-4}. The fast degrees of freedom, the positions of the electrons $\mathbf{r}_i$, and the slow degrees of freedom, the nuclear coordinates $\mathbf{R}_j$, are strongly coupled in the interaction V. Let us represent the set of electronic coordinates by $\mathbf{r} = \{\mathbf{r}_1, \mathbf{r}_2, \ldots, \mathbf{r}_n\}$ and the nuclear coordinates by $\mathbf{R} = \{\mathbf{R}_1, \mathbf{R}_2\}$. We assume (a nontrivial assumption) that we can find solutions for the eigenvalue problem

$$H_0(\mathbf{r}, \mathbf{R}) \psi_l(\mathbf{r}, \mathbf{R}) = W_l(\mathbf{R}) \psi_l(\mathbf{r}, \mathbf{R}) \tag{1443}$$

which for a fixed set of nuclear coordinates gives a set of electronic eigenvalues $W_0(\mathbf{R}), W_1(\mathbf{R}), \ldots$ labeled by l. This self-consistently assumes that the $\mathbf{R}$ values are not changing *drastically* (i.e., they are part of the quasistatic nuclear motions).

The averaging process corresponds to summing over the $\mathbf{r}$ dependence in the space spanned by the states $\psi_l(\mathbf{r}, \mathbf{R})$. Thus we can define an effective Hamiltonian, to first order in K_N, for each electronic state (which are widely separated on the scale of $k_B T$) by

$$H_E^s(\mathbf{R}) = W_l(\mathbf{R}) + \int d^3 r \, \psi_l(\mathbf{r}, \mathbf{R}) K_N \psi_l(\mathbf{r}, \mathbf{R}) + \mathcal{O}(K_N^2). \tag{1444}$$

Remembering that $H_E^s(\mathbf{R})$ is an operator, that $\psi_l(\mathbf{r}, \mathbf{R})$ is normalized, and that we can construct $\psi_l(\mathbf{r}, \mathbf{R})$ to be real, we can further reduce this result (see Problem 4.27) to obtain the final result for the effective Hamiltonian governing the nuclear motion:

$$H_{E,l}^s(\mathbf{R}) = K_N + u_l(\mathbf{R}) \tag{1445}$$

where

$$u_l(\mathbf{R}) = W_l(\mathbf{R}) + W_l'(\mathbf{R}), \tag{1446}$$

and $W_l(\mathbf{R})$ is the eigenvalue solution to Eq. (1443) and

$$W_l'(\mathbf{R}) = \sum_{j=1}^{N} \frac{\hbar^2}{2M_j} \int d^3 r (\nabla_{R_j} \psi_l(\mathbf{r}, \mathbf{R}))^2. \tag{1447}$$

It is clear that we have divided our original problem into a number of separate problems. In the first stage we must treat the eigenvalue problem Eq. (1443) to obtain $W_l(\mathbf{R})$ and $\psi_l(\mathbf{r}, \mathbf{R})$. Using $\psi_l(\mathbf{r}, \mathbf{R})$, we can then determine $W_l'(\mathbf{R})$ and the effective nuclear potential $u_l(\mathbf{R})$. For our purposes we can assume that for the electronic degrees of freedom, the first excited state lies several electron volts above the ground state and we need only consider the ground-state effective Hamiltonian $H_E^s = H_{E,0}^s$. The actual computation of $u_0(\mathbf{R})$ is very complicated. However, as shown in

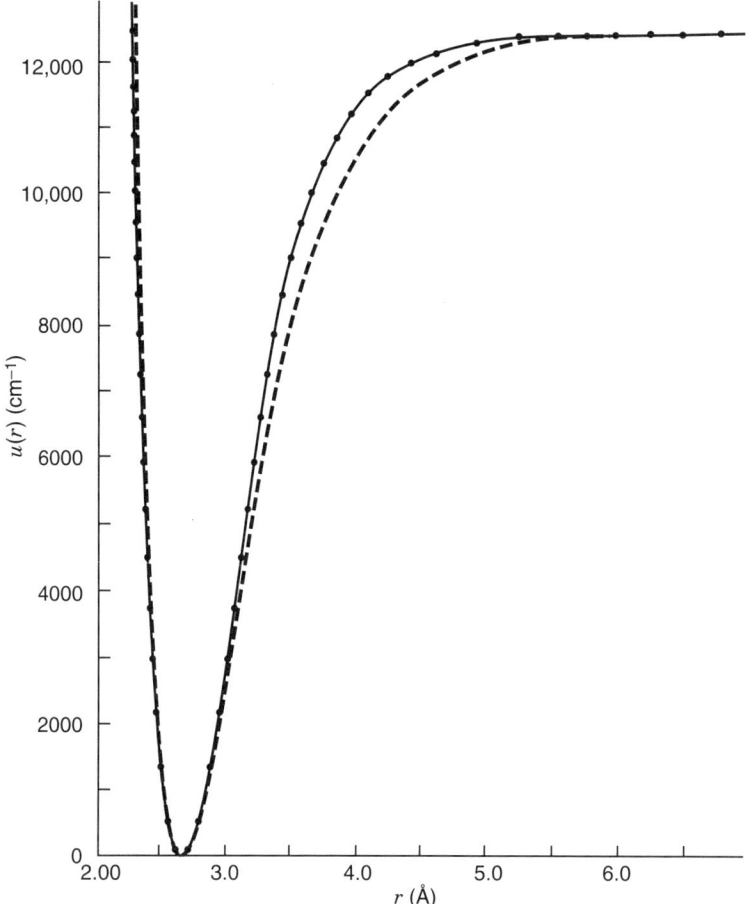

FIGURE 4.21 The internuclear potential energy curve for the ground state of I_2 as computed from ultraviolet spectroscopy. The dashed curve is a theoretical fit to the potential. [From R. D. Verma, *J. Chem. Phys.* **32**, 738 (1960).]

Fig. 4.21, one obtains potentials that have a deep minimum and can be parameterized by something like a Morse potential (see Problem 4.28). Notice that we have the physics here to determine the short-range repulsive part of the Lennard-Jones potential.

This means that our problem for the statistical mechanics of a diatomic molecule at room temperatures has been reduced to a solution of the Schrödinger equation for a two-body problem where the particles interact via a central potential $u_0(R)$. It is standard to separate this problem into one for the center-of-mass motion and one for the relative motion. The center-of-mass motion is what we referred to as the translational motion in Section 4.1 and treated in detail in Chapter 3. In dealing with the relative motion we need the energy eigenstates for the relative motion governed by the Schrödinger equation:

$$\left[-\frac{\hbar^2}{2\mu}\nabla_R^2 + u_0(R)\right]\psi(\mathbf{R}) = E\psi(\mathbf{R}) \tag{1448}$$

where, as usual, μ is the reduced mass for the two nuclei. We can write the Laplacian in the form

$$\nabla_R^2 = \frac{1}{R}\frac{\partial^2}{\partial R^2}R + \frac{\mathbf{L}^2}{\hbar^2 R^2} \tag{1449}$$

where $\mathbf{L}$ is the relative orbital angular momentum operator. It is a standard result [38] that we can find separated solutions of this Schrödinger equation of the form

$$\psi(\mathbf{R}) = Y_{lm}(\theta, \phi) f_{ln}(R) \tag{1450}$$

where the Y_{lm} are spherical harmonics and the radial component, f_{ln}, satisfies

$$\left[-\frac{\hbar^2}{2\mu}\frac{1}{R}\frac{\partial^2}{\partial R^2}R - \frac{1}{2\mu}\frac{\hbar^2 l(l+1)}{\hbar^2 R^2} + u_0(R)\right]f_{ln}(R) = Ef_{ln}(R). \tag{1451}$$

The next step is crucial is separating the problem into a decoupled problem between rotation and vibration. It is essential that the minimum of u_0 be sharp and deep. Thus it is a good approximation to assume that the two atoms are separated by the distance R_0 corresponding to the bottom of the well and determined by

$$\left(\frac{du_0(R)}{dR}\right)_{R=R_0} = 0. \tag{1452}$$

The potential can then be approximated by

$$u_0(R) = u_0(R_0) + \tfrac{1}{2}k(R - R_0)^2 + \cdots \tag{1453}$$

where the *force constant k* is defined by

$$k = u''(R_0). \tag{1454}$$

Using this result and that we should write $f_{ln}(R) = f_{ln}(R - R_0)$, it is straightforward to show (see Problem 4.28) that the relative energy eigenvalues for the diatomic system are given by

$$E_{n,l} = u(R_0) + \frac{\hbar^2}{2\mu}\frac{l(l+1)}{R_0^2} + \hbar\omega_0(n + \tfrac{1}{2}) \tag{1455}$$

which agrees with the result given by Eqs. (1118) and (1119), after we subtract the ground-state energy and identify the moment of inertia

$$I_0 = \mu R_0^2 \qquad (1456)$$

and the vibrational frequency

$$\omega_0 = \sqrt{\frac{k}{\mu}}. \qquad (1457)$$

Thus once we know the reduced mass μ, the distance of separation R_0, and the curvature of the potential at the bottom $k = u''(R_0)$, we know Θ_r and Θ_v, which characterize the theory in Section 4.1.3. Thus we have made contact with the underlying atomic physics in the problem.

4.10 COMPOSITE PARTICLES: DISSOCIATION AND IONIZATION

4.10.1 Introduction

In our treatment of chemical reactions in Chapter 2 we indicated that the key ingredient was the reaction formula of the form

$$A + B \leftrightarrow AB. \qquad (1458)$$

In the case of ideal constituents, the subsequent analysis is straightforward, as shown in Chapter 2, where we derived the law of mass action. When we look more closely and include interactions, however, the analysis is more difficult. How do we differentiate between the molecule AB and its dissociated pair A and B? Let us look at a simple example where we can understand the problem in detail.

4.10.2 Ionization of Atomic Hydrogen

Let us suppose that we have a gas of atomic hydrogen (for the sake of argument, let us ignore the presence of molecular hydrogen, which would only lead to further complications). From a more elementary point of view, this could be thought of as a gas of equal numbers of electrons and protons. These can combine to form hydrogen via the simple reaction

$$e + p \leftrightarrow H \qquad (1459)$$

where the inverse reaction is ionization. If we assume the system is sufficiently dilute that the constituents can be treated as noninteracting, we have a Hamiltonian of the form

$$H = \sum_{i=1}^{N_H} H_H^i + \sum_{i=1}^{N_e} H_e^i + \sum_{i=1}^{N_p} H_p^i \tag{1460}$$

where N_H, N_e, and N_p are the number of hydrogen atoms, electrons, and protons, respectively. We assume that the electrons and protons are point particles and that H_e^i and H_p^i are the associated kinetic energies and H_H^i is the Hamiltonian for the ith hydrogen atom. Remembering our results from Chapter 2, we associate a chemical potential μ_i with each component, and if the chemical reaction Eq. (1458) is written in the form

$$\sum_{i=1}^{3} b_i B_i = 0, \tag{1461}$$

with $B_1 = H$, $b_1 = 1$, $B_2 = p$, $b_2 = -1$, $B_3 = e$, and $b_3 = -1$, the chemical potentials must satisfy

$$\sum_{i=1}^{3} \mu_i b_i = 0. \tag{1462}$$

Since we assume no direct coupling between the three terms in the Hamiltonian, the thermodynamic potential for the entire system is given by

$$\begin{aligned}
\Omega &= -\beta^{-1} \ln \sum_{N_H=0}^{\infty} \frac{1}{N_H!} \operatorname{Tr}^H e^{-\beta(H_H - \mu_H N_H)} \sum_{N_p=0}^{\infty} \frac{1}{N_p!} \operatorname{Tr}^p e^{-\beta(H_p - \mu_p N_p)} \\
&\quad \times \sum_{N_e=0}^{\infty} \frac{1}{N_e!} \operatorname{Tr}^e e^{-\beta(H_e - \mu_e N_e)} \\
&= -\beta^{-1} \ln \left[\exp\left(\frac{V}{\lambda_H^3} e^{\beta \mu_H} Z_B\right) \exp\left(\frac{V}{\lambda_p^3} e^{\beta \mu_p}\right) \exp\left(\frac{V}{\lambda_e^3} e^{\beta \mu_e}\right) \right] \\
&= -\beta^{-1} V \left(\frac{e^{\beta \mu_H}}{\lambda_H^3} Z_B + \frac{e^{\beta \mu_p}}{\lambda_p^3} + \frac{e^{\beta \mu_e}}{\lambda_e^3} \right)
\end{aligned} \tag{1463}$$

where Z_B is the partition function associated with the internal states of hydrogen and we obtain the expected linear volume dependence from the translational degrees of freedom.

Let us return to our condition for equilibrium given by Eq. (1462). Clearly, the number of each type of particle is given by taking derivatives of Ω with respect to $\alpha_i = -\beta \mu_i$, so

$$N_H = \frac{V e^{\beta \mu_H}}{\lambda_H^3} Z_B \tag{1464}$$

$$N_p = \frac{V e^{\beta \mu_p}}{\lambda_p^3} \tag{1465}$$

$$N_e = \frac{V e^{\beta \mu_e}}{\lambda_e^3} \tag{1466}$$

where

$$\lambda_H = \left(\frac{2\pi \hbar^2}{(m_p + m_e) k_B T} \right)^{1/2} \tag{1467}$$

$$\lambda_p = \left(\frac{2\pi \hbar^2}{m_p k_B T} \right)^{1/2} \tag{1468}$$

$$\lambda_e = \left(\frac{2\pi \hbar^2}{m_e k_B T} \right)^{1/2} \tag{1469}$$

The pressure is given by

$$pV = -\Omega = V\beta^{-1} \left(\frac{e^{\beta \mu_H}}{\lambda_H^3} Z_B + \frac{e^{\beta \mu_p}}{\lambda_p^3} + \frac{e^{\beta \mu_e}}{\lambda_e^3} \right)$$

or

$$p = \beta^{-1}(n_H + n_p + n_e) \equiv \beta^{-1} n_T. \tag{1470}$$

The equilibrium condition, Eq. (1462),

$$\mu_H = \mu_p + \mu_e, \tag{1471}$$

becomes, after using Eqs. (1464), (1465), and (1466),

$$\ln n_H \lambda_H^3 Z_B^{-1} = \ln n_p \lambda_p^3 + \ln n_e \lambda_e^3. \tag{1472}$$

This can be rewritten as in the law of mass action:

$$\frac{n_H}{n_p n_e} = K \equiv \frac{\lambda_p^3 \lambda_e^3}{\lambda_H^3} Z_B \tag{1473}$$

where, explicitly,

$$\frac{\lambda_p^3 \lambda_e^3}{\lambda_H^3} = \left(\frac{2\pi\hbar^2}{k_B T \mu}\right)^{3/2} = (4\pi\beta R a_0^2)^{3/2}, \quad (1474)$$

$\mu = m_p m_e/(m_p + m_e)$ is the reduced mass, $a_0 = \hbar^2/\mu e^2$ is the Bohr radius, and $R = \mu e^4/2\hbar^2 = 13.6$ eV is Rydberg's constant.

Finally, if we started out with only hydrogen atoms, then

$$n_p = n_e \quad (1475)$$

due to ionization. We can then relate the equilibrium concentrations to the equilibrium temperature and pressure. From Eq. (1473) we have

$$n_H = K n_p^2 \quad (1476)$$

which can be used in Eq. (1470) to eliminate n_H and obtain

$$K n_p^2 + 2 n_p - p\beta = 0. \quad (1477)$$

This has the solution

$$n_p = \frac{1}{K}\left[(1 + p\beta K)^{1/2} - 1\right] \quad (1478)$$

and

$$n_H = \frac{1}{K}\left[p\beta K + 2(1 - (1 + p\beta K)^{1/2})\right]. \quad (1479)$$

The degree of ionization is defined by

$$\alpha = \frac{n_p}{n_p + n_H} \quad (1480)$$

and we easily find, using Eqs. (1478) and (1479), that

$$\alpha = \frac{1}{(1 + p\beta K)^{1/2}}. \quad (1481)$$

Up to this point, all the development is in parallel with the work in Chapter 2. All that is left is the evaluation of Z_B, which enters Eq. (1473) and then Eq. (1481), which is a sum over the internal hydrogen bound states. We, of course, know all the bound states of atomic hydrogen. If H_r is the relative Hamiltonian, for hydrogen

$$H_r = \frac{p^2}{2\mu} - \frac{e^2}{r} \tag{1482}$$

where μ is the reduced mass, then

$$H_r |nlm\rangle = E_n |nlm\rangle \tag{1483}$$

where n, l, and m are all the standard radial, orbital angular momentum, and z-component angular momentum quantum numbers, and the energy levels are given by

$$E_n = -\frac{R}{n^2} \tag{1484}$$

where R is Rydberg's constant, introduced above, and the principal quantum number takes on integer values, $n = 1, 2, \ldots$. The evaluation of Z_B is then, in principle, straightforward:

$$Z_B = \sum_{n=1}^{\infty} 2n^2 e^{-\beta E_n} = \sum_{n=1}^{\infty} 2n^2 e^{\beta R/n^2} \tag{1485}$$

where $2n^2$ is the degeneracy of the states, with energy E_n, including a factor of 2 for spin. We see immediately that the sum in Z_B does not converge, going as $\sum_n 2n^2$ for large n.

What has gone wrong? Clearly, we have made some type of error [39]. The first step in remedying our error is to remember [40] that our system is constrained to be in a box of volume V with linear dimensions L. We should recall that in the hydrogen atom the mean radius of a state with principal quantum number n is $a_0 n^2 \equiv L_H$, where $a_0 = \hbar^2/\mu e^2$ is the Bohr radius. Clearly, then, states where $L_H > L$ do not, in a rough sense, exist. This means that we can cut off the states for $n > n_m$ where n_m satisfies $a_0 n_m^2 = L$ or $n_m = (L/a_0)^{1/2}$. Then

$$Z_B = \sum_{n=1}^{n_m} 2n^2 e^{\beta R/n^2} = \sum_{n=1}^{n_m} 2n^2 + \sum_{n=1}^{n_m} 2n^2 (e^{\beta R/n^2} - 1)$$

$$= \sum_{n=1}^{n_m} 2n^2 + \sum_{n=1}^{n_m} 2n^2 \frac{\beta R}{n^2} + \sum_{n=1}^{\infty} 2n^2 \left(e^{\beta R/n^2} - 1 - \frac{\beta R}{n^2} \right) \tag{1486}$$

where, in the last term, which is a convergent sum, we have allowed $n_m \to \infty$. We have then

$$Z_B = \frac{n_m}{3} (2n_m^2 + 3n_m + 1) + 2\beta R n_m + f(\beta R) \tag{1487}$$

where

$$f(x) = \sum_{n=1}^{\infty} 2n^2 \left(e^{x/n^2} - 1 - \frac{x}{n^2} \right), \quad (1488)$$

and we see that the divergence is cut off in a finite system. It is clear that a more careful treatment of this problem is to solve for the bound-state spectrum of hydrogen *in a box*.

This result for Z_H is still not acceptable since it depends on the size of the system. Our description is still not correct in the thermodynamic limit. The reason is that we have assumed that all bound states for the noninteracting e, p, H system remain bound in the interacting system. Consider some bound pair in a state labeled by radial quantum number n. Then we know that the average radial separation of the electron and proton in that state is

$$r_n^2 = \langle r^2 \rangle_n = \int d^3 r \, r^2 \psi_{nlm}^*(\mathbf{r}) \psi_{nlm}(\mathbf{r}) = n^4 a_0^2 \quad (1489)$$

and increases with n. If we have densities $n_H \approx l_H^{-3}$, $n_p \approx l_p^{-3}$, where $r_n \approx l_H$ or l_p, it does not make sense to talk of a bound electron and proton since they are individually free to interact with the other e, p or H constituents which may approach much closer than r_n. Thus in an interacting system, only rather tightly bound states survive interactions and should be identified with the atomic or molecular system.

Let us return to the case of the ionization of atomic hydrogen. In this case we already have a clue as to how to proceed in the interacting system. In Section 4.8 we indicated that interactions between charged particles *in the fluid* are screened. This conclusion holds in the case of ionized hydrogen and it appears that we must solve for the bound-state spectrum of an electron and proton interacting via the Yukawa potential,

$$V(\mathbf{r}) = -\frac{q^2}{r} e^{-r/\lambda_D} \quad (1490)$$

where λ_D is the Debye screening length for this system. This is not a problem that we can solve analytically, but a simple variational calculation (see Problem 4.31) shows that for finite λ_D there are a finite number of bound states, and as λ_D decreases the number of bound states decreases until for small enough λ_D, there are no bound states. The condition to have at least one bound state is

$$\frac{2\mu \lambda_D^2}{\hbar^2} \frac{e^2}{\lambda_D} > 2 \quad (1491)$$

or

$$2 \frac{\lambda_D}{a_0} > 2. \quad (1492)$$

This makes good physical sense. If the screening length is shorter than the Bohr radius, we have no bound states. In principle, for a given λ_D we must solve for the bound-state spectrum, the number of states and their eigenvalues E_n. We know generally that there are only a finite number of states and that the sum Z_B is well behaved.

Let us assume that we are in a rather dilute system where $\lambda_D \gg a_0$ and the ground state is not strongly affected by screening $E_G = E_1 = -R$. Then at modest temperatures we expect

$$Z_B \approx 2e^{-\beta E_G} = 2e^{\beta R} \tag{1493}$$

where $R = \mu e^4/2\hbar^2$. At what range of temperature is our system highly ionized? From Eq. (1481) we have that the degree of ionization is

$$\alpha = (1 + p\beta K)^{-1/2} \tag{1494}$$

where, from Eqs. (1470), (1473), and (1474),

$$p\beta K = 2(4\pi)^{3/2} a_0^3 n_T (\beta R)^{3/2} e^{\beta R}. \tag{1495}$$

We see that α is a strong function of temperature and density. Since $R = 13.6\text{eV} \approx 1.6 \times 10^4 k_B \text{K}$, we see that at room temperatures the ionization is dominated by the *very* small $e^{-\beta R/2}$ term.

REFERENCES

[1] An early restricted version of this theorem is due to J. J. Waterston in 1846. This work was published in full in *Philos Trans. R. Soc. London A* **183**, 5 (1893). The theorem is typically attributed to J. C. Maxwell, *Philos Mag.* **19**, 19 (1860) in one of his famous works on kinetic theory. Classically, it was difficult to figure out how to count the internal degrees of freedom, and this was a major impetus for the development of quantum theory. See the discussion in S. G. Brush, *Kinetic Theory*, Vol. I, Pergamon Press, Oxford, 1965.

[2] D. A. McQuarrie, *Statistical Mechanics*, Harper & Row, New York, 1976, p. 95.

[3] L. D. Landau and E. M. Lifshitz, *Statistical Physics*, Pergamon Press, Oxford, 1980, p. 137

[4] R. W. Fowler and E. Guggenheim, *Statistical Thermodynamics*, Cambridge University Press, London, 1949, p. 99.

[5] For a more thorough discussion of the case of homonuclear diatomic molecules and the role of quantum statistics, see D. A. McQuarrie, *Statistical Mechanics*, Harper & Row, New York, 1976, p. 101.

[6] K. F. Bonhoeffer and P. Harteck, *Z. Phys. Chem. B* **4**, 113 (1926).

[7] D. A. McQuarrie, *Statistical Mechanics*, Harper & Row, New York, 1976, Chap. 8; L. D. Landau and E. M. Lifshitz, *Statistical Physics*, Pergamon Press, Oxford, 1980, p. 148.

REFERENCES 339

[8] S. G. Brush, *Statistical Physics and the Atomic Theory of Matter*, Princeton University Press, Princeton, N.J. 1983.

[9] J. C. Maxwell, *Philos. Mag.* **19**, 19 (1860).

[10] T. M. Reed and K. E. Gubbins, *Applied Statistical Mechanics*, McGraw-Hill, New York, 1973, p. 117.

[11] J. E. Lennard-Jones, *Proc. Phys. Soc. London* **43**, 461 (1931). In Section 4.9.3 we discuss the origins of this potential from an atomic point of view.

[12] J. E. Mayer, *J. Chem. Phys.* **5**, 67 (1937); S. F. Harrison and J. E. Mayer, *J. Chem. Phys.* **6**, 101 (1938).

[13] For a complete discussion, see R. Kubo, *J. Phys. Soc.* **17**, 110 (1962).

[14] K. Onnes and W. H. Keesom, "Die Zustandsgleichung," *Commun. Phys. Lab. Leiden Suppl.* **23** (1912).

[15] Our discussion here follows that of R. K. Pathria, *Statistical Mechanics*, Pergamon Press, Oxford, 1972, p. 264.

[16] L. Boltzmann, *Amsterdam Ber.* 477 (1899).

[17] F. H. Ree and W. G. Hoover, *J. Chem. Phys.* **40**, 939 (1964).

[18] In Chapter 1 we introduced the radial distribution function but did not explore the information contained in this quantity.

[19] L. S. Ornstein and F. Zernike, *Proc. Acad. Sci. Amsterdam* **17**, 793 (1914).

[20] J. K. Percus and G. J. Yevick, *Phys. Rev.* **110**, 1 (1958).

[21] M. S. Wertheim, *J. Math. Phys.* **5**, 643 (1964).

[22] P. Egelstaff, *An Introduction to the Liquid State*, Academic Press, London 1967.

[23] N. N. Ashcroft and J. Lekner, *Phys. Rev.* **145**, 83 (1966).

[24] E. A. Guggenheim, *J. Chem. Phys.* **13**, 253 (1945).

[25] In E. A. Guggenheim, *Themodyamics*, 6th ed., North-Holland, Amsterdam, 1977, he cites later experiments which give $p_c/n_c T_c = 0.288$ for xenon.

[26] A. Lenard and F. J. Dyson, *J. Math. Phys.* **9**, 698 (1968).

[27] E. H. Lieb and H. Narnhofer, *J. Stat. Phys.* **12**, 291 (1975); **14**, 465 (1976).

[28] P. Debye and E. Hückel, *Phys. Z.* **24**, 185 (1923); translation in *The Collected Papers of Peter J. W. Debye*, Interscience, New York, 1954.

[29] The first quantum mechanical calculation of the van der Waals force was by S. C. Wang, *Phys. Z.* **28**, 663 (1927), followed by a more accurate calculation by R. Eigenschitz and F. London, Z. Phys. **60**, 491 (1930).

[30] F. London, *Trans. Faraday Soc.* **33**, 8 (1937).

[31] H. Margeneau, *Rev. Mod. Phys.* **11**, 1 (1939).

[32] T. M. Reed and K. E. Gubbins, *Applied Statistical Mechanics*, McGraw-Hill, New York, 1973, p. 106.

[33] T. M. Reed and K. E. Gubbins, *Applied Statistical Mechanics*, McGraw-Hill, New York, 1973, p. 108; A. E. Kingston, *Phys. Rev. A.* **135** 1018 (1964).

[34] T. M. Reed and K. E. Gubbins, *Applied Statistical Mechanics*, McGraw-Hill, New York, 1973, p. 94.

[35] S. G. Brush, *Statistical Physics and the Atomic Theory of Matter*, Princeton University Press, Princeton, N.J. 1983, p. 212.

[36] M. Born and J. R. Oppenheimer, *Ann. Phys.* **84**, 457 (1927).
[37] The original calculation of this type was carried out by W. Heitler and F. London, *Z Phys.* **44**, 455 (1927).
[38] K. Gottfried, *Quantum Mechanics,* Vol. I, W. A. Benjamin, New York, 1966, p. 44.
[39] See the discussion in S. G. Brush, *Statistical Physics and the Atomic Theory of Matter,* Princeton University Press, Princeton, N.J. 1983, pp. 151 and 153. See also R. H. Fowler, *Philos. Mag.* **45**, 1 (1923); E. Fermi, *Collected Papers,* Vol. I, University of Chicago Press, Chicago, 1962, p. 118; and H. C. Urey, *Astrophy. S. J.* **59**, 1 (1924).
[40] S. J. Stickler, *J. Chem. Educ.* **43**, 364 (1966). In this article it is shown that for systems at room temperatures enclosed in a box, the contributions from the excited states are down by *many* orders of magnitude from those from the ground state.

PROBLEMS

4.1. Calculate the partition function in the grand canonical ensemble for a system of particles in an external nonlocal field described by the Hamiltonian

$$H = \sum_{i=1}^{N} \frac{\mathbf{p}_i^2}{2m} + \int d^3x\, u(\mathbf{x}) n(\mathbf{x})$$

where $n(\mathbf{x})$ is the particle density. Calculate the average density and pressure, and evaluate the chemical potential. Discuss the case where $u(\mathbf{x}) = -mgz$ is a gravitational potential.

4.2. Consider a Hamiltonian with a general quadratic form

$$H = \tfrac{1}{2}\sum_{ij} b_{ij} x_i x_j$$

with $i = 1, 2, \ldots, N$. If the matrix b has N positive eigenvalues and no negative ones, find the total average energy $\langle E \rangle$ when the system is at temperature $1/\beta$.

4.3. Consider a classical set of harmonic oscillators described by the Hamiltonian

$$H = \sum_{i=1}^{N} \frac{1}{2m} \mathbf{p}_i^2 + \frac{1}{2}\sum_{ij} \phi_{ij} \boldsymbol{\xi}_i \cdot \boldsymbol{\xi}_j$$

where i labels the N lattice sites located at $\mathbf{R}_i$. The potential energy can be written as

$$\tfrac{1}{2}\sum_{k_n} \phi(k_n) \boldsymbol{\xi}_{k_n} \cdot \boldsymbol{\xi}_{k_n}^*$$

where the sum is over reciprocal lattice vectors $\mathbf{k}_n$ and the Fourier transforms are defined by

$$\boldsymbol{\xi}_{k_n} = \frac{1}{\sqrt{N}} \sum_i e^{-i\mathbf{k}_n \cdot \mathbf{R}_i} \boldsymbol{\xi}_i$$

and

$$\phi(k_n) = \frac{1}{N}\sum_{ij} e^{-i\mathbf{k}_n\cdot(\mathbf{R}_i-\mathbf{R}_j)}\phi_{ij}.$$

Evaluate the average displacement squared,

$$r_0^2 = \frac{1}{N}\sum_{i=1}^{N}\langle(\xi_i)^2\rangle,$$

in the continuum approximation where, in d-dimensions,

$$\frac{1}{N}\sum_{k_n} \to a^d \int \frac{d^d k}{(2\pi)^d}$$

where a is the lattice spacing and

$$\phi(k) = \epsilon_o k^2 a^2(1+a^2 k^2).$$

Carry out the evaluation for both $d = 2$ and 3. (*Hint* : The Jacobian for change of variables $[\xi_i \to (\xi_{kn})]$ is a constant independent of the ξ's. An upper bound on the melting temperature T_m is given by $r_0 = a$. Determine this upper bound on T_m.

4.4. The rotational motion of a diatomic molecule is specified by the usual two angular variables θ and ψ and the corresponding canonically conjugate momenta P_θ and P_ψ. The rotational contribution to the kinetic energy is given by

$$K_{\text{rot}} = \frac{P_\theta^2}{2I} + \frac{1}{2I\sin^2\theta}P_\psi^2$$

(where I is the moment of inertia and the variables are in the ranges $0 < \psi < 2\pi, 0 < \theta < \pi, -\infty < P_\theta < \infty, -\infty < P_\psi < \infty$).

(a) Assuming that the Hamiltonian for the total system of N molecules is given by

$$H = \sum_i H_i^m$$

where

$$H_i^m = \frac{P_i^2}{2M} + K_{\text{rot}}^i,$$

P_i is the momentum of the center of mass, and M is the total mass of the molecule, compute the pressure, chemical potential, and specific heat (C_V) as functions of T, V, and the average particle number.

(b) Can you conceive of an experimental method for determining if a gas is monatomic or diatomic using this information?

4.5. Consider a gas of diatomic molecules. The molecules do not interact with each other, but the atoms in the molecules (each of mass M) interact with a potential energy

$$V(x) = \frac{k}{2}x^2 + ux^4$$

where x is the distance separating the two atoms in the molecule.

(a) Calculate the average energy per molecule in the thermodynamic limit. (You should be able to express your answer in terms of a simple one-dimensional integral).

(b) Discuss the limit $u \to 0$; the limit $k \to 0$. What do you obtain if both u and k go to zero?

4.6. Consider the Euler–MacLaurin summation formula:

$$\sum_{n=a}^{b} f(n) = \int_a^b dn\, f(n) + \frac{1}{2}[f(a) + f(b)]$$

$$+ \sum_{j=1}^{\infty}(-1)^j \frac{B_j}{(2j)!}\left[f^{(2j-1)}(a) - f^{(2j-1)}(b)\right]$$

where the B_j's are the Bernoulli numbers $B_1 = \frac{1}{6}$, $B_2 = \frac{1}{30}$, $B_3 = \frac{1}{42}$, …. Use this result to evaluate the sum Z_r given by Eq. (1124).

4.7. Show that the ratio of the number of *ortho*-hydrogen molecules to *para*-hydrogen molecules in equilibrium is the ratio of their contributions to the molecular partition function. Evaluate the ratio in the high- and low-temperature limits.

4.8. A gas of molecules (mass m) with a Maxwell distribution of velocities at temperature T is in a container bounded by a piston of area A. The piston is moving outward with a velocity u (small compared to $\sqrt{k_B T/m}$) and expanding the gas adiabatically. Show that because of the motion of the piston, each molecule that strikes the piston with velocity v at an angle of incidence Θ rebounds with a loss of kinetic energy of an amount $2mvu\cos\Theta\, (u \ll v)$. Show that the gas loses energy, per second, by an amount

$$-dE = pAu = p\,dV.$$

4.9. The Doppler formula for the observed frequency f from a source moving with velocity V_x along the line of sight to the observer is

$$f = f_o\left(1 + \frac{V_x}{c}\right)$$

where f_o is the frequency radiated when the source is at rest and c is the velocity of light.

(a) What is the distribution in frequency of a particular spectrum line radiated from a gas at temperature T?

(b) What is the variance $\langle (f - f_0)^2 \rangle$ of this radiated frequency? The square root of the variance is called the *width* of the line.

(c) Atomic hydrogen and atomic oxygen are both present in a hot gas. How much broader will the hydrogen lines be compared to oxygen lines of roughly the same frequency?

4.10. Consider a fluid system described by the Hamiltonian

$$H = \sum_{i=1}^{N} \frac{\mathbf{p}_i^2}{2m} + \frac{1}{2} \sum_{ij=1}^{N} V(\mathbf{r}_i - \mathbf{r}_j).$$

Working in the grand canonical ensemble, evaluate the quantity

$$\bar{f}(p) = \langle f(\mathbf{x}, \mathbf{p}) \rangle$$

where

$$f(\mathbf{x}, \mathbf{p}) = \sum_{i=1}^{N} \delta(\mathbf{x} - \mathbf{r}_i) \delta(\mathbf{p} - \mathbf{p}_i)$$

is the single-particle phase-space density.

4.11. The lattice gas is a very useful paradigm of an interacting system. The one-dimensional lattice gas has been used to understand, for example, the helix-coil transition of DNA molecules and the high-conductivity states of some one-dimensional conductors. This model consists of a ring of L sites, each of which may be either empty or occupied. In addition to the chemical potential, there is an interaction energy J for every pair of adjacent particles. A negative J constitutes an attraction, since it makes adjacent pairs more likely.

(a) By expanding the thermodynamic potential $\Omega = -pL$ in powers of the activity $a = \exp(\beta\mu)$, find the second virial coefficient b_2 of this gas as a function of J and the temperature β^{-1}. For what (negative) value of J does the second virial coefficient change sign?

(b) Show that the b_2 of this simple system may be expressed in terms of the number of adjacent pairs or the average energy $\langle E \rangle$.

4.12. Express the cluster coefficient B_4 in terms of the Q_Ns using the derivative expression given by Eq. (1214). Show that all of the contributions to B_4 come from connected graphs and reduce to the connected graphs shown in Fig. 4.9.

4.13. Consider a gas mixture of two types of particles. Assume that there is density n_A of particles of type A, which have mass m_A and can be treated as hard spheres with radius r_A. Similarly, there is a density n_B of particles of type B which have mass m_B and can be treated as hard spheres with radius r_B. Determine the equation of state $p = p(n_A, n_B, T)$ in a low-density expansion in

n_A and n_B. Keep terms, including those second order in the densities. You should be able to determine p explicitly in terms of the known parameters.

4.14. For given values of the density there exists a *Boyle temperature* for which the ideal gas law is valid even in the presence of interactions! Assuming that we are at low densities where our density expansion is valid and we can write the equation of state in the form

$$p = nk_B T - n^2 k_B T b_2(T)$$

where

$$b_2 = \int d^3 r (e^{-\beta V(r)} - 1)$$

and assuming that the interaction for this system is given by

$$V(r) = \begin{cases} \infty & r < r_0 \\ -V_0 & r_0 < r < r_1 \\ 0 & r > r_1, \end{cases}$$

find the Boyle temperature for this system as a function of V_0, r_0, and r_1.

4.15. Discuss the qualitative changes in C_p, α, and κ_T due to the first density correction to the ideal gas results. You may assume that the interaction for the system is of the form

$$V(r) = \begin{cases} \infty & r < r_0 \\ -u_0 \left(\dfrac{r_0}{r}\right)^m & r > r_0. \end{cases}$$

4.16. Work out the first density correction for the radial distribution function, given by Eq. (1316), for the case of hard spheres.

4.17. Show that

$$\left(\frac{\partial n}{\partial \mu}\right)_{T,V} = n^2 \kappa_T$$

where κ_T is the isothermal compressibility.

4.18. Show that the Percus–Yevick approximation for the radial distribution function, given by Eq. (1344), leads to the exact result for the direct correlation function at lowest order in the density.

4.19. Show that given the set of assumptions stated above Eq. (1352), the free energy can be written in the form given by Eq. (1352).

4.20. The law of rectilinear diameters states that near the liquid–gas coexistence curve, as we go to the critical point,

$$\frac{n_L + n_g}{2n_c} = 1 + \frac{3}{4}\left(1 - \frac{T}{T_c}\right).$$

Check this result for the van der Waals equation of state.

4.21. Using the virial theorem, in the form given by Eq. (1305), one can derive a result valid for Coulomb potentials:

$$p = \frac{1}{2}nk_BT + \frac{1}{3}\frac{E}{V}.$$

Generalize this result to the case of potentials that vary with a general power law $V \approx r^{-m}$.

4.22. Relate the two dimensionless parameters $\Gamma = \beta q^2/a$ and $\epsilon = \beta q^2/\lambda_D$ in the one-component plasma.

4.23. Starting with the perturbation theory result for the free energy given in Appendix N by Eq. (2820), take the low-temperature limit. Assume that in the low-temperature regime the physics is governed by the ground state $|0\rangle$ with energy eigenvalue E_G^0 separated from the first excited state $|1\rangle$ with energy eigenvalue E_1^0.

4.24. Carry out the algebra leading from Eq. (1424) to Eq. (1427).

4.25. Verify Eq. (1436) for the case where the ground states for the two atoms are isotropic.

4.26. In Chapter 5 we show that the atomic electric polarization can be written in the form

$$\alpha_E = \frac{2}{3}\sum_{m>0}\frac{|\langle m|\mu_E|0\rangle|^2}{E_m - E_0},$$

where m stands for the complete set of quantum numbers labeling the states of the decoupled set of closed-shell atoms. Find an upper bound on α_E in analogy to the development leading to Eq. (1437).

4.27. In working out the separation of electronic and nuclear degrees of freedom in Section 4.9.4, one is led to consider the operator

$$M_l(\mathbf{R}) = \int d^3r\, \psi_l(\mathbf{r},\mathbf{R}) K_N(\mathbf{R}) \psi_l(\mathbf{r},\mathbf{R}),$$

where K_N is the kinetic energy operator for the nucleus. Remembering that $\psi_l(\mathbf{r},\mathbf{R})$ is normalized and can be constructed to be real, show that

$$M_l(\mathbf{R}) = K_N(\mathbf{R}) + W_l'(\mathbf{R})$$

where

$$W_l' = \sum_{j=1}^N \frac{\hbar^2}{2M_j}\int d^3r\, [\nabla_{R_j}\psi_l(\mathbf{r},\mathbf{R})]^2.$$

4.28. Show how the energy spectrum given by Eq. (1455) follows from the radial Schrödinger equation Eq. (1451) under the assumptions of a sharp and deep potential $u(R)$.

4.29. When we treat the case of equilibrium for hydrogen atoms we run into problems when we try to compute directly the partition function for the bound states. In the case of an atom or molecule represented by a harmonic oscillator, we do not run into such problems. Explain why. Under what circumstances do we expect the oscillator model to be appropriate?

4.30. Assume, in our analysis of the degree of dissociation in Section 4.10, that we must keep two bound states, $E_1 = -R$ and $E_2 = -R/4$. Discuss quantitatively the range of temperatures over which the system is essentially dissociated $\alpha > 0.9$ and the range of temperatures over which the system is essentially a set of atoms $\alpha < 0.1$. Over what temperature range is our assumption that we need keep only E_1 valid?

4.31. Use a variational calculation to determine the number of bound states associated with the Yukawa potential given by Eq. (1490).

5 Equilibrium Properties of Dielectric and Magnetic Materials

5.1 INTRODUCTION

Electromagnetism arises in nature because elementary particles carry charge and intrinsic quantum mechanical spin. The general problem of the interaction of matter with electromagnetic radiation is a fundamental problem of great complexity. If we consider stable matter in thermal equilibrium interacting with static external electromagnetic fields, the possibilities are much more restricted. The case of weak, time-dependent external fields can be treated in nonequilibrium statistical mechanics using linear response theory. However, equilibrium requires that the electric and magnetic fields be static. Further, we cannot maintain thermal equilibrium when applying a static electric field in those cases where we have mobile charged particles. In a plasma (electrically neutral fluid with free charged particles) the charged particles will separate if a static electric field $\mathbf{E}$ is applied and the system will be taken far from equilibrium. Of course, similar questions arise in the study of charge transport in a metal where the applied electric field $\mathbf{E}$ induces a current $\mathbf{J}$. The application of a static magnetic field $\mathbf{B}$ to a system of charged particles is not incompatible with maintaining a stationary equilibrium state since $\mathbf{B}$ induces the particles (classically) to move in circular orbits. The equilibrium statistical mechanics of free charged particles in a static magnetic field is more subtle than we might guess and is treated in Section 5.2 after we have treated systems that are a bit simpler.

In charge-neutral materials there are four main considerations in determining their response to static external electric and magnetic fields:

1. On the atomic or molecular level we must decide whether the atom or molecule has a net magnetic or electric dipole moment. This question can be subtle since it requires characterizing the entire molecule by a *net* or coarse-grained property. If the molecule is nonmagnetic and nonpolar, its response properties are rather simple, and we discuss this case in Section 5.2.
2. If the molecule has a net magnetic or electric dipole moment, it is magnetic or polar, and in the case of dilute fluids, its response properties are well described by the Debye–Langevin theory of paramagnetism and dielectrics described in Section 5.5.

3. In concentrated materials, magnetic and electric dipoles interact via a long-range dipolar potential which leads to much of the physics associated with electro- and magnetostatics. The statistical mechanical treatment of this interaction is difficult. We present here a traditional treatment of these effects due to Lorentz and defer more complete treatment until Volume 2.

4. In the discussion upto this point there was an apparent symmetry between magnetic and dielectric properties. This symmetry is a bit of an illusion because of the existence of the exchange interaction in magnetic materials. As discussed in more detail below, because of intrinsic spin, magnetic moments interact with other magnetic moments with a component that is short ranged. In many systems this exchange interaction dominates the dipolar interaction and controls the nature of the magnetic ordering in the systems. It is a complicating factor in real materials that one must, on the longest-distance scales, include the dipolar interactions because, while they are weak, they are long ranged and couple many particles. The treatment of systems where there is a competition between the exchange and dipolar interactions will be discussed in Volume 2.

Magnetic and dielectric behavior are diverse and complicated phenomena. We begin with a rather general discussion and our detailed results are restricted to a number of oversimplified models. These models like the Ising and Heisenberg models, have a wide range of applicability to various problems in statistical physics, and as pointed out in more detail in Volume 2, many of the phenomena elucidated by their analysis are quite universal.

5.2 PERTURBATION THEORY

We begin our analysis with a rather general discussion of the response of a collection of charged particles, carrying spin, to weak static electric and magnetic fields. This will give some insight into the structure of the theory and allow us to establish some general results essentially independent of the details of the specific system of interest.

Restricting ourselves to the cases of constant $\mathbf{E}$ and $\mathbf{B}$ fields, the Hamiltonian governing [1] the motion of particles with coordinates $\mathbf{r}_i$, momentum $\mathbf{p}_i$, mass m_i, charge q_i and intrinsic spin $\mathbf{s}_i$ is given by

$$H = \sum_{i=1}^{N} \left\{ \frac{1}{2m_i} \left[\mathbf{p}_i - \frac{q_i}{c} \mathbf{A}(\mathbf{r}_i) \right]^2 + q_i \phi(\mathbf{r}_i) - \mu_B g_i \mathbf{s}_i \cdot \mathbf{B} \right\} + V \qquad (1496)$$

where g_i is the associated g-factor ($= 2$ for electrons), $\mu_B = e\hbar/2mc$ is the Bohr magneton, and the potential V includes all the pair interactions between particles, electrostatic and otherwise. For constant $\mathbf{B}$ and $\mathbf{E}$ fields, the vector potential $\mathbf{A}$ can be written as

$$\mathbf{A}(\mathbf{r}) = \tfrac{1}{2}(\mathbf{B} \times \mathbf{r}) \qquad (1497)$$

and the scalar potential ϕ can be expressed as

$$\phi(\mathbf{r}) = -\mathbf{E} \cdot \mathbf{r}. \tag{1498}$$

These choices for $\mathbf{A}$ and ϕ then generate the constant fields $\mathbf{B}$ and $\mathbf{E}$ via

$$\mathbf{B} = \nabla \times \mathbf{A} \tag{1499}$$
$$\mathbf{E} = -\nabla \phi. \tag{1500}$$

If we separate out the $\mathbf{E}$ and $\mathbf{B}$ dependence, Eq. (1496) can be rewritten as

$$H = H_0 - \boldsymbol{\mu}_E \cdot \mathbf{E} - \boldsymbol{\mu}_M \cdot \mathbf{B} + \sum_{i=1}^{N} \frac{q_i^2}{8m_i c^2} (\mathbf{B} \times \mathbf{r}_i)^2 \tag{1501}$$

where H_0 is the Hamiltonian in the absence of the fields,

$$H_0 = K + V, \tag{1502}$$

and K is the kinetic energy in the absence of electromagnetic forces. In Eq. (1501), the magnetic moment operator is

$$\boldsymbol{\mu}_M = \sum_{i=1}^{N} (\mu_B g_i \mathbf{s}_i + \mu_i \mathbf{L}_i), \tag{1503}$$

where

$$\mathbf{L}_i = \mathbf{r}_i \times \mathbf{p}_i \tag{1504}$$

is the orbital angular momentum operator,

$$\mu_i = \frac{q_i \hbar}{2 m_i c}, \tag{1505}$$

and the electric dipole operator is

$$\boldsymbol{\mu}_E = \sum_{i=1}^{N} q_i \mathbf{r}_i. \tag{1506}$$

We can make some general statements without specifying the system further. It is shown in Appendix N that for a quantum Hamiltonian of the form

$$H = H_0 + H_I, \tag{1507}$$

we can expand the free energy in a power series in the interaction part of the Hamiltonian H_I to obtain

$$F = -\beta^{-1} \ln \text{Tr}\, e^{-\beta H}$$
$$= F_0 + \langle H_I \rangle_0 - \frac{1}{2k_B T} \langle (\delta H_I)^2 \rangle_D - \frac{1}{2} \langle H_I^2 \rangle_s + \mathcal{O}(H_I^3) \tag{1508}$$

where

$$F_0 = -\beta^{-1} \ln \text{Tr}\, e^{-\beta H_0} \tag{1509}$$

is the free energy of the unperturbed system, $\langle \cdot \rangle_0$ indicates a thermal average over the probability operator $P_0 = e^{-\beta(H_0 - F_0)}$, and we have introduced the notation, for general operators A and B,

$$\langle \delta A \delta B \rangle_D = \sum_n P_n^0 (\delta A)_{nn} (\delta B)_{nn} \tag{1510}$$

$$(\delta A)_{nn} = A_{nn} - \sum_m P_m^0 A_{mm} \tag{1511}$$

where

$$P_n^0 = \frac{e^{-\beta E_n^0}}{Z_0} \tag{1512}$$

is just the equilibrium probability of the system being in the nth unperturbed energy eigenstate state $|\psi_n^0\rangle$:

$$H_0 |\psi_n^0\rangle = E_n^0 |\psi_n^0\rangle. \tag{1513}$$

Matrix elements of an operator A are defined by

$$A_{mn} = \langle \psi_m^0 | A | \psi_n^0 \rangle. \tag{1514}$$

We also introduce in Eq. (1508) the notation

$$\langle AB \rangle_s = \sum_{n \neq m} A_{nm} B_{mn} \frac{P_n^0 - P_m^0}{E_m^0 - E_n^0}, \tag{1515}$$

defined for two operators, A and B. For the case of present interest, the perturbation is given by

$$H_I = -\boldsymbol{\mu}_E \cdot \mathbf{E} - \boldsymbol{\mu}_M \cdot \mathbf{B} + \sum_{i=1}^N \frac{q_i^2}{8 m_i c^2} (\mathbf{B} \times \mathbf{r}_i)^2. \tag{1516}$$

Using this result in Eq. (1508), we can work out the free energy to second order in **E** and **B**, to obtain

$$F = F_0 - \langle \boldsymbol{\mu}_E \rangle_0 \cdot \mathbf{E} - \langle \boldsymbol{\mu}_M \rangle_0 \cdot \mathbf{B} + \sum_i \frac{q_i^2}{8 m_i c^2} \langle (\mathbf{B} \times \mathbf{r}_i)^2 \rangle_0$$
$$- \frac{1}{2 k_B T} \langle (\delta \boldsymbol{\mu}_E \cdot \mathbf{E} + \delta \boldsymbol{\mu}_M \cdot \mathbf{B})^2 \rangle_D$$
$$- \frac{1}{2} \langle (\boldsymbol{\mu}_E \cdot \mathbf{E} + \boldsymbol{\mu}_M \cdot \mathbf{B})^2 \rangle_s + \mathcal{O}(E, B)^3. \tag{1517}$$

It will simplify matters somewhat and will not change any of our basic conclusions if we assume that the average over the $\mathbf{r}_i$ in $\langle (\mathbf{B} \times \mathbf{r}_i)^2 \rangle_0$ is isotropic. We can then show (see Problem 5.2) that

$$\langle (\mathbf{B} \times \mathbf{r}_i)^2 \rangle_0 = \tfrac{2}{3} B^2 \langle r_i^2 \rangle_0. \tag{1518}$$

The average magnetization is given by

$$M_T^\alpha = -\frac{\partial F}{\partial B_\alpha} \tag{1519}$$

and the average polarization by

$$P_T^\alpha = -\frac{\partial F}{\partial E_\alpha}. \tag{1520}$$

We obtain immediately from Eq. (1517), valid to first order in **E** and **B**, that the average magnetization is given by

$$M_T^\alpha = \langle \mu_M^\alpha \rangle_0 - \sum_i \frac{q_i^2}{6 m_i c^2} \langle r_i^2 \rangle B_\alpha + \frac{1}{k_B T} \langle \{\delta H_1, \delta \mu_M^\alpha\} \rangle_D + \langle \{H_1, \mu_M^\alpha\} \rangle_s \tag{1521}$$

and the average polarization by

$$P_T^\alpha = \langle \mu_E^\alpha \rangle_0 + \frac{1}{k_B T} \langle \{\delta H_1, \delta \mu_E^\alpha\} \rangle_D + \langle \{H_1, \mu_E^\alpha\} \rangle_s, \tag{1522}$$

where we have introduced the notation

$$\{A, B\} = \tfrac{1}{2}(AB + BA) \tag{1523}$$

and

$$H_1 = \boldsymbol{\mu}_E \cdot \mathbf{E} + \boldsymbol{\mu}_M \cdot \mathbf{B}. \tag{1524}$$

In the limit of zero external fields, the average magnetization and polarization are given by

$$\mathbf{M}_T = \langle \boldsymbol{\mu}_M \rangle_0 \tag{1525}$$

and

$$\mathbf{P}_T = \langle \boldsymbol{\mu}_E \rangle_0. \tag{1526}$$

Thus a net magnetization or polarization corresponds to a nonzero average net magnetic or electric dipole moment. If $\mathbf{M}_T$ is nonzero, we have a ferromagnet; if $\mathbf{P}_T$ is nonzero, we have a ferroelectric. For the time being, we restrict the analysis to systems where, typically for reasons of symmetry,

$$\langle \boldsymbol{\mu}_M \rangle_0 = \langle \boldsymbol{\mu}_E \rangle_0 = 0. \tag{1527}$$

Let us turn now to the coefficients of the second-order terms in the free energy. The magnetic susceptibility in the limit as $\mathbf{E}$ and $\mathbf{B}$ go to zero is given by

$$\chi_M^{\alpha\beta} = \frac{\partial M_T^\alpha}{\partial B_\beta}$$
$$= -\delta_{\alpha\beta} \sum_i \frac{q_i^2}{6 m_i c^2} \langle \mathbf{r}_i^2 \rangle + \frac{1}{k_B T} \langle \{\delta\mu_M^\beta, \delta\mu_M^\alpha\} \rangle_D + \langle \{\mu_M^\beta, \mu_M^\alpha\} \rangle_s \tag{1528}$$

and the electric susceptibility is given by

$$\chi_E^{\alpha\beta} = \frac{\partial P_T^\alpha}{\partial E_\beta} = \frac{1}{k_B T} \langle \{\delta\mu_E^\beta, \delta\mu_E^\alpha\} \rangle_D + \langle \{\mu_E^\beta, \mu_E^\alpha\} \rangle_s. \tag{1529}$$

There is also a mixed susceptibility,

$$\chi_{EM}^{\alpha\beta} = \frac{\partial M_T^\alpha}{\partial E_\beta} = \frac{\partial P_T^\beta}{\partial B_\alpha} \tag{1530}$$

which, as discussed in Problem 5.4, will vanish if the unperturbed states have a definite parity.

In general, the susceptibilities are symmetric, since, for example,

$$\chi_M^{\alpha,\beta} = -\frac{\partial^2 F}{\partial B_\alpha \partial B_\beta} = \chi_M^{\beta,\alpha}. \tag{1531}$$

We do not focus here on the general dependence of $\chi_{M,E}^{\alpha\beta}$ on the vector labels α and β. We shall, however, return to this point in Volume 2. Here we focus on the total susceptibilities

$$\chi_M = \frac{1}{3}\sum_\alpha \chi_M^{\alpha\alpha}$$
$$= -\sum_i \frac{q_i^2}{6m_i c^2}\langle \mathbf{r}_i^2\rangle_0 + \frac{1}{3k_BT}\langle(\delta\boldsymbol{\mu}_M)^2\rangle_D + \frac{1}{3}\langle(\boldsymbol{\mu}_M)^2\rangle_s \quad (1532)$$

and

$$\chi_E = \frac{1}{3}\sum_\alpha \chi_E^{\alpha\alpha} = \frac{1}{3k_BT}\langle(\delta\boldsymbol{\mu}_E)^2\rangle_D + \frac{1}{3}\langle(\boldsymbol{\mu}_E)^2\rangle_s. \quad (1533)$$

There is considerable general information in these expressions for χ_M and χ_E. Note first that the factor multiplying the matrix elements in Eq. (1515) is positive:

$$\frac{P_n^0 - P_m^0}{E_m^0 - E_n^0} = \frac{1}{Z_0}\frac{e^{-\beta E_n^0} - e^{-\beta E_m^0}}{E_m^0 - E_n^0} \geq 0. \quad (1534)$$

The inequality follows if we write the factor

$$\frac{e^{-\beta E_n^0} - e^{-\beta E_m^0}}{E_m^0 - E_n^0} = e^{-\beta E_m^0}\frac{e^{\beta(E_m^0 - E_n^0)} - 1}{E_m^0 - E_n^0} \quad (1535)$$

for $E_m^0 - E_n^0 > 0$, or

$$\frac{e^{-\beta E_n^0} - e^{-\beta E_m^0}}{E_m^0 - E_n^0} = e^{-\beta E_n^0}\frac{e^{\beta(E_n^0 - E_m^0)} - 1}{E_n^0 - E_m^0} \quad (1536)$$

for $E_n^0 - E_m^0 > 0$. In each case, $(e^{\beta x} - 1)/x > 0$ for $x > 0$. Using Eqs. (1510) and (1515) in Eqs. (1532) and (1533), we have immediately that

$$\chi_E \geq 0 \quad (1537)$$

and

$$\chi_M + \sum_{i=1}^{N}\frac{q_i^2}{6m_i c^2}\langle \mathbf{r}_i^2\rangle_0 \geq 0. \quad (1538)$$

Here we see an important asymmetry between electric and magnetic properties. We cannot derive a stronger inequality [2] for the magnetic properties because of the *diamagnetic term*,

$$\chi_d \equiv -\sum_{i=1}^{N} \frac{q_i^2}{6m_i c^2} \langle \mathbf{r}_i^2 \rangle_0 \qquad (1539)$$

which is negative. This term is known as the *Larmor* or *Langevin diamagnetic susceptibility* [3]. If we write

$$\chi_M = \chi_p + \chi_d, \qquad (1540)$$

the *paramagnetic* contribution χ_p is formally analogous to χ_E with the replacement $\boldsymbol{\mu}_M \to \boldsymbol{\mu}_E$ and

$$\chi_p \geq 0. \qquad (1541)$$

5.3 NONPOLAR–NONMAGNETIC MATERIALS

5.3.1 Introduction

Let us now look at other specific possibilities. There are many cases where the basic constituents of matter (atoms or molecules) are sufficiently symmetric that they have no net electric or magnetic dipole moment in the absence of an applied field. Such systems are said to be nonpolar or nonmagnetic. All monatomic molecules (He, Ar, Kr, Xe) and all diatomic molecules (N_2, O_2, H_2, etc.) are nonpolar due to symmetry. Diatomic molecules of unlike atoms, for example HCl, are generally, like triatomic molecules, polars. Exceptions, due to symmetry, are CO_2, CCl_4, and CH_4. Similarly, nonmagnetic materials are closed-shell atoms which tend to be in a spin-zero ground state.

For nonpolar and nonmagnetic systems, we have the results

$$\langle \psi_n^0 | \boldsymbol{\mu}_E | \psi_n^0 \rangle = 0 \qquad (1542)$$

and

$$\langle \psi_n^0 | \boldsymbol{\mu}_M | \psi_n^0 \rangle = 0 \qquad (1543)$$

and the expressions for the susceptibilities reduce to

$$\chi_M = -\sum_i \frac{q_i^2}{6m_i c^2} \langle \mathbf{r}_i^2 \rangle_0 + \frac{1}{3} \langle (\boldsymbol{\mu}_M)^2 \rangle_s \qquad (1544)$$

and

$$\chi_E = \frac{1}{3} \langle (\boldsymbol{\mu}_E)^2 \rangle_s. \qquad (1545)$$

Note that the terms explicitly proportional to $(k_B T)^{-1}$ in Eqs. (1532) and (1533) do not contribute in this case.

This has still been rather general. Let us now go to the dilute limit, where we assume that each atom or molecule is well separated from all others. In this case the N-body wavefunction can be written as a product of the wavefunctions for each atom or molecule. Then (see Problem 5.5) Eq. (1544) reduces to

$$\chi_M = -N \sum_i \frac{q_i^2}{6m_i c^2} \langle r_i^2 \rangle_0 + \frac{N}{3} \langle (\mu_M)^2 \rangle_s, \qquad (1546)$$

and Eq. (1545) reduces to

$$\chi_E = \frac{N}{3} \langle (\mu_E)^2 \rangle_s, \qquad (1547)$$

where N is the total number of atoms or molecules and the averages given by Eq. (1515) are now for individual atoms or molecules. Thus i labels the various particles in each molecule in the first term in Eq. (1546). In making this reduction we have assumed that quantum statistics is playing no role. Otherwise, we cannot write the total wavefunction as a product wavefunction since it must be properly symmetrized or antisymmetrized (as, e.g., for the fermionic atom ^{3}He). When working in the dilute limit, we must keep in mind, the ideas developed in Chapter 4—that only the most tightly bound states retain their *atomic* identity when put into a concentrated system. Weakly bound atomic states will not be bound at all in a fluid, due to collisions and screening.

5.3.2 Molecular Polarizabilities

We can gain some feeling for the electric susceptibilities by looking at the simplest system: atomic hydrogen. In this case the electric dipole operator is

$$\mu_E = e\mathbf{r}_p - e\mathbf{r}_e = e\mathbf{r} \qquad (1548)$$

where $\mathbf{r}_p$ is the position of the proton, $\mathbf{r}_e$ the position of the electron, and $\mathbf{r}$ is their relative position. Since μ_E depends only on the relative position of the electron and proton, the only states contributing to the evaluation of Eq. (1547) are those associated with the relative motion of the atom. The translational degrees of freedom are not coupled into the determination of χ_E. It is conventional to call χ_E/N the *molecular polarizability* of the system,

$$\alpha_E = \frac{\chi_E}{N}, \qquad (1549)$$

and it is given in hydrogen by [using Eqs. (1515), (1547), and (1549)]

$$\alpha_E = \frac{1}{3} \sum_{n,l,m,m_s} \sum_{n',l',m',m'_s} \frac{1}{Z_0} \frac{e^{-\beta E_n^0} - e^{-\beta E_{n'}^0}}{E_{n'}^0 - E_n^0} |\langle nlmm_s|\boldsymbol{\mu}_E|n'l'm'm'_s\rangle|^2 \qquad (1550)$$

where the E_n^0 are the energy eigenstates, n and n' are the principal quantum numbers ($= 1, 2, 3, \ldots$), l and l' are total orbital angular momentum quantum numbers ($= 0, 1, \ldots, n-1$), m and m' are azimuthal angular momentum quantum numbers $-l \leq m \leq l$, $Z_0 = \sum_{n,l,m,m_s} e^{-\beta E_n^0}$ is the *relative* partition function, and

$$\langle n'l'm'm'_s|\boldsymbol{\mu}_E|nlmm_s\rangle = \int d^3r \, \psi^*_{n'l'm'}(\mathbf{r}) e\mathbf{r} \psi_{nlm}(\mathbf{r}) \delta_{m_s,m'_s} \qquad (1551)$$

are the matrix elements over the hydrogenic wavefunctions $\psi_{nlm}(\mathbf{r})$. It is imperative to keep in mind the restrictions on the sums over n and n' discussed in Chapter 4 and that $n \neq n'$. We should, in principle, sum only over the *tightly* bound states not disrupted by intermolecular collisions. In principle, α_E is temperature dependent. In practice, α_E is a very weak function of temperature. Consider the probability of the system being in energy state E_n^0:

$$\frac{e^{-\beta E_n^0}}{Z_0} = \frac{e^{-\beta E_n^0}}{\sum_{n',l',m',m'_s} e^{-\beta E_{n'}^0}}$$

$$= \frac{e^{\beta(E_1^0 - E_n^0)}}{1 + g_2 e^{\beta(E_1^0 - E_2^0)} + \cdots} \qquad (1552)$$

where g_2 is the degeneracy of the first excited state. Then since $E_n^0 = E_1^0/n^2$,

$$E_1^0 - E_2^0 = -\frac{3e^2}{8a_0} \approx -10 \, \text{eV}. \qquad (1553)$$

With room temperatures corresponding to $k_B T \approx 1/40$ eV, and for $n > 1$,

$$\beta|E_1^0 - E_n^0| \approx 400, \qquad (1554)$$

so

$$\frac{e^{-\beta E_n^0}}{Z_0} = \delta_{n,1} + \mathcal{O}(e^{-\beta|E_1^0 - E_2^0|}) \qquad (1555)$$

and the correction is very small at room temperatures. Using this information in Eq. (1550), we see that to *very* good accuracy we find that

$$\alpha_E = \frac{2}{3} \sum_{n,l,m,m_s,m'_s} \frac{1}{E_n^0 - E_1^0} |\langle 100 m'_s|\boldsymbol{\mu}_E|nlmm_s\rangle|^2 \qquad (1556)$$

TABLE 5.1 Experimental Results for the Electric Molecular Polarizability for the Noble Gases[a]

Atom	$\alpha_E \times 10^{-24}$ cm^3
He	0.201
Ne	0.390
Ar	1.62
Kr	2.46
Xe	3.99

Source: Data from Ref. 4.

[a]In some of the literature one finds the *molar* polarizability $\kappa = (4\pi/3)N_A\alpha$, where N_A is Avogadro's number. Early theoretical work is to be found in Ref. 5.

where the sum is over the excited states. As discussed in Problem 5.6, this can be worked out exactly, with the result

$$\chi_E = N\alpha_E = N\tfrac{9}{2}a_0^3 \tag{1557}$$

where the Bohr radius is $a_0 = \hbar^2/me^2$. If we consider

$$\frac{\chi_E}{V} = \alpha_E \frac{N}{V} = \alpha_E n \tag{1558}$$

we see that χ_E/V is typically very small for a gas since $\alpha_E \approx 0.5 \times 10^{-24}$ cm^3, $n \approx l^{-3}$, and $a_0/l \ll 1$. In Table 5.1 we give experimental results for molecular polarizabilities for nonpolar noble gas systems. In Problem 5.7 we discuss how much of the variation of α_E for noble gas systems can be explained by variation in the number of electrons Z.

5.3.3 Diamagnets

Turning to magnetic properties, we discuss nonmagnetic materials first. Hydrogen is magnetic because of the spin of the electron, so the simplest nonmagnetic materials are H$_2$ or helium, for which the *paramagnetic* contribution to the magnetic polarization can be written as

$$\alpha_M^p = \frac{\chi_p}{N} = \frac{1}{3}\langle(\mathbf{\mu}_M)^2\rangle_s. \tag{1559}$$

Assuming that the splitting to the first excited state is large, as in molecular hydrogen or helium, α_M^p can be evaluated, as in α_E, by taking matrix elements of the dipole moment between the ground and excited states. Thus, as in the sum in Eq. (1550) for α_E, the expression for α_M^p can be reduced to a single sum over excited states.

358   EQUILIBRIUM PROPERTIES OF DIELECTRIC AND MAGNETIC MATERIALS

However, if the ground state is assumed to be a spin singlet, then, as shown in Problem 5.8, $\langle 0|\boldsymbol{\mu}|m\rangle \cdot \langle m|\boldsymbol{\mu}|0\rangle = 0$. Then we have

$$\alpha_M^p = 0, \tag{1560}$$

up to corrections of order $e^{-\beta \Delta E}$, where ΔE is the splitting of the first excited state, and we know that this contribution is very small. For nonmagnetic materials the dominant contribution to the magnetic polarizability is from the diamagnetic term

$$\alpha_M = \frac{\chi_M}{N} = -\sum_i \frac{q_i^2}{6m_i c^2} \langle r_i^2 \rangle_0. \tag{1561}$$

For hydrogen, for example (see Problem 5.9),

$$\langle \mathbf{r}^2 \rangle_0 = 3a_0^2 \tag{1562}$$

and

$$\alpha_M = -\frac{e^2}{2mc^2} a_0^2 = -\frac{1}{2}\alpha^2 a_0^3 \tag{1563}$$

where $\alpha = e^2/\hbar c \approx \frac{1}{137}$ is the fine structure constant. So, in dilute systems, the magnitude of the susceptibility per unit volume,

$$\frac{\chi_M}{V} = \alpha_M n, \tag{1564}$$

is very small. We then have the general conclusions that for dilute nonpolar and nonmagnetic materials, the susceptibilities are essentially temperature independent and given by the appropriate molecular polarizabilities α_E and α_M. α_M is dominated by the diamagnetic contribution and is negative. Experimental values of α_M nonmagnetic materials are given in Table 5.2.

TABLE 5.2   Experimental Results for the Magnetic Molecular Susceptibility for the Noble Gases

Atom	$\alpha_M \times 10^{-6}$ mole^{-1}
He	-1.9
Ne	-7.2
Ar	-19.4
Kr	-28.0
Xe	-43.0

Source: Data from Ref. 6

5.4 PERMANENT DIPOLE MOMENTS

5.4.1 Coarse Graining

We turn next to dilute fluids composed of atoms or molecules carrying permanent electric or magnetic dipole moments. In the spirit of the discussion of coarse graining in Chapter 4, there is a range of temperatures for which we can think of such systems as a collection of point particles carrying a net electric or magnetic dipole moment. As stated by van Vleck [7] in the classic work in this area: "To find the permanent moment one must retain out of the total moment $\sum e_i \mathbf{r}_i$ only the constant part which remains on averaging over the internal motions of the electrons relative to the nuclei." In the case of a permanent magnetic moment, the key point is to realize that an atom or molecule, as a whole, has a net total angular momentum **J**. This requires averaging over the individual electronic and nuclear spins and orbital angular momenta.

5.4.2 Atomic Systems

Let us consider the coarse-grained treatment of a typical atomic system. The Hamiltonian describing an atomic system of n electrons interacting with a nucleus of charge Ze is given by [8]

$$H_0 = H_{\text{com}} + H_{\text{NR}} + H_{\text{LS}} \tag{1565}$$

where H_{com} is the kinetic energy of the center of mass of the atom, H_{NR} the nonrelativistic Hamiltonian describing the relative motion of the electrons and nucleus, and H_{LS} is the spin–orbit coupling which represents the main relativistic correction to H_{NR}. H_{NR} and H_{LS} are defined and discussed in detail in Ref. 8. We can construct the complete set of atomic states associated with H_0. If **L** is the total orbital angular momentum and **S** the total spin angular momentum for the atom, then

$$\mathbf{J} = \mathbf{L} + \mathbf{S} \tag{1566}$$

is the total angular momentum. The important point for our development here is that

$$[H_0, \mathbf{J}] = [H_0, L^2] = [H_0, S^2] = 0 \tag{1567}$$

and the eigenstates of H_0 can be labeled by eigenvalues of J^2, J_z, L^2, and S^2. Thus we have

$$H_0 |n, j, m_j, l, s\rangle = E(n) |n, j, m_j, l, s\rangle \tag{1568}$$

where, as usual, $J^2 = \hbar^2 j(j+1)$, and $j = 0, \frac{1}{2}, 1, \ldots$, $L^2 = \hbar^2 l(l+1)$, $l = 0, 1, 2, \therefore$, $S^2 = \hbar^2 s(s+1)$, $s = 0, \frac{1}{2}, 1, \ldots$, and $J_z = \hbar m_j$, $-j \leq m_j \leq j$, depending on the number of electrons. The quantum number n orders the energies by increasing values, and the $E(n)$ depend in general on j, m_j, l, and s.

We assume that we couple to this system with an external magnetic field B via a Zeeman term

$$H_B = -\frac{e}{2mc}\mathbf{L}\cdot\mathbf{B} - \frac{e}{mc}\mathbf{S}\cdot\mathbf{B}. \tag{1569}$$

The total Hamiltonian is given by

$$H = H_0 + H_B, \tag{1570}$$

and for modest values of the magnetic field we can treat H_B as a perturbation relative to H_0. In carrying out our coarse-graining procedure it is clear that we separate out from all other relative atomic degrees of freedom the *slow variables* associated with the conserved angular momenta and labeled by the quantum numbers j, m_j, l, and s. We do this by defining the sum over fast variables by

$$\text{Tr}^{\phi} A \to \text{Tr}^{j,m_j,l,s} A = \text{Tr}\delta_{j(j+1),J^2/\hbar^2}\delta_{s(s+1),S^2/\hbar^2}\delta_{l(l+1),L^2/\hbar^2}\delta_{m_j,J_z/\hbar} A \tag{1571}$$

which projects out of any average fixed values of the quantum numbers j, m_j, l, s. We average over all the other relative degrees of freedom.

Following the development in Appendix N, we have immediately that our effective Hamiltonian is given by

$$H_E(j, m_j, l, s) = H_{\text{com}} - \beta^{-1}\ln \text{Tr}^{j,m_j,l,s} e^{-\beta H_0} + \langle H_B\rangle_{j,m_j,l,s} + \mathcal{O}(B^2). \tag{1572}$$

The quantity $\langle H_B\rangle_{j,m_j,l,s}$ requires evaluating the average over H_B holding j, m_j, l, and s fixed. The key point is that the average of H_B depends only on the quantum numbers j, m_j, l, and s and not on remaining degrees of freedom. In particular, after using the Wigner–Eckhart theorem [8,9], one can show that

$$\langle H_B\rangle_{j,m_j,l,s} = -\mu_B g(j, l, s) m_j B \tag{1573}$$

where

$$g(j, l, s) = 1 + \frac{j(j+1) + s(s+1) - l(l+1)}{2j(j+1)} \tag{1574}$$

is the Lande [10] g-factor. Our effective Hamiltonian then reduces to

$$H_E(j, m_j, l, s) = H_{\text{com}} - \beta^{-1}\ln \text{Tr}^{j,m_j,l,s} e^{-\beta H_0} - \mu_B g(j, l, s) m_j B + \mathcal{O}(B^2). \tag{1575}$$

We next assume that the splitting Δ between the ground state and the first excited state is large compared to $k_B T$ and $\mu_B B$. We also assume that the ground state has a definite set of angular momentum quantum numbers J, L, S. Clearly, with these assumptions,

$$-\beta^{-1} \ln \text{Tr}^{j,m_j,l,s} e^{-\beta H_0} = E_G \delta_{j,J} \delta_{l,L} \delta_{s,S} \qquad (1576)$$

plus corrections going as $e^{-\beta \Delta}$. It is important that E_G does not depend on m_j. Thus we need only project out those angular momenta values corresponding to the ground state, and our temperature-independent effective Hamiltonian operator can be written as

$$H_E = H_{\text{com}} + E_G - \frac{\mu_B g(J, L, S) \mathbf{J} \cdot \mathbf{B}}{\hbar}. \qquad (1577)$$

Clearly, the term coupling the system to the external magnetic field has the physical interpretation in terms of an effective magnetic moment operator:

$$\boldsymbol{\mu}_{\text{eff}} = \frac{\mu_B g(J, L, S) \mathbf{J}}{\hbar} \qquad (1578)$$

with a magnetic moment in magnitude

$$\mu_{\text{eff}} = \mu_B g(J, L, S) \sqrt{J(J+1)}. \qquad (1579)$$

Thus the effective magnetic moment is determined once one knows the ground-state values of J, L, and S. These values can be determined by the empirically determined set of rules known as *Hund's rules* [11]. Table 5.3 gives the ground-state values of J, L, and S for the paramagnetic rare earth ions, where this theory works well. One finds good agreement between calculated and observed values of μ_{eff}. Clearly in the case where $J = 0$, the system is nonmagnetic.

There are several important points that we can draw from this analysis.

1. Coarse graining and the representation of a complex system by a simpler system require a separation of energy scales. Here we require the energy-level splitting $\Delta \gg \mu_B B$ if the perturbation expansion in H_B is to be valid, and if the new effective Hamiltonian is to be temperature independent, we also require that $\Delta \gg k_B T$. If these inequalities are satisfied, we can replace the atom or molecule with a point particle with total angular momentum $\mathbf{J}$ and an effective magnetic moment given by Eq. (1578).

2. Our first-order analysis leading to H_E neglects the diamagnetic contribution to χ_m (which is typically much smaller than the paramagnetic contribution) and the *van Vleck contribution*, given by

$$\frac{1}{3} \sum_{n \neq m} \frac{(P_n^o - P_m^o)}{E_m^o - E_n^o} |\boldsymbol{\mu}_M^{nm}|^2. \qquad (1580)$$

TABLE 5.3 Results for the Effective Magnetic Moments for the Rare Earth Compounds

Ion	S	L	J	g^a	Calculated[b]	Experimental
La^{3+}	0	0	0	—	0	0
Ce^{3+}	$\frac{1}{2}$	3	$\frac{5}{2}$	$\frac{6}{7}$	2.54	2.4
Pr^{3+}	1	5	4	$\frac{4}{5}$	3.58	3.5
Nd^{3+}	$\frac{3}{2}$	6	$\frac{9}{2}$	$\frac{8}{11}$	3.62	3.5
Pm^{3+}	2	6	4	$\frac{3}{5}$	2.68	—
Sm^{3+}	$\frac{5}{2}$	5	$\frac{5}{2}$	$\frac{2}{7}$	0.84	1.5
Eu^{3+}	3	3	0	—	0	3.4
Gd^{3+}	$\frac{7}{2}$	0	$\frac{7}{2}$	2	7.94	8.0
Tb^{3+}	3	3	6	$\frac{3}{2}$	9.72	9.5
Dy^{3+}	$\frac{5}{2}$	5	$\frac{15}{2}$	$\frac{4}{3}$	10.63	10.6
Ho^{3+}	2	6	8	$\frac{5}{4}$	10.60	10.4
Er^{3+}	$\frac{3}{2}$	6	$\frac{15}{2}$	$\frac{6}{5}$	9.59	9.5
Tm^{3+}	1	5	6	$\frac{7}{6}$	7.57	7.3
Yb^{3+}	$\frac{1}{2}$	3	$\frac{7}{2}$	$\frac{8}{7}$	4.54	4.5
Lu^{3+}	0	0	0	—	0	0

Source: Data from Ref. 12.
[a]g is related to S, L, and J by Eq.(1579).
[b]The calculated values are for the effective dipole magnitude given by Eq. (1579). More sophisticated calculations by van Vleck [13] and Frank [14] have accounted successfully for the poor agreement for Sm^{3+} and Eu^{3+}.

We obtain these contributions only if we determine the effective Hamiltonian to second order in B. As we shall see below, these contributions are not usually, quantitatively important, but we should keep their presence in mind.

3. In the case of large temperatures, we must include the second-order terms in the perturbation theory expansion in H_I and the effective Hamiltonian will become temperature dependent. In this regime the coarse-graining procedure may cease to be useful.

4. The effective Hamiltonian given by Eq. (1577) was derived for a dilute collection of atoms or molecules, which ignores the interactions between atoms or molecules. Calculation of the effective Hamiltonian for a concentrated paramagnetic system is quite difficult since it requires analysis of the overlap and the distortion of the wavefunctions for the distinct atoms. The resulting effective Hamiltonian is discussed in Section 5.9.1.

5. These coarse-graining ideas must be checked on a system-by-system basis. Unlike the case of the rare earth ions, in the iron group ions, due to the details of the electronic structure, the components of the orbital angular momentum **L** average to zero. In this case the effective moment is given by

TABLE 5.4 Experimental Values for Electric Dipole Moments[a]

Molecule	$\mu_E(D)$	Molecule	$\mu_E(D)$
CF	0.65	KBr	10.6
CN	1.4	KCl	10.3
CO	0.1098	LiF	6.4
CaCl	4.26	NaCl	9.0
ClO	1.24	PbO	4.6
CsCl	10.4	RbCl	10.5
HCl	1.1	SiO	3.1
HI	0.44		

Source: Data from Ref. 16.
[a] A debye is 10^{-18} cgs unit.

$\mu_{\text{eff}} = \mu_B \cdot 2\sqrt{S(S+1)}$. For further discussion, see Eyring, et al. [15]. For iron, where $S = 2$, this gives $\mu_{\text{eff}}/\mu_B = 5.92$, compared to the experimental value of 5.9.

The establishment of a permanent electric dipole moment is considerably more straightforward in dilute systems since it typically is a geometric effect in a nonsymmetric molecule. Thus, for example, in KCl we expect a permanent dipole pointing along the axis connecting the two ions of charge q with a moment given by qd, where d is simply the equilibrium distance of separation. Values of the dipole moments for various systems are given in Table 5.4.

At the thermodynamic level, electric and magnetic properties are closely analogous. At the atomic level they are very different. Although there are no magnetic charges, there are fundamental magnetic dipoles. Conversely, there are electric charges but no fundamental electric dipoles. Permanent magnetic dipoles are intrinsically quantum mechanical and quantized, while permanent electric dipoles are easy to visualize classically and are not essentially quantum mechanical in nature.

5.5 STATISTICAL MECHANICAL DEVELOPMENT FOR DIPOLES

Given the conclusions of Section 5.4, we can proceed to develop a statistical mechanical treatment for a collection of electric or magnetic dipoles $\boldsymbol{\mu}_i$ carried by point particles. An developed in Chapter 1, the equilibrium statistical mechanics of magnetic or dielectric systems proceeds according to the following steps: (1) enumeration of configurations, (2) specification of conserved variables, (3) specification of constraints, and (4) maximization of the entropy consistent with steps 1, 2, and 3. For the remainder of this section we shall refer only to magnetic systems, but it is clear that a completely analogous development holds for a system of electric dipoles.

We assume in our magnetic model that we can separate the translational and magnetic degrees of freedom. In the case where we treat the translational degrees of freedom classically, we assume that the configurations are specified by $\{q_i, \boldsymbol{\mu}_i\}$, where the q_i's are the phase-space coordinates of particle i and $\boldsymbol{\mu}_i$ is the magnetic moment of particle i. In addition to the usual conserved variables in a fluid, the magnetization, which is the total magnetic moment,

$$\mathbf{M}_T = \sum_{i=1} \boldsymbol{\mu}_i, \qquad (1581)$$

is often conserved. Even if $\mathbf{M}_T$ is not conserved, application of an external magnetic field $\mathbf{B}$ will act as a constraint on the system which forces a macroscopic magnetization $\langle \mathbf{M}_T \rangle$. Either way, $\mathbf{M}_T$ should be identified as a macrovariable. We can easily apply to this situation the analysis of the grand canonical ensemble developed in Chapter 1. Averages can now be written as

$$\langle A \rangle = \sum_{N=0}^{\infty} \frac{1}{N!} \mathrm{Tr}_q^N \mathrm{Tr}_M^N P_N[q_i, \boldsymbol{\mu}_i] A[q_i, \boldsymbol{\mu}_i] \qquad (1582)$$

where Tr_q^N is the sum over the q_i translational degrees of freedom, Tr_M^N the sum over the various possible dipole configurations, P_N the probability distribution, and A a function of the q_i's and $\boldsymbol{\mu}_i$'s. We write this as

$$\langle A \rangle = \mathrm{Tr}_M P A. \qquad (1583)$$

The equilibrium probability distribution follows by maximizing the entropy,

$$S = -k_B \mathrm{Tr}_M P \ln P \qquad (1584)$$

subject to the usual constraints: $\mathrm{Tr}_M P = 1$, $\mathrm{Tr}_M PH = E$, $\mathrm{Tr}_M PN = \bar{N}$, and the new constraint,

$$\mathrm{Tr}_M P \mathbf{M}_T = \langle \mathbf{M}_T \rangle. \qquad (1585)$$

In this development H is the Hamiltonian in the absence of the applied field and E is the associated average or *internal energy*. It should be quite clear from our analysis in Chapter 1 that the grand canonical probability distribution is now given by

$$P = \frac{e^{-\beta(H - \mu N) - \mathbf{F}_M \cdot \mathbf{M}_T}}{Z_M} \qquad (1586)$$

where

$$e^{-\beta \Omega} = Z_M \equiv \mathrm{Tr}_M \, e^{-\beta(H - \mu N) - \mathbf{F}_M \cdot \mathbf{M}_T}. \qquad (1587)$$

STATISTICAL MECHANICAL DEVELOPMENT FOR DIPOLES 365

Following now-familiar procedures, we identify the Lagrange multiplier (or conjugate force) as satisfying

$$k_B F_M^\alpha = \frac{\partial S}{\partial \langle M_T^\alpha \rangle}. \tag{1588}$$

The statement of conservation of energy in the entropy representation is given by

$$dS = \frac{dE}{T} - \frac{\mu}{T} d\bar{N} + \frac{p}{T} dV + k_B \mathbf{F}_M \cdot d\mathbf{M}_T. \tag{1589}$$

In our discussion of thermodynamics in Chapter 2, we pointed out that the force conjugate to the magnetization is related to the external magnetic field **B** by

$$\frac{\partial S}{\partial M_T^\alpha} = -\frac{B_\alpha}{T} = k_B F_M^\alpha \tag{1590}$$

and in the energy representation, where the energy is the fundamental thermodynamic potential, we have

$$\mathbf{g}_M = \mathbf{B} = \frac{\partial E}{\partial \mathbf{M}_T}. \tag{1591}$$

This means that the probability distribution for a magnetic system is of the form

$$P = \frac{e^{-\beta(H - \mu N - \mathbf{B} \cdot \mathbf{M}_T)}}{Z_M}. \tag{1592}$$

The entropy is given then by Euler's equation,

$$TS = E - \mathbf{B} \cdot \langle \mathbf{M}_T \rangle - \mu \bar{N} + pV. \tag{1593}$$

If we include the Zeeman term in the total Hamiltonian,

$$H_T = H - \mathbf{B} \cdot \mathbf{M}_T, \tag{1594}$$

Euler's equation takes the standard form

$$TS = E_T - \mu \bar{N} + pV. \tag{1595}$$

Magnetic properties can be obtained by taking derivatives of the grand potential with respect to B_α. The average magnetization is given by

$$M_T^\alpha(T, \mu, V) = -\frac{\partial \Omega}{\partial B^\alpha} = \langle M_T^\alpha \rangle. \tag{1596}$$

The magnetic susceptibility is given by a second derivative (see Problem 5.10),

$$\chi_{\alpha\nu}^M = -\frac{\partial^2 \Omega}{\partial B_\alpha \partial B_\nu} = \frac{\partial \langle M_\nu \rangle}{\partial B_\alpha} = \beta \langle \delta M_\alpha \delta M_\nu \rangle. \qquad (1597)$$

The total susceptibility is defined by

$$\chi_M = \frac{1}{3} \sum_\alpha \frac{\partial}{\partial B_\alpha} \langle M^\alpha \rangle$$

$$= \frac{\beta}{3} \langle (\delta \mathbf{M})^2 \rangle \geq 0. \qquad (1598)$$

Note, that because of the linear dependence of our model Hamiltonian H_T on $\mathbf{B}$, the diamagnetic contribution is not included in this development.

Just as in our previous discussion of fluids, there is structural information available for magnetic systems. If we define the magnetization density

$$\mathbf{M}(\mathbf{x}) = \sum_{i=1}^N \boldsymbol{\mu}_i \delta(\mathbf{x} - \mathbf{r}_i), \qquad (1599)$$

where $\mathbf{r}_i$ is the position of the ith moment, the magnetization correlation function is defined by

$$C_M(\mathbf{x}, \mathbf{x}') = \frac{1}{3} \sum_{\alpha=1}^3 \langle \delta M^\alpha(\mathbf{x}) \delta M^\alpha(\mathbf{x}') \rangle. \qquad (1600)$$

We can measure the magnetic structure factor,

$$C_M(\mathbf{k}) = \int \frac{d^3x \, d^3x'}{V} e^{-i\mathbf{k} \cdot (\mathbf{x}-\mathbf{x}')} C_M(\mathbf{x}, \mathbf{x}'), \qquad (1601)$$

by scattering thermal neutrons from our magnetic system. The analysis is similar to that for x-ray scattering, but the major difference is that light couples to the charge density (and therefore the particle density), while neutrons have a magnetic moment and will therefore scatter strongly from the magnetic moments in the system and thus reflect variations in both the particle and the magnetization densities.

5.6 FREE MOMENTS

5.6.1 Separation of Translational and Magnetic Degrees of Freedom

In the simplest magnetic system the magnetic moment of the ith atom is uncoupled from that of the jth atom. The total Hamiltonian for the system can then be written as a sum of of terms,

$$H_T = H + K_\mu - \mathbf{B} \cdot \mathbf{M}_T \tag{1602}$$

where H is the usual fluid Hamiltonian [17] and the last term is the usual Zeeman term. The K_μ term is a sum of terms,

$$K_\mu = \sum_i H_\mu^i, \tag{1603}$$

where H_μ^i represents possible fluctuations in the magnetic moment of the ith atom. This term is discussed below. In our simple permanent dipole model, $H_\mu^i = 0$. The Hamiltonian given by Eq. (1602) can be treated easily in the canonical ensemble, where

$$Z_M^N = \frac{1}{h^{3N} N!} \int dq_N \prod_{i=1}^N \mathrm{Tr}_M^i \, e^{-\beta H_T} \tag{1604}$$

where Tr_M^i is the sum over the magnetic moment degrees of freedom of the ith atom. Notice that the translational and magnetic degrees of freedom are not coupled and we can write

$$Z_M^N = Z_N (Q_M)^N \tag{1605}$$

where Z_N is the partition function for the translational degrees of freedom for the fluid, and the magnetic partition function for a single molecule,

$$Q_M = \mathrm{Tr}_M^i \, e^{-\beta(H_\mu^i - \mathbf{B} \cdot \boldsymbol{\mu}_i)}, \tag{1606}$$

is the same for each molecule. The free energy for the entire system is then given by

$$F_T = -\beta^{-1} \ln Z_M^N = F + F_M \tag{1607}$$

where F is the free energy for the fluid degrees of freedom. All the magnetic information is contained in the magnetic part of the free energy,

$$F_M = -\beta^{-1} N \ln Q_M. \tag{1608}$$

5.6.2 Classical Rigid Moments

Consider first the classical case where $\boldsymbol{\mu}$ is a fixed-length vector, $H_\mu^i = 0$, and the averaging is over the possible orientations of $\boldsymbol{\mu}$. The trace is given in this case by

$$\mathrm{Tr}_M = \int \frac{d\Omega_\mu}{4\pi} \tag{1609}$$

where $d\Omega_\mu$ is the solid angle associated with the direction of $\boldsymbol{\mu}$ and $\boldsymbol{\mu}_i^2 = \mu^2$ is fixed. Note that even though we use the word *classical* here, the underlying origins of $\boldsymbol{\mu}$ are quantum mechanical (indeed, $\mu \approx \mu_B \to 0$ as $\hbar \to 0$). $\boldsymbol{\mu}$ is classical in the sense that we ignore its quantization. If we use spherical polar coordinates and take $\mathbf{B}$ along the z-axis, we can evaluate Q_M directly:

$$Q_M = \int \frac{d\Omega_\mu}{4\pi} e^{\beta \mathbf{B}\cdot\boldsymbol{\mu}} = \int_0^{2\pi} \frac{d\phi}{4\pi} \int_0^\pi d\theta \sin\theta\, e^{\beta B\mu \cos\theta}$$
$$= \frac{1}{2}\int_{-1}^{+1} du\, e^{\beta B\mu u} \qquad (1610)$$

where we have introduced $u = \cos\theta$. Then

$$Q_M = \frac{1}{\beta B\mu} \sinh \beta B\mu \qquad (1611)$$

and

$$F_M = -\beta^{-1} N \ln\left(\frac{\sinh \beta B\mu}{\beta B\mu}\right). \qquad (1612)$$

The magnetization is given by

$$M_T^\alpha = -\frac{\partial F_M}{\partial B_\alpha} = \delta_{\alpha,z} N\mu L(x) \qquad (1613)$$

where $x = \beta B_z \mu$ and the Langevin function [18] is

$$L(x) = \frac{\partial}{\partial x} \ln\left(\frac{\sinh x}{x}\right) = \coth x - \frac{1}{x}. \qquad (1614)$$

We can make a number of interesting points concerning Eqs. (1613) and (1614). First we note that the average magnetization points in the direction of the magnetic field—the magnetic field lines up the moments. Next we observe that magnetic field and temperature enter only in the combination $\mu B/k_B T$, a ratio of energies. In Fig. 5.1, $L(x)$ is plotted as a function of x. We see that $L(x)$ is a monotonically increasing function of x which vanishes for small x as

$$L(x) = \frac{x}{3} - \frac{x^3}{45} + \mathcal{O}(x^5) \qquad (1615)$$

and goes to 1 for large x as

$$L(x) = 1 - \frac{1}{x} + \mathcal{O}(e^{-2x}). \qquad (1616)$$

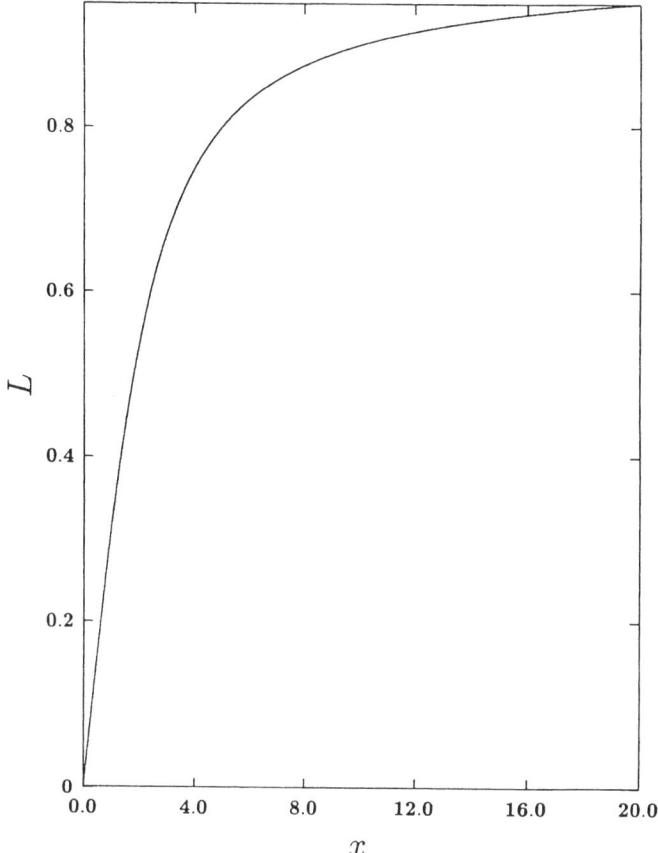

FIGURE 5.1 Langevin function versus scaled magnetic field.

If $x \ll 1$, which means the magnetic field energy is small compared to the thermal energy,

$$\langle \mathbf{M}_T \rangle = \frac{N\mu^2 \mathbf{B}}{3k_B T}. \tag{1617}$$

We see that there is zero net magnetization in the absence of an external magnetic field, and we have a paramagnet. If we compute the magnetic susceptibility, we obtain

$$\chi_M = \lim_{B \to 0} \frac{1}{3} \sum_\alpha \frac{\partial}{\partial B^\alpha} \langle M^\alpha \rangle = \frac{N\mu^2}{3k_B T}. \tag{1618}$$

It is conventional to write

$$\chi_M = \frac{C}{T} \qquad (16.19)$$

where $C = N\mu^2/3k_B$ is called the *Curie constant* [19], and this expression for χ_M is called the *Curie law for noninteracting spins*.

In the opposite limit, $x \gg 1$, we have either a large magnetic field or very low temperatures, and Eq. (16.13) reduces to

$$\langle \mathbf{M}_T \rangle = N\mu \hat{z}. \qquad (16.20)$$

Physically, this means either that we are at zero temperature and in the ground state, where all the spins line up, or we can interpret this result as having a magnetic field so strong that it *overpowers* the moments and forces them to line up. Note that zero temperature is quite special since the magnetization is nonzero even when $B \to 0$, and the magnetic susceptibility becomes infinite as $T \to 0$. We will come back to this point later.

5.6.3 Quantum Mechanical Theory

The generalization of the results of the preceding section to the quantum mechanical case requires choosing the magnetic moments in the Zeeman coupling to be given by the form established in Section 5.4.2:

$$\boldsymbol{\mu} = \mu_B g(J, L, S) \mathbf{J}. \qquad (19.21)$$

Then, in computing the contribution to the single-molecule magnetic partition function in Eq. (16.06), we take the external magnetic field along the z-direction and sum over the allowed quantum states of J_z. We find immediately that

$$Q_M = \mathrm{Tr}_M \, e^{\beta \boldsymbol{\mu} \cdot \mathbf{B}} = \mathrm{Tr}_M \, e^{\beta \mu_z B} = \sum_{m=-J}^{+J} e^{\beta \mu_B g m B}. \qquad (16.22)$$

If we let $x = e^{\beta \mu_B g B}$, we have

$$\sum_{m=0}^{J} x^m = \frac{1 - x^{J+1}}{1 - x} \qquad (16.23)$$

and

$$Q_M = -1 + \frac{1 - x^{J+1}}{1 - x} + \frac{1 - (x^{-1})^{J+1}}{1 - x^{-1}}. \qquad (16.24)$$

This can be rewritten, using standard identities for hyperbolic functions, as

$$Q_M = \frac{\sinh[(J+\tfrac{1}{2})y]}{\sinh(y/2)} \tag{1625}$$

where $y = \beta\mu_B g B$. The average magnetization per particle is given by

$$m_\alpha = \frac{\partial}{\partial B_\alpha}\left(\frac{F}{N}\right) = \delta_{\alpha z} g \mu_B L_B(y) \tag{1626}$$

where

$$\begin{aligned} L_B(y) &= \frac{1}{Q_M}\frac{\partial}{\partial y}Q_M \\ &= \left(J+\frac{1}{2}\right)\coth\left[\left(J+\frac{1}{2}\right)y\right] - \frac{1}{2}\coth\frac{y}{2} \end{aligned} \tag{1627}$$

is the Brillouin function [20] and is plotted versus y for various choices of J in Fig. 5.2. Also shown are comparisons with experimental data for paramagnetic salts, corresponding to different values of J. For small fields we easily find that

$$L_B(y) = \frac{y}{3}J(J+1) \tag{1628}$$

and

$$m_\alpha = \mu_B^2 g^2 \frac{J(J+1)}{3k_B T} B_\alpha. \tag{1629}$$

The zero-field susceptibility is given by

$$\chi_M = \mu_B^2 g^2 \frac{J(J+1)N}{3k_B T}. \tag{1630}$$

Note that Eqs. (1629) and (1630) are the same as the classical results given by Eqs. (1617) and (1618) if we introduce the effective moment

$$\mu_{\text{eff}}^2 = \mu_B^2 g^2(J,L,S) J(J+1), \tag{1631}$$

which agrees with Eq. (1579).

It is at this point that we can compare the diamagnetic contribution to the susceptibility with the Curie contribution given by Eq. (1630). The results of Problem 5.11 show that for room temperature or less, the Curie contribution dominates the diamagnetic contribution.

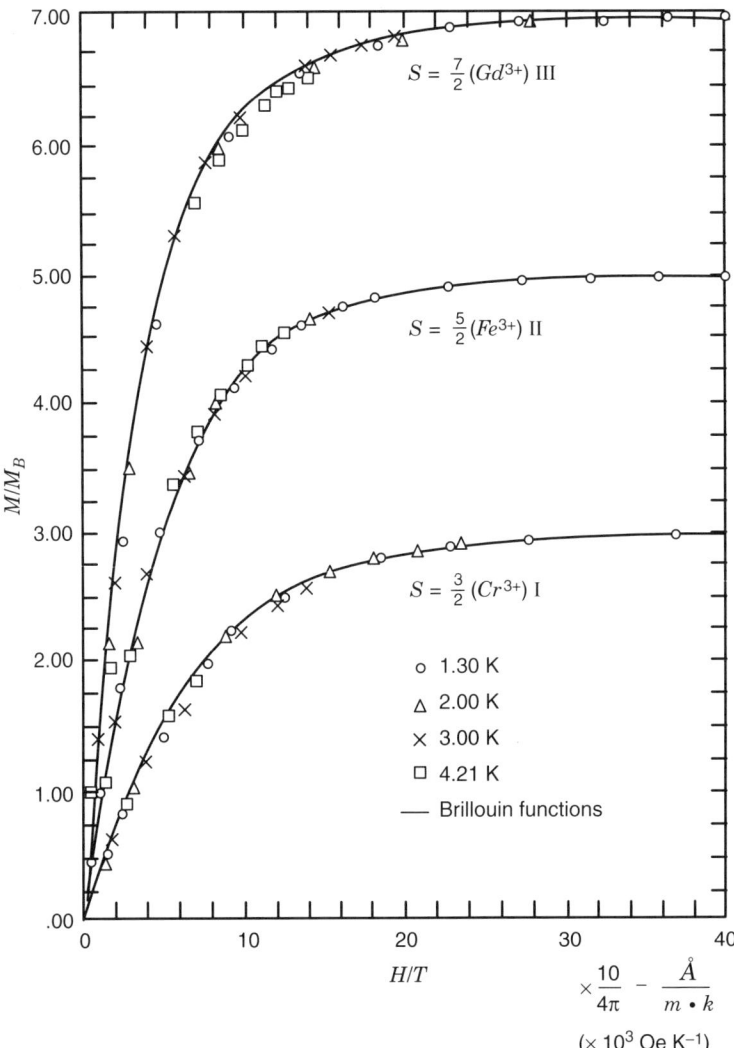

FIGURE 5.2 Magnetization curves of paramagnetic salts. I, potassium chromium alum; II, ferric ammonium alum; III, gadolinium sulfate octahydrate. [From S. Chikazumi, *Physics of Ferromagnetism*, Clarendon Press, Oxford, 1997; data from W. E. Henry, *Phys. Rev.* **88**, 559 (1952).]

For large y, low temperatures or large fields, one easily finds that the Brillouin function reduces to

$$L_B(y) = J - e^{-y} + \mathcal{O}(e^{-2y}), \tag{1632}$$

and the magnetization per particle saturates at the value

$$m = g\mu_B J. \tag{1633}$$

FREE MOMENTS 373

5.6.4 Fluctuating Dipole Model

A deficiency of the fixed-length model used in the last two calculations is that it neglects the van Vleck type of contribution to the susceptibilities. In the magnetic case this is not very important, since this contribution vanishes for nonmagnetic materials and is swamped by the Curie contribution for paramagnetic materials. For dielectrics, these contributions are much more important. Thus the model discussed in this section is more important for dielectric materials.

We now look at the consequences of associating with each atom or molecule an effective fluctuating dipole moment which contributes a piece to the Hamiltonian, which for an orientation $\hat{n}$ of its permanent dipole moment, is of the form

$$H^i_\mu = \tfrac{1}{2}\alpha_E^{-1}(\boldsymbol{\mu} - \boldsymbol{\mu}_E)^2 \tag{1634}$$

where $\boldsymbol{\mu}_E = \mu_E \hat{n}$. The physical significance of the parameters μ_E and α_E will emerge below. In carrying out the trace, we *sum* over all values of $\boldsymbol{\mu}$ (it can fluctuate in magnitude and direction) and over all angles $\hat{n}$:

$$\mathrm{Tr}_E = \int \frac{d\Omega_{\hat{n}}}{4\pi} \int \frac{d^3\mu}{(2\pi\alpha_E k_B T)^{3/2}}. \tag{1635}$$

We normalize the μ-integration such that the limit $\alpha_E \to 0$ returns us to the fixed-length model treated above. We then have that

$$Q_E = \int \frac{d\Omega_{\hat{n}}}{4\pi} \int \frac{d^3\mu}{(2\pi\alpha_E k_B T)^{3/2}} e^{-(\beta/2)\alpha_E^{-1}(\boldsymbol{\mu}-\boldsymbol{\mu}_E)^2} e^{+\beta\boldsymbol{\mu}\cdot\mathbf{E}}. \tag{1636}$$

Making the change of variables from $\boldsymbol{\mu}$ to

$$\mathbf{m} = \boldsymbol{\mu} - \boldsymbol{\mu}_E - \alpha_E \mathbf{E}, \tag{1637}$$

we have in the exponential in Eq. (1636),

$$\frac{\alpha_E^{-1}}{2}(\boldsymbol{\mu}-\boldsymbol{\mu}_E)^2 - \boldsymbol{\mu}\cdot\mathbf{E} = \frac{\alpha_E^{-1}}{2}m^2 - \boldsymbol{\mu}_E\cdot\mathbf{E} - \frac{\alpha_E E^2}{2}. \tag{1638}$$

The integral over $\mathbf{m}$ is easily carried out, to obtain

$$Q_E = \int \frac{d\Omega_{\hat{n}}}{4\pi} e^{\beta\boldsymbol{\mu}_E\cdot\mathbf{E}} e^{\beta\alpha_E E^2/2} = \frac{\sinh\beta E\mu_E}{\beta E\mu_E} e^{\beta\alpha_E E^2/2}. \tag{1639}$$

The first factor is the same as Eq. (1611) with $B \to E$ and $\mu \to \mu_E$. In direct analogy with magnetic development in Section 5.6.1, we have in the dielectric case for a set of noninteracting dipoles that the dielectric part of the free energy, F_E, for N molecules, is related to the dielectric partition function, Q_E, for a single molecule by

$$F_E = -\beta^{-1} N \ln Q_E, \qquad (1640)$$

and the average polarization is given by Eq. (1520):

$$P_\alpha = -\frac{\partial F_E}{\partial E_\alpha} = \beta^{-1} N \frac{\partial}{\partial E_\alpha} \ln Q_E. \qquad (1641)$$

Inserting Q_E, given by Eq. (1639), into this equation gives

$$P_\alpha = \delta_{\alpha_E, z} N [\mu_E L(x_0) + \alpha_E E] \qquad (1642)$$

where $x_0 = \beta E \mu_E$. The zero-field susceptibility is easily found to be given by

$$\chi_E = N \left(\frac{\mu_E^2}{3 k_B T} + \alpha_E \right). \qquad (1643)$$

It is clear that we can carry out the same analysis for a fluctuating magnetic dipole moment and obtain

$$\chi_M = N \left(\frac{\mu_M^2}{3 k_B T} + \alpha_M \right). \qquad (1644)$$

We discuss the interpretation of these results in the next section.

5.6.5 General Phenomenology of Dilute Paramagnets and Dielectrics

Let us recall from Section 5.2 that the magnetic and dielectric susceptibilities can be written in the forms given by Eqs. (1532) and (1533). As discussed in Problem 5.3, if there is a gap Δ between the ground-state energy and the first excited state, then for temperatures $k_B T \ll \Delta$, the van Vleck contribution corresponding to the average $\langle \cdot \rangle_s$ in Eqs. (1532) and (1533) are independent of temperature. Furthermore, as explored in Problem 5.1, if the average dipole moment is zero,

$$\langle \boldsymbol{\mu}_i \rangle = 0 \qquad (1645)$$

where $i = E$ or M, but if we define

$$\langle \boldsymbol{\mu}_i^2 \rangle \equiv \bar{\mu}_i^2(T) \qquad (1646)$$

then

$$\langle (\delta \boldsymbol{\mu}_i)^2 \rangle_D = \bar{\mu}_i^2(0) \qquad (1647)$$

for temperatures $k_B T \ll \Delta$. After noting that each contribution to the susceptibilities given by Eqs. (1532) and (1533) can be written as a product of N times a contribution per molecule, we find that each can be written in the Langevin–Debye form ($i = E$ and M) [21]

$$\frac{\chi_i}{N} = \frac{1}{3} \frac{\bar{\mu}_i^2}{k_B T} + \alpha_i \qquad (1648)$$

where $\bar{\mu}_i$ is the effective permanent dipole moment and α_i is the associated, approximately temperature independent molecular polarizability. α_i corresponds to the van Vleck contribution in the dielectric case and the van Vleck and diamagnetic contributions in the magnetic case. Experimentally, we determine $\bar{\mu}_i$ and α_i as shown in Tables 5.1 to 5.4. In nonpolar or nonmagnetic materials, $\bar{\mu}_i = 0$ and χ is temperature independent. α_M is negative due to the diamagnetic contribution, while α_E is positive. In magnetic systems we should treat a quantized moment, and the effective moment is given by Eq. (1631). This theoretical result (for given J, L, S) is compared with the experiment in Table 5.4 for a number of systems. The result given by Eq. (1648) is a clear experimental manifestation of the coarse-graining procedure discussed in Section 5.4.1. Thus in dilute systems we can determine the parameters entering the model through a measurement of the temperature dependence of χ at moderate temperatures. This is the typical interplay between the coarse-graining procedure, theory, and experiment. It is usually too difficult to carry out the coarse-graining procedure in a quantitative fashion, so theory is used to identify a few relevant parameters (e.g., μ and α) which are determined empirically.

5.6.6 Magnetic Structure Factor

It is straightforward to calculate the magnetic correlation function given by Eq. (1600) for free moments. Since for zero external field, the average of $\mathbf{M}(\mathbf{x})$ is zero, Eq. (1600) becomes

$$C_M(\mathbf{x}, \mathbf{x}') = \tfrac{1}{3} \langle \mathbf{M}(\mathbf{x}) \cdot \mathbf{M}(\mathbf{x}') \rangle$$

$$= \tfrac{1}{3} \sum_{i,j=1}^{N} \langle \boldsymbol{\mu}_i \cdot \boldsymbol{\mu}_j \delta(\mathbf{x} - \mathbf{r}_i) \delta(\mathbf{x}' - \mathbf{r}_j) \rangle. \qquad (1649)$$

The averages over different dipoles are independent:

$$\langle \boldsymbol{\mu}_i \cdot \boldsymbol{\mu}_j \rangle = \delta_{ij} \langle \boldsymbol{\mu}_i^2 \rangle. \qquad (1650)$$

Classically, we can identify $\langle \boldsymbol{\mu}_i^2 \rangle = \bar{\mu}^2$, whereas for the quantum mechanical model

$$\langle \boldsymbol{\mu}_i^2 \rangle = \langle g^2 \mu_B^2 \mathbf{J}^2 \rangle = g^2 \mu_B^2 J(J+1) = \bar{\mu}_M^2. \qquad (1651)$$

Inserting Eqs. (1650) and (1651) back into Eq. (1649) gives

$$C_M(\mathbf{x}, \mathbf{x}') = \frac{\bar{\mu}_M^2}{3} \sum_i \langle \delta(\mathbf{x} - \mathbf{r_i})\delta(\mathbf{x}' - \mathbf{r_i}) \rangle$$

$$= \frac{\bar{\mu}_M^2}{3} \delta(\mathbf{x} - \mathbf{x}')n \qquad (1652)$$

where n is the density of atoms (or molecules). The magnetic structure factor, defined by Eq. (1601), is then given by

$$C_M(\mathbf{k}) = \frac{\bar{\mu}_M^2 n}{3}. \qquad (1653)$$

In the case of noninteracting moments, $C_M(\mathbf{k})$ is independent of $\mathbf{k}$, and there is no structure. This will not be the case for interacting magnetic moments treated below. Comparing Eqs. (1653) and (1618) gives

$$C_M(\mathbf{k}) = k_B T \frac{\chi_M}{V} \qquad (1654)$$

where χ_M is the magnetic susceptibility. In Problem 5.12 you are asked to show that this result is also valid in the presence of an applied magnetic field.

5.7 FREE CHARGED PARTICLES IN A STATIC MAGNETIC FIELD

5.7.1 Classical Case

Let us look at a system that appears superficially much simpler than the atomic and molecular systems we investigated earlier in this chapter. Consider the statistical mechanics of a set of mutually noninteracting charges moving in a static magnetic field. In the classical case this system is rather trivial, but in the quantum mechanical case it is more involved, and some of the models studied previously help us work our way through the analysis.

Let us consider a system of N classical charged particles in a static external magnetic field. We assume that the configuration of the system is specified by the set of phase-space coordinates $(\mathbf{r}_i, \mathbf{p}_i)$. Each particle carries a charge $-e$ (we can think of electrons), and we suppose that these charges are embedded in a fixed array (lattice) of positive charges so that the system is, overall, electrically neutral, as in the one-component plasma treated earlier. The Hamiltonian for our system is given by

$$H = \sum_i \frac{1}{2m}\left(\mathbf{p}_i + \frac{e}{c}\mathbf{A}(\mathbf{r}_i)\right)^2 \qquad (1655)$$

where $\mathbf{A}(\mathbf{r}_i)$ is the vector potential evaluated at the position of the ith charged particle. The magnetic field is obtained from the vector potential using Eq. (1499).

The free energy is given in this case by

$$F_N = -\beta^{-1} \ln Z_N \tag{1656}$$

where the canonical partition function is given by

$$Z_N = \frac{1}{N!} \int \prod_i \left(d^3 r_i \frac{d^3 p_i}{(2\pi\hbar)^3} \right) e^{-\beta H}. \tag{1657}$$

By making the change of variables

$$\mathbf{p}_i \to \mathbf{P}_i = \mathbf{p}_i + \frac{e}{c}\mathbf{A}(\mathbf{r}_i), \tag{1658}$$

we obtain the new Hamiltonian,

$$H = \sum_i \frac{1}{2m} \mathbf{P}_i^2 \tag{1659}$$

and Z_N and F_N are independent of the magnetic field! The magnetic moment and susceptibility (as well as all higher-order derivatives of the free energy with respect to B) vanishes. This result, which clearly holds in the presence of an interaction potential which depends only on the particles' positions, known as *van Leeuwen's theorem* [22], states that there is no classical paramagnetism. The key aspect of this calculation centers around the independence of $\mathbf{p}_i$ and $\mathbf{r}_i$ in the classical partition function, as emphasized in Chapter 4.

5.7.2 Quantum Mechanical Case

In the quantum mechanical case we generalize the situation to treat particles with intrinsic spin $\mathbf{S}$ and consider the Hamiltonian for noninteracting particles:

$$H = \sum_i \frac{1}{2m} \left[\mathbf{p}_i + \frac{e}{c}\mathbf{A}(\mathbf{r}_i) \right]^2 - \sum_i \mu_B g_i \mathbf{S}_i \cdot \mathbf{B} \tag{1660}$$

where $\mathbf{r}_i$ and $\mathbf{p}_i$ are quantum mechanical position and momentum operators with the usual nonzero canonical commutation relations, g_i is the g-factor for particle i, and $\mathbf{S}_i$ is the spin-angular momentum operator for the ith electron.

We assume that the density of electrons and fixed ions is sufficiently small that we can consider all the particles as noninteracting. Interactions with the lattice or of electrons among themselves are important in solids or dense plasmas, but we treat the simpler situation here. We can make contact with the second-quantized formulation developed in Chapter 3 if we solve for the single-particle eigenstates. Thus we must solve the single-particle time-independent Schrödinger equation

$$\left\{\frac{1}{2m}\left[\mathbf{p}+\frac{e}{c}\mathbf{A}(\mathbf{r})\right]^2 - \mu_B g \mathbf{S}\cdot\mathbf{B}\right\}\Psi = E\Psi \tag{1661}$$

where Ψ is the wavefunction. If we assume that the applied magnetic field $\mathbf{B}$ is in the z-direction, the spin part of the Hamiltonian is diagonalized by writing

$$\Psi(\mathbf{r}) = \psi(\mathbf{r})\chi \tag{1662}$$

where χ is a spin eigenstate satisfying,

$$S^2\chi = \hbar^2 S(S+1)\chi \tag{1663}$$

and

$$S_z\chi = \hbar m_s \chi. \tag{1664}$$

The coordinate-dependent part of the wavefunction, ψ, satisfies

$$\frac{1}{2m}\left[\mathbf{p}+\frac{e}{c}\mathbf{A}(\mathbf{r})\right]^2\psi = \epsilon\psi \tag{1665}$$

where, after using Eq. (1664), we have

$$\epsilon = E + \mu_B g m_s B. \tag{1666}$$

The solution of Eq. (1665) is carried out most expeditiously in the Landau gauge, where $A_x = A_z = 0$ and $A_y = Bx$. Then, remembering that $\mathbf{p} = -i\hbar\nabla$, Eq. (1665) takes the form

$$\left(-\frac{\hbar^2}{2m}\nabla^2 - \frac{i\hbar eB}{mc}x\frac{\partial}{\partial y} + \frac{e^2 B^2}{2mc^2}x^2\right)\psi = \epsilon\psi. \tag{1667}$$

If we notice that the left-hand side depends only on powers of x, we see that we can find solutions of Eq. (1667) of the form

$$\psi(\mathbf{x}) = e^{ik_y y} e^{ik_z z} \phi(x). \tag{1668}$$

Substituting Eq. (1668) into Eq. (1667) gives

$$\left[-\frac{\hbar^2}{2m}\frac{\partial^2}{\partial x^2} + \frac{\hbar^2}{2m}(k_y^2 + k_z^2) + \frac{eB}{mc}\hbar k_y x + \frac{e^2 B^2}{2mc^2}x^2\right]\phi(x) = \epsilon\phi(x). \tag{1669}$$

If we let $X = x + \hbar k_y c / eB$, Eq. (1669) reads

$$\left(\frac{-\hbar^2}{2m}\frac{\partial^2}{\partial X^2} + \frac{e^2 B^2}{2mc^2}X^2\right)\phi(X) = \left(\epsilon - \frac{\hbar^2 k_z^2}{2m}\right)\phi(X). \tag{1670}$$

This is just the equation of a one-dimensional harmonic oscillator with energy eigenvalues

$$\epsilon = \left(n + \frac{1}{2}\right)\hbar\omega_c + \frac{\hbar^2 k_z^2}{2m} \qquad n = 0, 1, 2, \ldots \tag{1671}$$

where the cyclotron frequency, ω_c, is defined by

$$\omega_c = \frac{eB}{mc}. \tag{1672}$$

The motion associated with the x-direction is quantized into a set of energy levels. These *Landau levels* [23], labeled by n, are highly degenerate.

Although we could go on to construct the eigenfunctions $\phi_n(X)$ in detail, this will not be necessary for our purposes. All we need for our statistical mechanical calculation are the single-particle energy eigenstates, the ϵ_i in the notation of Chapter 3, and their degeneracy.

5.7.3 Density of States

The most awkward part of this analysis involves the determination of the density of states associated with various Landau levels. In the case of the free particles, we used box boundary conditions to count the number of states. We found the relationship

$$k_\alpha = n_\alpha \frac{2\pi}{L_\alpha} \qquad n_\alpha = 0, \pm 1, \pm 2, \ldots, \tag{1673}$$

for a box of dimension L_α in the αth direction. This allowed us to identify

$$\sum_{\mathbf{n}} f(\mathbf{n}) = V \int \frac{d^3 k}{(2\pi)^3} f(\mathbf{k}). \tag{1674}$$

The awkwardness in the present case arises because we have not really solved the problem we want to solve: the Schrödinger equation (1665) *in a box*. There is no problem dealing with the z-direction where we can use box normalization and obtain Eq. (1673) with $\alpha = z$. Sums over n_z can be evaluated in the usual way:

$$\sum_{n_z} f\left(\frac{k_z^2}{2m}\right) = L_z \int \frac{dk_z}{2\pi} f\left(\frac{k_z^2}{2m}\right). \tag{1675}$$

The problem is with the sum over k_y, which does not appear in the energy equation (1671). If we impose box normalization in the y-direction, we obtain

$$k_y = \frac{2\pi n_y}{L_y}, \quad n_y = 0, \pm 1, \pm 2, \ldots \quad (1676)$$

and without further consideration would conclude that n_y runs to $\pm\infty$. This sum is, however, cut off if we look at the behavior in the x-direction. We would like to impose box boundary conditions in the x-direction, but this would involve solving Eq. (1670), the one-dimensional harmonic oscillator Schrödinger equation, in the presence of boundaries. This is a particularly formidable task since the origin cannot be chosen arbitrarily to be at the center of the box. This is because the eigenfunction ϕ_n depends on the variable $x + \hbar k_y c/eB$. Thus as k_y marches through its allowed values, the center of the wavefunction ϕ_n changes. If we work with very large systems and our density of states is only used to evaluate quantities that fall off as $e^{-\beta\hbar\omega_c n}$ for large n, we can avoid this technical difficulty. Clearly, a minimum requirement for the use of box normalization requires that the origin of the eigenfunction $\phi_n(X)$ be inside the box. Thus we require that

$$-\frac{L_x}{2} \le \frac{\hbar k_y c}{eB} \le \frac{L_x}{2}. \quad (1677)$$

With $k_y = n_y(2\pi/L_y)$, this puts a bound on $|n_y|$ of

$$|n_y| \le \frac{eB}{\hbar c} \frac{L_x}{2} \frac{L_y}{2\pi} = n_y^{\max} \quad (1678)$$

and the total number of such states is

$$N_\Phi \equiv \sum_{n_y} = \frac{eB}{2\pi\hbar c} L_x L_y. \quad (1679)$$

This result has a direct physical interpretation. At a fundamental level the basic unit of magnetic flux is given by

$$\Phi_0 = \frac{hc}{e} = \frac{2\pi\hbar c}{e}. \quad (1680)$$

We see then that the total number of states in each Landau level given by Eq. (1679) can be rewritten as

$$N_\Phi = \frac{B L_x L_y}{\Phi_0} \quad (1681)$$

which is the total number of flux quanta penetrating the sample.

We can argue that it is not the center of the wave packet but the particle itself that must be confined in the box, and the particle, on average, is in an orbit of size [24]

$$a_n^2 = \frac{2c\hbar}{eB}\left(n+\frac{1}{2}\right) \quad (1682)$$

about the center of the wave packet. Then, for given fixed values of n and B, the fraction of particles near enough to the wall to experience boundary effects is

$$\frac{2a_n(L_x + L_y)}{L_x L_y}. \quad (1683)$$

The rest of the particles have *orbits* safely inside the box. If we first fix B and n and take the thermodynamic limit, the fraction of particles experiencing *edge effects* goes to zero and our determination of n_y^{max}, at least its dependence on BL_xL_y, seem secure. Note that the condition that the sum over n_y converge at large n ensures that the thermodynamic limit makes sense.

Although the basic dependence of n_y^{max} on B, L_x, and L_y has been established, we may not have confidence in the numerical coefficient. We adopt the following strategy. We have that

$$\sum_i f(\epsilon_i) = \sum_{n=0}^{\infty} \sum_{n_y} \sum_{n_z} f(\epsilon(n, k_z))$$

$$= \sum_{n_y} L_z \int \frac{dk_z}{2\pi} \sum_{n=0}^{\infty} f(\epsilon(n, k_z)). \quad (1684)$$

We assume that

$$N_\Phi = \sum_{n_y} = d_0 B L_x L_y \quad (1685)$$

and determine the numerical factor d_0 by demanding that calculation of the thermodynamic potential Ω reduce to that calculated for free particles as $B \to 0$.

Our problem is now precisely as in Chapter 3. We have a second quantized Hamiltonian of the form

$$H = \sum_{n=0}^{\infty} \sum_{n_y} \sum_{n_z} \sum_{m_s} \epsilon(n, k_z, m_s) a^+(n, n_y, n_z, m_s) a(n, n_y, n_z, m_s) \quad (1686)$$

where the single-particle energies are given by

$$\epsilon(n, k_z, m_s) = \hbar\omega_c\left(n + \frac{1}{2}\right) + \frac{\hbar^2 k_z^2}{2m} - \mu_B g m_s B \quad (1687)$$

382   EQUILIBRIUM PROPERTIES OF DIELECTRIC AND MAGNETIC MATERIALS

and the creation and annihilation operators, if the particles are electrons and $s = \frac{1}{2}$, correspond to fermions. We have immediately from Chapter 3 that the thermodynamic potential is given by

$$\Omega = -\beta^{-1} \sum_{n=0}^{\infty} \sum_{n_y} \sum_{n_z} \sum_{m_s} \ln(1 + e^{-\beta(\epsilon(n,k_z,m_s)-\mu)})$$

$$= -\beta^{-1} d_0 B L_x L_y L_z \sum_{n=0}^{\infty} \sum_{m_s} \int \frac{dk_z}{2\pi} \ln(1 + e^{-\beta(\epsilon(n,k_z,m_s)-\mu)}). \quad (1688)$$

5.7.4 Dilute Limit

We can gain some feeling for the problem and determine d_0 by looking first at the low-density limit. In that case, $e^{\beta\mu} \ll 1$, and we have, to leading order, that

$$\Omega = -\beta^{-1} d_0 BV \sum_{n=0}^{\infty} \int \frac{dk_z}{2\pi} e^{-\beta(\epsilon(n,k_z)-\mu)} \Omega_s \quad (1689)$$

where

$$\Omega_s = \sum_{m_s} e^{\beta \mu_B g m_s B}. \quad (1690)$$

The sum in Eq. (1690) is clearly given by Eq. (1625) to be

$$\Omega_s = \frac{\sinh[(s + 1/2)gx]}{\sinh(gx/2)}, \quad (1691)$$

where $x = \beta \mu_B B$. We have also that

$$\int_{-\infty}^{+\infty} \frac{dp_z}{2\pi\hbar} e^{-\beta p_z^2 / 2m} = \frac{1}{\lambda} \quad (1692)$$

where, as usual, $\lambda = \hbar(2\pi\beta/m)^{1/2}$ is the thermal de Broglie wavelength. We also need the sum

$$\sum_{n=0}^{\infty} e^{-2x(n+1/2)} = \frac{e^{-x}}{1 - e^{-2x}} = \frac{1}{2\sinh x}. \quad (1693)$$

Combining Eqs. (1691), (1692), and (1693) in Eq. (1689), we obtain

$$\Omega = -\beta^{-1} \frac{V e^{\beta\mu}}{\lambda^3} \frac{d_0 B \lambda^2}{2\sinh x} \Omega_s. \quad (1694)$$

Since as $B \to 0$, Ω is given by Eq. (223) [times a factor of $(2S+1)$ due to the electronic spin], and we can determine d_0 by requiring that these agree,

$$\lim_{B \to 0} \frac{d_0 B \lambda^2}{2 \sinh x} = 1, \tag{1695}$$

or

$$d_0 = \frac{e}{2\pi\hbar c} = \frac{1}{\Phi_0}. \tag{1696}$$

The degeneracy in this case associated with each level n is

$$N_\Phi = \frac{eB}{2\pi\hbar c} L_x L_y, \tag{1697}$$

in agreement with Eq. (1681). For general values of the magnetic field we have the thermodynamic potential,

$$\Omega = -\beta^{-1} \frac{V e^{\beta\mu}}{\lambda^3} \frac{x}{\sinh x} \Omega_s. \tag{1698}$$

The total number of particles is given as usual by

$$\bar{N} = \frac{\partial}{\partial \alpha}(\beta\Omega) = \frac{V e^{\beta\mu}}{\lambda^3} \frac{x}{\sinh x} \Omega_s. \tag{1699}$$

The magnetization is then given by

$$M_T^\alpha = -\frac{\partial \Omega}{\partial B_\alpha}$$

$$= \delta_{\alpha,z} \beta^{-1} \frac{\mu_B}{k_B T} \frac{V e^{\beta\mu}}{\lambda^3} \frac{\partial}{\partial x}\left[\frac{x}{\sinh x} \Omega_s(x)\right]. \tag{1700}$$

Eliminating the chemical potential in favor of the average number of particles, we obtain for the average magnetization

$$M_T^\alpha = \delta_{\alpha,z} \mu_B \bar{N} \frac{\sinh x}{x} \frac{1}{\Omega_s} \frac{\partial}{\partial x}\left[\frac{x}{\sinh x} \Omega_s(x)\right], \tag{1701}$$

which we can rewrite in the form

$$M_T^\alpha = \delta_{\alpha,z} \mu_B \bar{N} \left[-\frac{\partial}{\partial x} \ln \frac{\sinh x}{x} + \frac{\partial}{\partial x} \ln \Omega_s(x)\right]. \tag{1702}$$

Comparing this equation with Eq. (1614) and Eq. (1627), we obtain

$$M_T^\alpha = \delta_{\alpha,z} \mu_B \bar{N}[-L(x) + gL_B(gx)]. \tag{1703}$$

The first term in Eq. (1703) is proportional to the Langevin function, is diamagnetic in nature, and comes from the orbital part of the Hamiltonian. The positive term, which is proportional to the Brillouin function, comes directly from the intrinsic spin of the electron. The magnetic susceptibility in zero external field is clearly given in this case by

$$\chi_M = \frac{\bar{N}\mu_B^2}{3k_B T}\left[-1 + g^2 S(S+1)\right]. \tag{1704}$$

For spin-$\frac{1}{2}$ electrons, $g = 2$, $s = \frac{1}{2}$, and

$$\chi_M = \frac{\bar{N}\mu_B^2}{3k_B T}(-1+3) = \frac{2}{3}\frac{\bar{N}\mu_B^2}{k_B T}. \tag{1705}$$

The diamagnetic term is one-third of the paramagnetic term. Note that M_T^α and χ_M vanish as $\hbar \to 0$ (since $\mu_B \approx \hbar$), in agreement with the classical calculation at the beginning of this section.

5.7.5 Calculation for Arbitrary Density

We turn next to the calculation of the thermodynamic potential,

$$\Omega = -\beta^{-1}\frac{eB}{2\pi\hbar c}V \sum_{n=0}^{\infty}\sum_{m_s}\int\frac{dk_z}{2\pi}\ln(1 + e^{-\beta(\epsilon(n,k_z,m_s)-\mu)}), \tag{1706}$$

for arbitrary density. If we introduce the density of states,

$$D(\epsilon) = \frac{eB}{2\pi\hbar c}\sum_{n=0}^{\infty}\int\frac{dk_z}{2\pi}\delta\left(\epsilon - \epsilon(n) - \frac{\hbar^2 k_z^2}{2m}\right), \tag{1707}$$

where $\epsilon(n) = \hbar\omega_c(n + \frac{1}{2})$, we can then rewrite Eq. (1706) as

$$\Omega = -\beta^{-1}V\sum_{m_s}\int_0^{+\infty}d\epsilon D(\epsilon)\ln\left(1 + e^{-\beta(\epsilon-\mu-m_s g\mu_B B)}\right). \tag{1708}$$

The analysis of Ω is surprisingly involved and is facilitated by introduction of the auxiliary quantities

$$D_1(\epsilon) = \int_0^\epsilon dx\, D(x) \tag{1709}$$

and
$$D_2(\epsilon) = \int_0^\epsilon dx\, D_1(x). \tag{1710}$$

Since we have
$$\frac{\partial^2}{\partial \epsilon^2} D_2(\epsilon) = D(\epsilon), \tag{1711}$$

we can write
$$\Omega = -\beta^{-1} V \sum_{m_s} \int_0^{+\infty} d\epsilon\, \frac{\partial^2}{\partial \epsilon^2} D_2(\epsilon) \ln(1 + e^{-\beta(\epsilon - \mu - m_s g \mu_B B)}). \tag{1712}$$

If we integrate by parts twice, remembering that $D_1(0) = D_2(0) = 0$, we obtain
$$\Omega = V \sum_{m_s} \int_0^{+\infty} d\epsilon\, D_2(\epsilon) \frac{\partial}{\partial \epsilon} n(\epsilon) \tag{1713}$$

where, as usual for fermions,
$$n(\epsilon) = \frac{1}{e^{\beta(\epsilon - \mu - m_s g \mu_B B)} + 1}. \tag{1714}$$

This formulation is convenient since $D_2(\epsilon)$ is better behaved than $D(\epsilon)$ and because $\partial n(\epsilon)/\partial \epsilon$ approaches a δ-function at low temperatures. The next step in the derivation is to evaluate $D_1(\epsilon)$ and $D_2(\epsilon)$:

$$\begin{aligned}D_1(\epsilon) &= \frac{eB}{hc} \sum_{n=0}^{\infty} \int \frac{dk_z}{2\pi} \int_0^\epsilon dx\, \delta\left(x - \epsilon(n) - \frac{\hbar^2 k_z^2}{2m}\right) \\ &= \frac{eB}{hc} \sum_{n=0}^{\infty} \int \frac{dk_z}{2\pi} \theta\left[\epsilon - \epsilon(n) - \frac{\hbar^2 k_z^2}{2m}\right]. \end{aligned} \tag{1715}$$

Inserting this result in Eq. (1710), we obtain

$$\begin{aligned}D_2(\epsilon) &= \frac{eB}{hc} \sum_{n=0}^{\infty} \int \frac{dk_z}{2\pi} \int_0^\epsilon dx\, \theta\left[x - \epsilon(n) - \frac{\hbar^2 k_z^2}{2m}\right] \\ &= \frac{eB}{hc} \sum_{n=0}^{\infty} \int \frac{dk_z}{2\pi} \left(\epsilon - \epsilon(n) - \frac{\hbar^2 k_z^2}{2m}\right) \theta\left(\epsilon - \epsilon(n) - \frac{\hbar^2 k_z^2}{2m}\right) \\ &= \frac{eB}{hc} \sum_{n=0}^{\infty} \frac{2}{3\pi} \left(\frac{2m}{\hbar^2}\right)^{1/2} |\epsilon - \epsilon(n)|^{3/2} \theta(\epsilon - \epsilon(n)). \end{aligned} \tag{1716}$$

386 EQUILIBRIUM PROPERTIES OF DIELECTRIC AND MAGNETIC MATERIALS

Defining $x = \epsilon/\hbar\omega_c$, we can write

$$D_2(\epsilon) = \frac{2}{3\pi} \frac{eB}{hc} \left(\frac{2m}{\hbar^2}\right)^{1/2} (\hbar\omega_c)^{3/2} S(x) \tag{1717}$$

where

$$S(x) = \sum_{n=0}^{\infty} \theta(x - n - \tfrac{1}{2})(x - n - \tfrac{1}{2})^{3/2}. \tag{1718}$$

The evaluation of $S(x)$ is rather involved and is discussed in Problem 5.14, where the result, valid for large x or small B, is determined:

$$S(x) = \frac{2}{5} x^{5/2} - \frac{\sqrt{x}}{16} + \cdots \tag{1719}$$

Putting this result for $S(x)$ back into Eq. (1717), we find after rearrangement that

$$D_2(\epsilon) = \bar{D}\left[\frac{2}{5} \epsilon^{5/2} - \frac{(\mu_B B)^2}{4} \sqrt{\epsilon} + \cdots\right] \tag{1720}$$

where

$$\bar{D} = \frac{1}{6\pi^2} \left(\frac{2m}{\hbar^2}\right)^{3/2}. \tag{1721}$$

Since we are ultimately interested in the zero-field magnetic susceptibility, we need keep only terms up to $\mathcal{O}(B^2)$ in Ω and $D_2(\epsilon)$. We can, therefore, drop the higher-order terms in Eq. (1720) which come in at order $B^{5/2}$, and obtain

$$\Omega = V\bar{D} \sum_{m_s} \int_0^{+\infty} d\epsilon \left[\frac{2}{5} \epsilon^{5/2} - \frac{(\mu_B B)^2}{4} \epsilon^{1/2}\right] \frac{\partial}{\partial \epsilon} \left[\frac{1}{e^{\beta(\epsilon - \mu - m_s g \mu_B B)} + 1}\right] \tag{1722}$$

where all of the B dependence is displayed explicitly. It is easy to see that the zero-field magnetization vanishes. The zero-field magnetic susceptibility,

$$\chi_M = -\left(\frac{\partial^2 \Omega}{\partial B^2}\right)_{B=0}, \tag{1723}$$

has two contributions,

$$\chi_M = \chi_M^p + \chi_M^d \tag{1724}$$

where

$$\chi_M^d = V\bar{D} \sum_{m_s} \int_0^{+\infty} d\epsilon \, \frac{2\mu_B^2}{4} \epsilon^{1/2} \frac{\partial}{\partial \epsilon} n(\epsilon) \tag{1725}$$

comes from the $\mathcal{O}(B^2)$ contribution to $D_2(\epsilon)$, and

$$\chi_M^p = -V\bar{D} \sum_{m_s} \int_0^{\infty} d\epsilon \left(\frac{2}{5}\right) \epsilon^{5/2} \left[\frac{\partial^3}{\partial B^2 \partial \epsilon} [e^{\beta(\epsilon - \mu - m_s g \mu_B B)} + 1]^{-1}\right]\bigg|_{B=0}, \tag{1726}$$

which clearly comes from the Zeeman term in the original Hamiltonian. After doing the sum over m_s, χ_M^d is reduced to

$$\chi_M^d = V\mu_B^2 \frac{\bar{D}}{2}(2S+1) \int_0^{+\infty} d\epsilon \sqrt{\epsilon} \frac{\partial n(\epsilon)}{\partial \epsilon}. \tag{1727}$$

Using the identity

$$\frac{\partial}{\partial B}\left[e^{\beta(\epsilon - \mu - m_s g \mu_B B)} + 1\right]^{-1} = -m_s g \mu_B \frac{\partial}{\partial \epsilon}\left[e^{\beta(\epsilon - \mu - m_s g \mu_B)} + 1\right]^{-1}, \tag{1728}$$

Eq. (1726) can be written as

$$\chi_M^p = -V\bar{D} \sum_{m_s} \int_0^{\infty} d\epsilon \left(\frac{2}{5}\right) \epsilon^{5/2} m_s^2 g^2 \mu_B^2 \frac{\partial^3 n(\epsilon)}{\partial \epsilon^3}. \tag{1729}$$

Integrating by parts twice, we find that

$$\chi_M^p = -\frac{3}{2} V\bar{D} \sum_{m_s} m_s^2 g^2 \mu_B^2 \int_0^{\infty} d\epsilon \sqrt{\epsilon} \frac{\partial n(\epsilon)}{\partial \epsilon}. \tag{1730}$$

The sum over m_s is given by

$$\sum_{m_s=-S}^{S} m_s^2 = \frac{S(S+1)(2S+1)}{3}. \tag{1731}$$

We then have

$$\chi_M^p = -V\mu_B^2 \frac{\bar{D}}{2} S(S+1)(2S+1) g^2 \int_0^{\infty} d\epsilon \sqrt{\epsilon} \frac{\partial n(\epsilon)}{\partial \epsilon}. \tag{1732}$$

Comparing Eqs. (1727) and (1732), we see that they share the same temperature-dependent integral. Taking the ratio, we obtain the very nice result

$$\frac{\chi_M^d}{\chi_M^p} = -\frac{1}{S(S+1)g^2} \tag{1733}$$

and for electrons ($S = \frac{1}{2}, g = 2$)

$$\frac{\chi_M^d}{\chi_M^p} = -\frac{1}{3}. \tag{1734}$$

This result is correct for all temperatures and densities, including the dilute limit.

In all of our previous calculations, χ_M^p was either constant or showed Curie law behavior. Here we obtain more complicated behavior. We already know that χ_M^p shows Curie law behavior at high temperatures [see Eq. (1705)]. At zero temperature,

$$\frac{\partial n(\epsilon)}{\partial \epsilon} = -\delta(\epsilon - \epsilon_F), \tag{1735}$$

and in zero external field Eq. (1722) reduces to

$$\Omega = -V\left(\tfrac{2}{5}\right)\bar{D}\epsilon_F^{5/2}(2S+1). \tag{1736}$$

From Eq. (1736) we can compute the average particle number,

$$\langle N \rangle = -\frac{\partial \Omega}{\partial \epsilon_F} = V\bar{D}\epsilon_F^{3/2}(2S+1). \tag{1737}$$

Using Eqs. (1735) and Eq. (1737) in Eqs. (1727) and (1732), we obtain

$$\chi_M^p = \langle N \rangle \frac{S(S+1)}{2} g^2 \frac{\mu_B^2}{\epsilon_F}$$

$$\chi_M^d = -\langle N \rangle \frac{\mu_B^2}{\epsilon_F}. \tag{1739}$$

For electrons we have the *Pauli paramagnetism* [25],

$$\chi_M^p = \langle N \rangle \frac{3}{2} \frac{\mu_B^2}{\epsilon_F} \tag{1740}$$

and the *Landau diamagnetism* [23],

$$\chi_M^d = -\frac{\langle N \rangle}{2} \frac{\mu_B^2}{\epsilon_F}. \tag{1741}$$

Both give nonzero contributions to the magnetic susceptibilities for zero temperature. Clearly, χ_M^p must decrease with increasing temperature and eventually cross over to Curie behavior at large temperatures.

The effects we have been looking at above have been weak-field effects. In the case of stronger fields there is a very interesting phenomenon that has been very useful in studying the properties of solids. One can show that the magnetization is a periodic function of $\epsilon_F/\mu_B B$. These oscillations of the magnetization, known as the *De Haas–Van Alphen effect* [26], are associated with moving Landau levels past ϵ_F. This leads to a practical way of measuring ϵ_F. For a detailed discussion, see the book by Shoenberg [27].

One can go on to study the properties of electrons in a magnetic field in quasi–two dimensional systems. This leads one to the fascinating phenomena of the integer quantum Hall effect discovered by von Klitzing in 1980 and the fractional quantum Hall effect discovered by Tsui, Stömer, and Gossard in 1982. Both discoveries have been rewarded with Nobel prizes and collectively represent one of the most remarkable condensed-matter phenomena discovered in the second half of the twentieth century [28].

5.8 INTERACTING ELECTRIC DIPOLE MOMENTS

5.8.1 Dipole–Dipole Interaction

Thus far we have treated the simplest of magnetic and dielectric systems—noninteracting dipoles. Even for relatively dilute fluids we must begin to account for the effects due to the interaction of dipoles on different atoms or molecules. We concentrate first on the dielectric case. The form of the electric dipole interaction can be worked out easily by starting with the Coulomb interaction between the four charges shown in Figure 5.3. In this system the pair of charges e_1 and e_3 are constrained to be a distance a apart and form one dipole, while the charges e_2 and e_4, which form a second dipole, are also constrained to be a distance a apart. We are in a

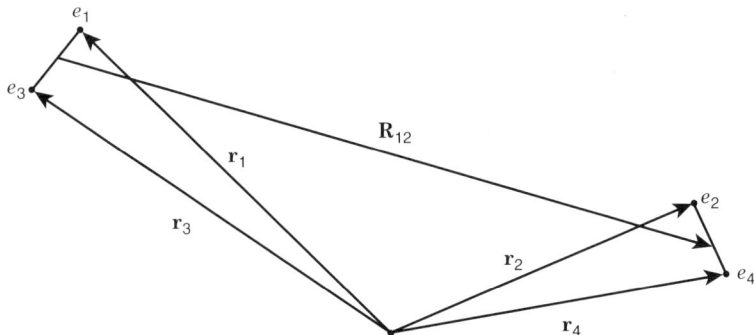

FIGURE 5.3 Geometry describing the interaction between two electric dipoles.

dilute regime where we can assume that on average the distance of separation of the centers of mass of the two dipoles is much larger than a:

$$|\mathbf{R}_{12}| \gg a. \tag{1742}$$

Now, the interaction between the two dipoles can easily be derived from the potential energy for four interacting charges,

$$V_4 = \sum_{i \neq j=1}^{4} \frac{e_i e_j}{|\mathbf{r}_i - \mathbf{r}_j|}, \tag{1743}$$

by subtracting off the energy of the bound pairs (13) and (24), and then expanding in inverse powers of the separation of the center of masses R_{12}. We obtain (see Problem 5.15), to lowest order in powers of R_{12}^{-1},

$$V(12) = \frac{3\boldsymbol{\mu}_1 \cdot \hat{R}_{12} \boldsymbol{\mu}_2 \cdot \hat{R}_{12} - \boldsymbol{\mu}_1 \cdot \boldsymbol{\mu}_2}{|\mathbf{R}_{12}|^3} \tag{1744}$$

where $\boldsymbol{\mu}_1$ is the dipole moment of pair 1 and $\boldsymbol{\mu}_2$ is the dipole moment of pair 2. We can write this in the form

$$V(12) = \sum_{\alpha,\beta} \mu_1^\alpha \mu_2^\beta \Gamma_{\alpha\beta}(\mathbf{R}_1 - \mathbf{R}_2) \tag{1745}$$

where

$$\Gamma_{\alpha\beta}(\mathbf{x}) = \frac{3\hat{x}_\alpha \hat{x}_\beta - \delta_{\alpha\beta}}{|\mathbf{x}|^3}. \tag{1746}$$

The interaction potential for a collection of N dipoles is

$$V_{\mathrm{DP}} = \tfrac{1}{2} \sum_{\alpha,\beta} \sum_{ij} \mu_i^\alpha \mu_j^\beta \Gamma_{\alpha\beta}(\mathbf{R}_i - \mathbf{R}_j). \tag{1747}$$

The total Hamiltonian is then of the form

$$H = K + V_R + V_{\mathrm{DP}} - \mathbf{E} \cdot \mathbf{P}_T \tag{1748}$$

where K is the kinetic energy, V_R a short-range repulsive contribution which restricts the atoms or molecules from coming too close together, and $\mathbf{E}$ the applied electric field which couples to the total dipole moment $\mathbf{P}_T = \sum_i \boldsymbol{\mu}_i$.

A full treatment of a system of interacting dipoles is a challenging problem because of the long-range nature of the interaction. Indeed, this model forms the microscopic basis for macroscopic electrostatics of dielectric materials. A careful treatment of this model requires a discussion of a variety of subtle topics such as boundary effects and separation of applied and total fields. Thus one has to worry, for example, that in general a uniform external field does not generate a uniform local field in the dielectric. This topic will be discussed in some detail in Volume 2. Here we develop a rather conventional treatment due to Lorentz [29]. In some ways this treatment raises almost as many questions as it answers.

5.8.2 Mean-Field Theory

The Hamiltonian given by Eq. (1748) is rather complicated since it involves the interaction of the translational degrees of freedom and the dipole moments. Although we could go on to formulate a density expansion to obtain the corrections to the free-dipole limit, this approach, as with a plasma, is of limited usefulness without further elaboration because of the long-range nature of the interaction potential. To understand the issues, let us consider the simplified Hamiltonian

$$H = K + V_{\text{DP}} - \mathbf{E} \cdot \mathbf{P}_T. \tag{1749}$$

We focus here on an approximation technique, of some historical importance, which allows us to treat the long-range part of the interaction. The basic idea is to assume that the net result of the interactions with other dipoles on a given dipole can be evaluated in terms of an effective or mean field.

Let us introduce the polarization field

$$\mathbf{P}(\mathbf{x}) = \sum_i \boldsymbol{\mu}_i \delta(\mathbf{x} - \mathbf{r}_i) \tag{1750}$$

so that

$$V_{\text{DP}} = \tfrac{1}{2} \int d^3x \int d^3x' \sum_{\alpha,\beta} P_\alpha(\mathbf{x}) \Gamma_{\alpha\beta}(\mathbf{x} - \mathbf{x}') P_\beta(\mathbf{x}'). \tag{1751}$$

We then write

$$V_{\text{DP}} = \tfrac{1}{2} \int d^3x \int d^3x' \sum_{\alpha,\beta} [\langle P_\alpha(\mathbf{x}) \rangle + \delta P_\alpha(\mathbf{x})] \Gamma_{\alpha\beta}(\mathbf{x} - \mathbf{x}') [\langle P_\beta(\mathbf{x}') \rangle + \delta P(\mathbf{x}')] \tag{1752}$$

where

$$\delta P_\alpha(\mathbf{x}) = P_\alpha(\mathbf{x}) - \langle P_\alpha(\mathbf{x}) \rangle \tag{1753}$$

is the fluctuation in the polarization about an average value. The main approximation we make is to assume that we can ignore the fluctuation term in Eq. (1752)

392 EQUILIBRIUM PROPERTIES OF DIELECTRIC AND MAGNETIC MATERIALS

proportional to $\delta P_\alpha(\mathbf{x})\delta P_\beta(\mathbf{x}')$. Then the dipolar part of the interaction is given by

$$V_{\text{DP}} = V_0 - \int d^3x \sum_\alpha \delta P_\alpha(\mathbf{x}) E_\alpha^M(\mathbf{x}) \tag{1754}$$

where V_0 is the constant term proportional to $\langle P_\alpha(\mathbf{x})\rangle\langle P_\beta(\mathbf{x}')\rangle$, and

$$E_\alpha^M(\mathbf{x}) = -\int d^3x' \sum_\beta \Gamma_{\alpha\beta}(\mathbf{x} - \mathbf{x}')\langle P_\beta(\mathbf{x}')\rangle \tag{1755}$$

is the field acting at $\mathbf{x}$ due to the polarization of the other moments in the system. We see that the local electric field $E_\alpha^M(\mathbf{x})$ is proportional to the average polarization $\mathbf{P}$, but care must be used in evaluation of the integral. The reason for the need for this care is the long-range nature of the dipole–dipole interaction ($\approx r^{-3}$). We evaluate the integral using an idea due to Lorentz [29]. First split the integral into two parts: the contribution, $\mathbf{E}^{M,1}$ from all the dipoles outside some spherical region of radius R enclosing the dipole of interest (see Fig. 5.4) and the contribution $\mathbf{E}^{M,2}$ from those dipoles inside this sphere. The local electric field $\mathbf{E}^{M,1}$ can then be evaluated as the electric field due to a fixed constant polarization surrounding a sphere. This is a standard problem in electrostatics and is solved by recognizing that the constant polarization sets up a surface charge density

$$Q(\mathbf{x}) = -\langle \mathbf{P}\rangle \cdot \hat{x}\delta(|\mathbf{x}| - R). \tag{1756}$$

The resulting electric field in the sphere is given by

$$\mathbf{E}^{M,1}(\mathbf{x}) = \int d^3x' Q(\mathbf{x}') \frac{\mathbf{x} - \mathbf{x}'}{|\mathbf{x} - \mathbf{x}'|^3}$$

$$= -\int d^3x' \sum_\beta \langle P_\beta\rangle \hat{x}'_\beta \delta(|\mathbf{x}'| - R) \frac{\mathbf{x} - \mathbf{x}'}{|\mathbf{x} - \mathbf{x}'|^3}. \tag{1757}$$

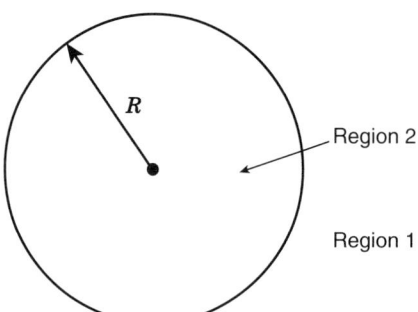

FIGURE 5.4 Geometry for Lorentz calculation for dielectric material.

We are interested in the behavior at the origin, where

$$E_\alpha^{M,1}(0) = -\int d^3x' \sum_\beta \langle P_\beta\rangle \hat{x}'_\beta \delta(|\mathbf{x}'|-R)\left(\frac{-R\hat{x}'_\alpha}{R^3}\right)$$

$$= \sum_\beta \langle P_\beta\rangle \int d\Omega' \hat{x}'_\beta \hat{x}'_\alpha = \sum_\beta \langle P_\beta\rangle \tfrac{1}{3}\delta_{\alpha\beta}4\pi$$

$$= \frac{4\pi}{3}\langle P_\alpha\rangle. \tag{1758}$$

The next part of the argument is probably the weakest point. We assume, due to symmetry and averaging, that $\mathbf{E}^{M,2}(0) = 0$. Lorentz justified this by putting the nearby dipoles on a cubic lattice and assuming that all are oriented in the same direction, and found that they averaged to zero. Clearly, the prefactor of $\mathbf{P}$ in Eq. (1758) depends on the shape of the domains surrounding the selected dipole. Let us assume with Lorentz that $\mathbf{E}^{M,2}(\mathbf{0}) = 0$, and then

$$\mathbf{E}^M = \frac{4\pi}{3}\langle \mathbf{P}\rangle. \tag{1759}$$

We see then that the mean field is uniform, and comparing Eqs. (1755) and (1759), we see that we have in effect evaluated the ill-defined integral

$$\int d^3x' \Gamma_{\alpha\beta}(\mathbf{x}-\mathbf{x}') = -\frac{4\pi}{3}\delta_{\alpha\beta}. \tag{1760}$$

We must now determine $\langle \mathbf{P}\rangle$ self-consistently as the polarization due to the total electric field,

$$\mathbf{E}_T = \mathbf{E} + \mathbf{E}^M = \mathbf{E} + \frac{4\pi}{3}\langle \mathbf{P}\rangle. \tag{1761}$$

It is to be noted here that $\mathbf{E}_T$ is the electric field due to both the external charge and the bound surface charge one obtains for a system with dipolar interactions. The next step is to return to Eq. (1754) and see that we can self-consistently solve for the average polarization. We note (see Problem 5.16) that the polarization is just the average of the dipole density,

$$\langle \mathbf{P}\rangle = \left\langle \sum_{i=1}^N \boldsymbol{\mu}_i \delta(\mathbf{x}-\mathbf{r}_i)\right\rangle = n\langle \boldsymbol{\mu}_i\rangle, \tag{1762}$$

where n is the density of dipoles. In calculating the average over the dipole moments, the translational degrees of freedom decouple and the average can be taken with respect to the Hamiltonian,

$$H = -\int d^3x \mathbf{P}(\mathbf{x}) \cdot \mathbf{E}_T = -\sum_{i=1}^{N} \boldsymbol{\mu}_i \cdot \mathbf{E}_T. \tag{1763}$$

If we assume classical dipoles with length μ_0, the average polarization is given by a calculation identical to that carried out in Section 5.F.2 and leads to the result,

$$\langle \mathbf{P} \rangle = n \langle \boldsymbol{\mu}_i \rangle = n \hat{E}_T \mu_0 L(\beta E_T \mu_0) \tag{1764}$$

where L is the Langevin function given by Eq. (1614). If we assume that the total field E_T is weak, we can expand Eq. (1764) to obtain

$$\langle \mathbf{P} \rangle = n\chi_0 \mathbf{E}_T = n\chi_0 \left(\mathbf{E} + \frac{4\pi}{3} \langle \mathbf{P} \rangle \right) \tag{1765}$$

where

$$\chi_0 = \frac{\mu_0^2}{3k_B T} \tag{1766}$$

is the noninteracting susceptibility per molecule. We solve Eq. (1765) for the average polarization,

$$\langle \mathbf{P} \rangle = \frac{n\chi_0 \mathbf{E}}{1 - (4\pi n/3)\chi_0}. \tag{1767}$$

Recall for dielectric materials that the displacement field $\mathbf{D}$ is defined by

$$\mathbf{D} = \mathbf{E} + 4\pi \langle \mathbf{P} \rangle \tag{1768}$$

and proportional to the applied field with a proportionality constant given by the dielectric constant

$$\mathbf{D} = \epsilon \mathbf{E}. \tag{1769}$$

Since the polarization is proportional to the field, we can combine Eqs. (1767), (1768), and (1769) to solve for the dielectric constant with the result:

$$\epsilon = 1 + \frac{4\pi n \chi_0}{1 - 4\pi n \chi_0/3}. \tag{1770}$$

It is convenient to rewrite this in the form

$$\frac{\epsilon - 1}{\epsilon + 2} = \frac{4\pi n \chi_0}{3} \quad (1771)$$

which is known as the *Clausius–Mossotti relation* [30]. We expect it to work only for low densities. Indeed, the result can be established for nonpolar molecules where we replace χ_0 with α_E. For further development and refinement, see Frölich [31]. See Problem 5.17 for a discussion of the associated *polarization catastrophe*.

5.9 MAGNETIC LATTICE MODELS

5.9.1 Heisenberg Model

We know from everyday experience that solids exhibit interesting magnetic properties. Permanent bar magnets are a common example. The microscopic origins of the interaction between dipoles in solids is complicated and varies from system to system. There are some features that are common to a large number of systems and can be expressed in terms of a relatively simple model. We assume that the primary contributor to the magnetic properties of a large class of solids is due to a net magnetic moment of molecules that form the crystalline lattice characteristic of solids. As a first approximation, we can assume that the molecules are fixed on a lattice and ignore their oscillation about their equilibrium positions. The molecules are thus a fixed distance a from their nearest neighbor. As discussed in some detail in Chapter 6, we can label the lattice sites by a set of lattice vectors $\mathbf{R}(\mathbf{n})$, where $\mathbf{n}$ is a triad of integers. Typically, we suppress the label $\mathbf{n}$. We then assume that the molecule at lattice site $\mathbf{R}$ has a net magnetic moment $\boldsymbol{\mu}(\mathbf{R})$ (we will also call this a *spin*). We then assume, following work by Heisenberg [32], that the Hamiltonian describing this system is

$$H = -\frac{1}{2} \sum_{\mathbf{R} \neq \mathbf{R}'} J(\mathbf{R}, \mathbf{R}') \boldsymbol{\mu}(\mathbf{R}) \cdot \boldsymbol{\mu}(\mathbf{R}') - \mathbf{B} \cdot \sum_{\mathbf{R}} \boldsymbol{\mu}(\mathbf{R}) \quad (1772)$$

where we have included the Zeeman term [33] in the Hamiltonian. The nontrivial term is proportional to $\boldsymbol{\mu}(\mathbf{R}) \cdot \boldsymbol{\mu}(\mathbf{R}')$ and gives the interaction of a spin on site $\mathbf{R}$ with a spin on site $\mathbf{R}'$. This interaction (in a rough sense) results from a calculation of the wavefunction overlap between adjacent molecules. The spin dependence of this overlap is controlled by quantum statistical effects. For this reason this interaction between the spins, $J(\mathbf{R}, \mathbf{R}')$, is known as the *exchange interaction*. In principle, we should be able to compute $J(\mathbf{R}, \mathbf{R}')$ given the type of molecules and the lattice structure. In practice, a quantitative determination of $J(\mathbf{R}, \mathbf{R}')$ is very difficult, and typically, $J(\mathbf{R}, \mathbf{R}')$ is modeled in terms of a few parameters that are determined empirically.

We should note that one also always has the long-range dipolar interaction in the magnetic case. Typically, dipolar interactions, when viewed on microscopic spatial scales, are weak compared to exchange forces. Therefore, we focus here on the important physics associated with the exchange interaction not present in the electric case. The very interesting case of the competition between exchange and dipolar interactions that occur on larger length scales because of the long-range interaction of the dipolar interactions in ferromagnetic materials will be discussed in Volume 2.

The *exchange interaction* is as much a guiding principle as a quantitative *law* describing the interaction between permanent moments. The origins of the Heisenberg Hamiltonian has been studied intensely by theorists for many years. A very significant "effort went into calculating the magnitudes of the exchange constants J_{ij} from atomistic considerations. In spite of all this effort, and perhaps because of what it revealed, it was not possible to elevate exchange to the rank of a universal principle..." [34]. This is another way of indicating that microscopic derivations of coarse-grained models can be difficult. Thus, in detail, one may need to modify the Heisenberg model. For example, the role of anisotropy is discussed in Section 5.13. For a discussion of these points, see Mattis [35] and Herring [36]. For our purposes, as we discuss below, the exchange coupling can be determined empirically.

There are some general properties of the exchange coupling $J(\mathbf{R}, \mathbf{R}')$ that we can specify. First we note, due to translational invariance, that it is a function only of the relative positions of the sites $\mathbf{R}$ and $\mathbf{R}'$,

$$J(\mathbf{R}, \mathbf{R}) = J(|\mathbf{R} - \mathbf{R}'|), \tag{1773}$$

and decays to zero rapidly as $|\mathbf{R} - \mathbf{R}'|$ increases. This means that spins interact strongly only with their nearest neighbors and possibly their next-nearest neighbors. It is also quite significant that J may be positive or negative (spins may attract or repel their neighbors). Let us consider the physical difference. If two neighboring spins are aligned in the same direction, parallel, their interaction is $H^P_{12} = -J_{12}\mu^2$. If the spins are aligned antiparallel, their interaction is $H^A_{12} = J_{12}\mu^2$. If the interaction is positive, $J_{12} > 0$, then

$$H^P_{12} = -|J_{12}\mu^2| < 0 \tag{1774}$$

and

$$H^A_{12} = |J_{12}\mu^2| > 0. \tag{1775}$$

Since a system at low temperatures acts to *lower* its energy, for $J_{12} > 0$ the system *wants* to line spins up parallel. Clearly, for $J_{12} < 0$ we have

$$H^P_{12} > 0 \tag{1776}$$

and

$$H_{12}^A < 0 \tag{1777}$$

and the spins want to be antiparallel. The situation where $J_{12} > 0$, and the spins line up, clearly corresponds to a ferromagnetic magnetic state. We study this situation first. Later, we return to the case $J_{12} < 0$.

5.9.2 Mean-Field Theory Solution

It is clear for a system defined on a lattice that it makes physical sense to work in the canonical ensemble with a fixed number of lattice sites. The partition function can then be written as

$$Z_M^N = \prod_\mathbf{R} \int \frac{d\Omega_\mathbf{R}}{4\pi} e^{-\beta H} \tag{1778}$$

where $d\Omega_\mathbf{R}$ is the integral over all angles associated with the spin $\boldsymbol{\mu}(\mathbf{R})$ which, for simplicity, we treat as a classical fixed-length vector as in Section 5.6.2. The total magnetization is again given by

$$\langle M_T^\alpha \rangle = -\frac{\partial F}{\partial B^\alpha} = \beta^{-1} \frac{\partial}{\partial B^\alpha} \ln Z_M^N. \tag{1779}$$

Consider first the noninteracting case $J(\mathbf{R} - \mathbf{R}') = 0$, so

$$H = -\mathbf{B} \cdot \sum_\mathbf{R} \boldsymbol{\mu}(\mathbf{R}), \tag{1780}$$

and the system reduces to a collection of free moments as treated in Section 5.6.2. The Hamiltonian is a sum of terms from individual lattice sites, so

$$Z_M^N = \prod_\mathbf{R} \int \frac{d\Omega_\mathbf{R}}{4\pi} e^{\beta \mathbf{B} \cdot \boldsymbol{\mu}(\mathbf{R})} = (Q_M)^N \tag{1781}$$

where Q_M is given by Eq. (1611). The average magnetization is given by Eq. (1613),

$$\langle \mathbf{M}_T \rangle = N\mu \hat{z} L(\beta B \mu), \tag{1782}$$

where μ is the fixed length of the spin and L is the Langevin function given by Eq. (1614).

We now want to look at the corrections to the $J = 0$ result due to interactions. As in the fluid case, when we include interactions, we cannot in general solve the problem (with two exceptions we discuss later). We must therefore resort to some

EQUILIBRIUM PROPERTIES OF DIELECTRIC AND MAGNETIC MATERIALS

approximation technique. The simplest approximation that makes sense for our interacting system is to assume that the interaction of all the other spins with the spin at site **R** can be replaced by the average interaction due to those spins:

$$\frac{1}{2}\sum_{\mathbf{R},\mathbf{R}'} J(\mathbf{R},\mathbf{R}')\boldsymbol{\mu}(\mathbf{R})\cdot\boldsymbol{\mu}(\mathbf{R}') \to \sum_{\mathbf{R},\mathbf{R}'} J(\mathbf{R},\mathbf{R}')\boldsymbol{\mu}(\mathbf{R})\cdot\langle\boldsymbol{\mu}(\mathbf{R}')\rangle. \tag{1783}$$

We note that the average value of a spin on site **R**′, in a translationally invariant system, is just the average magnetization per spin and independent of the particular site label **R**′:

$$\langle\boldsymbol{\mu}(\mathbf{R}')\rangle = \frac{\langle\mathbf{M}_T\rangle}{N} \equiv \mathbf{m}. \tag{1784}$$

With these approximations, the Hamiltonian can then be written as

$$H = -\sum_{\mathbf{R},\mathbf{R}'} J(\mathbf{R},\mathbf{R}')\boldsymbol{\mu}(\mathbf{R})\cdot\mathbf{m} - \mathbf{B}\cdot\sum_{\mathbf{R}} \boldsymbol{\mu}(\mathbf{R}). \tag{1785}$$

Because of the translational invariance of the interaction,

$$\sum_{\mathbf{R}'} J(\mathbf{R}-\mathbf{R}') = \sum_{\mathbf{R}'} J(|\mathbf{R}'|) \equiv \bar{J} \tag{1786}$$

and

$$H = -\bar{J}\mathbf{m}\cdot\sum_{\mathbf{R}} \boldsymbol{\mu}(\mathbf{R}) - \mathbf{B}\cdot\sum_{\mathbf{R}} \boldsymbol{\mu}(\mathbf{R}). \tag{1787}$$

We see immediately that the term $\bar{J}\mathbf{m}$ serves as an effective average internal magnetic field,

$$\mathbf{B}_I = \bar{J}\mathbf{m}. \tag{1788}$$

Because our approximation introduces an average or mean field, it is called a *mean-field theory* [37]. Note the significant difference between the mean field theory here, where we have a short-range interaction, and the case of a dipolar interaction treated in the preceding section. In the case of the exchange interaction, the sum $\sum_{\mathbf{R}} J(\mathbf{R})$ is very well behaved. In the case of the long-range dipolar interaction, we had the sensitive considerations leading to Eq. (1760). In the mean-field approximation, the Hamiltonian given by Eq. (1787) can be written as

$$H = -\mathbf{B}_T \cdot \sum_{\mathbf{R}} \boldsymbol{\mu}(\mathbf{R}) \tag{1789}$$

where the *total* field is given by

$$\mathbf{B}_T = \mathbf{B}_I + \mathbf{B}. \tag{1790}$$

This problem is then formally equivalent to the noninteracting case and we obtain without further work that the average magnetization is given by

$$\langle \mathbf{M}_T \rangle = N\mu \hat{z} L(\beta B_T \mu) \tag{1791}$$

where L is again the Langevin function given by Eq. (1614). After we substitute $\mathbf{B}_T = \mathbf{B} + \bar{J}\mathbf{m}$ into Eq. (1791), we obtain (assuming that $\mathbf{B} = B\hat{z}$)

$$\mathbf{m} = \mu \hat{z} L[\beta \mu (B + \bar{J} m_z)] \tag{1792}$$

which is a transcendental equation for $\mathbf{m}$. The first thing we note is that $\mathbf{m}$, as in the noninteracting case, is in the z-direction. It is therefore convenient to write

$$\mathbf{m} = \hat{z} \mu m_0, \tag{1793}$$

and we arrive at the self-consistent equation,

$$m_0 = L(\beta \mu B + \bar{J}\beta \mu^2 m_0), \tag{1794}$$

to be solved for m_0. Let us first consider this equation for $B = 0$ and look for solutions of

$$m_0 = L(\bar{J}\beta \mu^2 m_0). \tag{1795}$$

If we define $x = \bar{J}\beta \mu^2 m_0$, then Eq. (1795) takes the form

$$\frac{x}{\bar{J}\beta \mu^2} = L(x). \tag{1796}$$

We see that the general shape of $L(x)$ (as shown in Fig. 5.1) is approximately linear for small $x (\approx x/3)$ and then bends over for larger x. If the coefficient x is larger than $\frac{1}{3}$, then, as we see in Fig. 5.5, there is only one solution to Eq. (1796): $x = 0$ or $m_0 = 0$. We then have for

$$\frac{1}{\bar{J}\beta \mu^2} > \frac{1}{3} \tag{1797}$$

or

$$k_B T > \frac{\bar{J}\mu^2}{3} \tag{1798}$$

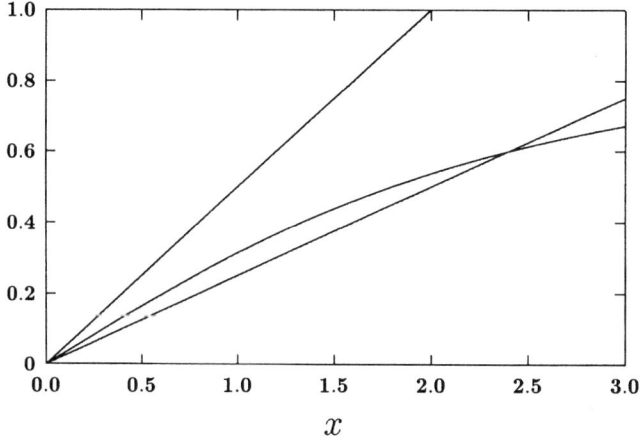

FIGURE 5.5 Search for solutions to mean-field theory equation of state. Curved line is the Langevin function as a function of $x = m_0/t$, where m_0 is scaled spontaneous magnetization defined by Eq. (1795) and $t = k_B T/\bar{J}\mu^2$ is a dimensionles temperature. The equation of state is given by $tx = L(x)$. The top curve corresponds to $tx = x/3$ and gives no solutions for nonzero x. For $t < \frac{1}{3}$, as for the lower straight line corresponding to $tx = x/4$, one obtains nonzero solutions.

that $m_0 = 0$. On the other hand, for $1/\bar{J}\beta\mu^2 < \frac{1}{3}$ or $k_B T < \bar{J}\mu^2/3$, we can have two solutions: $m_0 = 0$ and $m_0 \neq 0$. Although our equations allow for two possible values of the average magnetization, nature will pick one of these as the physical solution. We can determine *the* equilibrium solution by returning to the stability conditions discussed in Chapter 2. We are now discussing a system at constant temperature, volume, and spin number. Under these circumstances, the equilibrium configuration is that configuration that lowers the free energy,

$$F_M = -\beta^{-1} \ln Z_M^N. \tag{1799}$$

We have from our previous calculation for free spins [see Eq. (1612)], that

$$F_M = -\beta^{-1} N \ln \frac{\sinh \beta B_T \mu}{\beta B_T \mu}. \tag{1800}$$

In zero external field this reduces to

$$F_M = -\beta^{-1} N \ln \frac{\sinh \beta \bar{J}\mu^2 m_0}{\beta \bar{J}\mu^2 m_0}. \tag{1801}$$

If $m_0 = 0$, then

$$F_M = -\beta^{-1} N \ln 1 = 0. \tag{1802}$$

If $m_0 \neq 0$, we must investigate the quantity $\ln(\sinh x/x)$. If we remember that the Taylor series expansion for the hyperbolic sine is

$$\sinh x = \sum_{n=0}^{\infty} \frac{x^{2n+1}}{(2n+1)!}, \tag{1803}$$

we have

$$\frac{\sinh x}{x} = \sum_{n=0}^{\infty} \frac{x^{2n}}{(2n+1)!} \geq 1, \tag{1804}$$

and

$$\ln \frac{\sinh x}{x} > 0. \tag{1805}$$

We see then that for $m_0 \neq 0$,

$$F_M = -\beta^{-1} N \ln \frac{\sinh \beta \bar{J} \mu^2 m_0}{\beta \bar{J} \mu^2 m_0} < 0, \tag{1806}$$

and the $m_0 \neq 0$ solution is the most stable solution since it leads to a lower free energy than for $m_0 = 0$. Let us summarize the situation. For temperatures greater than $T_c \equiv \bar{J}\mu^2/3k_B$, there is no net magnetization. However, for $T < T_c$, the system prefers to have a net magnetization. Near $T = T_c$, the average magnetization will be small and we can assume that $\beta \bar{J} \mu^2 m_0/3$ is small. Expanding Eq. (1795) in powers of m_0, we obtain

$$m_0 = \frac{T_c}{T} m_0 - \frac{3}{5} \left(\frac{T_c}{T}\right)^3 m_0^3 + \mathcal{O}(m_0)^5, \tag{1807}$$

where we have used

$$\beta \frac{\bar{J}\mu^2}{3} = \frac{T_c}{T}. \tag{1808}$$

If we ignore the terms of $\mathcal{O}(m_0^5)$ and solve for the nonzero solution for m_0, we obtain

$$m_0 = \left[\frac{5}{3}\left(\frac{T}{T_c}\right)^2 \epsilon\right]^{1/2} \tag{1809}$$

where

$$\epsilon = \frac{T_c - T}{T_c}. \tag{1810}$$

402 EQUILIBRIUM PROPERTIES OF DIELECTRIC AND MAGNETIC MATERIALS

We see that the critical index β, as defined in Chapter 2 and telling us how $m \to 0$ as $T \to T_c$, is given by

$$\beta = \tfrac{1}{2} \tag{1811}$$

in agreement with the Landau theory.

Notice, as in our treatment of Landau theory in Chapter 2, that we have symmetry breaking for $T < T_c$. In this case the rotational symmetry at high temperatures, in the absence of an applied field, is broken since the average magnetization points in a particular direction for $T < T_c$.

For low temperatures, x becomes large in Eq. (1614),

$$L(\bar{J}\beta\mu^2 m_0) \to \pm 1 \tag{1812}$$

(depending on the sign of m_0), $m_0 = \pm 1$, and all the spins are lined up:

$$\langle \mathbf{M} \rangle = \hat{z} N \mu. \tag{1813}$$

The solution of Eq. (1794) for all temperatures is shown in Fig. 5.6 for a variety of values of the applied field.

Let us look next at the case of a nonzero external field but for $T = T_c$, where $\beta_c \bar{J} \mu^2 = 3$,

$$m_0 = L(b + 3m_0) \tag{1814}$$

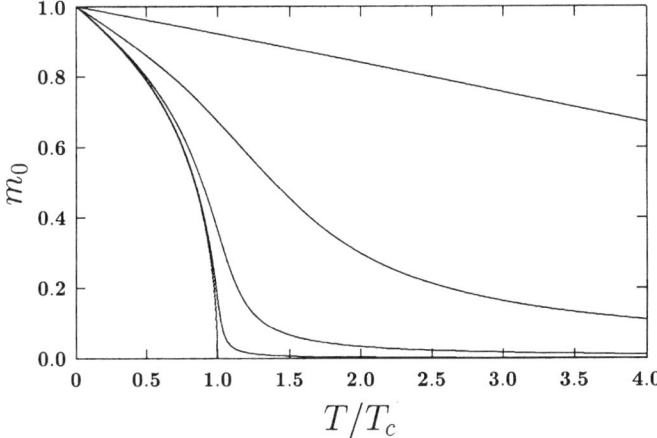

FIGURE 5.6 Scaled magnetization density from mean-field theory as a function of scaled temperature. Different curves currespond to different values of the scaled magnetic field. From bottom to top one has $\mu B/k_B T_c = 0, 0.01, 0.1, 1.0,$ and 10.0.

with $b \equiv \beta_c \mu B$. Again, assuming that m_0 and b are small, we expand Eq. (1814) in powers of $(b + 3m_0)$ to obtain

$$m_0 = \frac{b}{3} + m_0 - \frac{1}{45}(b + 3m_0)^3, \qquad (1815)$$

neglecting higher-order terms. After canceling terms, Eq. (1815) reduces to

$$b = \tfrac{1}{15}(b + 3m_0)^3. \qquad (1816)$$

Solving for m_0, we easily find that

$$m_0 = \tfrac{1}{3}[(15b)^{1/3} - b]. \qquad (1817)$$

Looking at small b, we find that

$$m_0 = \left(\tfrac{5}{9}b\right)^{1/3} \qquad (1818)$$

and the critical index δ is, as in the Landau theory (Chapter 2), given by

$$\delta = 3. \qquad (1819)$$

Let us now look at the magnetic susceptibility for our system. Differentiating Eq. (1791) with respect to B, the susceptibility is given by

$$\chi_M = N\mu \frac{\partial}{\partial B} L(\beta\mu B + \beta\mu \bar{J} m_z). \qquad (1820)$$

If we let

$$x = \beta\mu B + \beta\mu \bar{J} m_z \qquad (1821)$$

and use the chain rule for differentiation, we obtain

$$\chi_M = N\mu \frac{\partial L(x)}{\partial x} \frac{\partial x}{\partial B} \qquad (1822)$$

and

$$\frac{\partial x}{\partial B} = \beta\mu + \frac{\beta\mu \bar{J} \chi_M}{N}. \qquad (1823)$$

Using Eq. (1823) in Eq. (1822), we have a linear equation for χ_M, which we can solve to obtain

$$\chi_M = [1 - \beta\mu^2 \bar{J}\alpha(x)]^{-1} N\beta\mu^2 \alpha(x) \qquad (1824)$$

where

$$\alpha(x) = \frac{\partial L(x)}{\partial x} = \frac{\partial}{\partial x}\left(\coth x - \frac{1}{x}\right) = 1 - \coth^2 x + \frac{1}{x^2}. \tag{1825}$$

In particular, we are interested in the susceptibility for zero external field where $x = \beta\mu^2 \bar{J} m_0$. For $T > T_c$, $m_0 = 0$ and for small x,

$$\alpha(x) = \frac{1}{3} - \frac{x^2}{15} + \mathcal{O}(x^4), \tag{1826}$$

so $\alpha(0) = \frac{1}{3}$, and

$$\chi_M = \left(1 - \frac{\beta\mu^2 \bar{J}}{3}\right)^{-1} \frac{N\beta\mu^2}{3}. \tag{1827}$$

If we remember that

$$\frac{\beta \bar{J}\mu^2}{3} = \frac{T_c}{T} \tag{1828}$$

and the Curie constant is defined as

$$C = \frac{N\mu^2}{3k_B}, \tag{1829}$$

we obtain the *Curie–Weiss law* [37] for the magnetic susceptibility:

$$\chi_M = \frac{C}{T} \frac{1}{1 - T_c/T} = \frac{C}{T - T_c} \tag{1830}$$

for $T \geq T_c$. We see then for $T \gg T_c$ that

$$\chi_M \approx \frac{C}{T} \tag{1831}$$

and we recover the Curie law. Physically, this means that at high temperatures, the spins are essentially independent—they rotate primarily due to thermal agitation rather than interacting with their neighbors. As we lower the temperature, however, we see that χ_M will increase and will *diverge* as $T \to T_c$. The divergence of χ_M as $T \to T_c$ is characterized by the critical index γ [see Eq. (675)], which takes the value 1, again as in the Landau theory. This divergence of χ_M at $T = T_c$ is very similar to the divergence of the isothermal compressibility κ_T near the liquid–gas critical point. There is a critical point in the magnetic phase diagram at the Curie

TABLE 5.5 Examples of Ferromagnets and Their Transition Temperatures

Ferromagnet	T_c(K)	Ferromagnet	T_c(K)
Fe	1043	MnBi	630
CO	1388	MnSb	587
Ni	627.2	Dy	88
Gd	292.5	$CrBr_3$	32.56
CrO_2	386.5	EuS	16.5
MnAs	318	EuO	69

Source: Data from Ref. 38.

temperature T_c. From an empirical point of view we can use the transition temperature as a mechanism for determining the effective exchange coupling. In the mean field approximation, assuming that we know the magnetic moment μ and there are z nearest neighbors $\bar{J} = zJ$, we can estimate

$$J = \frac{3k_B T_c}{z\mu^2}. \tag{1832}$$

In Table 5.5 we list some typical ferromagnets with values for T_c.

Consider next χ_M for $T < T_c$. Here we must be careful. Let us calculate $1 - \beta\mu^2 \bar{J}\alpha(x)$ for T slightly less than T_c. In this case x is small but not zero and

$$1 - \beta\mu^2 \bar{J}\alpha(x) = 1 - \frac{3T_c}{T}\left[\frac{1}{3} - \frac{x^2}{15} + \mathcal{O}(x^4)\right]$$

$$= 1 - \frac{T_c}{T} + \frac{T_c}{5T}(\bar{J}\beta\mu^2 m_0)^2 - \mathcal{O}(\bar{J}\beta\mu^2 m_0)^4. \tag{1833}$$

The key point is that the third term vanishes as $(T_c - T)$ and is of the same order as the first two terms:

$$\frac{T_c}{5T}(\bar{J}\beta\mu^2 m_0)^2 = \frac{T_c}{5T}\left(\frac{3T_c}{T}\right)^2 \frac{5}{3}\left(\frac{T}{T_c}\right)^2 \frac{T_c - T}{T_c} = 3\left(\frac{T_c - T}{T}\right) \tag{1834}$$

so

$$1 - \beta\mu^2 \bar{J}\alpha(x) = 2\left(\frac{T_c - T}{T}\right), \tag{1835}$$

and

$$\chi_M = \frac{C}{T}\frac{T}{2(T_c - T)} = \frac{C}{2(T_c - T)} \tag{1836}$$

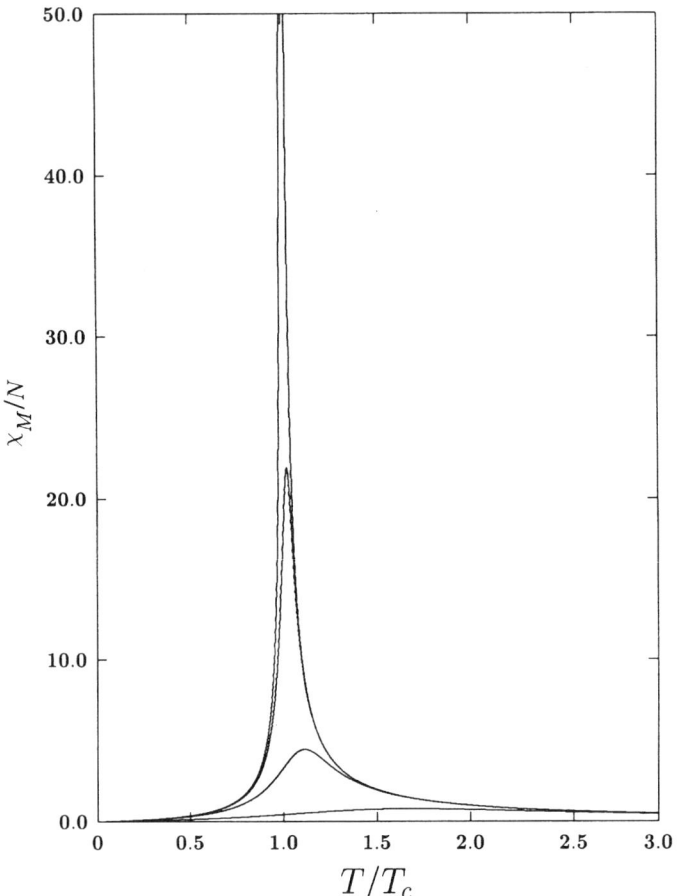

FIGURE 5.7 Magnetic susceptibility for a ferromagnet from mean-field theory as a function of scaled temperature. Different curves correspond to different values of the scaled magnetic field. From top to bottom one has $\mu B/k_B T_c = 0, 0.01, 0.1$, and 1.0.

for T near but below T_c. Note that χ_M is still positive, as is required for thermodynamic stability. It is left as a problem (5.19) to work out the behavior of χ_M as $T \to 0$. The full temperature dependence of χ_M as a function of T for various applied fields has the rather spectacular behavior shown in Fig. 5.7. This behavior is observed in inelastic neutron scattering experiments from magnetic systems such as nickel.

5.9.3 Antiferromagnetism

We have thus far restricted the analysis to a positive exchange, $J > 0$. What if $J < 0$? We can still carry out the mean field theory and obtain the equation for the average magnetization density as a simple modification of Eq. (1795),

$$m_0 = L(-|\bar{J}|\beta\mu^2 m_0) \tag{1837}$$

and for the susceptibility,

$$\chi_M = \left[1 + \beta\mu^2|\bar{J}|\alpha(x)\right]^{-1} N\beta\mu^2 \alpha(x) \tag{1838}$$

where $x = -|\bar{J}|\beta\mu^2 m_0$. Since $L(-x) = -L(x)$, we have

$$m_0 = -L(|\bar{J}|\beta\mu^2 m_0). \tag{1839}$$

Clearly, the only solutions for Eqs. (1838) and (1839) are

$$m_0 = 0 \tag{1840}$$

and

$$\chi_M = \left(1 + \frac{\beta\mu^2|\bar{J}|}{3}\right)^{-1} \frac{N\beta\mu^2}{3}. \tag{1841}$$

There is no ferromagnetic phase transitions for a Heisenberg magnetic with negative coupling J! Remember, however, that unlike the ferromagnet, when J is negative, the systems wants to order in an antiparallel manner:

$$\rightarrow \leftarrow \rightarrow \leftarrow \rightarrow \leftarrow \rightarrow \leftarrow \tag{1842}$$

Systems that have negative J values and undergo this antiparallel ordering at low temperatures are known as *anti ferromagnets*. This suggests that we look at our system in the following way. Assume that our spin system is made up of two interpenetrating sublattices; thus the A's in Fig. 5.8 form one sublattice and the B's form another lattice. Now suppose that the system is in a completely antiparallel configuration, as in Fig. 5.9. Look at just the B sublattice. The spins on this sublattice

A	B	A	B	A	B	
B	A	B	A	B	A	
A	B	A	B	A	B	
B	A	B	A	B	A	
A	B	A	B	A	B	
B	A	B	A	B	A	**FIGURE 5.8** Sublattice designations for a square lattice.

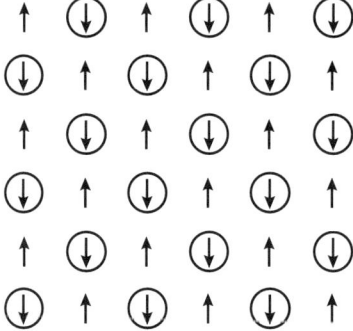

FIGURE 5.9 Sublattice ordering for an antiferromagnetic. Spins on sublattice B are circled.

are ordered. Similarly, the spins on the *A* sublattice are ordered. Let us define the magnetization for the *B* sublattice as

$$\mathbf{M}_B = \sum_{\mathbf{R} \in B} \boldsymbol{\mu}(\mathbf{R}) \tag{1843}$$

and for the *A* sublattice as

$$\mathbf{M}_A = \sum_{\mathbf{R} \in A} \boldsymbol{\mu}(\mathbf{R}). \tag{1844}$$

The exchange interaction will, assuming nearest-neighbor interactions only, couple the two sublattices. Again using mean-field theory, we obtain the coupled set of equations

$$m_B = L(\beta \mu B - \beta \mu^2 |\bar{J}| m_A) \tag{1845}$$

$$m_A = L(\beta \mu B - \beta \mu^2 |\bar{J}| m_B), \tag{1846}$$

where m_A and m_B are defined by

$$\langle \mathbf{M}_B \rangle = \hat{z} \frac{N}{2} \mu m_B \tag{1847}$$

and

$$\langle \mathbf{M}_A \rangle = \hat{z} \frac{N}{2} \mu m_A. \tag{1848}$$

In the case of zero applied field, the coupled equations (1845) and (1846) have the solution

$$m_A = -m_B = m_0 \tag{1849}$$

where

$$m_0 = L(|\bar{J}|\beta\mu^2 m_0) \tag{1850}$$

has solutions completely equivalent to the ferromagnetic problem. We see in this case that the total magnetization is zero since

$$\langle \mathbf{M}_T \rangle = \langle \mathbf{M}_A \rangle + \langle \mathbf{M}_B \rangle = \hat{z}\frac{N}{2}\mu(m_A + m_B)$$
$$= \hat{z}\frac{N}{2}\mu(m_0 - m_0) = 0. \tag{1851}$$

This is, however, a new *order parameter*, known as the *staggered magnetization*,

$$\langle \mathbf{N} \rangle \equiv \langle \mathbf{M}_A \rangle - \langle \mathbf{M}_B \rangle = \hat{z}\frac{N}{2}\mu(m_0 + m_0)$$
$$= \hat{z}N\mu m_0 \tag{1852}$$

which is nonzero. The nonzero value of $\langle \mathbf{N} \rangle$ corresponds to a different type of spontaneous symmetry breaking and a different phase transition. The transition temperature in this case is known as the *Néel temperature* [39]. Numerically, the Néel temperature is given in mean-field theory by the same expression as for ferromagnets except that we take the magnitude of J:

$$k_B T_N = \frac{z|J|\mu^2}{3}. \tag{1853}$$

It is interesting to solve Eqs. (1845) and (1846) for m_A and m_B in the presence of a field. In Fig. 5.10 we present the solutions for m_A and m_B as a function of temperature for various values of the applied field. Unlike in the ferromagnetic case, the applied magnetic field does *not* destroy the phase transition. Instead, the Neel temperature is decreased with increased applied field. Conceptually, there is a change in the nature of the phase transition for the case of nonzero external field. In the case of an applied field, one no longer breaks rotational symmetry as one passes through the transition since it has been broken by the applied field. However, the translational invariance on the lattice is broken when the sublattices become inequivalent, and one sublattice is more aligned with the field than the other. A key observation in this case is that the uniform applied external field couples, via the Zeeman term, directly to the order parameter, the total magnetization in the ferromagnetic case, but it does not couple directly to the order parameter in the antiferromagnetic case. It turns out that antiferromagnets are more prevalent in nature than ferromagnets (see Table 5.6 for examples).

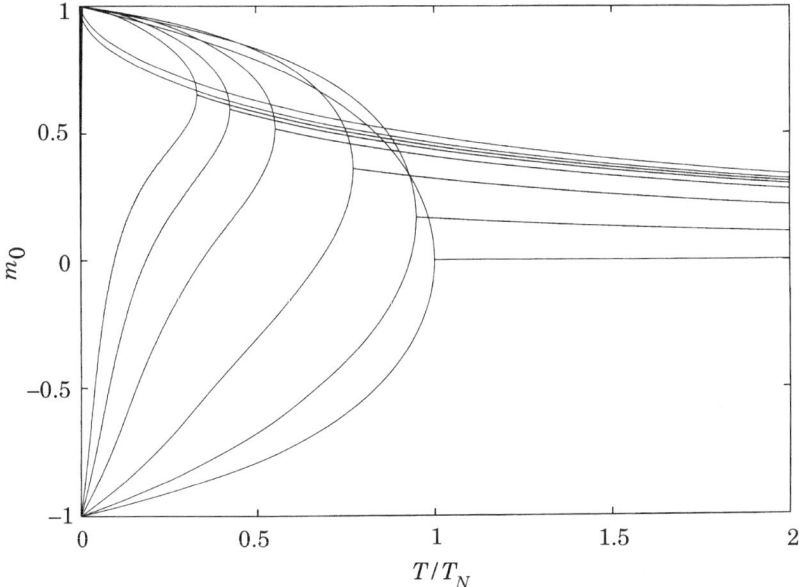

FIGURE 5.10 Solution for the sublattice ordering m_0 in mean-field theory for an antiferromagnet as a function of temperature. Different curves correspond to different applied magnetic fields. In the region of the curve for $T > T_N$, as one moves from bottom to top the applied field $\mu B / k_B T_c$ runs through values 0.01, 1.0, 2.0, 2.6, 2.8, 2.9, 3.0, and 3.2. For the last two values, unlike for lower values of the applied field, one does not have a breaking of the symmetry between the two sublattices.

TABLE 5.6 Examples of Antiferromagnets and Their Transition Temperatures

Antiferromagnet	$T_N(K)$	Antiferromagnet	$T_N(K)$
MnO	116	FeO	198
MnS	160	$CoCl_2$	25
MnTe	307	CoO	291
MnF_2	67	$NiCl_2$	50
FeF_2	79	NiO	525
$FeCl_2$	24	Cr	308

Source: Data from Ref. 40.

5.9.4 Magnetic Structure

In the case of noninteracting spins, we found [see Eq. (1654)] that the magnetic structure function was independent of wavenumber:

$$C_{\alpha\beta}(\mathbf{q}) = C_{\alpha\beta}(\mathbf{q} = 0). \tag{1854}$$

For interacting spins in general and in mean-field theory in particular, this is no longer true. In the classical regime we can calculate $C_{\alpha\beta}(\mathbf{q})$, the Fourier transform of

$$C_{\alpha\beta}(\mathbf{R},\mathbf{R}') = \langle \delta\mu_\alpha(\mathbf{R})\delta\mu_\beta(\mathbf{R}')\rangle, \tag{1855}$$

where

$$\delta\mu_\alpha(\mathbf{R}) = \mu_\alpha(\mathbf{R}) - \langle\mu_\alpha(\mathbf{R})\rangle, \tag{1856}$$

using the following formal device. Assume that the magnetic system of interest is acted upon by an external magnetic field $\mathbf{B}(\mathbf{R})$ which varies from site to site. The Zeeman term in the Hamiltonian becomes $-\sum_\mathbf{R} \mathbf{B}(\mathbf{R}) \cdot \boldsymbol{\mu}(\mathbf{R})$. It is then left as a problem (5.22) to prove the identities:

$$M_\alpha(\mathbf{R}) = \langle\mu_\alpha(\mathbf{R})\rangle = \beta^{-1}\frac{\partial}{\partial B_\alpha(\mathbf{R})}\ln Z_M^N(B) \tag{1857}$$

and

$$C_{\alpha\beta}(\mathbf{R},\mathbf{R}') = \beta^{-1}\frac{\partial}{\partial B_\beta(\mathbf{R}')}M_\alpha(\mathbf{R}) \tag{1858}$$

where

$$Z_M^N(B) = \left[\prod_\mathbf{R}\int\frac{d\Omega(\mathbf{R})}{4\pi}\right]e^{-\beta H}, \tag{1859}$$

and

$$H = -\tfrac{1}{2}\sum_{\mathbf{R},\mathbf{R}'}J(\mathbf{R},\mathbf{R}')\boldsymbol{\mu}(\mathbf{R})\cdot\boldsymbol{\mu}(\mathbf{R}') - \sum_\mathbf{R}\mathbf{B}(\mathbf{R})\cdot\boldsymbol{\mu}(\mathbf{R}). \tag{1860}$$

Mean-field theory is carried out in much the same fashion as in the case of a homogeneous field, and the Hamiltonian H is written in terms of an effective field:

$$H = -\sum_\mathbf{R}\mathbf{B}_T(\mathbf{R})\cdot\boldsymbol{\mu}(\mathbf{R}) \tag{1861}$$

where the total field is again the sum of the external field and the local field due to the neighbors:

$$\mathbf{B}^T(\mathbf{R}) = \mathbf{B}(\mathbf{R}) + \sum_{\mathbf{R}'}J(\mathbf{R},\mathbf{R}')\mathbf{M}(\mathbf{R}'). \tag{1862}$$

It is important to realize at this stage that the external field $\mathbf{B}(\mathbf{R})$ breaks translational invariance and $\mathbf{M}(\mathbf{R})$ depends on $\mathbf{R}$. Despite this difference, the computation of $\mathbf{M}(\mathbf{R})$ follows in the same fashion as in the homogeneous case, and assuming that

$$\mathbf{B}(\mathbf{R}) = \hat{z}B(\mathbf{R}), \tag{1863}$$
$$\mathbf{M}(\mathbf{R}) = \hat{z}M_z(\mathbf{R}), \tag{1864}$$

we find that

$$M_z(\mathbf{R}) = \mu L(\beta\mu B_z^T(\mathbf{R})). \tag{1865}$$

Taking the derivative of Eq. (1865) with respect to $B_z(\mathbf{R}')$ and using the identity Eq. (1858), we obtain

$$\beta C_{zz}(\mathbf{R}, \mathbf{R}') = \frac{\partial}{\partial B_z(\mathbf{R}')}[\mu L(\beta\mu B_z^T(\mathbf{R}))]. \tag{1866}$$

Letting $x = \beta\mu B_z^T(\mathbf{R})$, we have, again using the chain rule for differentiation,

$$\beta C_{zz}(\mathbf{R}, \mathbf{R}') = \mu\frac{\partial L(x)}{\partial x}\frac{\partial}{\partial B_z(\mathbf{R}')}\left[\beta\mu B_z(\mathbf{R}) + \beta\mu\sum_{\mathbf{R}''} J(\mathbf{R},\mathbf{R}'')M_z(\mathbf{R}'')\right]$$

$$= \beta\mu^2\frac{\partial L(x)}{\partial x}\left[\delta_{\mathbf{R},\mathbf{R}'} + \sum_{\mathbf{R}''}\beta J(\mathbf{R},\mathbf{R}'')\beta C_{zz}(\mathbf{R}'',\mathbf{R}')\right]. \tag{1867}$$

We now assume that the external field is uniform, $B(\mathbf{R}) = B$, so the system becomes translationally invariant. We can then solve Eq. (1867) for the structure factor using Fourier transforms:

$$C(\mathbf{q}) = \frac{1}{N}\sum_{\mathbf{R},\mathbf{R}'} e^{+i\mathbf{q}\cdot(\mathbf{R}-\mathbf{R}')} C_{zz}(\mathbf{R},\mathbf{R}') \tag{1868}$$

$$J(\mathbf{q}) = \frac{1}{N}\sum_{\mathbf{R}\mathbf{R}'} e^{+i\mathbf{q}\cdot(\mathbf{R}-\mathbf{R}')} J(\mathbf{R},\mathbf{R}') = \sum_{\mathbf{R}} e^{+i\mathbf{q}\cdot\mathbf{R}} J(\mathbf{R}). \tag{1869}$$

We easily obtain from Eq. (1867),

$$C(\mathbf{q}) = \frac{\mu^2\alpha(x)}{1 - \beta\mu^2\alpha(x)J(\mathbf{q})} \tag{1870}$$

where $x = \beta\mu B_z^T$ and $\alpha(x)$ is given by Eq. (1825). Let us further restrict the discussion to zero external field and $T \geq T_c$. In this case, $\alpha(0) = \frac{1}{3}$ and

$$C(\mathbf{q}) = \frac{\mu^2}{3}\frac{1}{1 - (T_c/T)[J(\mathbf{q})/\bar{J}]} \tag{1871}$$

where $\bar{J} = J(\mathbf{q} = 0)$ and we identify the Curie temperature using Eq. (1828). Let us further specialize to the case of nearest-neighbor interactions on a cubic lattice. We easily find that

$$J(\mathbf{q}) = 2J \sum_{\alpha=1}^{3} \cos q_\alpha a \qquad (1872)$$

where a is the lattice spacing. Then $\bar{J} = 6J$ (with $z = 6$), and

$$C(\mathbf{q}) = \frac{\mu^2}{3} \frac{T}{T - (T_c/3) \sum_{\alpha=1}^{3} \cos q_\alpha a}. \qquad (1873)$$

In Fig. 5.11, $C(\mathbf{q})$ is plotted versus qa for $\mathbf{q} = \hat{z}q$ and various values of $T/T_c \geq 1$. We see that for very high temperatures, $C(\mathbf{q})$ approaches its noninteracting structureless value. However, as T/T_c drops, we see a buildup of structure for $\mathbf{q}$ near zero and multiples of $2\pi/a$. For $T = T_c$ we find that $C(\mathbf{q})$ diverges as $1/q^2$ as $\mathbf{q} \to 0$. We study the significance of this wavenumber dependence in detail in Volume 2.

In the case of antiferromagnets ($J \to -|J|$), a similar analysis shows that

$$C(\mathbf{q}) = \frac{\mu^2}{3} \frac{T}{T + (T_c/3) \sum_{\alpha=1}^{3} \cos q_\alpha a}. \qquad (1874)$$

In this case, as shown in Fig. 5.12, at $T = T_N$, $C(\mathbf{q})$ diverges near those *antiferromagnetic points* where all three directions $q_\alpha a$ are an odd multiple of π. In Problem 5.25 the physical significance of this result is explored.

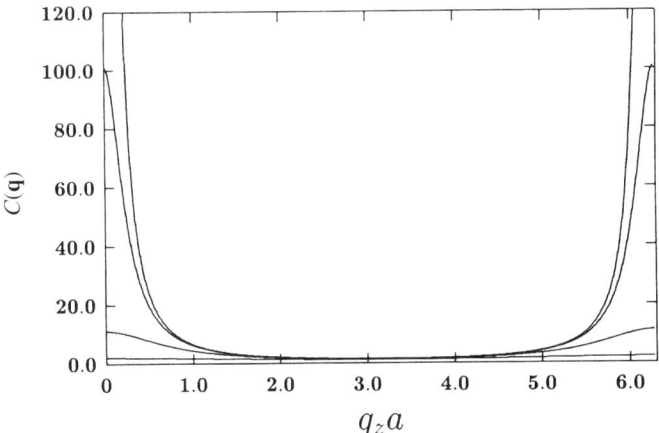

FIGURE 5.11 Magnetic structure factor versus wavenumber for a ferromagnet within mean-field theory. Plotted for $q_x = q_y = 0$ and curves with decreasing peak amplitudes correspond to temperatures $T/T_c = 1, 1.01, 1.1$, and 2.0.

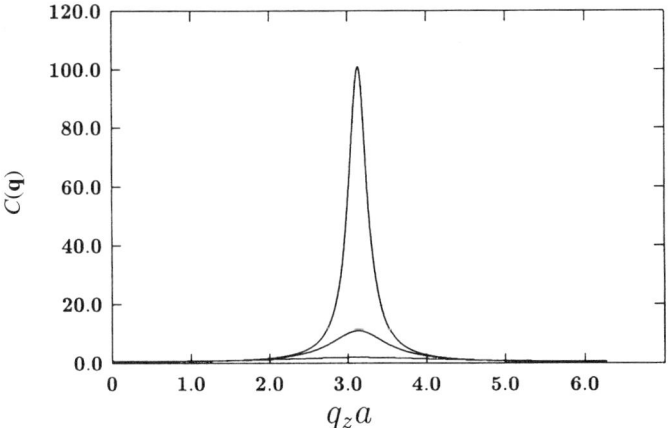

FIGURE 5.12 Magnetic structure factor versus wavenumber for an antiferromagnet within mean-field theory. Plotted for $q_x a = q_y a = \pi$ and from top to bottom for temperatures $T/T_N = 1.01$, 1.1, and 2.0. The peak is at the antiferromagnetic point $q_z a = \pi$.

5.9.5 Magnetic Models

Thus far we have concentrated on the Heisenberg Hamiltonian given by Eq. (1772). More generally, we can have an anisotropic Hamiltonian of the form

$$H = -\frac{1}{2} \sum_{\alpha} \sum_{\mathbf{R},\mathbf{R}'} J_\alpha(\mathbf{R}, \mathbf{R}') \mu_\alpha(\mathbf{R}) \mu_\alpha(\mathbf{R}'). \tag{1875}$$

In the isotropic limit J_α is independent of α. In many cases of interest the J_α's are dependent on α. In such cases we obtain *easy* axis or planes of orientation. Consider first the case $J_z \gg J_x, J_y > 0$. Clearly, the system will be inclined to order in the z-direction in order to lower the energy. Thus, to a first approximation, we can ignore the x and y components of $\boldsymbol{\mu}(\mathbf{R})$ and treat the simplified model,

$$H = -\frac{1}{2} \sum_{\mathbf{R},\mathbf{R}'} J_z(\mathbf{R}, \mathbf{R}') \mu_z(\mathbf{R}) \mu_z(\mathbf{R}'). \tag{1876}$$

Known as the *Ising model*, this is among the most studied models in all of theoretical physics. It was named after Ernst Ising [41], a graduate student of Wilhelm Lenz, who invented the model in 1925. The key simplifying feature of this model is that the magnetic moments can take on only two distinct values,

$$\mu_z(\mathbf{R}) = \pm \mu \tag{1877}$$

where μ is the magnitude of the moment. If we write

$$\mu_z(\mathbf{R}) = \mu \sigma(\mathbf{R}), \tag{1878}$$

the variable $\sigma(\mathbf{R})$ takes on the values ± 1 and the Hamiltonian can be rewritten as

$$H = -\tfrac{1}{2}\sum_{\mathbf{R},\mathbf{R}'} J(\mathbf{R},\mathbf{R}')\sigma(\mathbf{R})\sigma(\mathbf{R}') \tag{1879}$$

where

$$J(\mathbf{R},\mathbf{R}') = \mu^2 J_z(\mathbf{R},\mathbf{R}'). \tag{1880}$$

The Ising model has wider applicability than a simple description of magnetism. It serves as a model for any system that has two competing states on a lattice site. One example is a lattice gas where the density at lattice site $\mathbf{R}$ is given by

$$n(\mathbf{R}) = \tfrac{1}{2}[1 + \sigma(\mathbf{R})]. \tag{1881}$$

If $\sigma(\mathbf{R}) = 1, n(\mathbf{R}) = 1$ and the site is occupied; if $\sigma(\mathbf{R}) = -1, n(\mathbf{R}) = 0$ and the site is unoccupied. The interactions $J(\mathbf{R},\mathbf{R}')$ are then given by evaluating the interparticle potential at separations corresponding to different relative orientations on the lattice.

Returning to Eq. (1875), the next situation of interest is when there is an easy plane of orientation. In this case $J_x = J_y \gg J_z$, and the moments want to align in the x–y plane. This model, known as the x–y or *planar model*, looks like the Heisenberg model except that $\boldsymbol{\mu}(\mathbf{R})$ is now a two-dimensional vector, $\boldsymbol{\mu}(\mathbf{R}) = [\mu_x(\mathbf{R}), \mu_y(\mathbf{R})]$.

We shall proceed for the remainder of this chapter to analyze the Ising model in some detail. The same techniques can also be applied to the x–y and Heisenberg models. It should be understood that magnetic systems are very rich and there are a variety of other effects that come into play beyond the exchange interaction. Because magnets are embedded in solids, the symmetry of the lattice must be taken into account, which can introduce anisotropies (couplings between various directions). The magnetic degrees of freedom can couple to the vibrations (oscillations) of the lattice. There are also the effects of impurities and imperfections, which can have a strong influence on the magnetic properties of a solid.

5.10 SOLUTIONS FOR THE ISING MODEL

5.10.1 Introduction

The Ising Hamiltonian in the presence of an external field can be written as

$$H = -\tfrac{1}{2}\sum_{\mathbf{R},\mathbf{R}'} J(\mathbf{R},\mathbf{R}')\sigma(\mathbf{R})\sigma(\mathbf{R}') - B\sum_{\mathbf{R}}\sigma(\mathbf{R}) \tag{1882}$$

where we have absorbed a factor of the magnitude of the magnetic moment, μ, into B, the external magnetic field. This is a nontrivial interacting statistical mechanical

416　EQUILIBRIUM PROPERTIES OF DIELECTRIC AND MAGNETIC MATERIALS

model and we have learned that in general we must use approximation techniques in treating such models. The Ising model is an exception to this rule. As indicated below, this model can be solved rather completely in one and two dimensions. Let us look at the much simpler one-dimensional case first.

5.10.2　One-Dimensional Case

Let us consider the one-dimensional Ising model in the case of nearest-neighbor interactions, where the Hamiltonian reduces to

$$H = -J \sum_l \sigma_l \sigma_{l+1} - B \sum_l \sigma_l \tag{1883}$$

where l is an integer labeling the lattice sites. For convenience, we assume periodic boundary conditions $\sigma_1 = \sigma_{N+1}$, so the spins are on a circle. The canonical ensemble partition function can now be written

$$Z(B,T) = \sum_{\sigma_1=\pm 1} \sum_{\sigma_2} \cdots \sum_{\sigma_N} \exp\left[\sum_{l=1}^{N} \beta(J\sigma_l\sigma_{l+1} + B\sigma_l)\right]. \tag{1884}$$

In evaluating Z it is convenient to write

$$\sum_{l=1}^{N} \sigma_l = \tfrac{1}{2} \sum_{l=1}^{N} (\sigma_l + \sigma_{l+1}) \tag{1885}$$

which follows from a use of $\sigma_1 = \sigma_{N+1}$. We then have

$$Z(B,T) = \sum_{\sigma_1=\pm 1} \sum_{\sigma_2} \cdots \sum_{\sigma_N} \exp\left\{\sum_{l=1}^{N} \beta\left[J\sigma_l\sigma_{l+1} + \frac{B}{2}(\sigma_l + \sigma_{l+1})\right]\right\}. \tag{1886}$$

Note that the Boltzmann factor can be written in a product form:

$$\exp\left\{\beta \sum_{l=1}^{N} \left[J\sigma_l\sigma_{l+1} + \frac{B}{2}(\sigma_l + \sigma_{l+1})\right]\right\}$$
$$= \exp\left\{\beta\left[J\sigma_1\sigma_2 + \frac{B}{2}(\sigma_1 + \sigma_2) + J\sigma_2\sigma_3 + \frac{B}{2}(\sigma_2 + \sigma_3) + \cdots\right]\right\}$$
$$= \exp\left\{\beta\left[J\sigma_1\sigma_2 + \frac{B}{2}(\sigma_1 + \sigma_2)\right]\right\} \exp\left\{\beta\left[J\sigma_2\sigma_3 + \frac{B}{2}(\sigma_2 + \sigma_3)\right]\right\} \times \cdots$$
$$\tag{1887}$$

Let us introduce the notation

$$\langle \sigma_i | T | \sigma_j \rangle \equiv \exp\left\{ \beta \left[J \sigma_i \sigma_j + \frac{B}{2}(\sigma_i + \sigma_j) \right] \right\} \quad (1888)$$

so that

$$Z(B, T) = \sum_{\sigma_1} \sum_{\sigma_2} \cdots \sum_{\sigma_N} \langle \sigma_1 | T | \sigma_2 \rangle \langle \sigma_2 | T | \sigma_3 \rangle \cdots \langle \sigma_{N-1} | T | \sigma_N \rangle \langle \sigma_N | T | \sigma_1 \rangle. \quad (1889)$$

Now we observe that $\langle \sigma_i | T | \sigma_j \rangle$ is simply a symmetric 2×2 matrix with matrix elements

$$\langle +1 | T | +1 \rangle = e^{\beta(J+B)} \quad (1890)$$

$$\langle -1 | T | -1 \rangle = e^{\beta(J-B)} \quad (1891)$$

$$\langle +1 | T | -1 \rangle = \langle -1 | T | +1 \rangle = e^{-\beta J}. \quad (1892)$$

In dealing with these matrices, it is extremely useful to introduce an abstract vector space. The basic definitions are discussed in Appendix J. The idea is that for every symmetric matrix we can introduce an operator and a set of basis states, and the *average* of this operator between a set of basis states gives the matrix. Thus the matrix $\langle i | A | j \rangle$ can be thought of as a matrix element of an abstract operator A, where $|i\rangle$ and $|j\rangle$ are the basis states. These basis states are assumed to be complete,

$$\sum_i |i\rangle\langle i| = 1 \quad (1893)$$

and the orthonormal

$$\langle i | j \rangle = \delta_{ij}. \quad (1894)$$

Going back to the one-dimensional Ising model, we can introduce the operator T and the basis set $|\sigma_i\rangle$, which for a given i can be $|+\rangle$ or $|-\rangle$. This basis set is complete:

$$1 = \sum_{\sigma_i = \pm 1} |\sigma_i\rangle\langle\sigma_i| \quad (1895)$$

and orthonormal:

$$\langle \sigma_i | \sigma'_j \rangle = \delta_{ij} \delta_{\sigma,\sigma'}. \quad (1896)$$

We can then write

$$Z(B,T) = \sum_{\sigma_1}\sum_{\sigma_3}\cdots\sum_{\sigma_N}\langle\sigma_1|T\sum_{\sigma_2}|\sigma_2\rangle\langle\sigma_2|T\sigma_3\rangle\cdots\langle\sigma_{N-1}|T|\sigma_N\rangle\langle\sigma_N|T|\sigma_1\rangle$$

$$= \sum_{\sigma_1}\sum_{\sigma_3}\cdots\sum_{\sigma_N}\langle\sigma_1|T^2|\sigma_3\rangle\cdots\langle\sigma_{N-1}|T|\sigma_N\rangle\langle\sigma_N|T|\sigma_1\rangle, \quad (1897)$$

where we have used the completeness condition

$$\sum_{\sigma_2}|\sigma_2\rangle\langle\sigma_2| = 1. \quad (1898)$$

We can use this result $N-1$ times to obtain

$$Z(B,T) = \sum_{\sigma_1}\langle\sigma_1|T^N|\sigma_1\rangle \equiv \operatorname{Tr}T^N. \quad (1899)$$

We obtain the trace of a 2×2 matrix raised to the Nth power. If we can find the eigenstates $|\lambda_\mu\rangle$ of T with eigenvalues λ_μ,

$$T|\lambda_\mu\rangle = \lambda_\mu|\lambda_\mu\rangle, \quad (1900)$$

we can trace over this complete set of states,

$$Z(B,T) = \sum_\mu \langle\lambda_\mu|T^N|\lambda_\mu\rangle \quad (1901)$$

$$= \sum_\mu \lambda_\mu^N \langle\lambda_\mu|\lambda_\mu\rangle \quad (1902)$$

and we can choose our normalization such that

$$\langle\lambda_\mu|\lambda_\mu\rangle = 1, \quad (1903)$$

to obtain

$$Z(B,T) = \sum_\mu \lambda_\mu^N. \quad (1904)$$

This evaluation of the partition function has been reduced to the determination of the eigenvalues of a 2×2 matrix. This method, expressing the partition function in terms of a product of matrices, is called the *transfer matrix method*. The eigenvalues are determined as solutions to the linear equation

$$\langle\nu|T|\lambda\rangle = \lambda\langle\nu|\lambda\rangle, \quad (1905)$$

which we can rewrite as

$$\sum_\alpha (\langle \nu|T|\alpha\rangle - \lambda \delta_{\alpha,\nu})\langle \alpha|\lambda\rangle = 0. \tag{1906}$$

This homogeneous equation has solutions only for choices of λ obtained as solutions to the determinantal equation

$$\det(\langle \nu|T|\alpha\rangle - \lambda \delta_{\alpha,\nu}) = 0. \tag{1907}$$

Evaluating this 2×2 determinant, we obtain

$$(e^{\beta(J+B)} - \lambda)(e^{\beta(J-B)} - \lambda) - e^{2\beta J} = 0 \tag{1908}$$

or

$$\lambda^2 - 2\lambda e^{\beta J}\cosh \beta B + 2\sinh 2\beta J = 0. \tag{1909}$$

Equation (1909) has the solutions

$$\lambda_\pm = e^{\beta J}[\cosh \beta B \pm (\cosh^2 \beta B - 2e^{-2\beta J}\sinh 2\beta J)^{1/2}]. \tag{1910}$$

Since

$$\cosh^2 \beta B - 2e^{-2\beta J}\sinh 2\beta J = \sinh^2 \beta B + e^{-4\beta J}, \tag{1911}$$

we can rewrite Eq. (1910) in the form

$$\lambda_\pm = e^{\beta J}[\cosh \beta B \pm (\sinh^2 \beta B + e^{-4\beta J})^{1/2}]. \tag{1912}$$

Clearly, $\lambda_+ > \lambda_-$, and since we can write

$$Z(B,T) = \lambda_+^N \left[1 + \left(\frac{\lambda_-}{\lambda_+}\right)^N\right], \tag{1913}$$

and $1 \gg (\lambda_-/\lambda_+)^N$ for large N, we have

$$Z(B,T) \approx \lambda_+^N. \tag{1914}$$

Finite-size corrections are explored in Problem 5.30. The free energy is given in the thermodynamic limit by

$$\frac{F}{N} = \lim_{N\to\infty} -\frac{\beta^{-1}}{N}\ln Z(B,T) = -\beta^{-1}\ln \lambda_+$$
$$= -J - \beta^{-1}\ln[\cosh \beta B + (\sinh^2 \beta B + e^{-4\beta J})^{1/2}]. \tag{1915}$$

It is left to Problem 5.33 to work out the various thermodynamic properties of the system.

1. The magnetization per spin:

$$\frac{M}{N} = -\frac{1}{N}\left(\frac{\partial F}{\partial B}\right)_T = \frac{\sinh\beta B}{(\sinh^2\beta B + e^{-4\beta J})^{1/2}}. \quad (1916)$$

2. The magnetic susceptibility in zero external field:

$$\chi_M = \left(\frac{\partial M}{\partial B}\right)_{B=0} = N\beta e^{2\beta J}. \quad (1917)$$

3. The energy per spin in zero external field:

$$\frac{E}{N} = \frac{\partial}{\partial \beta}\left(\frac{\beta F}{N}\right) = -J\tanh\beta J. \quad (1918)$$

4. The specific heat in zero external field:

$$C_V = \left(\frac{dE}{dT}\right)_{B=0} = \frac{k_B N(\beta J)^2}{\cosh^2 \beta J}. \quad (1919)$$

The thermal properties E and C_V are smooth functions of temperature. The energy goes to a constant at low temperatures,

$$\lim_{T \to 0} E = -JN \quad (1920)$$

corresponding to a coherent lining up of the spins each contributing energy $-J$. The magnitude of the energy decreases as the temperature rises, going as

$$E = -\frac{J^2}{k_B T}N \quad (1921)$$

for large T. The decrease in E corresponds to the spins becoming completely uncorrelated at high temperatures. The specific heat, shown in Fig. 5.13, vanishes exponentially for small temperatures and as $(J/k_B T)^2$ for large temperatures.

Turning to the magnetic properties, we see that the magnetization increases with magnetic field as in Fig. 5.14. For a fixed nonzero temperature

$$\lim_{B \to \infty} \frac{M}{N} = \pm 1 \quad (1922)$$

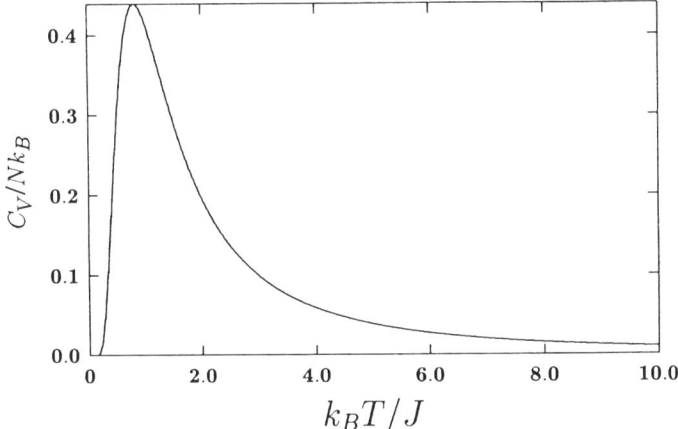

FIGURE 5.13 Specific heat for the one-dimensional Ising model versus reduced temperature in zero external field.

and

$$\lim_{B \to 0} \frac{M}{N} = 0. \tag{1923}$$

A very strong magnetic field overwhelms the thermal motion of the spins and lines them up. Of more immediate interest is the result [Eq. (1923)] which says that there is no magnetization at nonzero temperatures for the one-dimensional Ising model. In the zero-temperature limit,

$$\lim_{T \to 0} \frac{M}{N} = \pm 1 \tag{1924}$$

and the spins will line up. The magnetic susceptibility per spin in zero field is plotted in Fig. 5.15. At high temperatures this reproduces the Curie law. In this case the phase transition is at zero temperature. χ_M is finite for all $T > 0$. It is left as a fairly difficult problem to show that one can also work out the spatial correlations for the one-dimensional Ising model (see Problem 5.31).

The lack of a phase transition at nonzero ($T > 0$) temperatures in the one-dimensional Ising model is a special case of a general theorem [42]: One-dimensional systems with short-range forces can undergo a phase transition only at $T = 0$. This is discussed in some detail in Volume 2.

5.10.3 Ising Model in Higher Dimensions

The two-dimensional Ising model is qualitatively different from the one-dimensional Ising model. Whereas the one-dimensional Ising model shows no ferromagnetism for nonzero temperatures, the two-dimensional system does. Peierls

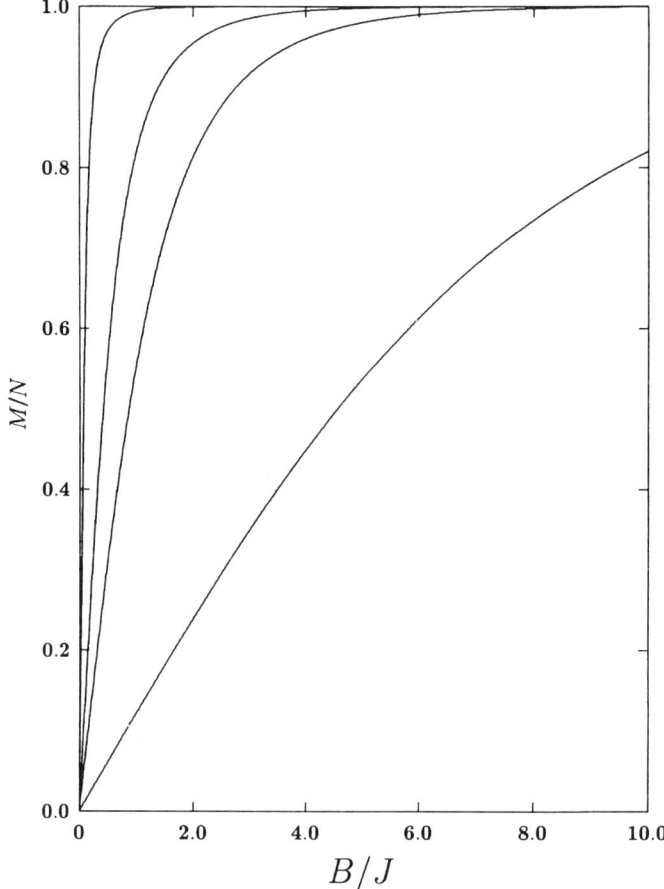

FIGURE 5.14 Magnetization per spin for the one-dimensional Ising model versus reduced applied field for temperatures (from bottom to top) $k_BT/J = 10.0$, 3.0, 2.0, and 1.0.

[43] first showed in 1936 that the two-dimensional Ising model has a ferromagnetic phase transition. In 1941, Kramers and Wannier [44] determined the transition temperature rigorously, *assuming* that there existed a phase transition point. In 1944, Onsager [45] opened a new door for the study of the statistical mechanics of phase transitions by solving the two-dimensional Ising model rigorously and proving the existence of a phase transition. This was a monumental event because, before Onsager, it was not known whether a rigorous theory could show a phase transition.

The various methods developed for solving the two-dimensional Ising model are mathematically very sophisticated and are not easily generalizable to other problems or higher dimensions. We shall not, therefore, pause here to go through the solutions, which are discussed in the book by McCoy and Wu [46]. We shall, instead, summarize and discuss the results of the solutions for a square lattice.

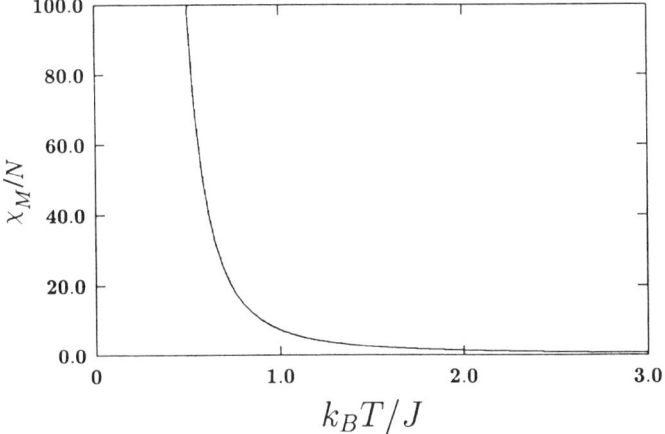

FIGURE 5.15 Magnetic susceptibility per spin for the one-dimensional Ising model in zero external field versus reduced temperature.

In Onsager's original 1944 paper, he found the free energy per spin for the two-dimensional nearest-neighbor Ising model on a square lattice to be given by

$$\frac{F}{N} = -\beta^{-1} \frac{1}{N} \ln Z$$
$$= -\beta^{-1} \left[\ln(2\cosh 2K) + \frac{1}{2\pi} \int_0^{2\pi} d\phi \ln \frac{1 + (1 - \kappa^2 \sin^2 \phi)^{1/2}}{2} \right] \quad (1925)$$

where $K = \beta J$, and

$$\kappa = 2 \frac{\sinh 2K}{\cosh^2 2K}. \quad (1926)$$

From the free energy, he was able to compute the energy per spin (see Fig. 5.16),

$$\frac{E}{N} = -J \coth 2K \left[1 \pm \frac{2}{\pi} (1 - \kappa^2)^{1/2} J(\kappa) \right], \quad (1927)$$

where $\pm$ means temperatures above and below the transition temperature, and T_c is determined by

$$e^{-2\beta_c J} = \sqrt{2} - 1. \quad (1928)$$

The quantity $J(\kappa)$ in Eq. (1927), defined by

$$J(\kappa) = \int_0^{\pi/2} \frac{d\phi}{(1 - \kappa^2 \sin^2 \phi)^{1/2}}, \quad (1929)$$

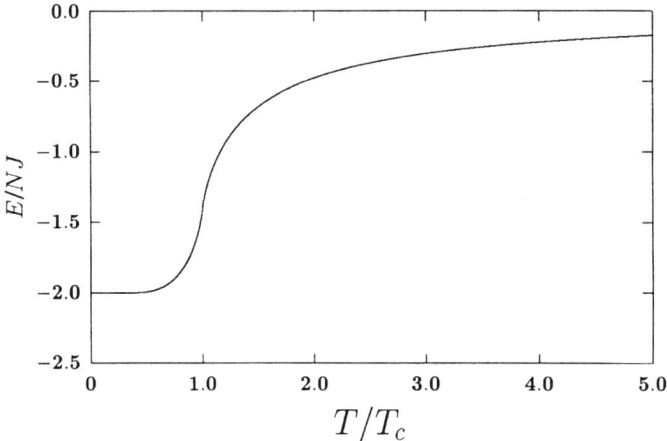

FIGURE 5.16 Onsager solution for the average energy per spin for the ferromagnetic two-dimensional Ising model of a square lattice in zero external field.

is the complete elliptic integral of the first kind. For values of T near T_c, we have

$$J(\kappa) \approx \ln|T - T_c| \tag{1930}$$

so

$$\frac{E}{N} \approx \ln(T - T_c)\ln|T - T_c| \tag{1931}$$

and the specific heat is given by (see Problem 5.34)

$$\frac{C_V}{N} \approx \ln|T - T_c|. \tag{1932}$$

Thus the specific heat is logarithmically divergent at the critical temperature.

Onsager's solution in 1944 gave only the thermal properties that follow from the free energy, $F = F(T)$, evaluated in zero external field. The results for the spontaneous magnetization per spin became known [47] later, when it was shown

$$m(0, T) = \begin{cases} 0 & T > T_c \\ \left(1 - \dfrac{1}{\sinh^4 2K}\right)^{1/8} & T < T_c. \end{cases} \tag{1933}$$

For temperatures just below T_c this reduces to

$$m(0, T) \approx \left[8\sqrt{2}\left(\frac{J}{k_B T_c}\right)\frac{T_c - T}{T_c}\right]^{1/8} \tag{1934}$$

and the critical index $\beta = \frac{1}{8}$. The result for the magnetization was reported [48] by Onsager at a conference at Cornell in 1948, but the detailed proof was given by Yang [47] in 1952. Many other [49] properties of the two-dimensional Ising model are known (for triangular and hexagonal lattices as well as square lattices) exactly. We note here, and discuss in detail in Volume 2, that the exact value of β given by Eq. (1934) does not agree with the Landau theory or mean field theory, which give $\beta = \frac{1}{2}$. Therefore, mean field theory is only an approximate theory for critical phenomena.

No one has yet obtained an exact solution for the three-dimensional Ising model. However, using various approximation methods, we know a great deal about the three-dimensional case. One of the primary approximation tools is discussed in the next section.

5.11 HIGH-TEMPERATURE EXPANSIONS

We develop in this section a very general approximation method that can be used on a variety of lattice models. Let us consider the nearest-neighbor Ising model where the Hamiltonian can be written as

$$H = -JU - BM \tag{1935}$$

where M is a dimensionless version of the total magnetization,

$$M = \sum_{\mathbf{R}} \sigma(\mathbf{R}), \tag{1936}$$

and

$$U = \frac{1}{2} \sum_{\mathbf{R},\mathbf{R}'} V(\mathbf{R}, \mathbf{R}') \sigma(\mathbf{R}) \sigma(\mathbf{R}') \tag{1937}$$

is a dimensionless interaction with

$$V(\mathbf{R}, \mathbf{R}') = \sum_{\boldsymbol{\delta}} \delta_{\mathbf{R}', \mathbf{R}+\boldsymbol{\delta}} \tag{1938}$$

and the $\boldsymbol{\delta}$ are the lattice vectors connecting a site to its z nearest neighbors. Thus, for a two-dimensional square lattice, $(\delta_x, \delta_y) = a(1,0), a(-1,0), a(0,1)$, and $a(0,-1)$, where a is the lattice spacing.

The partition function can then be written in the form

$$Z(K, h) = \left(\prod_{\mathbf{R}} \sum_{\sigma(\mathbf{R})} \right) e^{KU(\sigma) + hM(\sigma)} \tag{1939}$$

where

$$K = \frac{J}{k_B T} \qquad (1940)$$

and

$$h = \frac{B}{k_B T}. \qquad (1941)$$

Written in this form, we see that for fixed N, the partition function depends on the dimensionless variables K and h. One very successful approach to this problem (and also for the x–y and Heisenberg models) has been to develop expansions in powers of K. Since K is small for high temperatures, these and related expansions have come to be called *high-temperature expansions*. The ideas behind these expansions and the basic techniques are closely related to the development of low-density expansions discussed in Chapter 4.

As in the development of low-density expansions, we recognize that we should not expand $Z(K, h)$ directly but expand the associated potential:

$$\Omega(K, h) = \ln Z(K, h) = \sum_{n=0}^{\infty} \frac{K^n}{n!} \Omega^{(n)}(h) \qquad (1942)$$

where

$$\Omega^{(n)}(h) = \left(\frac{\partial^n}{\partial K^n} \Omega(K, h)\right)_{K=0}. \qquad (1943)$$

It is useful, in evaluating the $\Omega^{(n)}(h)$, to introduce the probability distribution $P_0[\sigma]$ corresponding to $K = 0$. The key point is that $P_0[\sigma]$ factorizes into a product of terms

$$P_0[\sigma] = \prod_{\mathbf{R}} P_0[\mathbf{R}, \sigma] \qquad (1944)$$

where

$$P_0[\mathbf{R}, \sigma] = \frac{e^{h\sigma(\mathbf{R})}}{2 \cosh h}. \qquad (1945)$$

Introducing averages over $P_0[\sigma]$,

$$\langle A[\sigma] \rangle_0 = \left(\prod_{\mathbf{R}} \sum_{\sigma(\mathbf{R})}\right) P_0[\sigma] A[\sigma], \qquad (1946)$$

we easily obtain the averages at a site

$$\langle \sigma^{2n}(\mathbf{R}) \rangle_0 = 1$$
$$\langle \sigma^{2n+1}(\mathbf{R}) \rangle_0 = \tanh h \equiv m_0 \qquad (1948)$$

where n is an integer. Using Eqs. (1942), (1943), and (1939), we easily obtain

$$\Omega^{(0)} = \ln Z(0, h), \qquad (1949)$$
$$\Omega^{(1)} = \langle U \rangle_0, \qquad (1950)$$
$$\Omega^{(2)} = \langle U^2 \rangle_0 - \langle U \rangle_0^2, \qquad (1951)$$
$$\Omega^{(3)} = \langle U^3 \rangle_0 - 3\langle U^2 \rangle_0 \langle U \rangle_0 + 2\langle U \rangle_0^3, \qquad (1952)$$

and so on. Note again the clustering structure for the $\Omega^{(n)}$. While $\langle U^n \rangle_0 \approx N^n$, we expect each $\Omega^{(n)}$ to be linear in N, the total number of spins.

The evaluation of the $\langle U^n(\sigma) \rangle_0$ and $\Omega^{(n)}$ is technically different than in the low-density expansions. In that case we had translational invariance but difficult overlap integrals to evaluate. In the case of the Ising model the evaluation of the *trace*, sums over the σ's, is trivial since the average over $\langle \rangle_0$ factorizes into the averages over independent sites. However, the treatment of the lattice structure became the challenge as one goes to higher order. Consider the lowest-order term,

$$\langle U \rangle_0 = \tfrac{1}{2} \sum_{\mathbf{R},\mathbf{R}'} V(\mathbf{R}, \mathbf{R}') \langle \sigma(\mathbf{R})\sigma(\mathbf{R}') \rangle_0. \qquad (1953)$$

Since $V(\mathbf{R}, \mathbf{R}')$ requires that $\mathbf{R} \neq \mathbf{R}'$, $\langle \sigma(\mathbf{R})\sigma(\mathbf{R}') \rangle_0$ factorizes into the product $\langle \sigma(\mathbf{R}) \rangle_0 \langle \sigma(\mathbf{R}') \rangle_0$. We therefore obtain

$$\langle U \rangle_0 = \frac{1}{2} \sum_{\mathbf{R},\mathbf{R}'} V(\mathbf{R}, \mathbf{R}') m_0^2 = \frac{Nz}{2} m_0^2 \qquad (1954)$$

since

$$\sum_{\mathbf{R}'} V(\mathbf{R}, \mathbf{R}') = z, \qquad (1955)$$

where z is the number of nearest neighbors. We then have

$$\Omega^{(1)}(h) = N \frac{z}{2} m_0^2. \qquad (1956)$$

Turning to $\Omega^{(2)}$, we restrict ourselves, for simplicity, to the case $h = 0$. The case $h \neq 0$ is discussed in Problem 5.37. For $h = 0$, $\langle U \rangle_0 = 0$,

$$\Omega^{(2)} = \langle U^2(\sigma) \rangle_0 \qquad (1957)$$

and

$$\langle U^2(\sigma)\rangle_0 = \frac{1}{4}\sum_{R_1}\sum_{R_2}\sum_{R_3}\sum_{R_4} V(\mathbf{R}_1,\mathbf{R}_2)V(\mathbf{R}_3,\mathbf{R}_4)\langle\sigma(\mathbf{R}_1)\sigma(\mathbf{R}_2)\sigma(\mathbf{R}_3)\sigma(\mathbf{R}_4)\rangle_0. \tag{1958}$$

Since the V's require that $\mathbf{R}_1 \neq \mathbf{R}_2$ and $\mathbf{R}_3 \neq \mathbf{R}_4$, we easily obtain

$$\langle\sigma(\mathbf{R}_1)\sigma(\mathbf{R}_2)\sigma(\mathbf{R}_3)\sigma(\mathbf{R}_4)\rangle_0 = \delta_{\mathbf{R}_1,\mathbf{R}_3}\delta_{\mathbf{R}_2,\mathbf{R}_4} + \delta_{\mathbf{R}_1,\mathbf{R}_4}\delta_{\mathbf{R}_2\mathbf{R}_3} \tag{1959}$$

since, for example, $\sigma^2(\mathbf{R}_1) = 1$. Equation (1958) then reduces to

$$\langle U^2(\sigma)\rangle_0 = \frac{1}{4}\sum_{R_1}\sum_{R_2} 2V^2(\mathbf{R}_1,\mathbf{R}_2) \tag{1960}$$

having used the fact that $V(\mathbf{R}_1,\mathbf{R}_2) = V(\mathbf{R}_2,\mathbf{R}_1)$. Then, using Eq. (1938), Eq. (1960) becomes

$$\langle U^2(\sigma)\rangle_0 = \frac{1}{2}\sum_{R_1}\sum_{R_2}\sum_{\delta}\sum_{\delta'} \delta_{\mathbf{R}_2,\mathbf{R}_1+\delta}\delta_{\mathbf{R}_2,\mathbf{R}_1+\delta'}. \tag{1961}$$

We can do the sum over $\mathbf{R}_2$ easily to obtain

$$\sum_{R_2} \delta_{\mathbf{R}_2,\mathbf{R}_1+\delta}\delta_{\mathbf{R}_2,\mathbf{R}_1+\delta'} = \delta_{\delta,\delta'}. \tag{1962}$$

The result is independent of $\mathbf{R}_1$, so the sum over $\mathbf{R}_1$ gives the expected factor of N and

$$\Omega^{(2)}(0) = \langle U^2(\sigma)\rangle_0 = \frac{N}{2}\sum_{\delta}\sum_{\delta'} \delta_{\delta,\delta'} = \frac{Nz}{2}. \tag{1963}$$

We could then proceed to compute physical quantities such as the average energy and specific heat which follow by taking derivatives of Ω with respect to temperature. We concentrate here instead on the magnetic properties which follow from taking derivatives with respect to the magnetic field:

$$M = -\frac{\partial}{\partial B}\beta^{-1}\ln Z = -\frac{\partial\Omega}{\partial h} = -\sum_{n=0}^{\infty}\frac{K^n}{n!}\frac{\partial}{\partial h}\Omega^{(n)}(h) \tag{1964}$$

gives the magnetization, and the magnetic susceptibility is given by

$$\chi = -\frac{\partial M}{\partial B} = \beta\sum_{n=0}^{\infty}\frac{K^n}{n!}\frac{\partial^2\Omega^{(n)}(h)}{\partial^2 h^2}. \tag{1965}$$

The zero-field magnetization vanishes since all of the $\partial \Omega^{(n)}(h)/\partial h$ vanish as $h \to 0$ because of the odd number of spins in the zeroth-order averages. This is not true for χ and we can write

$$\chi = \beta \sum_{n=0}^{\infty} K^n \tilde{\chi}^{(n)} \qquad (1966)$$

as $h \to 0$, where

$$\tilde{\chi}^{(n)} = \frac{1}{n!} \left(\frac{\partial^2 \Omega^{(n)}(h)}{\partial h^2} \right)_{h=0}. \qquad (1967)$$

There are two paths for evaluating the $\tilde{\chi}^{(n)}$. One can formally take the derivatives with respect to h, let $h \to 0$, and evaluate the subsequent average over $P_0[\sigma]$. For $n = 0$ this leads to

$$\tilde{\chi}^{(0)} = \left\langle \left[\sum_{\mathbf{R}} \sigma(\mathbf{R}) \right]^2 \right\rangle_0$$
$$= \sum_{\mathbf{R}} \sum_{\mathbf{R}'} \langle \sigma(\mathbf{R}) \sigma(\mathbf{R}') \rangle_0 = N, \qquad (1968)$$

since in zero external field,

$$\langle \sigma(\mathbf{R}) \sigma(\mathbf{R}') \rangle_0 = \delta_{\mathbf{R},\mathbf{R}'}. \qquad (1969)$$

The second path is to evaluate the averages leading to $\Omega^{(n)}(h)$ for arbitrary h, take the h derivatives, and then set $h \to 0$. For $n = 1$ we have $\Omega^{(1)}(h)$ from Eq. (1956), and

$$\tilde{\chi}^{(1)} = \lim_{h \to 0} \frac{\partial^2}{\partial h^2} N \frac{z}{2} m_0^2 = Nz. \qquad (1970)$$

The $\tilde{\chi}^{(n)}$ have been evaluated to rather large order n for a variety of lattices. It turns out to be convenient to write the series in terms of the variable

$$u = \tanh K \qquad (1971)$$

rather than K. The variable u enters naturally into the development since we can write

$$e^{K\sigma(\mathbf{R})\sigma(\mathbf{R}')} = \cosh K + \sigma(\mathbf{R})\sigma(\mathbf{R}') \sinh K$$
$$= \cosh K [1 + u\sigma(\mathbf{R})\sigma(\mathbf{R}')] \qquad (1972)$$

and as discussed in Problem 5.36, the partition function can be written for $h = 0$ in the form

$$Z(K) = (\cosh K)^{Nz/2} f(u) \tag{1973}$$

where f is a function of u alone. In any event, since K can be expanded in powers of u, we can write

$$\chi = \sum_{n=0}^{\infty} \chi^{(n)} u^n \tag{1974}$$

and the $\chi^{(n)}$ are related to the $\tilde{\chi}^{(n)}$. The $\chi^{(n)}$ are listed for a simple cubic lattice in Table 5.7. The main interest in developing series of this length has been to look at the critical phenomena in these systems. We assume that near the Curie temperature,

$$\chi = \tilde{\chi}_0 \epsilon^{-\gamma} \tag{1975}$$

where

$$\epsilon = \frac{T - T_c}{T_c}. \tag{1976}$$

If we restrict ourselves to the region near T_c, one can write

$$u(T) = u(T_c) + (T - T_c) \frac{\partial u}{\partial T}\bigg|_{T=T_c} + \mathcal{O}(T - T_c)^2 \tag{1977}$$

or

$$\frac{T - T_c}{T_c} = \frac{\cosh^2 K_c}{K_c u_c} \left[1 - \frac{u(T)}{u_c} \right]. \tag{1978}$$

TABLE 5.7 High-Temperature Expansion Coefficients for the Magnetic Susceptability for a Simple Cubic Lattice

n	$\tilde{\chi}_n$	n	$\tilde{\chi}_n$
1	6	10	8,306,862
2	30	11	38,975,286
3	150	12	182,265,822
4	726	13	852,063,558
5	3,510	14	3,973,784,886
6	16,710	15	18,527,532,310
7	79,494	16	86,228,667,894
8	375,174	17	401,225,391,222
9	1,769,686		

Source: Data from Ref. 50.

Thus we have

$$\chi = \chi_0 \left(1 - \frac{u}{u_c}\right)^{-\gamma} \tag{1979}$$

for u near u_c, where

$$\chi_0 = \left(\frac{\cosh^2 K_c}{K_c u_c}\right)^{-\gamma} \tilde{\chi}_0. \tag{1980}$$

Expanding Eq. (1979) in powers of u/u_c, we obtain

$$\chi = \chi_0 \left[1 + \gamma \frac{u}{u_c} + \gamma \frac{\gamma+1}{2}\left(\frac{u}{u_c}\right)^2 + \cdots \right.$$
$$\left. + \frac{1}{n!}\gamma(\gamma+1)\cdots(\gamma+n-1)\left(\frac{u}{u_c}\right)^n + \cdots\right]$$
$$= \sum_{n=0}^{\infty} \chi^{(n)} u^n. \tag{1981}$$

Taking the ratios of successive coefficients, we obtain

$$q_n = \frac{\chi^{(n)}}{\chi^{(n-1)}} = \frac{\gamma(\gamma+1)\cdots(\gamma+n-1)(n-1)!u_c^{n-1}}{n!u_c^n \gamma(\gamma+1)\cdots(\gamma+n-2)}$$
$$= \frac{\gamma+n-1}{nu_c}. \tag{1982}$$

If we plot q_n versus $1/n$, we expect, if χ has a dominant powerlaw divergence, to obtain a straight line with intercept at $1/n \to 0$ given by $1/u_c$ and with slope $(\gamma-1)/u_c$. Thus we can estimate both $1/u_c$ and γ. In Fig. 5.17 we plot q_n versus $1/n$ for the case of a simple cubic lattice. We see some *noise* associated with the points corresponding to small n. Linear least squares fits to the data give $u_c = 1/4.58$ and the critical exponent in the range $1.23 \leq \gamma \leq 1.26$. Considerably more elaborate techniques for analyzing such series have been developed [51], and essentially every physical property and lattice type have been studied.

5.12 MONTE CARLO SIMULATIONS

5.12.1 Basic Technique

In Chapter 1 we discussed the molecular dynamics method for direct numerical simulation of an N-body system. Using this approach, we can prepare a time sequence of phase-space configurations, q, generated by forward-stepping Newton's laws. In the long-time limit this sequence will generate averages compatible with

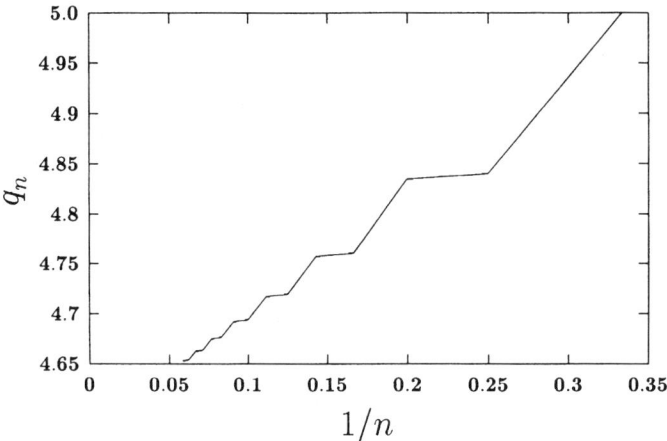

FIGURE 5.17 Ratio of high-temperature expansion coefficients for the magnetic susceptibility versus $1/n$, where n is the order of the expansion.

those of the microcanonical ensemble. Using Nosé dynamics we can generate configurations compatible with the canonical ensemble. We turn now to a complementary numerical simulational technique known as the *Monte Carlo method*. In this rather general method we again generate a sequence of configurations but with two major differences from the molecular dynamics method. The evolution is not given by a deterministic equation of motion where the configuration at the next time step is a known function of the present configuration. There is now a statistical or probabilistic aspect to the dynamics—there will be a probability that the system will move to a given new configuration. The second difference, which gives rise to the first, is that it is assumed that the system of interest is in contact with an external heat bath. This heat bath is assumed to be in thermal equilibrium at some temperature T_B and the dynamics of the system of interest are arranged to drive that system toward being in equilibrium with the bath at the temperature T_B.

This method can be developed rather generally for a variety of systems. Let us focus on the case of Ising models, where the ideas can easily be made explicit. Let us first discuss how one carries out a Monte Carlo simulation of an Ising model and then come back and discuss some of the formal underpinnings of the method.

Ising models differ from fluid models in their discrete nature. The Ising spins are located at discrete lattice sites and take on discrete values (± 1). The configuration space for N Ising spins is therefore 2^N-dimensional. Let us assume that we have an Ising system in a particular configuration $\{\sigma\}_0 = \{\sigma(\mathbf{R}_1), \sigma(\mathbf{R}_2), \ldots \sigma(\mathbf{R}_N)\}$. In a particular form of the Monte Carlo method we forward step this system in time via the following steps:

1. Randomly choose a site $\mathbf{R}_i$.
2. Compute the change in energy of the system $\Delta E(\mathbf{R}_i)$ if the spin $\sigma(\mathbf{R}_i)$ at site $\mathbf{R}_i$ is flipped $[\sigma(\mathbf{R}_i) \to -\sigma(\mathbf{R}_i)]$.

3. Compute the probability W [which depends on $\Delta E(\mathbf{R}_i)$], that $\sigma(\mathbf{R}_i)$ should be flipped; we discuss appropriate choices for W below.
4. Generate a random number r between 0 and 1.
5. If $W \geq r$, flip $\sigma(\mathbf{R})$. If $W < r$, do not flip $\sigma(\mathbf{R})$.

This gives us a new spin configuration $\{\sigma\}_1 = \{\sigma(\mathbf{R}_1), \sigma(\mathbf{R}_2), \ldots, \sigma'(\mathbf{R}_i), \ldots, \sigma(\mathbf{R}_N)\}$, where $\sigma'(\mathbf{R}_i)$ is the value of the spin at site $\mathbf{R}_i$ after this procedure. We then repeat this procedure many times and obtain a sequence of configurations $\{\sigma\}_0$, $\{\sigma\}_1, \{\sigma\}_2, \ldots, \{\sigma\}_n$. Given a sequence of configurations, we can monitor any observable. For example, the magnetization per spin for a given configuration n is given by

$$m_n = \frac{1}{N} \sum_{\mathbf{R}} \sigma_n(\mathbf{R}) \tag{1983}$$

where $\sigma_n(\mathbf{R})$ is the value of the spin at site $\mathbf{R}$ for configuration n. Similarly, the magnetic susceptibility per spin is given in the paramagnetic phase by

$$\chi_n = \frac{1}{k_B T} \tilde{\chi}_n \tag{1984}$$

$$\tilde{\chi}_n = \frac{1}{N} \left[\sum_{\mathbf{R}} \sigma_n(\mathbf{R}) \right]^2. \tag{1985}$$

Averages are given by summing over configurations:

$$m = \frac{1}{n+1} \sum_{j=0}^{n} m_j \tag{1986}$$

$$\tilde{\chi} = \frac{1}{n+1} \sum_{j=0}^{n} \tilde{\chi}_j. \tag{1987}$$

It can be seen that this procedure is well suited for programming on a computer. The use of integer arithmetic for the Ising spins can be evaluated very quickly for rather large systems.

5.12.2 Detailed Balance

The main unknown in this procedure is the flipping probability W. We must develop a procedure for identifying those probabilities that will drive a system into equilibrium at some temperature T_B and keep it there. There are many possible ways of constructing such a Monte Carlo procedure. Our approach is to assume that the probability that the system is in configuration $\{\sigma\}$, $P_{n+1}[\sigma]$, after $(n+1)$ Monte

Carlo steps, is proportional to the probability distribution at the preceding step:

$$P_{n+1}[\sigma] = \sum_{\{\sigma'\}} W_n[\sigma|\sigma']P_n[\sigma'] \tag{1988}$$

where

$$\sum_{\{\sigma\}} = \prod_{\mathbf{R}} \left(\sum_{\sigma'(\mathbf{R})=\pm 1} \right) \tag{1989}$$

is the sum over all possible configurations. $W_n[\sigma|\sigma']$ is just the probability that the system in configuration $\{\sigma'\}$ at step n is taken to $\{\sigma\}$ in the next step. Since $P_n[\sigma]$ is a probability distribution, we have the normalization

$$\sum_{\{\sigma\}} P_{n+1}[\sigma] = 1 = \sum_{\{\sigma\}} \sum_{\{\sigma'\}} W_n[\sigma|\sigma']P_n[\sigma'] \tag{1990}$$

which is satisfied by

$$\sum_{\{\sigma\}} W_n[\sigma|\sigma'] = 1. \tag{1991}$$

The form of $W_n[\sigma|\sigma']$ is fixed by choosing a spin-flipping mechanism. Clearly, there are many other types of *dynamics* that we could introduce. We will be satisfied with developing the single-spin-flip case here. $W_n[\sigma|\sigma']$ consists of two obvious terms: one for when the spin chosen at random flips and another term for those times when it does not flip. We have immediately the general form

$$W_n[\sigma|\sigma'] = \frac{1}{N}\sum_{\mathbf{R}'} \{\hat{f}_\sigma(\mathbf{R}')\delta_{\sigma,\sigma'}W[\Delta E(\sigma',\mathbf{R}')] + \delta_{\sigma,\sigma'}(1 - W[\Delta E(\sigma',\mathbf{R}')])\} \tag{1992}$$

where

$$\delta_{\sigma,\sigma'} = \prod_{\mathbf{R}} \delta_{\sigma(\mathbf{R}),\sigma'(\mathbf{R})} \tag{1993}$$

sets the values of the old configuration $\{\sigma'\}$ equal to the values of the new spin configuration $\{\sigma\}$ at every lattice point. The operator $\hat{f}_\sigma(\mathbf{R}')$ flips the spin at site $\mathbf{R}'$ in the configuration σ:

$$\hat{f}_\sigma(\mathbf{R}')A[\sigma(\mathbf{R}_1),\sigma(\mathbf{R}_2),\ldots,\sigma(\mathbf{R}'),\ldots,\sigma(\mathbf{R}_N)]$$
$$= A[\sigma(\mathbf{R}_1),\sigma(\mathbf{R}_2),\ldots,-\sigma(\mathbf{R}'),\ldots,\sigma(\mathbf{R}_N)]. \tag{1994}$$

Since the flipped spin at site $\mathbf{R}'$ is chosen at random, we sum over all sites and divide by the total number of sites. It is left to Problem 5.39 to show that this choice for $W_n[\sigma, \sigma']$ satisfies Eq. (1991). Putting Eq. (1992) back into the *master equation*, Eq. (1988), and doing the sum over $\{\sigma'\}$, we obtain

$$P_{n+1}[\sigma] - P_n[\sigma] = \frac{1}{N}\sum_{\mathbf{R}'}\{\hat{f}_\sigma(\mathbf{R}')[W(\Delta E(\sigma, \mathbf{R}'))P[\sigma]] - W(\Delta E(\sigma, \mathbf{R}')P[\sigma]\}. \tag{1995}$$

We can use this equation to choose W by requiring that the flipping probability $W(\Delta E(\sigma, \mathbf{R}'))$ be such that the system is driven toward the *stationary* (independent of n) *equilibrium state* at temperature T_B:

$$P_{n+1}[\sigma] = P_n[\sigma] = \frac{e^{-\beta H[\sigma]}}{Z} = P_E[\sigma] \tag{1996}$$

where Z is the normalizing partition function and $\beta = 1/k_B T_B$. The master equation reduces in this case to

$$0 = \frac{1}{N}\sum_{\mathbf{R}'}\{\hat{f}_\sigma(\mathbf{R}')[W(\Delta E(\sigma, \mathbf{R}'))P_E[\sigma]] - W(\Delta E(\sigma, \mathbf{R}'))P_E[\sigma]\}. \tag{1997}$$

It is clear that this condition must be satisfied at each site so we have

$$\hat{f}_\sigma(\mathbf{R})\{W(\Delta E(\sigma, \mathbf{R}))P_E[\sigma]\} = W(\Delta E(\sigma, \mathbf{R}))P_E[\sigma], \tag{1998}$$

which is known as the *condition of detailed balance*. Similar conditions hold for processes other than spin-flip dynamics. In this case detailed balance requires that the product $W[\Delta E(\sigma, \mathbf{R})]P_E[\sigma]$ be independent of the spin at site $\mathbf{R}$. In the case of the nearest-neighbor Ising model, the Hamiltonian is given by

$$H[\sigma] = -\frac{J}{2}\sum_{\mathbf{R}'}\sum_{\delta'}\sigma(\mathbf{R}')\sigma(\mathbf{R}' + \delta) \tag{1999}$$

where δ connects nearest neighbors. $H[\sigma]$ depends on the particular spin at site $\mathbf{R}$ such that

$$H[\sigma] = -J\sigma(\mathbf{R})\sum_\delta \sigma(\mathbf{R} + \delta) + H[\mathbf{R}, \sigma] \tag{2000}$$

where $H[\mathbf{R}, \sigma]$ does not depend on $\sigma(\mathbf{R})$. Note that the change in energy in flipping the spin at site $\mathbf{R}$ is given by

$$\Delta E(\sigma, \mathbf{R}) = 2J\sigma(\mathbf{R})\sum_\delta \sigma(\mathbf{R} + \delta) \tag{2001}$$

so

$$H[\sigma] = -\frac{\Delta E(\sigma, \mathbf{R})}{2} + H[\mathbf{R}, \sigma]. \tag{2002}$$

We can therefore, using Eqs. (1996) and (2002), rewrite the detailed balance condition, Eq. (1998), in the form

$$W[-\Delta E(\sigma, \mathbf{R})]e^{-\beta \Delta E(\sigma, R)/2} = W[\Delta E(\sigma, \mathbf{R})]e^{\beta \Delta E(\sigma, R)/2}. \tag{2003}$$

The local statement of detailed balance for spin-flip dynamics can be written in the form

$$\frac{W[\Delta E(\sigma, \mathbf{R})]}{W[-\Delta E(\sigma, \mathbf{R})]} = e^{-\beta \Delta E(\sigma, R)}. \tag{2004}$$

There are various choices for W which are compatible with this condition. We discuss two forms for W that are used widely. The first choice is physically appealing and was introduced in 1953 by Metropolis, Rosenbluth, Rosenbluth, Teller, and Teller [52], who published the first calculation of this type. They chose

$$W(\Delta E) = \begin{cases} 1 & \text{if } \Delta E \leq 0 \\ e^{-\beta \Delta E} & \text{if } \Delta E \geq 0 \end{cases} \tag{2005}$$

which clearly satisfies Eq. (2004). The physical motivation is clear since the system always lowers its energy if it can. Another choice is to assume the *Glauber form*:

$$W(\Delta E) = \frac{1}{2}\left[1 - \tanh\left(\frac{\beta \Delta E}{2}\right)\right] \tag{2006}$$

which also satisfies Eq. (2004). This choice is very convenient for the case of the one-dimensional Ising model [53] since the resulting master equation can be solved for $P_n[\sigma]$ analytically.

5.12.3 Computation of Averages

The power of the Monte Carlo method is that it is not restricted to a particular range of system parameters—high temperatures or low density. The disadvantages of the system are that we are restricted to finite sizes, and we must take great care with the averaging process. In this section we use the method to compute for a situation where we know the solution to the problem. We treat the one-dimensional Ising model and compute the equilibrium susceptibility and magnetization density. We carry out a Monte Carlo simulation for a one-dimensional Ising model using Glauber

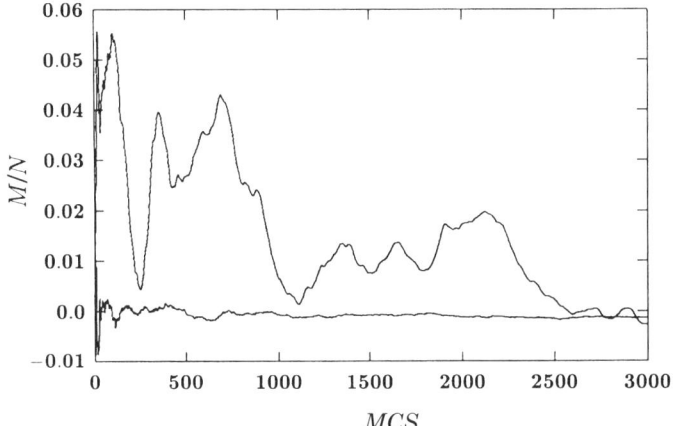

FIGURE 5.18 Total magnetization per spin versus Monte Carlo steps per spin (MCS) for a one-dimensional Ising model spin-flip dynamics. The system consists of 1000 spins. The lower, less fluctuating curve corresponds to a very high temperature ($u = 0$); the other fluctuating curve corresponds to a run at a rather low temperature ($u = 0.8$).

spin-flip *dynamics*, given by Eq. (2006). We assume a system of N spins satisfying periodic boundary conditions

$$\sigma(N + 1) = \sigma(1). \tag{2007}$$

Initially, the value of each spin is chosen at random. In the first set of simulations (see Fig. 5.18) a system of 1000 spins is treated for the set of parameters $u = \tanh(\beta J) = 0.0$ and $u = 0.8$ with $h = 0.0$. Since we have zero external magnetic field the average magnetization density should be zero. We see in Fig. 5.18 that as the sequence of Monte Carlo steps increases, the average magnetization as defined by Eq. (1986) approaches zero. Notice that the appropriate unit of *time* associated with the sequence of Monte Carlo steps is the *Monte Carlo steps per spin* (MCS). This unit can be used to compare the state of evolution toward equilibrium of systems of different-sized systems. Thus, on average, in one Monte Carlo step per spin, we try to flip each spin in the system once. In Fig. 5.19 we show the results for the magnetic susceptibility, defined by Eq. (1987), as a function of MCS for $u = 0, 0.2, 0.4, 0.6$, and 0.8. Notice the substantial fluctuations $\tilde{\chi}$. In this case we have from Eq. (1917) that

$$\tilde{\chi} = \frac{1 + u}{1 - u} \tag{2008}$$

and for example, $\tilde{\chi} = 4$ for $u = 0.6$. The simulation is approaching the equilibrium value as the system evolves. Notice that the result given by Eq. (2008) is true only for an infinite system. Indeed, we can calculate χ as a function of N analytically using

438 EQUILIBRIUM PROPERTIES OF DIELECTRIC AND MAGNETIC MATERIALS

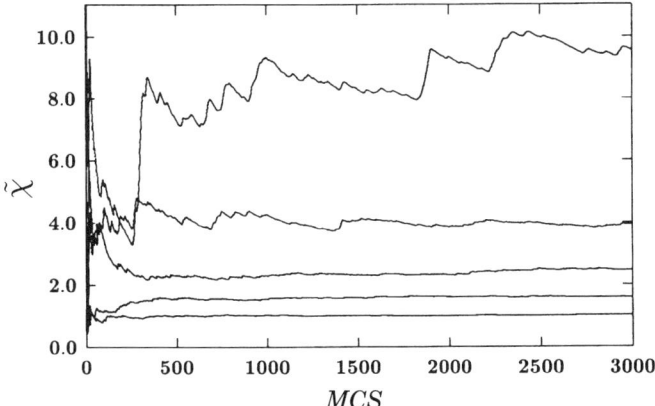

FIGURE 5.19 Averaged scaled magnetic susceptibility versus Monte Carlo steps per spin for a one-dimensional Ising model for various final temperatures. The initial state is assumed to be at infinite temperature ($u_I = 0.0$), while the final temperatures, moving from the bottom to the top of the plot, correspond to $u_F = 0.0, 0.2, 0.4, 0.6,$ and 0.8. These results for long times should correspond to the exact equilibrium results $\tilde{\chi} = (1 + u_F)/(1 - u_F)$.

Eq. (1904). Numerically, we should determine χ for various values of N and then use some extrapolation formula to take the $N \to \infty$ limit. These finite-sized limitations become more severe as we investigate systems with correlations over large distances. An important example is the case of a system near its critical point, where the correlation length is limited to the size of a finite system even at the transition temperature.

Clearly, the method discussed here is very general and can be applied to higher dimensions, interactions with longer range, and so on. For a more complete discussion, see Ref. 54.

5.13 TRANSITION FROM CLASSICAL TO MODERN THEORY OF CRITICAL PHENOMENA

In earlier chapters we discussed critical phenomena for a variety of systems: liquid–gas, ferromagnets, antiferromagnets, and systems described by the Landau theory discussed in Chapter 2. In our specific treatments of the van der Waals equation for the liquid–gas case, the Curie–Weiss theory for ferro-and antiferromagnets, and Landau theory, we found that each can be expressed in terms of an order parameter whose average, M, shows bifurcation behavior below the critical point and an associated susceptibility which blows up at the critical point. In each case we can introduce the set of critical indices [55] given in Table 5.8. We list in Table 5.9 the values of the critical indices given in each case by mean field theory. More generally, we can introduce an order parameter for any phase transition and introduce the equivalent set of critical indices. Over the first part of the twentieth century there was

TABLE 5.8 Definitions of Critical Indices

Quantity	Definition of Index
$M(T, 0)$	ϵ^β
$M(T_c, B)$	$B^{1/\delta}$
$\chi(T, 0)$	$\epsilon^{-\gamma}$

TABLE 5.9 Values of Critical Indices in Mean-Field Theory and for the Two-Dimensional Ising Model

Exponent	Mean Field Theory	Ising Two-Dimensional Model
β	$\frac{1}{2}$	$\frac{1}{8}$
δ	3	15
γ	1	$\frac{7}{4}$

a feeling that the mean-field theories were essentially exact. However, this conventional wisdom was overthrown with Onsager's solution of the two-dimensional Ising model, where he found the exponent $\beta = \frac{1}{8}$, not the $\frac{1}{2}$ given by mean field theory. The values of the critical indices for the two-dimensional Ising model are also listed in Table 5.9.

The central problem in the theory of critical phenomena has been to calculate these critical indices for a given system and, in turn, determine their systematic variation from system to system. During the 1950s and 1960s, experimental data and series expansion results using the methods of Sections 5.11 and 5.12 began to accumulate for these critical indices. It is easiest to summarize the results of this work in terms of a generalization of the Heisenberg model given by

$$H = -\frac{J}{2} \sum_{\langle \mathbf{R},\mathbf{R}' \rangle} \sum_{\alpha=1}^{n} \mu_\alpha(\mathbf{R}) \mu_\alpha(\mathbf{R}') \tag{2009}$$

where $\mathbf{R}$ and $\mathbf{R}'$ label nearest-neighbor lattice sites on a d-dimensional lattice, J the interaction between nearest-neighbor spins, and $\mu_\alpha(\mathbf{R})$ the αth component of an n-component spin on lattice site $\mathbf{R}$. This n-vector model can be taken to be quantum mechanical or classical. Clearly, for $n = 1$ this is the Ising model, for $n = 2$ this is the planar or x–y model, and for $n = 3$ we return to the Heisenberg model.

In Tables 5.10 to 5.12 results of extensive high-temperature expansion studies are presented for the critical indices as a function of spatial dimensionality d, the vector nature of the order parameter n, lattice structure, and quantum nature of the spins. The indices do depend on d and n, but do not appear to depend on the particular lattice type, the magnitude of the spin, or quantum effects. All of our best information now indicates that to a first approximation, for systems with short-range interactions, the indices depend basically only on two properties of the system: the spatial dimensionality and the vector nature of the order parameter.

TABLE 5.10 Variation of Ising Model Critical Indices with Spatial Dimension

Exponent	Two-Dimensional	Three-Dimensional	Four-Dimensional
β	$\frac{1}{8}$	0.325	1.00
δ	15	5	3
γ	$\frac{7}{4}$	1.30	1.0

TABLE 5.11 Variation of n-Vector Model Critical Indices with n in Three Dimensions

n	γ	β	δ
1	1.25	0.303	5.0
2	1.32	0.33	5.0
3	1.38	0.345	5.0
∞	2.0	0.5	5.0

Source: Data from Ref. 56.

TABLE 5.12 Values for γ for Various Lattices and Spin Values

Lattice/Spin Values	$\frac{1}{2}$	1	$\frac{3}{2}$	$\frac{5}{2}$	∞
Face-centered cubic	1.43 ± 0.04	1.39 ± 0.02	1.38 ± 0.01	1.37 ± 0.01	1.38 ± 0.02
Body-centered cubic	1.39 ± 0.04	1.39 ± 0.02	1.38 ± 0.01	1.36 ± 0.01	1.37 ± 0.01
Simple cubic	1.42 ± 0.02	1.39 ± 0.07	1.38 ± 0.04	1.37 ± 0.04	1.38 ± 0.02

Source: Data from Ref. 57.

These results are summarized by the hypothesis of *universality* [58]: All phase-transition problems can be divided into a small number of classes, depending on the dimensionality of the system and the symmetries of the ordered state. Within each class, all phase transitions have identical behavior in the critical region; only the names of the variables are changed.

This was the state of the art near the end of the 1960s. The main question was *why* one had universality! This leads to the modern theory of critical phenomena [59] and renormalization group theory. This theory, which led to the awarding of a Nobel prize to Kenneth Wilson in 1982, builds on the ideas of scaling [60] and coarse graining and shows how universality follows for a wide range of microscopically very different systems and offered a method for computing the critical indices as a function of d and n. The ideas behind the renormalization group are discussed fully in Volume 2.

REFERENCES

[1] K. Gottfried, *Quantum Mechanics*, Vol. I, W. A. Benjamin, New York, 1966, Chap. VIII.
[2] One can formulate these inequalities in terms of the dielectric constant and magnetic permeability as discussed by L. D. Landau and E. M. Lifshitz, *Statistical Physics*, 3rd

ed., Pergamon Press, Oxford, 1980, p. 95. This point is developed in some detail in Volume 2.

[3] P. Langevin, *Ann. Chim. Phys.* **5**, 70 (1905).

[4] C. Cuthbertson and M. Cuthbertson, *Proc. R. Soc. A* **84**, 13 (1911).

[5] L. Pauling, *Proc. R. Soc. London A* **114**, 181 (1927).

[6] R. Kubo and T. Nagamiya, eds., *Solid State Physics*, McGraw-Hill, New York, 1969, p. 439.

[7] J. H. van Vleck, *The Theory of Electric and Magnetic Susceptibilities*, Oxford University Press, London, 1932, p. 27.

[8] R. J. Elliott, "Magnetic phase transitions," in *Magnetic Phase Transitions*, ed. M. Ausloos and R. J. Elliott, Springer-Verlag, Berlin, 1983.

[9] One lesson to learn about the coarse-graining procedure is that it requires dealing with the fast degrees of freedom in explicit detail if one is to express the new effective parameters explicitly in terms of parameters characterizing the more microscopic description.

[10] A. Lande, *Z. Phys.* **15**, 189 (1923).

[11] For a simple discussion of Hund's rules, see H. Eyring, D. Henderson, B. J. Stover, and E. M. Eyring, *Statistical Mechanics and Dynamics*, 2nd ed., Wiley, New York, 1982, p. 330.

[12] H. Eyring, D. Henderson, B. J. Stover, and E. M. Eyring, *Statistical Mechanics and Dynamics*, 2nd ed., Wiley, New York, 1982, p. 334.

[13] J. H. van Vleck, *The Theory of Electric and Magnetic Susceptibilities*, Oxford University Press, London, 1932, p. 244.

[14] A. Frank, *Phys. Rev.* **39**, 119 (1932).

[15] H. Eyring, D. Henderson, B. J. Stover, and E. M. Eyring, *Statistical Mechanics and Dynamics*, 2nd ed., Wiley, New York, 1982, p. 336.

[16] A. Radzig and B. Smirnov, *Reference Data on Atoms, Molecules and Ions*, Springer-Verlag, New York, 1985.

[17] As long as the magnetic moments for different atoms or molecules are uncorrelated, it does not matter, with respect to magnetic properties whether or not H in Eq. (1602), has the usual nonmagnetic pairwise interactions.

[18] P. Langevin, *J. Phys.* **4**, 678 (1905).

[19] P. Curie, *Ann. Chim. Phys.* **5**, 289 (1895).

[20] L. Brillouin, *J. Phys.* **8**, 74 (1927).

[21] There is an instructive discussion by J. H. van Vleck in *The Theory of Electric and Magnetic Susceptibilities*, Oxford University Press, London, 1932, of Eq. (1648), where he refers to as the Langevin–Debye formula. Similarly, there is background discussion by P. Debye in *Polar Molecules*, Dover, New York, 1929, p. 30.

[22] J. H. van Leeuwen, *J. Phys.* **2**, 361 (1921).

[23] L. D. Landau, "Diamagnetismus der Metalle," *Z. Phys.* **65**, 629 (1930); English translation in *Collected Papers of L. D. Landau*, Pergamon Press, Elmsford, N. Y. 1965.

[24] The dependence of the average *orbit* distance for a harmonic oscillator on the energy quantum number n is worked out, for example, by K. Gottfried in *Quantum Mechanics*, Vol. I, W. A. Benjamin, New York, 1966, p. 259.

[25] W. Pauli, "Über Gasentartung and Paramagnetismus," *Z. Phys*, **41**, 81 (1927).

[26] W. J. De Haas and P. M. Van Alphen, *Leiden Commun.* **212** (1931); J. M. Luttinger, *Phys. Rev.* **121**, 1251 (1961).

[27] D. Shoenberg, *The Fermi Surface*, Wiley, New York, 1960, pp. 74–83.

[28] S. M. Girvin, *The Quantum Hall Effect: Novel Excitations and Broken Symmetries*, cond-mat/9907002, to be published by Springer-Verlag and Les Editions de Physique.

[29] For further discussion of Lorentz's treatment of electric dipoles, see H. Eyring, D. Henderson, B. J. Stover, and E. M. Eyring, *Statistical Mechanics and Dynamics*, 2nd ed., Wiley, New York, 1982. The original references are given by H. Lorentz, *Ann. Phys.* **9**, 641 (1880); L. Lorenz, *Ann. Phys.* **11**, 70 (1880); and H. Lorentz, *The Theory of Electrons*, Dover, New York, 1952.

[30] R. Clausius, *Die Mechanische Wärmelehre*, Vol. II, Vieweg-Verlag, Brunswick, Germany, 1879 p. 94; O. Mossotti, *Mem. Math. Fis. Modena* **24**(II), 49 (1850).

[31] H. Fröhlich, *Theory of Dielectrics*, 2nd ed., Oxford University Press, New York, 1958.

[32] W. Heisenberg, *Z. Phys.* **49**, 619 (1928). See also P. A. M. Dirac, *Proc. R. Soc. A* **112**, 661 (1926).

[33] P. Zeeman, *Nature* **35**, 347 (1897).

[34] D. C. Mattis, *The Theory of Magnetism*, Harper & Row, New York, 1965, p. 32.

[35] D. C. Mattis, *The Theory of Magnetism*, Harper & Row, New York, 1965, Chap. 2.

[36] C. Herring, "Direct exchange between well separated atoms," in *Magnetism*, Vol. 2B, ed. G. T. Rado and H. Suhl, Academic Press, New York, 1966.

[37] P. Weiss [*J. Phys.* **6**, 661 (1907)] introduced the very important idea of a mean or molecular field.

[38] C. Kittel, *Introduction to Solid State Physics*, 7th ed, Wiley, New York, 1996, p. 445.

[39] L. Néel, *Ann. Phys. (Paris)* **17**, 64 (1932); *J. Phys. Radium* **3**, 160 (1932).

[40] C. Kittel, *Introduction to Solid State Physics*, 7th ed., Wiley, New York, 1996, p. 465.

[41] E. Ising, *Z. Phys.* **31**, 253 (1925).

[42] L. D. Landau and E. M. Lifshitz, *Statistical Physics*, 3rd ed, Pergamon Press, Oxford, 1980, p. 537.

[43] R. Peierls, *Proc. Cambridge Philos. Soc.* **32**, 471 (1936), first showed that the two-dimensional Ising model has a ferromagnetic phase transition.

[44] In 1941, H. A. Kramers and G. H. Wannier [*Phys. Rev.* **60**, 253, 263 (1941)] determined the transition temperature rigorously.

[45] In 1944, Onsager solved the two-dimensional Ising model on a square lattice: L. Onsager, *Phys. Rev.* **65**, 117 (1944). See also B. Kaufmann, *Phys. Rev.* **76**, 1232 (1949).

[46] B. M. McCoy and T. T. Wu, *The Two-Dimensional Ising Model*, Harvard University Press, Cambridge, Mass., 1973.

[47] C. N. Yang, *Phys. Rev.* **85**, 809 (1952).

[48] L. Onsager at a conference at Cornell in 1949 [*Nuovo Cimento Suppl.* **6**, 261 (1949)].

[49] See Ref. 46.

[50] C. Domb and M. S. Green, *Phase Transitions and Critical Phenomena*, Vol. 3, Academic Press, London, 1974, p. 381.

[51] An introduction is given by H. E. Stanley, *Introduction to Phase Transitions and Critical Phenomena*, Oxford University Press, New York, 1971. Considerably more elaborate techniques for analyzing such series have been developed, and essentially every physical property and lattice type have been studied and are discussed in Ref. 50.

[52] N. Metropolis, A. W. Rosenbluth, M. Rosenbluth, A. H. Teller, and E. Teller, *J. Chem. Phys.* **27**, 1087 (1953).

[53] This choice is very convenient for the case of the one-dimensional Ising model since the resulting master equation can be solved for $P_n[\sigma]$ analytically: R. J. Glauber, *J. Math. Phys. (N.Y.)* **4**, 294 (1963).

[54] H. Gould and J. Tobochnik, *An Introduction to Computer Simulation Methods*, Part 2, Addison-Wesley, Reading, Mass., 1988, p. 483; K. Binder, ed., *Monte Carlo Methods in Statistical Physics*, Springer-Verlag, Berlin, 1979: K. Binder, ed., *Application of the Monte Carlo Method in Statistical Physics*, Springer-Verlag, Berlin, 1998.

[55] There are at least two more critical indices, η and ν, which control the spatial correlations near the critical point and which are essential in the modern theory discussed in Volume 2.

[56] H. E. Stanley, *Introduction to Phase Transitions and Critical Phenomena*, Oxford University Press, New York, 1971.

[57] D. S. Ritchie and M. E. Fisher, *Phys. Rev. B* **5**, 2668 (1972).

[58] L. P. Kadanoff et al., *Rev. Mod. Phys.* **39**, 615 (1967).

[59] S. Ma, *Modern Theory of Critical Phenomena*, W. A. Benjamin, New York, 1976.

[60] L. P. Kadanoff, *Physics (N.Y.)* **2**, 263 (1966); B. Widom, *J. Chem. Phys.* **43**, 3989 (1965).

PROBLEMS

5.1. Consider the quantum mechanical average defined in Eq. (1510) with $A = B$ in the case of low temperatures. Suppose that the ground-state energy E_G, which is separated from the next energy state by a gap Δ, is degenerate with a set of g ground states $|g\rangle$. There are two cases of particular interest.

(a) Suppose that the matrix elements of the operator A, between the degenerate ground states, are independent of the particular state,

$$\langle g|A|g'\rangle = A_0$$

for all $|g\rangle$ and $|g'\rangle$. In this case, which includes as a special case a nondegenerate ground state, shows that as $T \to 0$, $\langle (\delta A)^2 \rangle_D \to 0$ exponentially.

(b) Suppose that $\langle A \rangle = 0$; then show that

$$\lim_{T \to 0} \langle (\delta A)^2 \rangle_D = \frac{1}{g}\sum_g \langle g|A^2|g\rangle.$$

5.2. Show for an isotropic system that

$$\langle (\mathbf{B} \times \mathbf{r}_i)^2 \rangle = \tfrac{2}{3} B^2 \langle r_i^2 \rangle.$$

5.3. Show that if there is a gap Δ between the ground-state energy and the excited states, the van Vleck type of contribution, given by Eq. (1515), has a nonzero low-temperature limit.

5.4. Write out the microscopic expression for the mixed susceptibility

$$\chi_{ME}^{\alpha\beta} = \frac{\partial M_\alpha}{\partial E_\beta}.$$

Show that this quantity vanishes in zero fields if the unperturbed states have a definite parity.

5.5. Show that if the total wavefunction is written as a product of molecular wavefunctions, Eqs. (1544) and (1545) reduce to Eqs. (1546) and (1547) respectively.

5.6. Compute the molecular polarizability, α_E, for atomic hydrogen by evaluating the sum over matrix elements in Eq. (1556).

5.7. How much of the measured variation in α_E for noble gas systems can be explained by the variation with the number of electrons Z?.

5.8. If the ground state $|0\rangle$ is a magnetic spinsinglet, show that

$$\langle 0|\boldsymbol{\mu}|m\rangle \cdot \langle m|\boldsymbol{\mu}|0\rangle = 0$$

where m is any excited state.

5.9. Show that $\langle \mathbf{r}_i^2 \rangle = 3a_0^2$ for atomic hydrogen.

5.10. The expression given by Eq. (1597) for the magnetic susceptibility,

$$\chi_{\alpha\nu}^M = \beta \langle \delta M_\alpha\, \delta M_\nu \rangle,$$

does not generally hold for quantum systems. State under what circumstances this expression does hold for quantum systems.

5.11. Compare the Curie and diamagnetic contributions to the magnetic susceptibility and indicate under what conditions the Curie contribution dominates.

5.12. Determine the magnetic correlation function

$$C_M(\mathbf{x}, \mathbf{x}') = \tfrac{1}{3}\langle \delta\mathbf{M}(\mathbf{x}) \cdot \delta\mathbf{M}(\mathbf{x}') \rangle$$

for a set of noninteracting quantum mechanical magnetic moments in the presence of a uniform external magnetic field.

5.13. Consider a set of *tightly bound* noninteracting electrons governed by the Hamiltonian

$$H = \sum_{l,\sigma}(\epsilon_l - \mu B\sigma)a_{l,\sigma}^+ a_{l,\sigma}$$

where l labels the set of electron levels in the absence of B and $\sigma = \pm 1$ labels the spin states. Assume that one knows the density of states

$$D(\epsilon) = \sum_l \delta(\epsilon - \epsilon_l).$$

Determine the zero-field magnetic susceptibility. Evaluate this quantity in the zero-temperature limit.

5.14. Consider the sum

$$S(x) = \sum_{n=0}^{\infty} \theta\left(x - n - \tfrac{1}{2}\right)\left(x - n - \tfrac{1}{2}\right)^{3/2}$$

which occurs in our treatment of electrons in a magnetic field and which we want to evaluate for large x. This evaluation is facilitated by the use of the Poisson sum formula,

$$\sum_{n=-\infty}^{+\infty} \phi(n) = \sum_{l=-\infty}^{+\infty} \int_{-\infty}^{+\infty} dy\, \phi(y) e^{2i\pi ly}.$$

First show that for general x we can write

$$S(x) = \tfrac{1}{2}\left(x - \tfrac{1}{2}\right)^{3/2} + f\left(x - \tfrac{1}{2}\right)$$

where

$$f(x) = \tfrac{1}{2} \sum_{n=-\infty}^{+\infty} \theta(x - |n|)(x - |n|)^{3/2}.$$

Using the Poisson sum formula, show that

$$f(x) = \frac{2}{5} x^{5/2} + 2 \sum_{l=1}^{\infty} \int_0^x dz\, z^{3/2} \cos 2\pi l(x - z).$$

Using two integrations by parts, this can be rewritten in the form

$$f(x) = \frac{2}{5} x^{5/2} + \frac{x^{1/2}}{8} - \frac{3}{8\pi^2} \sum_{l=1}^{\infty} \frac{1}{l^2} \int_0^x \frac{dy}{y^{1/2}} \cos[2\pi l(y - x)].$$

[Hint:

$$\sum_{l=1}^{\infty} \frac{1}{l^2} = \frac{\pi^2}{6}.\bigg]$$

For large x one can drop the oscillatory term. Show that in this case that $S(x)$ is given by Eq. (1719).

5.15. Starting with the expression for the interaction of four changes,

$$V_4 = \sum_{i \neq j=1}^{4} \frac{e_i e_j}{|\mathbf{r}_i - \mathbf{r}_j|},$$

where $|e_i| = e$, show that the interaction between two rigid dipoles (where the changes are separated by a distance a) is given by

$$V(\mathbf{R}) = \frac{3\boldsymbol{\mu}_1 \cdot \hat{R}\hat{\mu}_2 \cdot \hat{R} - \boldsymbol{\mu}_1 \cdot \boldsymbol{\mu}_2}{R^3}$$

where $\boldsymbol{\mu}_1$ and $\boldsymbol{\mu}_2$ are the two dipole moments and $\mathbf{R}$ is the separation of their centers of mass. Assume that $R \gg a$.

5.16. Assume that one has the mean-field theory Hamiltonian

$$H = \sum_{i=1}^{N} \boldsymbol{\mu}_i \cdot \mathbf{E}_T,$$

where $\boldsymbol{\mu}_i$ is the magnetic moment of the ith molecule and the total field $\mathbf{E}_T$ is uniform. If the polarization density is given by

$$\mathbf{P}(\mathbf{x}) = \sum_{i=1}^{N} \boldsymbol{\mu}_i \delta(\mathbf{x} - \mathbf{r}_i)$$

show that the average is given by

$$\langle \mathbf{P}(\mathbf{x}) \rangle = n \langle \boldsymbol{\mu}_i \rangle.$$

5.17. Starting with the Clausius–Mossotti relation, given by Eq. (1771), show that it predicts that there is a temperature, T_{cat}, where the dielectric constant blows up. Assuming an electric dipole moment $\mu \approx 1D$ and a density of 10^{23}cm^{-3}, estimate T_{cat}. This divergence is not seen experimentally. This is an indication that the use of the Lorentz field breaks down for concentrated polar systems.

5.18. Determine the value of the Curie temperature in mean field theory for the quantum Heisenberg system where $\boldsymbol{\mu}(\mathbf{R})$ in Eq. (1772) is given by the operator defined by Eq. (1621).

5.19. Starting with the mean-field theory expression, Eq. (1824), for the magnetic susceptibility for a ferromagnetic system, work out the low-temperature limit for χ_M.

5.20. Starting with the basic mean-field equations

$$m_A = L(\beta \mu B - \beta \mu^2 |\bar{J}| m_B)$$

$$m_B = L(\beta \mu B - \beta \mu^2 |\bar{J}| m_A),$$

describing the sublattice ordering is an anti ferromagnet, calculate the magnetic susceptibility

$$\chi = \mu \frac{N}{2} \frac{\partial}{\partial B}(m_A + m_B)$$

for the one-phase regime above the Néel temperature.

5.21. It is clear from Fig. 5.10 that for large enough fields one loses the antiferromagnetic transition. This means that for a given temperature there is an upper critical magnetic field above which the transition disappears. Alternatively, we have that the Neel temperature depends on the applied field: $T_N(B)$. Working with the correlation function expression given by Eq. (1870), which is valid for $T > T_N(B)$, find the expression determining $T_N(B)$. Work out $T_N(B) = T_N(0)(1 + A[\mu B/k_B T_N(0)]^2 + \cdots)$ for small $\mu B/k_B T_N(0)$ and determine the constant A.

5.22. Starting with the partition function given by Eq. (1859), show that the identities given by Eqs. (1857) and (1858) follow.

5.23. Analyze the magnetic structure factor given by Eq. (1870) in the low-temperature limit.

5.24. Identify the antiferromagnetic wavenumbers q_o for a magnet ordering on a body-centered cubic lattice.

5.25. In the case of an antiferromagnet defined on a simple cubic lattice it was shown [see Eq. (1874)] that the magnetic correlation function $C(\mathbf{q})$ blows up at the Neel temperature when all three components of $q_\alpha a$ are odd multiples of π. Show that this is a manifestation of the result that the Fourier transform of the magnetization density reduces to the staggered magnetization for just these antiferromagnetic points.

5.26. Develop mean-field theory for the Ising model governed by the Hamiltonian

$$H = -\frac{J}{2} \sum_{\mathbf{R} \neq \mathbf{R}'} \sigma(\mathbf{R})\sigma(\mathbf{R}') - B \sum_{\mathbf{R}} \sigma(\mathbf{R})$$

where $J > 0$, the sums over $\mathbf{R}$ and $\mathbf{R}'$ in the interaction term are over nearest neighbors, and each spin takes on only values of $+1$ or -1. Determine the zerofield magnetization and magnetic susceptibility near the Curie temperature.

5.27. Calculate the average energy for a Heisenberg magnet in the mean-field approximation. Derive from this result the specific heat at constant volume. What is the form of the specific heat near the Curie temperature?

5.28. Consider a set of N Ising spins $\sigma_i = \pm 1$. Determine the probability $P[M]$ of finding the system with a total magnetization ($\sum_i \sigma_i$) equal to M at infinite temperature and zero external magnet field.

(a) What is the probability of all the spins being lined up?

(b) What is the probability of the total magnetization equaling zero? Evaluate the leading behavior of $P[0]$ in the large N limit.

5.29. Compute the entropy for the one-dimensional Ising model in zero external field. Find the high- and low-temperature limits. Interpret these results in terms of the information-theoretic definition of entropy.

5.30. Determine the leading-order correction as $N \to \infty$ to the free energy for the one-dimensional Ising model. Show that finite-size corrections are more important for low temperatures.

5.31. Using the transfer matrix method, determine the spin–spin correlation function,

$$C(n) = \langle \sigma(n+i)\sigma(i) \rangle \tag{2010}$$

for a one-dimensional Ising model in zero external field. Your final result for spins separated by n lattice sites can be put in the simple form

$$C(n) = u^{|n|}$$

where $u = \tanh(J/k_B T)$.

5.32. Given that the spin–spin correlation function for a one-dimensional Ising model is as determined in Problem 5.21, determine the correlation length,

$$\xi^2 = 2 \frac{\sum_{n=-\infty}^{+\infty} a^2 n^2 C(n)}{\sum_{n=-\infty}^{+\infty} C(n)}.$$

Compare this correlation length with the definition ξ_T defined by

$$C(n) \equiv e^{-a|n|/\xi_T}. \tag{2011}$$

5.33. Starting with the free energy for the one-dimensional Ising model given by Eq. (1915), find the magnetization per spin, magnetic susceptibility, average energy, and specific heat in a zero external field.

5.34. Show that as $T \to T_c^+$, Onsager's solution for the average energy per spin, given by Eq. (1927), leads to a specific heat that can be written in the form

$$\frac{C_V}{N} = A \ln|T - T_c| + B. \tag{2012}$$

Determine the constants A and B.

5.35. Determine the first nontrivial correction for the low-temperature magnetization for the two-dimensional Ising model starting with Eq. (1933).

5.36. Show that the partition function for a three-dimensional Ising model can be written in the form

$$Z(K) = (\cosh K)^{Nz/2} f(u)$$

where f is a function of $u = \tanh(J/k_B T)$.

5.37. Evaluate $\Omega^{(2)}(h)$, defined by Eq. (1951), in the case of a nonzero magnetic field. Show that this leads to the $\mathcal{O}(K^2)$ contribution to the susceptibility.

5.38. Analyze the high-temperature expansion data for $\tilde{\chi}_n$ and obtain your own estimates for u_c and γ.

5.39. Show that the flipping probability given by Eq. (1992) satisfies the normalization condition given by Eq. (1991).

6 Statistical Mechanics of Solids

6.1 INTRODUCTION

6.1.1 What Is a Solid?

Most substances form solids at sufficiently low temperature and high pressures. Thus, in the phase diagram shown in Fig. 2.7 for a simple noble gas such as neon, one has a fluid (gas or liquid) at high temperatures and low pressures, a solid at low temperatures and high pressures, and phase transitions connecting the various phases. Of interest to us here is the solid phase, which is fundamentally different from the fluid phases. From the discussion in Chapter 1 we know that the distinguishing factor of a solid is that it is rigid! It is this property that allows one to hold a pencil on one end and write on paper with the other. In contrast to a solid, we walk easily through gases and swim through liquids.

This macroscopic solid property is easily understood in terms of the fundamental microscopic characteristic of a solid. In a typical solid the atoms are arranged on a periodic lattice as shown in Fig. 6.1a, and that lattice forms a crystal. The atoms sitting on a lattice site find it energetically unfavorable to leave the vicinity of a lattice site. If one applies a force to a solid sample, unlike in a fluid, where the atoms can conveniently *get out of the way*, the atoms in the solid are difficult to displace, and the result is resistance to the force applied. As we discuss below, the energetic stability of a solid with atoms ordered on a periodic lattice typically comes from a balancing of the attractive long-range and repulsive short-range forces [1] for a system of particles at high density. Thus the chosen crystal structure is the energy-efficient packing mechanism.

As previously discussed, one of the main and intriguing differences between the fluid and solid phases is that there is a fundamental change in symmetry. The gas and liquid phases have complete translational and rotational symmetry. There are no preferred positions or directions in space. For example, in the fluid phases the average density is a uniform constant $\langle n(\mathbf{x}) \rangle = \bar{n}$. In a solid, however, since the atoms form a periodic lattice, the average density, $\langle n(\mathbf{x}) \rangle = \bar{n}(\mathbf{x})$, depends on position $\mathbf{x}$ and is periodic:

$$\bar{n}(\mathbf{x}) = \bar{n}(\mathbf{x} + \mathbf{R}) \tag{2013}$$

where $\mathbf{R}$ is a *lattice vector* connecting any points on the lattice. Solids have reduced symmetry properties compared to the complete translational invariance in fluids. As pointed out in Chapter 1, the canonical ensembles are translationally invariant, so,

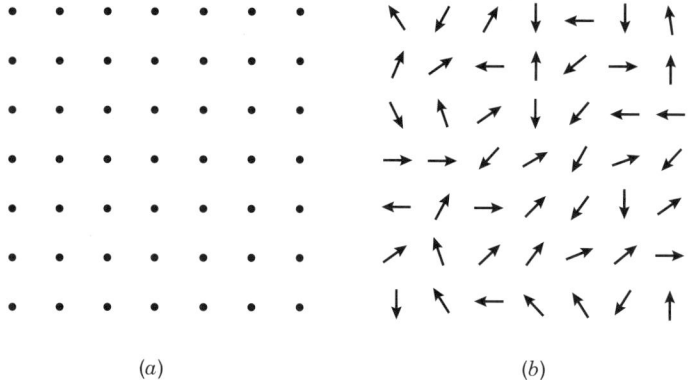

FIGURE 6.1 Schematic of a perfect classical solid: (*a*) for $T = 0$ particles are at rest lattice site; (*b*) for $T > 0$ particles jiggle about their equilibrium positions.

for example, the grand canonical ensemble cannot, without modification, be used to describe the solid state.

In developing a theory for the solid state and for the phase transitions into the solid state, one is confronted by the substantial technical problem of treating the strong interactions among many particles. One possible approach to developing a theory of solids is to extend the theories we have for describing fluids (discussed in Chapter 4) across the fluid–solid phase transition. One can attempt to extend the perturbation theory methods developed for low-density fluids into these much higher-density regimes. However, it turns out to be a very difficult and delicate enterprise to determine which of the possible periodic lattices a system will choose as one crosses the coexistence curve separating fluid and solid. Consequently, an approach based on a low-density expansion is problematic.

We can, however, gain considerable information about the solid if instead of approaching the problem from the high-temperature side, we start from the low-temperature side. This approach is built on the assumption that at sufficiently low temperatures, the system does form a solid. In a simple classical picture the ground state corresponds to a set of atoms at rest and the low-temperature corrections correspond to small lattice vibrations about these equilibrium positions. Although this picture is very useful and practically applicable in many cases, there are situations where it fails:

1. If quantum mechanically zero-point fluctuations are sufficiently large, as in helium, the system will not form a solid even at zero temperature (for sufficiently low pressure).
2. If a system is one- or two-dimensional, it will not form a conventional solid. In two dimensions one has a very subtle situation, which is discussed in Volume 2.
3. As we raise the temperature, the range of the displacements about the equilibrium positions will grow. The system will eventually undergo a displacive

phase transition (form a different lattice) or melt. It is the possibility of phase transitions in the solid between different lattice structures which makes it difficult to answer the question of the lattice type near the coexistence curve. The lattice near the liquid–solid coexistence curve need not be the same as the low-temperature lattice structure.

6.1.2 Building Up a Theory of Solids

Solids can be complex and diverse objects. The properties of insulators, semiconductors, and metals differ significantly as to their electronic properties. In this chapter we restrict ourselves to the much simpler problem of simple insulating solids. The problem of understanding the electronic structure in solids is very interesting and challenging but is beyond the scope of the discussion here.

A direct theoretical assault on a general solid-state system is essentially hopeless. What is done is to build up, successively, effect by effect, an aesthetically appealing and practical theory of solids. Thus we start with a theory for the simplest type of solid (an insulating noble gas solid) and then add additional effects to build up the theory. It is useful here to give a brief outline of the structure of this procedure.

Atomic Motion

Classical Harmonic Oscillator. The most elementary picture (see Fig. 6.1) of a solid in its ground state consists of a set of classical (closed-shell) atoms at rest on an array of periodic lattice points. This is, of course, a zero-temperature picture. At nonzero temperatures the atoms will begin to *jiggle* or oscillate. For low temperatures it will be a good approximation to assume that these vibrations are small enough that each oscillator looks like a harmonic oscillator. Thus to a first approximation an insulating solid looks like a collection of classical harmonic oscillators. Later, when we solve for the properties of a set of harmonic oscillators, we will find that this picture leads us to the idea of correlated or collective vibrations on a lattice. These correlated vibrations travel over long distances on the lattice and are known as *phonons*. These are the *elementary excitations* on the lattice.

Quantum Mechanical Harmonic Oscillators. An obvious refinement of this model is to make the system quantum mechanical. At sufficiently low temperatures the classical picture will break down and the system should be quantized. Quantum effects include zero-point motion and the need to use quantum (Bose–Einstein) statistics. This will also involve *quantizing* the elementary excitations (phonons) in the system much as photons are quantized in going from classical to quantum electromagnetism.

These *harmonic systems* have properties that depend, to some extent, on the particular lattice structure of the solid and the complications arising from treating molecular and compositional solids (e.g., NaCl) can be treated using the methods of lattice dynamics.

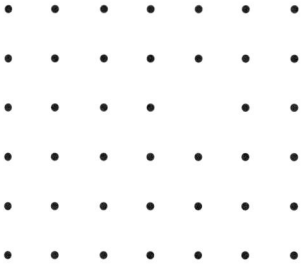

FIGURE 6.2 Point defect (vacancy) in a classical solid at $T = 0$.

Beyond the Harmonic Approximation. Although we limit our discussions here to the harmonic approximation, it has its limitations. As we raise the temperature of a solid, the atomic oscillations will be of increased amplitude and we must include anharmonic corrections. When the displacements become large enough, of course, the solid will melt.

Defects. A second very important correction to the harmonic, perfect crystal, model is the natural occurrence of defects. Crystals are not perfect. The simplest example (see Fig. 6.2) of a defect in a solid is a *point defect*, where an atom is simply missing from the lattice. The missing atom may show up elsewhere as an *interstitial*. There are other types of imperfections, *grain boundaries, dislocations*, and so on, which are important for understanding the mechanical properties of solids. A discussion of defect structures is given in Volume 2.

6.2 LATTICE STRUCTURES

The mathematical theory of crystal lattices is an old and elaborate discipline. We will be satisfied here with a brief overview absent the rigorous mathematics available in texts [2] on the subject. The basic ideas are rather straightforward.

An ideal crystal is constructed by the infinite repetition in space of identical structural units. In the simplest crystals (noble gases, copper, aluminum, and alkali metals) the structural unit is a single atom. The lattice formed by these units is called a Bravais lattice [3]. The structural unit, which may contain many atoms or molecules, is called a basis. The Bravais lattice plus a basis give the atomic crystal structure. For now, let us focus on the various Bravais lattices.

A Bravais lattice is specified by a set of basis vectors $\mathbf{a}_\alpha$ and a set of integers (n_1, n_2, n_3) via

$$\mathbf{R}(n_1, n_2, n_3) = n_1 \mathbf{a}_1 + n_2 \mathbf{a}_2 + n_3 \mathbf{a}_3 = \sum_{\alpha=1}^{3} n_\alpha \mathbf{a}_\alpha. \qquad (2014)$$

If there are N_α sites along side α, the integers n_α are in the range

$$-\frac{N_\alpha}{2} \leq n_\alpha \leq \frac{N_\alpha}{2}. \tag{2015}$$

The set of integers $\{n_\alpha\}$ define a set of lattice points that fill space, and any two lattice points can be connected by some set $\{n_\alpha\}$. The $\mathbf{a}_\alpha$ form a parallelepiped known as a *primitive unit cell*. The primitive unit cells fill all space. In two dimensions we have the general lattice shown in Fig. 6.3. Clearly, this primitive unit cell is space filling.

The generalization of this result to three dimensions is straightforward. This primitive basis is defined by the length of the $\mathbf{a}_\alpha$ and the three angles α, β, and γ shown in Fig. 6.4. From a mathematical point of view, this is the generic case. However, in the real world, for reasons we discuss below, nature forms lattices with higher symmetry. Thus one can investigate those lattices that have some degree of rotational symmetry or inversion symmetry. Let us focus on the two-dimensional case and address the question: Are there rotations in the plane that leave the lattice invariant? In Problem 6.1 it is shown that the only lattice structures that are invariant under a set of rotations are those where $a_1 = a_2$ and

$$\theta = \frac{2\pi}{n} \tag{2016}$$

and $n = 3, 4$, and 6. The case $n = 3$ generates a honeycomb lattice, $n = 4$ generates a square lattice and $n = 6$ gives a triangular lattice (see Fig. 6.5). More specifically, the square lattice is a Bravais lattice with

$$\mathbf{a}_x = a\hat{x}, \qquad \mathbf{a}_y = a\hat{y}, \tag{2017}$$

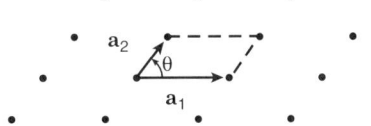

FIGURE 6.3 Primitive lattice vectors for generic two-dimensional lattice.

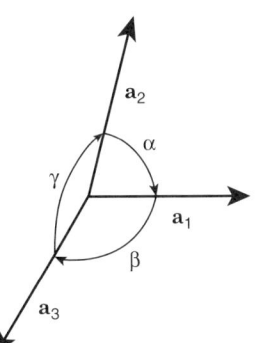

FIGURE 6.4 Primitive lattice vectors for generic three-dimensional lattice.

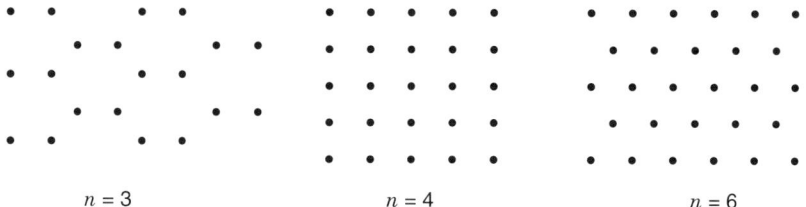

$n = 3$ $n = 4$ $n = 6$

FIGURE 6.5 Special lattices for two-dimensional systems: $n = 3$ corresponds to the honeycomb lattice, $n = 4$ to the square lattice, and $n = 6$ to a triangular lattice.

the triangular lattice is a Bravais lattice with one possible choice of lattice vectors given by

$$\mathbf{a}_1 = a\hat{x}, \qquad \mathbf{a}_2 = a\left(\frac{1}{2}\hat{x} + \frac{\sqrt{3}}{2}\hat{y}\right), \tag{2018}$$

and the honeycomb lattice is not a Bravais lattice. It is, instead, a triangular lattice with a two-site basis.

In three dimensions there are 14 Bravais lattices, which are discussed in standard texts on crystallography [4]. We shall typically discuss the three Bravais lattices that form the cubic group. The basis vectors for the *simple cubic* (sc) *lattice* are given by

$$\mathbf{a}_x = a\hat{x}, \qquad \mathbf{a}_y = a\hat{y}, \qquad \mathbf{a}_z = a\hat{z}.$$

The basis vectors for the *face-centered cubic* (fcc) *lattice*, given by

$$\begin{aligned}
\mathbf{a}_1 &= \frac{a}{2}(\hat{y} + \hat{z}), \\
\mathbf{a}_2 &= \frac{a}{2}(\hat{z} + \hat{x}), \\
\mathbf{a}_3 &= \frac{a}{2}(\hat{x} + \hat{y}),
\end{aligned} \tag{2019}$$

are shown in Fig. 6.6, and some selected elements that form an fcc lattice are Ar, Ag, Al, Au, Cu, Kr, Ne, Ni, Pb, and Xe. The basis vectors for the *body-centered cubic*

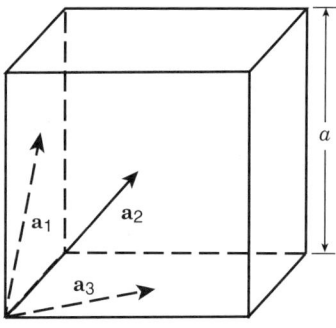

FIGURE 6.6 Primitive lattice vectors for fcc system.

(bcc) *lattice* given by

$$\mathbf{a}_1 = \frac{a}{2}(\hat{y} + \hat{z} - \hat{x}),$$
$$\mathbf{a}_2 = \frac{a}{2}(\hat{z} + \hat{x} - \hat{y}), \qquad (2020)$$
$$\mathbf{a}_3 = \frac{a}{2}(\hat{x} + \hat{y} - \hat{z}),$$

are shown in Fig. 6.7, and some selected elements that form a bcc lattice are Ba, Cr, Cs, Fe, K, Li, Na, and Rb.

As discussed below, we must also consider regular hexagonal lattices formed by stacking triangular lattices in layers separated by a distance c. Such a lattice is a non-Bravais lattice formed by alternating planes of triangular lattices if the Bravais lattice of one set of planes is given by

$$\mathbf{R} = n_1\mathbf{a}_1 + n_2\mathbf{a}_2 + n_3\mathbf{a}_3 \qquad (2021)$$

where

$$\mathbf{a}_1 = a\hat{x} \qquad (2022)$$
$$\mathbf{a}_2 = \frac{a}{2}(\hat{x} + \sqrt{3}\,\hat{y}) \qquad (2023)$$
$$\mathbf{a}_3 = c\hat{z}. \qquad (2024)$$

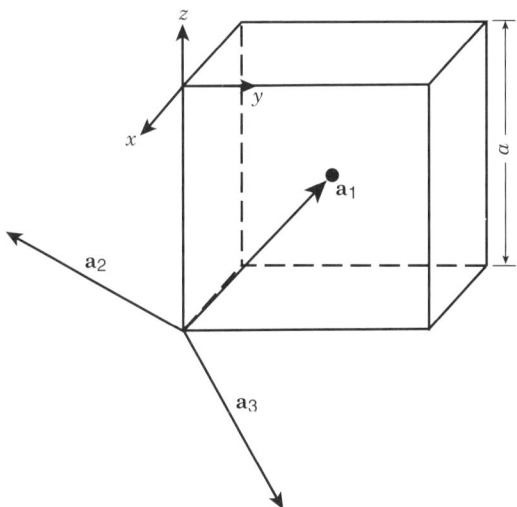

FIGURE 6.7 Primitive lattice for bcc system.

Then an atom on an alternating site is given by

$$\mathbf{d} = \tfrac{1}{3}(\mathbf{a}_1 + \mathbf{a}_2) + \tfrac{1}{2}\mathbf{a}_3. \tag{2025}$$

We are particularly interested in *hexagonal close-packed* (hcp) *lattice*. For close packing one requires that $c = (\sqrt{8}/3)\, a$. This guarantees that there are 12 nearest neighbors on the hcp lattice, the same as for fcc. There are six nearest neighbors for a simple cubic lattice and eight nearest neighbors for bcc. Examples of elements that form an hcp lattice are Be, Ce, He, Ho, La, Mg, Nd, Os, Ru, and Tb.

If one goes beyond the Bravais lattices and adds a basis, one can consider the symmetries for a general crystal structure. These depend on the symmetry of the basis and of the Bravais lattice. There turn out to be 230 different *space groups* [5].

6.3 GROUND-STATE LATTICE STRUCTURE

Let us begin our analysis of perfect crystalline solids with an investigation of the *classical* ground-state properties of the noble gases: neon, argon, krypton, and xenon. We should exclude helium from the discussion since helium atoms are so light that a classical treatment is inappropriate. (We return to this later.) We will assume that the noble gas atoms are described by the Hamiltonian

$$H = \sum_{i=1}^{N} \frac{\mathbf{p}_i^2}{2m} + \frac{1}{2} \sum_{i \neq j=1}^{N} V(\mathbf{r}_i - \mathbf{r}_j) \tag{2026}$$

where there are N particles of mass m. Particle i is located at position $\mathbf{r}_i$, with momentum $\mathbf{p}_i$ and the potential energy is a sum of pair potentials. We assume, for definiteness, that each pair interacts via the Lennard-Jones potential

$$V(r) = 4\epsilon \left[\left(\frac{\sigma}{r}\right)^{12} - \left(\frac{\sigma}{r}\right)^{6} \right] \tag{2027}$$

where r is the distance separating the centers of mass of the atoms. The status of this potential was discussed in Chapter 4. The parameters ϵ and σ can be determined by low-density gas-phase experiments and are listed in Table 4.3.

The ground state for a classical solid consists of a system of particles at rest (zero kinetic energy) with positions, $\mathbf{R}(\mathbf{n})$, on a regular lattice specified by the set of points given by Eq. (2014). How does the system choose the particular lattice structure found in its ground state? The answer is simply that the lattice is that lattice, assuming that there is an atom located at each lattice site, which minimizes the ground-state energy. For nonzero temperatures the physically chosen lattice is that lattice which minimizes the free energy.

The ground-state energy is just the potential energy

$$U_0 = \frac{1}{2} \sum_{\mathbf{R} \neq \mathbf{R}'} V(\mathbf{R} - \mathbf{R}'). \tag{2028}$$

After using the translational invariance satisfied by the pair potential, this reduces to

$$U_0 = \frac{N}{2} \sum_{\mathbf{R} \neq 0} V(\mathbf{R}). \tag{2029}$$

The ground-state energy per particle is given by

$$u_0 = \frac{U_0}{N} = \frac{1}{2} \sum_{\mathbf{R} \neq 0} V(\mathbf{R}) \tag{2030}$$

and is an intensive quantity (independent of N for large N).

We now consider the minimization process where we select the lattice which gives the lowest value for u_0. Thus we must select the $\mathbf{a}_\alpha$ in Eq. (2014) which minimizes Eq. (2030). Let us expand the $\mathbf{a}_\alpha$ in terms of Cartesian coordinates,

$$\mathbf{a}_\alpha = \sum_i a_\alpha^i \hat{x}_i. \tag{2031}$$

Then the a_α^i are the independent variables in the minimization. Inserting Eq. (2031) in Eq. (2014), we obtain

$$\mathbf{R} = \sum_\alpha n_\alpha \sum_i a_\alpha^i \hat{x}_i = \sum_i R_i(\mathbf{n}) \hat{x}_i \tag{2032}$$

where

$$R_i(\mathbf{n}) = \sum_\alpha n_\alpha a_\alpha^i \tag{2033}$$

and

$$R^2(\mathbf{n}) = \sum_i R_i^2(\mathbf{n}). \tag{2034}$$

Minimizing the energy density u_0 with respect to the a_α^i gives

$$\frac{\partial u_0}{\partial a_\alpha^i} = \frac{1}{2} \sum_\mathbf{n} V'(R(\mathbf{n})) \frac{\partial R(\mathbf{n})}{\partial a_\alpha^i} = 0. \tag{2035}$$

However, from Eq. (2034),

$$2R(\mathbf{n})\frac{\partial R(\mathbf{n})}{\partial a_\alpha^i} = 2R_i(\mathbf{n})n_\alpha, \tag{2036}$$

or

$$\frac{\partial R(\mathbf{n})}{\partial a_\alpha^i} = R_i(\mathbf{n})\frac{n_\alpha}{R(\mathbf{n})}. \tag{2037}$$

Inserting this result back into Eq. (2035), the minimization condition becomes

$$\sum_{\mathbf{n}} V'(R(\mathbf{n}))R_i(\mathbf{n})\frac{n_\alpha}{R(\mathbf{n})} = 0. \tag{2038}$$

If we multiply by a_α^j and sum over α, we obtain the basic equation that must be satisfied in determining the ground-state lattice structure:

$$\sum_{\mathbf{n}} \frac{V'(R(\mathbf{n}))}{R(\mathbf{n})} R_i(\mathbf{n})R_j(\mathbf{n}) = 0. \tag{2039}$$

Later we obtain this result in another, complementary way.

We can get a better feeling for the process of lattice selection by considering the two-dimensional case. We start with the generic case of a Bravais lattice with lattice vectors:

$$\mathbf{R} = n_1\mathbf{a}_1 + n_2\mathbf{a}_2 \tag{2040}$$

where

$$\mathbf{a}_1^2 = a^2 \tag{2041}$$
$$\mathbf{a}_2^2 = b^2 \tag{2042}$$

and

$$\mathbf{a}_1 \cdot \mathbf{a}_2 = ab\cos\alpha. \tag{2043}$$

If we break the lattice vectors up into orthogonal components, then (with $\mathbf{a}_1 = a\hat{x}$)

$$R_x = an_1 + bn_2\cos\alpha \tag{2044}$$

and

$$R_y = bn_2\sin\alpha. \tag{2045}$$

Taking the orthogonal components of Eq. (2039), we obtain three equations:

$$\sum_n \frac{V'(R)}{R} (an_1 + bn_2 \cos\alpha)^2 = 0 \tag{2046}$$

$$\sum_n \frac{V'(R)}{R} bn_2 \sin\alpha (an_1 + bn_2 \cos\alpha) = 0 \tag{2047}$$

and

$$\sum_n \frac{V'(R)}{R} b^2 n_2^2 \sin^2\alpha = 0. \tag{2048}$$

Assuming that $\sin\alpha \neq 0$, we see that these equations are equivalent to the set

$$\sum_n \frac{V'(\mathbf{R})}{R} n_1^2 = 0 \tag{2049}$$

and

$$\sum_n \frac{V'(\mathbf{R})}{R} n_1 n_2 = 0 \tag{2050}$$

and

$$\sum_n \frac{V'(\mathbf{R})}{R} n_2^2 = 0. \tag{2051}$$

How can we satisfy all three equations simultaneously? Clearly, if $\alpha = \pi/2$ and $\mathbf{a}_1 \cdot \mathbf{a}_2 = 0$, then $\mathbf{R}^2(\mathbf{n}) = a^2 n_1^2 + b^2 n_2^2$, $\mathbf{R}(-\mathbf{n}) = -\mathbf{R}(\mathbf{n})$ and

$$\sum_n \frac{V'(R)}{R} n_1 n_2 = 0 \tag{2052}$$

by symmetry. If we choose $a_1 = a_2 = a$, the remaining two equations, Eqs. (2049) and (2051), can be replaced by a single equation,

$$\sum_n \frac{V'(R)}{R} R^2 = 0. \tag{2053}$$

This result is equivalent to Eq. (2039) if we have inversion symmetry $\mathbf{R}(-\mathbf{n}) = -\mathbf{R}(\mathbf{n})$. Therefore, we have shown for the generic set of Bravais lattices that the energy is minimized by the square lattice. This energy must then be compared with the two exceptional symmetric lattices (with rotational symmetry). The triangular lattice [with shells with six nearest neighbors (sixfold coordination)] and the honeycomb lattice (a nonBravais lattice with a shell of three nearest neighbors and threefold coordination).

Our calculation is therefore reduced to one where we compare the ground-state energy of the rotationally invariant lattices. We then note that all of the two-dimensional lattices that are invariant under rotations (honeycomb, square, and triangular) can be organized in terms of a shell structure with sites on shell α an equal distance R_α from a central site and with m_α equivalent neighbors (see Table 6.1). The ground-state energy can then be written as

$$u_0 = \frac{1}{2} \sum_\alpha m_\alpha V(R_\alpha). \tag{2054}$$

For these lattices there is a single characteristic length that we can take as the nearest-neighbor distance a,

$$R_\alpha = a l_\alpha \tag{2055}$$

If we minimize Eq. (2054) with respect to a and multiply by a, we obtain

$$\sum_\alpha m_\alpha V'(R) R_\alpha = 0, \tag{2056}$$

which is equivalent to Eq. (2053) written in terms of shell variables.

TABLE 6.1 Shell Structure for Various Two-Dimensional Lattices

Shell (α)	l_α^2	m_α
	Square	
1	1	4
2	2	4
3	4	4
4	5	8
	Triangular	
1	1	6
2	3	6
3	4	6
	Honeycomb	
1	1	3
2	3	6
3	6	3

GROUND-STATE LATTICE STRUCTURE

Let us apply this result to the case of a Lennard-Jones potential. The derivative of the potential is given by

$$V'(R) = 4\epsilon\left[-12\left(\frac{\sigma}{R}\right)^{12}\frac{1}{R} + 6\left(\frac{\sigma}{R}\right)^{6}\frac{1}{R}\right], \tag{2057}$$

and using Eq. (2055), the minimization condition, Eq. (2056), reduces to

$$\sum_\alpha m_\alpha \left[2\left(\frac{\sigma}{a}\right)^{12}\frac{1}{l_\alpha^{12}} - \left(\frac{\sigma}{a}\right)^{6}\frac{1}{l_\alpha^{6}}\right] = 0, \tag{2058}$$

where we have canceled an overall multiplicative factor of $-24\epsilon/a$. If we define the dimensionless sums,

$$A_p = \sum_{\alpha \neq 0} m_\alpha \frac{1}{l_\alpha^p}, \tag{2059}$$

the minimization condition reduces to

$$2\left(\frac{\sigma^2}{a^2}\right)^{6} A_{12} - \left(\frac{\sigma^2}{a^2}\right)^{3} A_6 = 0,$$

or

$$a = \sigma \left(\frac{2A_{12}}{A_6}\right)^{1/6}. \tag{2060}$$

This gives the nearest-neighbor distance which minimizes the ground-state energy for a particular lattice type. The information about the lattice type is contained in the sums A_p. The energy per particle, given by Eq. (2054), can be written as

$$\begin{aligned}
u_0 &= \frac{1}{2}\sum_\alpha m_\alpha 4\epsilon \left[\left(\frac{\sigma}{a}\right)^{12}\frac{1}{l_\alpha^{12}} - \left(\frac{\sigma}{a}\right)^{6}\frac{1}{l_\alpha^{6}}\right] \\
&= 2\epsilon\left[\left(\frac{\sigma}{a}\right)^{12} A_{12} - \left(\frac{\sigma}{a}\right)^{6} A_6\right].
\end{aligned} \tag{2061}$$

Substituting for a/σ from Eq. (2060), we find that the minimum energy for each lattice type is of the form

$$\begin{aligned}
u_0 &= 2\epsilon\left[\left(\frac{A_6}{2A_{12}}\right)^{2} A_{12} - \frac{A_6}{2A_{12}} A_6\right] \\
&= -\frac{\epsilon A_6^2}{2 A_{12}}.
\end{aligned} \tag{2062}$$

Since $A_{12} > 0$, u_0 is negative and we have a minimum. Notice that the information about the lattice type is contained in A_6 and A_{12}, which enter because of our choice of the Lennard-Jones potential.

The sums A_p converge after a few terms. The needed quantities m_α and l_α are listed in Table 6.1 for the three regular lattice types. We then have, for example, for the square lattice

$$A_p = 4 + \frac{4}{2^{p/2}} + \frac{4}{2^p} + \frac{8}{5^{p/2}} + \cdots. \tag{2063}$$

More specifically, for $p = 6$,

$$A_6 = 4\left(1 + \frac{1}{2^3} + \frac{1}{2^6} + \frac{2}{5^3} + \cdots\right) = 4.626\ldots, \tag{2064}$$

and for $p = 12$,

$$A_{12} = 4\left(1 + \frac{1}{2^6} + \frac{1}{2^{12}} + \frac{2}{5^6} + \cdots\right) = 4.0636\ldots. \tag{2065}$$

The various results for the three lattices are listed in Table 6.2. We see that A_p is determined, to a first approximation, by the number of nearest neighbors, z,

$$A_6 \approx A_{12} \approx z, \tag{2066}$$

and

$$u_0 \approx -\frac{\epsilon}{2} z \tag{2067}$$

and clearly the triangular lattice gives the lowest energy. Systems interacting by a Lennard-Jones potential in two dimensions will form triangular lattices.

In the three-dimensional case the development is similar but more involved. It seems clear that treatment of the generic case will lead to the conclusion that the simple cubic case gives a lower ground-state energy than any general six-fold

TABLE 6.2 Sums for Regular Lattices in Two Dimensions

Lattice	A_6	A_{12}	A_6^2/A_{12}
□	4.626	4.0636	5.27
△	6.3158	6.0084	6.64
H	3.2361	3.009	3.49

coordinated Bravais lattice. As in two dimensions, one must then look at more symmetric cases with higher coordination. This certainly leads one to consider the cubic lattices. We must also include the hcp lattices in the analysis. If one has a simple radial pair potential, then to a first approximation the key information is the number of nearest neighbors. Fcc and hcp lattices have 12 nearest neighbors, bcc lattices have eight nearest neighbors, and simple cubic lattices have six nearest neighbors.

In complete analogy with the two-dimensional case, the energy is again given in terms of the A_p's by Eq. (2062). Again, using the shell technique developed for two dimensions, we obtain the converged results shown in Table 6.3. Comparison with experiment is shown in Table 6.4.

Again to a first approximation, the results are determined by counting the number of nearest neighbors. It is, however, extremely difficult to distinguish between the energies of an fcc and an hcp system. It is found experimentally that the noble gas systems form an fcc lattice. Clearly, we need a more sophisticated theoretical model to distinguish between these two lattices.

The determination of the lattice structure for the case of a noble gas is relatively simple because the system contains a single component and the potentials of interaction are reasonably well known. More generally, the situation is less favorable. In the case, for example, of an ionic crystal we have the additional complication that there is effective dissociation between the gas molecular phase and the ionic solid phase. For example, for salt (NaCl), unlike for a noble gas, the ions in the solids are in a different state than the gas-phase neutral salt. In many cases the interaction potential is not well known. One has multiple interactions between

TABLE 6.3 Sums for Three-Dimensional Lattices

	sc	bcc	fcc	hcp
A_6	8.40	12.2533	14.45392	14.45489
A_{12}	6.20	9.11418	12.13188	12.13229
$(2A_{12}/A_6)^{1/6}$	1.181	1.179	1.09017	1.09
$A_6^2/2A_{12}$	5.690	8.236	8.610199	8.611107

Source: Data from Ref. 6.

TABLE 6.4 Comparison of Ground-State Properties Obtained from Theory with Experiment

	Ne	Ar	Kr	Xe
a (Å) exptl.	3.13	3.75	3.99	4.33
$a = 1.09\sigma$	2.99	3.71	3.98	4.34
u_0(eV) exptl.	−0.02	−0.08	−0.11	−0.17
$u_0 = -8.6\epsilon$	−0.027	−0.089	−0.120	−0.172
B_0 (10^{10} dyn/cm) exptl.	1.1	2.7	3.5	3.6
$B_0 = 75\epsilon/\sigma^3$	1.81	3.18	3.46	3.81

Source: Data from Ref. 7.

multiple constituents. One has both short (as in argon) and long-range Coulomb contributions to the potentials. The short-range *core* contributions depend on the parameters peculiar to constituents (e.g., core size). In these problems the Coulomb energy is the dominant new element. Thus one is faced with the calculation of the Madelung energy [8,9], which arises when performing the sums A_p, as defined for the noble gas case, but for $p = 1$. The evaluation of A_1 involves subtleties that lead one to organize the calculation so that one maintains overall charge neutrality. A complete description of the problem of lattice formation and the associated *cohesion energy* in ionic solids is given by Born and Huang [10]. In more complicated situations, such as in covalent crystals and metals, the electronic charge distribution is spread out in the crystal, and we must take the electronic properties explicitly into account.

6.4 LATTICE VIBRATIONS

6.4.1 Harmonic Hamiltonian

Once we have determined the lattice structure that gives the minimum value of the ground-state energy, we can investigate other properties of the low-temperature solid. The first step is to look at the form of the Hamiltonian for a system near the ground state.

We consider the simplest situation of a single-component system of particles with mass m. At nonzero temperatures the atoms oscillate about their equilibrium lattice positions labeled by **R**. The instantaneous position of the particles is given by

$$\mathbf{r}(\mathbf{R}) = \mathbf{R} + \mathbf{u}(\mathbf{R}), \qquad (2068)$$

where $\mathbf{u}(\mathbf{R})$ is the deviation from equilibrium. The momentum of the particle whose equilibrium position is denoted by **R** is $\mathbf{p}(\mathbf{R})$. The Hamiltonian for this system is the sum of the kinetic energy,

$$K = \sum_{\mathbf{R}} \frac{\mathbf{P}(\mathbf{R})^2}{2m}, \qquad (2069)$$

and the potential energy, given by

$$\begin{aligned} U &= \tfrac{1}{2} \sum_{\mathbf{R} \neq \mathbf{R}'} V(\mathbf{r}(\mathbf{R}) - \mathbf{r}(\mathbf{R}')) \\ &= \tfrac{1}{2} \sum_{\mathbf{R} \neq \mathbf{R}'} V(\mathbf{R} - \mathbf{R}' + \mathbf{u}(\mathbf{R}) - \mathbf{u}(\mathbf{R}')). \end{aligned} \qquad (2070)$$

A major simplification occurs in the problem if we assume that the atomic fluctuations $\mathbf{u}(\mathbf{R})$ are small relative to the spacing between equilibrium sites. We can then expand U in a power series in $\mathbf{u}(\mathbf{R})$:

$$U = U_0 + \frac{1}{2}\sum_{\mathbf{R}\neq\mathbf{R}'}\sum_\alpha [u_\alpha(\mathbf{R}) - u_\alpha(\mathbf{R}')]\frac{\partial}{\partial r_\alpha}V(\mathbf{r})\Big|_{\mathbf{r}=\mathbf{R}-\mathbf{R}'}$$
$$+ \frac{1}{4}\sum_{\mathbf{R}\neq\mathbf{R}'}\sum_{\alpha,\beta}[u_\alpha(\mathbf{R}) - u_\alpha(\mathbf{R}')][u_\beta(\mathbf{R}) - u_\beta(\mathbf{R}')]$$
$$\times \left(\frac{\partial^2}{\partial r_\alpha \partial r_\beta}V(\mathbf{r})\right)\Big|_{\mathbf{r}=\mathbf{R}-\mathbf{R}'} \tag{2071}$$

where U_0 is the ground-state energy computed earlier and we neglect higher-order anharmonic terms. Let us look at the term linear in $\mathbf{u}$:

$$U_1 = \frac{1}{2}\sum_{\mathbf{R}\neq\mathbf{R}'}\sum_\alpha \left(\frac{\partial V(r)}{\partial r_\alpha}\right)_{\mathbf{r}=\mathbf{R}-\mathbf{R}'}[\mathbf{u}(\mathbf{R}) - \mathbf{u}(\mathbf{R}')]_\alpha$$
$$\equiv \sum_\mathbf{R}\mathbf{u}(\mathbf{R})\cdot\mathbf{F}(\mathbf{R}) \tag{2072}$$

where

$$F_\alpha(\mathbf{R}) = \sum_{\mathbf{R}'}\left(\frac{\partial V(r)}{\partial r_\alpha}\right)_{\mathbf{r}=\mathbf{R}-\mathbf{R}'}$$
$$= \sum_{\mathbf{R}'}\frac{V'(|\mathbf{R}-\mathbf{R}'|)}{|\mathbf{R}-\mathbf{R}'|}(\mathbf{R}-\mathbf{R}')_\alpha = \sum_{\mathbf{R}'}\frac{V'(|\mathbf{R}'|)}{|\mathbf{R}'|}(-R'_\alpha) \tag{2073}$$

is the average force on each particle. Physically, if we are to have equilibrium, this force must vanish. Mathematically, this force vanishes if the lattice has inversion symmetry [11]. Therefore,

$$U_1 = 0, \tag{2074}$$

and the Hamiltonian is given in this approximation by

$$U = U_0 + U_H. \tag{2075}$$

The harmonic contribution can be written as

$$U_H = \frac{1}{4}\sum_{\mathbf{R}\neq\mathbf{R}'}\sum_{\alpha,\beta}[u_\alpha(\mathbf{R}) - u_\alpha(\mathbf{R}')][u_\beta(\mathbf{R}) - u_\beta(\mathbf{R}')]\phi_{\alpha\beta}(\mathbf{R}-\mathbf{R}') \tag{2076}$$

where

$$\phi_{\alpha\beta}(\mathbf{r}) = \frac{\partial}{\partial r_\alpha}\frac{\partial}{\partial r_\beta}V(r). \tag{2077}$$

It is a matter of simple rearrangement to rewrite Eq. (2076) in the more convenient form

$$U_H = \tfrac{1}{2} \sum_{\mathbf{R} \neq \mathbf{R}'} \sum_{\alpha,\beta} u_\alpha(\mathbf{R}) u_\beta(\mathbf{R}') D_{\alpha\beta}(\mathbf{R} - \mathbf{R}') \tag{2078}$$

where the *dynamical matrix* is defined by

$$D_{\alpha\beta}(\mathbf{R} - \mathbf{R}') = \delta_{\mathbf{R},\mathbf{R}'} \sum_{\mathbf{R}''} \phi_{\alpha\beta}(\mathbf{R} - \mathbf{R}'') - \phi_{\alpha\beta}(\mathbf{R} - \mathbf{R}'). \tag{2079}$$

6.4.2 Normal Modes of Vibration

Further progress is facilitated by the use of Born–von Kármán [12] periodic boundary conditions for the crystal. We require that displacements satisfy the periodicity condition

$$\mathbf{u}(\mathbf{R} + N_\alpha \mathbf{a}_\alpha) = \mathbf{u}(\mathbf{R}) \tag{2080}$$

for each of the three primitive lattice vectors $\mathbf{a}_\alpha$, where N_α are large integers satisfying

$$N = N_1 N_2 N_3 \tag{2081}$$

where N is the total number of atoms. This periodicity is ensured if we write

$$\mathbf{u}(\mathbf{R}) = \frac{1}{N} \sum_{\mathbf{k}} \mathbf{u}(\mathbf{k}) e^{i\mathbf{k}\cdot\mathbf{R}} \tag{2082}$$

where the wavenumbers $\mathbf{k}$ are of the restricted form

$$\mathbf{k} = \sum_{\alpha=1}^{3} \frac{m_\alpha}{N_\alpha} \mathbf{b}_\alpha \tag{2083}$$

where the m_α are integers and the $\mathbf{b}$'s are the reciprocal lattice vectors

$$\mathbf{b}_1 = \frac{2\pi(\mathbf{a}_2 \times \mathbf{a}_3)}{\mathbf{a}_1 \cdot (\mathbf{a}_2 \times \mathbf{a}_3)} \tag{2084}$$

$$\mathbf{b}_2 = \frac{2\pi(\mathbf{a}_3 \times \mathbf{a}_1)}{\mathbf{a}_1 \cdot (\mathbf{a}_2 \times \mathbf{a}_3)} \tag{2085}$$

$$\mathbf{b}_3 = \frac{2\pi(\mathbf{a}_1 \times \mathbf{a}_2)}{\mathbf{a}_1 \cdot (\mathbf{a}_2 \times \mathbf{a}_3)} \tag{2086}$$

which satisfy

$$\mathbf{b}_\alpha \cdot \mathbf{a}_\beta = 2\pi \delta_{\alpha\beta}. \tag{2087}$$

Then $\mathbf{k} \cdot \mathbf{a}_\alpha N_\alpha = m_\alpha 2\pi$, which guarantees that Eq. (2082) satisfies Eq. (2080). Since

$$\sum_\mathbf{R} e^{+i(\mathbf{k}-\mathbf{k}')\cdot\mathbf{R}} = N\delta_{\mathbf{k},\mathbf{k}'}, \tag{2088}$$

we can invert Eq. (2082) to obtain

$$\mathbf{u}(\mathbf{k}) = \sum_\mathbf{R} e^{-i\mathbf{k}\cdot\mathbf{R}} \mathbf{u}(\mathbf{R}). \tag{2089}$$

Using the Fourier representation for $u_\alpha(\mathbf{R})$ and $u_\beta(\mathbf{R}')$ in Eq. (2078), we obtain, after straightforward manipulations,

$$U_H = \frac{1}{2N} \sum_\mathbf{k} \sum_{\alpha,\beta} u_\alpha(\mathbf{k}) u_\beta^*(\mathbf{k}) D_{\alpha\beta}(\mathbf{k}) \tag{2090}$$

where we have introduced the Fourier representation

$$D_{\alpha\beta}(\mathbf{k}) = \sum_\mathbf{R} D_{\alpha\beta}(\mathbf{R}) e^{-i\mathbf{k}\cdot\mathbf{R}}. \tag{2091}$$

Since U_H is real and $D_{\alpha\beta}(\mathbf{R})$ is symmetric under $\mathbf{R} \to -\mathbf{R}$, we see that $D_{\alpha\beta}(\mathbf{k})$ is a real, symmetric, 3×3 matrix. It therefore can be diagonalized by a set of eigenvectors $\epsilon_\alpha^i(\mathbf{k})$ such that

$$\sum_\beta D_{\alpha\beta}(\mathbf{k}) \epsilon_\beta^i(\mathbf{k}) = \lambda_i(\mathbf{k}) \epsilon_\alpha^i(\mathbf{k}) \tag{2092}$$

where i takes on three values and, as usual, we can show that the eigenvalues λ_i are real. If the harmonic approximation is to be valid and the system thermodynamically stable, the eigenvalues must be positive. We will return to this point in the next section. Clearly, the eigenvectors for a real symmetric matrix can be constructed to be real, complete

$$\sum_\alpha \epsilon_\alpha^i(\mathbf{k}) \epsilon_\alpha^j(\mathbf{k}) = \delta_{i,j}, \tag{2093}$$

and orthonormal

$$\sum_i \epsilon_\alpha^i(\mathbf{k}) \epsilon_\beta^i(\mathbf{k}) = \delta_{\alpha\beta}. \tag{2094}$$

If we expand the Fourier transforms of the displacements in this complete set,

$$u_\alpha(\mathbf{k}) = \sum_i \epsilon^i_\alpha(\mathbf{k})\tilde{u}_i(\mathbf{k}), \tag{2095}$$

then

$$\tilde{u}_i(\mathbf{k}) = \sum_\alpha \epsilon^i_\alpha(\mathbf{k})u_\alpha(\mathbf{k}), \tag{2096}$$

and the potential energy can be put in the diagonal form

$$U_H = \tfrac{1}{2}N \sum_{\mathbf{k},i} \tilde{u}_i(\mathbf{k})\lambda_i(\mathbf{k})\tilde{u}_i^*(\mathbf{k}). \tag{2097}$$

The eigenvalues, $\lambda_i(\mathbf{k})$, are related to the normal modes of oscillation in the solid. We can see this by looking at the classical equation of motion for atoms with mass m:

$$\begin{aligned} m\ddot{u}_\alpha(\mathbf{R},t) &= -\frac{\partial}{\partial u_\alpha(\mathbf{R},t)} U_H \\ &= -\sum_{\mathbf{R}'}\sum_\beta D_{\alpha\beta}(\mathbf{R}-\mathbf{R}')u_\beta(\mathbf{R}',t). \end{aligned} \tag{2098}$$

Fourier transforming over $\mathbf{R}$ gives immediately

$$m\ddot{u}_\alpha(\mathbf{k},t) = -\sum_\beta D_{\alpha\beta}(\mathbf{k})u_\beta(\mathbf{k},t). \tag{2099}$$

If we then insert the eigenfunction expansion for $u_\alpha(\mathbf{k},t)$ given by Eq. (2095), use Eq. (2092), and matrix multiply by $\epsilon^i_\alpha(\mathbf{k})$, we obtain the equation of motion,

$$m\ddot{\tilde{u}}_i(\mathbf{k},t) = -\lambda_i(\mathbf{k})\tilde{u}_i(\mathbf{k},t). \tag{2100}$$

After defining the eigenfrequencies

$$\omega_i^2(\mathbf{k}) = \frac{\lambda_i(\mathbf{k})}{m}, \tag{2101}$$

we obtain the oscillating solutions

$$\tilde{u}_i(\mathbf{k},t) = \tilde{u}_i(\mathbf{k},0)\cos(\omega_i(\mathbf{k})t) + \frac{\dot{\tilde{u}}_i(\mathbf{k},0)}{\omega_i(\mathbf{k})}\sin(\omega_i(\mathbf{k})t). \tag{2102}$$

This clearly displays the normal-mode structure of the system.

6.4.3 Example: Cubic Crystal with Nearest-Neighbor Interactions

In general, the dynamic matrix, $D_{\alpha\beta}(\mathbf{k})$, is a complicated quantity depending on the specific lattice and form of the potential. To gain some feeling for its structure, let us restrict ourselves here to the simple case of particles on a cubic lattice, with lattice spacing a, interacting with a radial potential that acts only out to a range of the first six nearest neighbors. For a radial potential, $\phi_{\alpha\beta}(\mathbf{r})$, defined by Eq. (2077), can be written as

$$\phi_{\alpha\beta}(\mathbf{r}) = A(r)\hat{r}_\alpha \hat{r}_\beta + E(r)\delta_{\alpha\beta} \tag{2103}$$

where

$$A(r) = V''(r) - \frac{V'(r)}{r} = r\left(\frac{d}{dr}\frac{V'(r)}{r}\right) \tag{2104}$$

and

$$E(r) = \frac{V'(r)}{r}. \tag{2105}$$

For this very simple model the condition determining the minimum ground-state energy, Eq. (2053), reduces to $V'(a)a = 0$. Our model interaction, evaluated for lattice vectors $\mathbf{R}$, can be written in the form

$$\phi_{\alpha\beta}(\mathbf{R}) = \sum_{\mathbf{a}} \delta_{\mathbf{R},\mathbf{a}} A \hat{a}_\alpha \hat{a}_\beta \tag{2106}$$

where $A = A(a) = V''(a)$ and the $\mathbf{a}$ are the six nearest neighbors $(a,0,0)$, $(-a,0,0)$, $(0,a,0)$, $(0,-a,0)$, $(0,0,a)$, $(0,0,-a)$. We then have, from Eq. (2079), that

$$D_{\alpha\beta}(\mathbf{R}) = \delta_{\mathbf{R},0} 6A\delta_{\alpha\beta} - \sum_{\mathbf{a}} \delta_{\mathbf{R},\mathbf{a}} A \hat{a}_\alpha \hat{a}_\beta. \tag{2107}$$

Taking the Fourier transform of Eq. (2107) gives

$$D_{\alpha\beta}(\mathbf{k}) = 4\delta_{\alpha\beta} A \sin^2 \frac{k_\alpha a}{2}. \tag{2108}$$

Clearly, the eigenvectors are given by

$$\epsilon_\alpha^i(\mathbf{k}) = \delta_{\alpha i} \tag{2109}$$

where the three choices for i correspond to the three orthogonal directions, and

$$\lambda_i(\mathbf{k}) = D_i(\mathbf{k}) = 4A \sin^2 \frac{k_i a}{2}. \tag{2110}$$

We will return to these results later.

6.4.4 Long-Wavelength Limit

Let us concentrate next on the long-wavelength limit of our lattice dynamical system. We see from our example in the preceding section that the eigenvalues vanish for small wavenumbers. In this section we want to show that this is a general feature of crystalline systems with important physical consequences. The general analysis is based on some general properties of the dynamical matrix. We have from Eq. (2079) and the symmetries of $\phi_{\alpha\beta}(\mathbf{R})$ [$\phi_{\alpha\beta}(\mathbf{R}) = \phi_{\beta\alpha}(\mathbf{R})$, $\phi_{\alpha\beta}(-\mathbf{R}) = \phi_{\alpha\beta}(\mathbf{R})$] that

$$D_{\alpha\beta}(\mathbf{R}) = D_{\beta\alpha}(\mathbf{R}) \tag{2111}$$

$$D_{\alpha\beta}(-\mathbf{R}) = D_{\alpha\beta}(\mathbf{R}). \tag{2112}$$

We also have from the definition, Eq. (2079), that

$$\sum_{\mathbf{R}} D_{\alpha\beta}(\mathbf{R}) = 0. \tag{2113}$$

Using these properties, one can show (see Problem 6.3) that the Fourier transform of $D_{\alpha\beta}(\mathbf{R})$, Eq. (2091), can be put in the form

$$D_{\alpha\beta}(\mathbf{k}) = -2 \sum_{\mathbf{R}} D_{\alpha\beta}(\mathbf{R}) \sin^2\left(\frac{\mathbf{k}\cdot\mathbf{R}}{2}\right). \tag{2114}$$

Substituting Eq. (2079) into Eq. (2114) gives the general result:

$$D_{\alpha\beta}(\mathbf{k}) = 2 \sum_{\mathbf{R}} \phi_{\alpha\beta}(\mathbf{R}) \sin^2\left(\frac{\mathbf{k}\cdot\mathbf{R}}{2}\right). \tag{2115}$$

In the long-wavelength limit this general expression reduces to

$$D_{\alpha\beta}(\mathbf{k}) = \tfrac{1}{2} \sum_{\mathbf{R}} \phi_{\alpha\beta}(\mathbf{R})(\mathbf{k}\cdot\mathbf{R})^2, \tag{2116}$$

and the dynamical matrix vanishes as $\mathcal{O}(k^2)$ as $k \to 0$. Inserting this result back into Eq. (2090), we have the long-wavelength contribution to the potential energy,

$$U_H = \frac{1}{2N} \sum_{\mathbf{k}} \sum_{\alpha\beta} u_\alpha(\mathbf{k}) u_\beta^*(\mathbf{k}) \frac{1}{2} \sum_{\mathbf{R}} \phi_{\alpha\beta}(\mathbf{R})(\mathbf{k}\cdot\mathbf{R})^2. \tag{2117}$$

This expression suggests that we introduce the elastic tensor

$$C_{\alpha\mu\beta\nu} = \frac{n_0}{2} \sum_{\mathbf{R}} R_\mu \phi_{\alpha\beta}(\mathbf{R}) R_\nu \tag{2118}$$

where $n_0 = N/V$. In the long-wavelength limit the dynamical matrix can be written as

$$D_{\alpha\beta}(\mathbf{k}) = \frac{1}{n_0} \sum_{\mu\nu} C_{\alpha\mu\beta\nu} k_\mu k_\nu. \qquad (2119)$$

All of the detailed information about the particular lattice and potential are contained in the elastic tensor. The long-wavelength contribution to the harmonic part of the interaction potential is then given by

$$U_H = \frac{1}{2Nn_0} \sum_{\mathbf{k}} \sum_{\alpha\beta\mu\nu} k_\mu u_\alpha(\mathbf{k}) k_\nu u_\beta^*(\mathbf{k}) C_{\alpha\mu\beta\nu}. \qquad (2120)$$

For wavelengths long compared to the lattice spacing, we can express the displacement in terms of a continuous field,

$$\mathbf{u}(\mathbf{k}) = n_0 \int d^3x\, e^{-i\mathbf{k}\cdot\mathbf{x}} \mathbf{u}(\mathbf{x}) \qquad (2121)$$

and the sum over wavenumbers in Eq. (2120) can be written in the form

$$\frac{1}{Nn_0} \sum_{\mathbf{k}} k_\mu k_\nu n_0 \int d^3x\, e^{-i\mathbf{k}\cdot\mathbf{x}} u_\alpha(\mathbf{x}) n_0 \int d^3x'\, e^{-i\mathbf{k}\cdot\mathbf{x}'} u_\beta(\mathbf{x}')$$

$$= \frac{n_0}{N} \int d^3x\, u_\alpha(\mathbf{x}) \int d^3x'\, u_\beta(\mathbf{x}') \nabla_x^\mu \nabla_{x'}^\nu \sum_{\mathbf{k}} e^{-i\mathbf{k}\cdot(\mathbf{x}-\mathbf{x}')}$$

$$= \frac{1}{V} \int d^3x\, u_\alpha(\mathbf{x}) \int d^3x'\, u_\beta(\mathbf{x}') \nabla_x^\mu \nabla_{x'}^\nu V\delta(\mathbf{x}-\mathbf{x}')$$

$$= \int d^3x\, \nabla_\mu u_\alpha(\mathbf{x}) \nabla_\nu u_\beta(\mathbf{x}). \qquad (2122)$$

In the last line we have carried out two integrations by parts. The long-wavelength contribution to the potential energy is then given by

$$U_H = \tfrac{1}{2} \int d^3x \sum_{\alpha\beta\mu\nu} C_{\alpha\mu\beta\nu} \nabla_\mu u_\alpha(\mathbf{x}) \nabla_\nu u_\beta(\mathbf{x}). \qquad (2123)$$

Inserting the result for $\phi_{\alpha\beta}(\mathbf{R})$, given by Eq. (2103), into Eq. (2118), and remembering Eq. (2039), leads to the expression for the elastic tensor,

$$C_{\alpha\mu\beta\nu} = \frac{n_0}{2} \sum_{\mathbf{R}} A(R) \hat{R}_\alpha \hat{R}_\beta R_\mu R_\nu, \qquad (2124)$$

valid for a single component solid with a radial pair potential. Since the elastic tensor is invariant under the exchange $\mu \leftrightarrow \alpha$ and $\nu \leftrightarrow \beta$, we can write the final expression for the long-wavelength contribution to the potential energy as

$$U_H = \tfrac{1}{2} \int d^3x \sum_{\alpha\beta\mu\nu} C_{\alpha\mu\beta\nu} U_{\alpha\mu}(\mathbf{x}) U_{\beta\nu}(\mathbf{x}) \tag{2125}$$

where, as in Chapter 1, we have reintroduced the strain tensor

$$U_{\alpha\mu}(\mathbf{x}) = \tfrac{1}{2}\left[\nabla_\mu u_\alpha(\mathbf{x}) + \nabla_\alpha u_\mu(\mathbf{x})\right]. \tag{2126}$$

Later in this chapter we return to a more general discussion of the long-wavelength properties of solids and the role of elastic constants in understanding the mechanical properties of solids.

6.4.5 Thermal Excitations: Classical Theory

Even though we know that it is proper to develop a quantum mechanical theory for the low-temperature properties of a solid, it is useful and instructive to work out the results of the associated classical theory. This will serve as a check on the quantum theory and will also allow us to develop some theoretical methods that are of some general use.

Consider then the Hamiltonian for our solid

$$H = U_0 + K + U_H \tag{2127}$$

where U_0 is the ground-state energy, K the kinetic energy, and U_H the harmonic part of the interaction potential. The partition function in the canonical ensemble for our perfect harmonic crystal, with N sites, is given by

$$\begin{aligned} Z_N &= \int \prod_{\mathbf{R}} \left[\frac{d^3 p(\mathbf{R})}{(2\pi\hbar)^3} d^3 u(\mathbf{R})\right] e^{-\beta H} \\ &= e^{-\beta U_0} \int \prod_{\mathbf{R}} \left[\frac{d^3 p(\mathbf{R})}{(2\pi\hbar)^3} d^3 u(\mathbf{R})\right] e^{-\beta(K+U_H)}. \end{aligned} \tag{2128}$$

In the current situation where the N atoms have formed a lattice, there is no ambiguity in counting which atom is at which site. The atoms are fixed and labeled so that there is no need to divide by $N!$. Alternatively, there are exactly $N!$ ways of distributing the N atoms on the lattice sites, which cancels the factor of $N!$ in the denominator in the canonical ensemble.

Since our system corresponds to a large collection of harmonic oscillators, the temperature dependence can be extracted through an appeal to the equipartition theorem, or equivalently, by rescaling all the momenta $\mathbf{p}(\mathbf{R})$ and displacements $\mathbf{u}(\mathbf{R})$ by a factor of $\beta^{-1/2}$. After the rescaling we easily find that Eq. (2128) can be rewritten as

$$Z_N = e^{-\beta U_0} \beta^{-3N} \tilde{Z}_N \tag{2129}$$

where $\tilde{Z}_N$ is independent of temperature. The free energy is given by

$$\begin{aligned} F_N &= -\beta^{-1} \ln Z_N \\ &= U_0 + 3N\beta^{-1} \ln \beta - \beta^{-1} \ln \tilde{Z}_N. \end{aligned} \tag{2130}$$

The average energy then follows:

$$E = \frac{\partial}{\partial \beta}(\beta F_N) = U_0 + 3Nk_B T \tag{2131}$$

and the specific heat is the expected Dulong–Petit law [13]:

$$C_V = 3Nk_B. \tag{2132}$$

It is in treating the low-temperature behavior of the specific heats where the classical theory fails [14] most severely. We return to this point later when discussing the quantum theory.

It turns out that we can compute the partition function and free energy quite generally in the harmonic approximation. Let us write Eq. (2128) as

$$Z_N = Z_K Z_U e^{-\beta U_0} \tag{2133}$$

where the contribution of the kinetic energy is the same as for the ideal gas case:

$$Z_K = \left(\int_{-\infty}^{+\infty} \frac{dp}{2\pi\hbar} e^{-\beta p^2/2m} \right)^N = \lambda^{-3N}, \tag{2134}$$

and the potential energy contribution is given by

$$Z_U = \int \left[\prod_{\mathbf{R}} d^3 u(\mathbf{R}) \right] e^{-\beta U_H}. \tag{2135}$$

It is at this stage that we can understand why all the eigenvalues $\lambda_i(\mathbf{k})$ must be positive. If we change integration variables to the set $\tilde{u}_i(\mathbf{k})$ where U_H is diagonal [see Eq. (2097)], the integrals over $d\tilde{u}_i(\mathbf{k})$ will converge for large $|\tilde{u}_i(\mathbf{k})|$ only if $\lambda_i(\mathbf{k}) > 0$.

The determination of Z_U is an integration over a classical Gaussian probability distribution. The mathematics of this situation is treated in Appendix O. We find there that the potential contribution to the free energy is given by

$$F_U = -\beta^{-1} \ln Z_U$$
$$= \frac{3N}{2}\beta^{-1} \ln 2\pi\beta + \frac{k_B T}{2} \sum_{\mathbf{k}} \sum_i \ln \lambda_i(\mathbf{k}). \qquad (2136)$$

The total free energy in the classical harmonic approximation is then given by

$$F = U_0 + 3N\beta^{-1} \ln 2\pi\beta + \frac{k_B T}{2} \sum_{\mathbf{k}} \sum_i \ln \lambda_i(\mathbf{k}). \qquad (2137)$$

The temperature dependence is simple, as indicated above. The mechanical properties are governed at low temperatures by the ground-state energy. While the pressure vanishes at zero temperature, the bulk modulus (or inverse compressibility)

$$B = -V\left(\frac{\partial p}{\partial V}\right)_{T,N} = V\left(\frac{\partial^2 U_0}{\partial V^2}\right)_{T,N} \qquad (2138)$$

characterizes the ground state. We discuss this quantity in more detail below. As developed in Problem 6.4, we can show in the nearest-neighbor model of Section 6.4.3, where $U_0 = \frac{1}{2} \cdot 6NV(a)$, that

$$B = \frac{n_0}{3} V''(a) a^2. \qquad (2139)$$

6.4.6 Spatial Correlations: Classical Treatment

A property of considerable interest in a solid is the magnitude and correlation of lattice vibrations. Clearly, when the average displacements grow with increasing temperature to a magnitude on the order of a lattice spacing, we expect that the system will melt. It is of interest then to consider the average displacement correlation functions

$$G_{\alpha\beta}(\mathbf{R}, \mathbf{R}') = \langle u_\alpha(\mathbf{R}) u_\beta(\mathbf{R}') \rangle. \qquad (2140)$$

This quantity can be worked out without further approximation in the case of a harmonic Hamiltonian. The details of working out averages over a *Gaussian* probability distribution are given in Appendix O. We will refer to the results as needed here. The displacement correlation function can be expressed in terms of the Fourier transform,

$$G_{\alpha\beta}(\mathbf{R}, \mathbf{R}') = \frac{1}{N} \sum_{\mathbf{k}} G_{\alpha\beta}(\mathbf{k}) e^{+i\mathbf{k}\cdot(\mathbf{R}-\mathbf{R}')}, \qquad (2141)$$

which is worked out in Appendix O, using Eq. (2875), with the result

$$G_{\alpha\beta}(\mathbf{k}) = k_B T \sum_i \frac{\epsilon^i_\alpha(\mathbf{k})\epsilon^i_\beta(\mathbf{k})}{\lambda_i(\mathbf{k})} \qquad (2142)$$

where the $\lambda_i(\mathbf{k})$ are the eigenvalues of the dynamical matrix discussed earlier in this chapter. This expression is valid for all wavenumbers, but it is interesting to look at the long-wavelength limit where

$$\lambda_i(\mathbf{k}) \approx k^2 f_i(\hat{k}), \qquad (2143)$$

and

$$G_{\alpha\beta}(\mathbf{k}) = \frac{kT}{k^2} \sum_i \frac{\epsilon^i_\alpha(\mathbf{k})\epsilon^i_\alpha(\mathbf{k})}{f_i(\hat{k})} \qquad (2144)$$

and the Fourier components diverge as $k \to 0$. This is a very general phenomena in systems with a broken continuous symmetry. Such collective effects are known as Nambu–Goldstone modes [15]. These modes enter into the discussion of the average displacement at a given site, which is given by

$$\begin{aligned}\langle \mathbf{u}^2(\mathbf{R}) \rangle &= \sum_\alpha G_{\alpha\alpha}(\mathbf{R},\mathbf{R}) = \frac{1}{N}\sum_{\mathbf{k},\alpha} G_{\alpha\alpha}(\mathbf{k}) \\ &= \frac{k_B T}{N} \sum_{\mathbf{k},\alpha}\sum_i \frac{\epsilon^i_\alpha(\mathbf{k})\epsilon^i_\alpha(k)}{\lambda_i(\mathbf{k})} \\ &= \frac{k_B T}{N} \sum_{\mathbf{k},i} \frac{1}{\lambda_i(\mathbf{k})}.\end{aligned} \qquad (2145)$$

It is clear from this result that the amplitude of the lattice vibrations increases as $\sqrt{k_B T}$. This means that at high enough temperatures the displacement begins to approach a substantial fraction of the lattice spacing and one expects that the solid will melt. The Lindemann criterion [16] suggests that a system will melt when the amplitude of vibration is about 0.2 to 0.25 of the lattice spacing.

One of the main implications of this result for $\langle \mathbf{u}^2 \rangle$ comes from looking at the long-wavelength contribution to the displacement squared as a function of dimensionality. Since $\lambda_i(\mathbf{k}) \approx k^2$ for small wavenumbers, we have for general dimensionality [17] d that there is a contribution

$$\langle \mathbf{u}^2(\mathbf{R}) \rangle \approx \int_0^\Lambda k^{d-1} \frac{dk}{k^2} + \cdots \qquad (2146)$$

where Λ is a wavenumber cutoff below which $\lambda_i(\mathbf{k})$ can be replaced by its small-wavenumber limit, and $+\cdots$ indicates contributions from higher wavenumbers. Inspection of the integral in Eq. (2146) shows that it blows up at the lower limit for $d \leq 2$. The implication of this very general result is that conventional harmonic solids cannot exist in two or less spatial dimensions. If the solid is set up, the fluctuations are sufficiently strong to destroy it. Two dimensions is the marginal case where one can have a transition to a new phase without conventional long-range order. As developed by Halperin and Nelson [18] and Young [19], one has a Kosterlitz–Thouless [20] phase. This fascinating phenomenon is discussed in some detail in Volume 2.

Let us turn next to the more directly measurable quantities associated with the particle density and it fluctuations. Consider first the average density for a harmonic solid,

$$\bar{n}(\mathbf{x}) = \langle n(\mathbf{x}) \rangle, \tag{2147}$$

where, since we can identify a particle with lattice sites,

$$n(\mathbf{x}) = \sum_{\mathbf{R}} \delta(\mathbf{x} - \mathbf{r}(\mathbf{R})). \tag{2148}$$

Assuming Born–Von Kármán boundary conditions, we have the Fourier representation

$$f(\mathbf{x}) = \frac{1}{N} \sum_{\mathbf{k}} f(\mathbf{k}) e^{+i\mathbf{k}\cdot\mathbf{x}}, \tag{2149}$$

with the inverse transform

$$f(\mathbf{k}) = n_0 \int d^3 x \, e^{-i\mathbf{k}\cdot\mathbf{x}} f(\mathbf{x}) \tag{2150}$$

where $n_0 = N/V$. Substituting Eq. (2150) back into Eq. (2149) and using $f(\mathbf{x}) = \delta(\mathbf{x} - \mathbf{r}(\mathbf{R}))$ inside the integral, we obtain the representation

$$\delta(\mathbf{x} - \mathbf{r}(\mathbf{R})) = \frac{1}{V} \sum_{\mathbf{k}} e^{+i\mathbf{k}\cdot[\mathbf{x}-\mathbf{r}(\mathbf{R})]}. \tag{2151}$$

Inserting this result back into Eq. (2147) gives

$$\bar{n}(\mathbf{x}) = \frac{1}{V} \sum_{\mathbf{R}} \sum_{\mathbf{k}} e^{+i\mathbf{k}\cdot(\mathbf{x}-\mathbf{R})} \langle e^{-i\mathbf{k}\cdot\mathbf{u}(\mathbf{R})} \rangle. \tag{2152}$$

It is left as a problem (6.5) to show, using the techniques developed in Appendix O, that we have the result

$$\langle e^{-\mathbf{k}\cdot\mathbf{u}(\mathbf{R})} \rangle = \exp\left[-\tfrac{1}{2}\sum_{\beta,\mu} k_\beta k_\mu G_{\beta\mu}(\mathbf{R},\mathbf{R})\right]. \tag{2153}$$

The on-site lattice correlation function, $G_{\beta\mu}(\mathbf{R},\mathbf{R}) = G_{\beta\mu}(\mathbf{0})$, is given by Eq. (2145) and the average density by

$$\bar{n}(\mathbf{x}) = \frac{1}{V}\sum_{\mathbf{k}}\sum_{\mathbf{R}} e^{i\mathbf{k}\cdot(\mathbf{x}-\mathbf{R})}\exp\left[-\tfrac{1}{2}\sum_{\alpha\mu} k_\alpha k_\mu G_{\alpha\mu}(\mathbf{0})\right]. \tag{2154}$$

We have, however, that

$$\sum_{\mathbf{R}} e^{-i\mathbf{k}\cdot\mathbf{R}} = N\sum_{\mathbf{Q}} \delta_{\mathbf{k},\mathbf{Q}}, \tag{2155}$$

where the $\mathbf{Q}$'s are the reciprocal lattice vectors

$$\mathbf{Q} = \sum_{\alpha=1}^{s} m_\alpha \mathbf{b}_\alpha, \tag{2156}$$

where the $\mathbf{b}_\alpha$ are defined by Eq. (2084) and the m_α are integers. The average density can therefore be written as

$$\bar{n}(\mathbf{x}) = n_0 \sum_{\mathbf{Q}} e^{i\mathbf{Q}\cdot\mathbf{x}} \exp\left[-\tfrac{1}{2}\sum_{\alpha\mu} Q_\alpha Q_\mu G_{\alpha\mu}(\mathbf{0})\right]. \tag{2157}$$

At zero temperature in the classical theory the vibrations go to zero, $G_{\alpha\mu}(\mathbf{0}) = 0$, and the average density is given by

$$\bar{n}(\mathbf{x}) = \sum_{\mathbf{R}} \delta(\mathbf{x}-\mathbf{R}). \tag{2158}$$

We have δ-function peaks at the *rest* positions. At nonzero temperatures these peaks are broadened by the Debye–Waller factor [21] e^{-W}, where $W = \tfrac{1}{2}\sum_{\alpha,\mu} Q_\alpha Q_\mu G_{\alpha\mu}(\mathbf{0})$. We discuss this further at the end of the next section.

Consider next the density–density correlation function

$$S(\mathbf{x},\mathbf{x}') = \langle n(\mathbf{x})n(\mathbf{x}')\rangle \tag{2159}$$

and its Fourier transform, which is the structure factor:

$$S(\mathbf{k}) = \frac{1}{V}\int d^3x\, d^3x'\, e^{i\mathbf{k}\cdot(\mathbf{x}-\mathbf{x}')} S(\mathbf{x},\mathbf{x}') = \frac{1}{V}\sum_{\mathbf{R},\mathbf{R}'}\langle e^{i\mathbf{k}\cdot[\mathbf{r}(\mathbf{R})-\mathbf{r}(\mathbf{R}')]}\rangle$$

$$= \frac{1}{V}\sum_{\mathbf{R},\mathbf{R}'} e^{i\mathbf{k}\cdot(\mathbf{R}-\mathbf{R}')}\langle e^{i\mathbf{k}\cdot[\mathbf{u}(\mathbf{R})-\mathbf{u}(\mathbf{R}')]}\rangle. \qquad (2160)$$

The Gaussian average needed here is clearly related to that encountered in Eq. (2152). It is left to Problem 6.6 to show that the needed average is given by

$$\langle e^{i\mathbf{k}\cdot[\mathbf{u}(\mathbf{R})-\mathbf{u}(\mathbf{R}')]}\rangle = \exp\left\{-\sum_{\alpha,\mu} k_\alpha k_\mu [G_{\alpha\mu}(0) - G_{\alpha\mu}(\mathbf{R}-\mathbf{R}')]\right\} \qquad (2161)$$

and

$$S(\mathbf{k}) = \frac{1}{V}\sum_{\mathbf{R},\mathbf{R}'} e^{+i\mathbf{k}\cdot(\mathbf{R}-\mathbf{R}')}\exp\left\{-\sum_{\alpha\mu} k_\alpha k_\mu [G_{\alpha\mu}(0) - G_{\alpha\mu}(\mathbf{R}-\mathbf{R}')]\right\}. \qquad (2162)$$

Although it is tempting to stop here and analyze the consequences of this formal development, we delay this discussion until after development of the associated quantum theory.

6.5 QUANTUM MECHANICAL THEORY

6.5.1 Quantization

Proper treatment of the low-temperature properties of solids requires development of a quantum mechanical theory. The quantization of the classical theory involves replacing the momenta $\mathbf{p}(\mathbf{R})$ and displacements $\mathbf{u}(\mathbf{R})$ with operators that satisfy the usual canonical commutation relations:

$$[u_\alpha(\mathbf{R}), p_\mu(\mathbf{R}')] = i\hbar \delta_{\alpha\mu}\delta_{\mathbf{R},\mathbf{R}'} \qquad (2163)$$

$$[u_\alpha(\mathbf{R}), u_\mu(\mathbf{R}')] = [p_\alpha(\mathbf{R}), p_\mu(\mathbf{R}')] = 0. \qquad (2164)$$

At the harmonic level we have a quantum mechanical Hamiltonian given by Eqs. (2127), (2069), and (2078), where $\mathbf{u}$ and $\mathbf{p}$ are operators. If we introduce Fourier transforms as in Eq. (2089), we easily obtain

$$H = U_0 + \frac{1}{2N}\sum_{\mathbf{k}}\sum_{\alpha,\mu}\left[p_\alpha(-\mathbf{k})\frac{1}{m}p_\mu(\mathbf{k})\delta_{\alpha\mu} + u_\alpha(-\mathbf{k})D_{\alpha\mu}(\mathbf{k})u_\mu(\mathbf{k})\right]. \qquad (2165)$$

QUANTUM MECHANICAL THEORY

If we then change to the *normal-mode coordinates*,

$$\tilde{u}_i(\mathbf{k}) = \sum_\alpha \epsilon^i_\alpha(\mathbf{k}) u_\alpha(\mathbf{k}) \qquad (2166)$$

$$\tilde{p}_i(\mathbf{k}) = \sum_\alpha \epsilon^i_\alpha(\mathbf{k}) p_\alpha(\mathbf{k}), \qquad (2167)$$

this takes the form

$$H = U_0 + \frac{1}{2} N \sum_{\mathbf{k},i} \left[\tilde{p}_i(-\mathbf{k}) \frac{1}{m} \tilde{p}_i(\mathbf{k}) + \tilde{u}_i(-\mathbf{k}) \lambda_i(\mathbf{k}) \tilde{u}_i(\mathbf{k}) \right]. \qquad (2168)$$

The commutation relations satisfied by the transformed operators (see Problem 6.7) are

$$[\tilde{u}_i(\mathbf{k}), \tilde{p}_j(\mathbf{k}')] = i\hbar N \delta_{\mathbf{k},\mathbf{k}'} \delta_{ij} \qquad (2169)$$

and

$$[\tilde{u}_i(\mathbf{k}), \tilde{u}_j(\mathbf{k}')] = [\tilde{p}_i(\mathbf{k}), \tilde{p}_j(\mathbf{k}')] = 0. \qquad (2170)$$

Matters simplify further if we introduce the creation and annihilation operators,

$$a_i(\mathbf{k}) = [2\hbar m \omega_i(\mathbf{k}) N]^{-1/2} [m \omega_i(\mathbf{k}) \tilde{u}_i(\mathbf{k}) + i\tilde{p}_i(\mathbf{k})] \qquad (2171)$$

$$a_i^+(\mathbf{k}) = [2\hbar m \omega_i(\mathbf{k}) N]^{-1/2} [m \omega_i(\mathbf{k}) \tilde{u}_i(-\mathbf{k}) - i\tilde{p}_i(-\mathbf{k})] \qquad (2172)$$

and $\omega_i(\mathbf{k})$ is to be determined. Note immediately that a and a^+ satisfy the appropriate canonical commutation relations

$$[a_i(\mathbf{k}), a_j^+(\mathbf{k}')] = \delta_{\mathbf{k},\mathbf{k}'} \delta_{ij} \qquad (2173)$$

$$[a_i(\mathbf{k}), a_j(\mathbf{k}')] = [a_i^+(\mathbf{k}), a_j^+(\mathbf{k}')] = 0. \qquad (2174)$$

Clearly, then, inverting Eqs. (2171) and (2172) gives

$$\tilde{p}_i(\mathbf{k}) = i[m\hbar \omega_i(\mathbf{k}) N]^{1/2} [a_i^+(-\mathbf{k}) - a_i(\mathbf{k})] \qquad (2175)$$

and

$$\tilde{u}_i(\mathbf{k}) = \left[\frac{\hbar N}{2m \omega_i(\mathbf{k})} \right]^{1/2} [a_i(\mathbf{k}) + a_i^+(-\mathbf{k})]. \qquad (2176)$$

Substituting for $\tilde{p}_i$ and $\tilde{u}_i$ in the Hamiltonian, Eq. (2168), leads to the expression

$$H = U_0 + \frac{1}{2}\sum_{\mathbf{k},i}\left\{-\frac{m\hbar\omega_i(\mathbf{k})}{2m}[a_i^+(-\mathbf{k})a_i^+(\mathbf{k}) + a_i(\mathbf{k})a_i(-\mathbf{k}) - a_i^+(-\mathbf{k})a_i(-\mathbf{k})\right.$$

$$- a_i(\mathbf{k})a_i^+(\mathbf{k})] + \lambda_i(\mathbf{k})\frac{\hbar}{2m\omega_i(\mathbf{k})}[a_i^+(-\mathbf{k})a_i^+(\mathbf{k}) + a_i(\mathbf{k})a_i(-\mathbf{k})$$

$$\left. + a_i^+(-\mathbf{k})a_i(-\mathbf{k}) + a_i(\mathbf{k})a_i^+(\mathbf{k})]\right\}. \tag{2177}$$

The terms proportional to $a_i^+(-\mathbf{k})a_i^+(\mathbf{k}) + a_i(\mathbf{k})a_i(-\mathbf{k})$ vanishes if we choose

$$-m\frac{\hbar\omega_i(\mathbf{k})}{2m} + \frac{\lambda_i(\mathbf{k})\hbar}{2m\omega_i(\mathbf{k})} = 0 \tag{2178}$$

or

$$\omega_i^2(\mathbf{k}) = \frac{\lambda_i(\mathbf{k})}{m} \tag{2179}$$

as in the classical theory. After noting that $\omega_i(-\mathbf{k}) = \omega_i(-\mathbf{k})$, we see that the Hamiltonian then takes the form

$$H = U_0 + \frac{1}{2}\sum_{\mathbf{k},i}\hbar\omega_i(\mathbf{k})[a_i^+(\mathbf{k})a_i(\mathbf{k}) + a_i(\mathbf{k})a_i^+(\mathbf{k})]. \tag{2180}$$

Using the commutation relations for the creation and annihilation operators, this can be rewritten in the standard form

$$H = U_0 + U_{zp} + H_p \tag{2181}$$

where the *zero-point energy* is given by

$$U_{zp} = \sum_{\mathbf{k},i}\frac{\hbar}{2}\omega_i(\mathbf{k}), \tag{2182}$$

and the fluctuating contribution by

$$H_p = \sum_{\mathbf{k},i}\hbar\omega_i(\mathbf{k})a_i^+(\mathbf{k})a_i(\mathbf{k}). \tag{2183}$$

6.5.2 Phonons

This Hamiltonian H_p is precisely of the quadratic form discussed in Chapter 3. We also see immediately the similarity of H_p to the Hamiltonian describing electromagnetic radiation. In that case photons are created and destroyed with energy $\hbar\omega(k) = c\hbar k$, where c is the speed of light. The index i labeling $a_i^+(\mathbf{k})$ corresponded to the polarization in that case. In solids the excitations with energy $\hbar\omega_i(\mathbf{k})$ are called phonons [22]. They are quantized lattice vibrations. In direct analogy with the electromagnetic case, we can identify, in the long-wavelength limit, a *speed of sound*

$$c_i(\hat{k}) = \lim_{k \to 0} \frac{\omega_i(\mathbf{k})}{k} = \lim_{k \to 0} \left(\frac{\lambda_i(\mathbf{k})}{k^2 m} \right)^{1/2}. \tag{2184}$$

Since we showed earlier that $\lambda_i(\mathbf{k}) \approx \mathcal{O}(k^2)$ as $k \to 0$, the limit in Eq. (2184) is well behaved. $c_i(\hat{k})$ gives the speed with which a long-wavelength normal mode propagates through the lattice in a particular direction.

Given the quantum mechanical harmonic Hamiltonian given by equation (2183) describing the set of N atoms, we first want to work out the thermodynamic properties. The free energy in the canonical ensemble is given by

$$F_N = -\beta^{-1} \ln Z_N \tag{2185}$$

where the quantum generalization of Eq. (2128) is

$$Z_N = \text{Tr } e^{-\beta H} \tag{2186}$$

and the trace is over the phonon degrees of freedom. Since we can have an arbitrary number of phonons, we must use the grand canonical ensemble (with respect to the phonon degrees of freedom) with a zero chemical potential (as for photons). We then have, with $H_N^G = U_0 + U_{\text{zp}}$,

$$F_N = -\beta^{-1} \ln \text{Tr } e^{-\beta(H_N^G + H_p)}$$
$$= H_N^G + \Omega_p \tag{2187}$$

where

$$\Omega_p = -\beta^{-1} \ln \text{Tr } e^{-\beta H_p}. \tag{2188}$$

Since this problem is precisely of the type discussed in Chapter 3, we have, without further work, that

$$\Omega_p = \beta^{-1} \sum_{\mathbf{k},i} \ln(1 - e^{-\beta\hbar\omega_i(\mathbf{k})}). \tag{2189}$$

Since $\hbar\omega_i(\mathbf{k}) \geq 0$, we can write this in the convenient form

$$\Omega_p = \beta^{-1} \sum_{i,\mathbf{k}} \int_0^{+\infty} d\epsilon \, \delta(\epsilon - \hbar\omega_i(\mathbf{k})) \ln\left(1 - e^{-\beta\hbar\omega_i(\mathbf{k})}\right)$$

$$= \beta^{-1} V \int_0^{+\infty} d\epsilon \, D(\epsilon) \ln\left(1 - e^{-\beta\epsilon}\right) \qquad (2190)$$

where the phonon density of states is given by

$$D(\epsilon) = \frac{1}{V} \sum_{\mathbf{k},i} \delta(\epsilon - \hbar\omega_i(\mathbf{k})). \qquad (2191)$$

We obtain immediately, using $E = \partial(\beta F_N)/\partial\beta$, that the total energy of the system is

$$E = U_0 + U_{zp} + V \int_0^\infty d\epsilon \, D(\epsilon) \epsilon n(\epsilon) \qquad (2192)$$

where, as usual for massless bosons, the Planck distribution,

$$n(\epsilon) = (e^{\beta\epsilon} - 1)^{-1}, \qquad (2193)$$

enters the calculation. Note that the zero-point energy equation (2182) can also be expressed as an integral over the density of states,

$$U_{zp} = \frac{V}{2} \int_0^\infty d\epsilon \, D(\epsilon) \epsilon. \qquad (2194)$$

The introduction of the density of states neatly separates the mode structure of the lattice (which is a mechanics problem) from the thermal properties, which are characterized by the Planck distribution. As discussed above, the determination of $\lambda_i(\mathbf{k})$, $\omega_i(\mathbf{k})$, and the density of states is a complicated question in general and forms one of the basis for the field of lattice dynamics [23]. We see that we need methods for treating $\omega_i(\mathbf{k})$ and $D(\epsilon)$.

6.5.3 Spatial Structure

We next consider the quantum generalization of the classical results for the equilibrium spatial correlations in the system. As in the classical case [see Eq. (2152)], we consider the average density

$$\bar{n}(\mathbf{x}) = \frac{1}{V} \sum_{\mathbf{k}} \sum_{\mathbf{R}} e^{+i\mathbf{k}\cdot(\mathbf{x}-\mathbf{R})} \langle e^{-i\mathbf{k}\cdot\mathbf{u}(\mathbf{R})} \rangle \qquad (2195)$$

and the structure factor [see Eq. (2160)],

$$S(\mathbf{k}) = \frac{1}{V} \sum_{\mathbf{R},\mathbf{R}'} e^{+i\mathbf{k}\cdot(\mathbf{R}-\mathbf{R}')} \langle e^{+i\mathbf{k}\cdot(u(\mathbf{R})-u(\mathbf{R}'))} \rangle. \tag{2196}$$

These quantities can be explicitly evaluated in the quantum case if we use the very neat result

$$\langle e^A \rangle = \exp\left(\tfrac{1}{2}\langle A^2 \rangle\right) \tag{2197}$$

where the operator A is linear in $\mathbf{u}(\mathbf{R})$ and $\mathbf{p}(\mathbf{R})$. This result is worked out in Problem 6.9. Using Eq. (2197) to evaluate the averages in Eqs. (2195) and (2196), we obtain

$$\langle e^{+i\mathbf{k}\cdot\mathbf{u}(\mathbf{R})} \rangle = \exp\left[-\tfrac{1}{2}\sum_{\alpha,\mu} k_\alpha k_\mu G_{\alpha\mu}(0)\right] \tag{2198}$$

and

$$\langle e^{+i\mathbf{k}\cdot[\mathbf{u}(\mathbf{R})-\mathbf{u}(\mathbf{R}')]} \rangle = \exp\left[-\sum_{\alpha,\mu} k_\alpha k_\mu [G_{\alpha,\mu}(0) - G_{\alpha\mu}(\mathbf{R}-\mathbf{R}')]\right], \tag{2199}$$

where $G_{\alpha\mu}$ is the quantum mechanical correlation function

$$G_{\alpha\mu}(\mathbf{R}-\mathbf{R}') = \langle u_\alpha(\mathbf{R})u_\mu(\mathbf{R}')\rangle. \tag{2200}$$

We see then that the analysis is formally identical to the classical case, Eqs. (2153) and (2161), except that we must evaluate the quantum mechanical correlation functions $G_{\alpha\mu}(\mathbf{R}-\mathbf{R}')$. We have most of the tools necessary for this analysis. The first step in this calculation is to express the correlation function in terms of the normal modes:

$$G_{\alpha\mu}(\mathbf{R}-\mathbf{R}') = \frac{1}{N}\sum_{\mathbf{k}} e^{+i\mathbf{k}\cdot(\mathbf{R}-\mathbf{R}')} G_{\alpha\mu}(\mathbf{k}), \tag{2201}$$

where

$$G_{\alpha\mu}(\mathbf{k}) = \sum_{ij} \epsilon_\alpha^i(\mathbf{k})\epsilon_\mu^j(\mathbf{k})\tilde{G}_{ij}(\mathbf{k}), \tag{2202}$$

and

$$\tilde{G}_{ij}(\mathbf{k}) = \frac{1}{N}\langle \tilde{u}_i(\mathbf{k})\tilde{u}_j(-\mathbf{k})\rangle. \tag{2203}$$

Eliminating the $\tilde{u}_i(\mathbf{k})$ in terms of the creation and annihilation operators using Eq. (2176) we obtain

$$\tilde{G}_{ij}(\mathbf{k}) = \frac{\hbar}{2m\omega_i^{1/2}(\mathbf{k})\omega_j^{1/2}(\mathbf{k})} \langle [a_i(\mathbf{k}) + a_i^+(-\mathbf{k})][a_j(-\mathbf{k}) + a_j^+(\mathbf{k})]\rangle$$

$$= \frac{\hbar}{2m\omega_i^{1/2}(\mathbf{k})\omega_j^{1/2}(\mathbf{k})} \langle 2a_i^+(\mathbf{k})a_j(-\mathbf{k}) + \delta_{ij}\rangle$$

$$= \frac{\hbar \delta_{ij}}{2m\omega_i(\mathbf{k})}[1 + 2n_i(\mathbf{k})] \qquad (2204)$$

where we have used

$$\langle a_i(\mathbf{k})a_j(-\mathbf{k})\rangle = \langle a_i^+(-\mathbf{k})a_j^+(\mathbf{k})\rangle = 0 \qquad (2205)$$

and

$$n_i(\mathbf{k}) = \langle a_i^+(\mathbf{k})a_i(-\mathbf{k})\rangle = (e^{\beta\hbar\omega_i(\mathbf{k})} - 1)^{-1}. \qquad (2206)$$

Then, for example, the on-site oscillations are given by

$$\langle \mathbf{u}^2(\mathbf{R})\rangle = \frac{1}{N}\sum_{\mathbf{k}}\sum_i \frac{\hbar}{2m\omega_i(\mathbf{k})}[1 + 2n_i(\mathbf{k})]. \qquad (2207)$$

In the classical limit

$$n_i(\mathbf{k}) = \frac{1}{\beta\hbar\omega_i(\mathbf{k})} \qquad (2208)$$

and this fluctuation in the displacement field reduces to

$$\lim_{\hbar \to 0}\langle \mathbf{u}^2(\mathbf{R})\rangle = \frac{k_B T}{N}\sum_{\mathbf{k}}\sum_i \frac{1}{m\omega_i^2(\mathbf{k})}$$

$$= \frac{k_B T}{N}\sum_{\mathbf{k}}\sum_i \frac{1}{\lambda_i(\mathbf{k})}, \qquad (2209)$$

in agreement with Eq. (2145). In terms of the density of states, Eq. (2207) can be rewritten as

$$\langle \mathbf{u}^2(\mathbf{R})\rangle = n_0^{-1}\int_0^{+\infty} d\epsilon \frac{\hbar^2}{2m\epsilon}[1 + 2n(\epsilon)]D(\epsilon). \qquad (2210)$$

The Debye–Waller factor, which enters the structure factor and is given by

$$W \equiv \frac{1}{2}\sum_{\alpha\mu} k_\alpha k_\mu G_{\alpha\mu}(0)$$
$$= \frac{1}{2}\sum_{\alpha\mu} k_\alpha k_\mu \frac{1}{N}\sum_{\mathbf{k}',ij} \epsilon^i_\alpha(\mathbf{k}')\epsilon^j_\mu(\mathbf{k}')\tilde{G}_{ij}(\mathbf{k}'), \quad (2211)$$

cannot be reduced simply to an integral over the density of states.

The main differences between the classical and quantum theories are the zero-point energy and the related zero-point fluctuation in the displacement. At zero temperature one has the nonzero result

$$\langle \mathbf{u}^2(\mathbf{R}) \rangle = n_0^{-1} \int_0^\infty d\epsilon \, \frac{\hbar^2}{2m\epsilon} D(\epsilon). \quad (2212)$$

6.5.4 The Debye Theory

The difficulty in completing our calculations and obtaining quantitative results is due to the complexity, in general, of the eigenvalues $\lambda_i(\mathbf{k})$ and the density of states. It is useful and instructive to develop an approximate theory of the lattice vibrations, due to Peter Debye, which preserves the basic physics of the situation. The Debye theory [24] makes two basic approximations:

1. Replace $\omega_i(\mathbf{k})$ by its long-wavelength limit,

$$\omega_i(\mathbf{k}) = c_i(\hat{k})k, \quad (2213)$$

over the entire Brillouin zone.

2. The Brillouin zone is replaced by a spherical zone of radius k_D. Then, for some function $f(\mathbf{k})$, we have

$$\frac{1}{V}\sum_{\mathbf{k}} f(\mathbf{k}) = \int \frac{d^3k}{(2\pi)^3} \Theta(k_D - k) f(\mathbf{k}). \quad (2214)$$

The Debye density of states is given by using Eq. (2214) in Eq. (2191):

$$D_D(\epsilon) = \sum_i \int \frac{d^3k}{(2\pi)^3} \theta(k_D - k)\delta(\epsilon - \hbar c_i(\hat{k})k)$$
$$= \int \frac{d^3k}{(2\pi)^3} \theta(k_D - k)[\delta(\epsilon - \hbar c_L k) + 2\delta(\epsilon - \hbar c_T k)] \quad (2215)$$
$$= \frac{\epsilon^2}{2\pi^2\hbar^3}\left(\frac{1}{c_L^3} + \frac{2}{c_T^3}\right)\theta(\hbar c k_D - \epsilon), \quad (2216)$$

where c_L and c_T are the longitudinal and transverse speeds of sound. For small ϵ, written in terms of frequency $\nu = \epsilon/h$, this is known as the Rayleigh–Jeans density of states [25]. It is very convenient to define the average speed of sound, c, via

$$\frac{3}{c^3} = \frac{1}{c_L^3} + \frac{2}{c_T^3}. \tag{2217}$$

Notice that we have made the additional approximation in going from Eq. (2215) to Eq. (2216) that the upper cutoff is given by $\hbar c k_D$ is all directions.

The wavenumber cutoff k_D is determined by requiring that the exact sum rule

$$\int_0^\infty d\epsilon\, D(\epsilon) = \frac{1}{V}\sum_{\mu,\mathbf{k}} = \frac{3N}{V} = 3n_0 \tag{2218}$$

be satisfied by $D_D(\epsilon)$. This gives the constraint

$$\int_0^\infty d\epsilon\, \frac{3}{2\pi^2(\hbar c)^3} \epsilon^2 \Theta(\hbar c k_D - \epsilon) = \frac{k_D^3}{2\pi^2} = 3n_0. \tag{2219}$$

Introducing the Debye temperature Θ_D via,

$$k_B \Theta_D \equiv \hbar c k_D, \tag{2220}$$

the density of states Eq. (2216) can be rewritten as

$$D_D(\epsilon) = \frac{9n_0}{k_B \Theta_D} \left(\frac{\epsilon}{k_B \Theta_D}\right)^2 \theta(k_B \Theta_D - \epsilon). \tag{2221}$$

Clearly, the Debye energy, $k_B \Theta_D$, plays a role similar to the Fermi energy for degenerate Fermi systems. We can obtain Θ_D experimentally by fits to the low-temperature specific heat [see Eq. (2230) below] and these values are given in Table 6.5.

Table 6.5 Debye Temperature for a Selection of Solids

Element	Θ_D (K)	Element	Θ_D (K)
Ne	75	Rb	56
Ar	92	Cs	38
Kr	72	C	2230
Xe	64	Si	645
Li	344	Ge	374
Na	158	Sn	200
K	91	Pb	105

Source: Data from Ref. 26.

Given the Debye model, we can go on to compute many of the interesting physical quantities. Consider first the harmonic contribution to the average energy. Inserting Eq. (2221) in Eq. (2192) gives

$$E_H = 9N \int_0^{k_B \Theta_D} \frac{d\epsilon}{k_B \Theta_D} \left(\frac{\epsilon}{k_B \Theta_D}\right)^2 (e^{\beta\epsilon} - 1)^{-1}. \tag{2222}$$

Changing variables to $x = \beta\epsilon$ and defining $t = T/\Theta_D$, we obtain

$$E_H = Nk_B \Theta_D t^4 \int_0^{1/t} \frac{dx\, x^3}{e^x - 1}. \tag{2223}$$

We plot $E_H/Nk_B\Theta_D \equiv f_H(t)$ in Fig. 6.8. The limits $t \gg 1$ and $\ll 1$ can be worked out analytically: For high temperatures $t \gg 1$,

$$f_H(t) = \frac{t}{3} + \mathcal{O}(1). \tag{2224}$$

For low temperatures, $t \ll 1$,

$$f(t) = \frac{\pi^4}{15} t^4 + \mathcal{O}(t^5). \tag{2225}$$

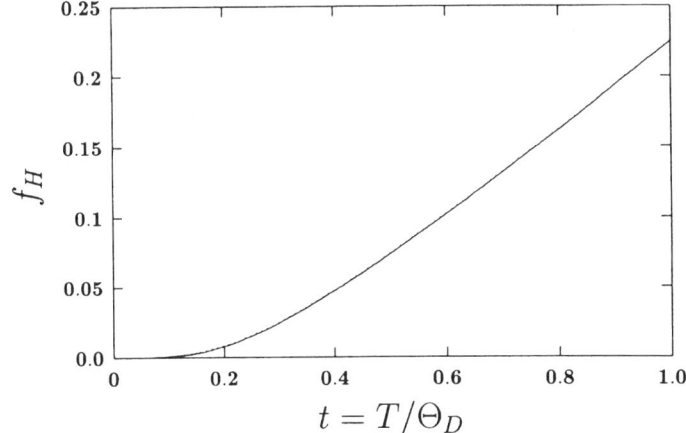

FIGURE 6.8 Scaled harmonic contribution to the average energy for a Debye solid versus reduced temperature.

The zero-point energy is given by

$$U_{zp} = V \int_0^\infty d\epsilon \frac{\epsilon}{2} D(\epsilon)$$
$$= \frac{9N}{2} \int_0^\infty d\epsilon \left(\frac{\epsilon}{k_B \Theta_D}\right)^3 \theta(k_B \Theta_D - \epsilon)$$
$$= \frac{9N}{2} (k_B \Theta_D) \int_0^1 dx x^3 = \frac{9}{8} N(k_B \Theta_D). \quad (2226)$$

Then, in the harmonic and Debye approximations, the energy per particle is given by

$$\frac{E}{N} = u_o + \frac{9}{8} k_B \Theta_D + 9 k_B \Theta_D f(t) \quad (2227)$$

where u_0 is given by Eq. (2030). The specific heat is given by

$$C_V = \left(\frac{\partial E}{\partial T}\right)_{N,V} = 9 N k_B \frac{df(t)}{dt} \quad (2228)$$

and $C_V/3Nk_B$ is plotted as a function of t in Fig. 6.9. We see, using Eq. (2224), that in the high-temperature limit, $f'(t) = 1/3$ and we regain, as expected, the classical

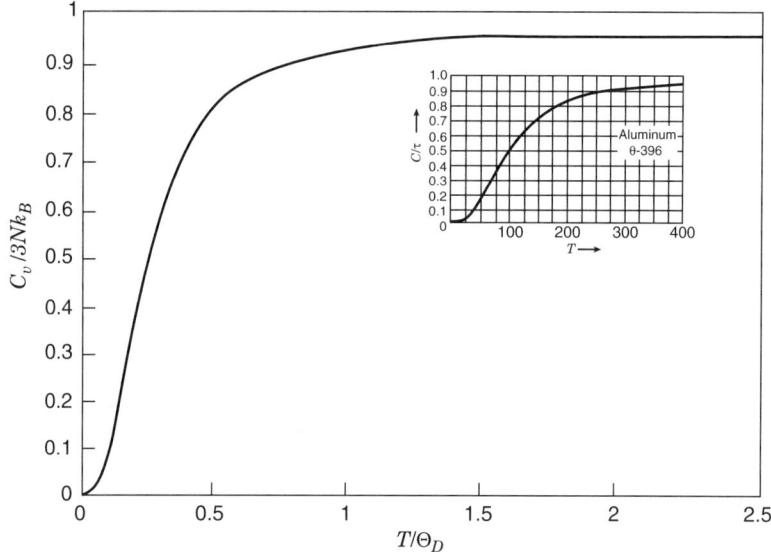

FIGURE 6.9 Debye theory result for the specific heat versus reduced temperature. Inset is comparison of this result with experimental results for aluminum taken from the original paper by Debye. The theory is sufficiently good that the experimental points are obscured by the theoretical curve.

result
$$C_V = 3Nk_B, \quad T \gg \Theta_D. \tag{2229}$$

In the low-temperature regime, using Eq. (2225), we have $f'(t) = 4\pi^4 t^3/15$, and

$$C_V = \tfrac{12}{5}\pi^4 Nk_B t^3, \quad T \ll \Theta_D. \tag{2230}$$

Thus the specific heat shows a T^3 behavior at low temperatures. Empirically, we can determine Θ_D by setting the coefficient of the NT^3 term in C_V equal to $12\pi^4/5\Theta_D^3$. This is the procedure used to obtain the results in Table 6.5.

We see from Eq. (2226) that the zero-point motion gives a quantum correction to the ground-state energy. In Table 6.6 we list the quantitative estimates for u_0, given in Table 6.4 and the ratio $|u_{zp}/u_0|$, where $u_{zp} = U_{zp}/N$ is given in the Debye theory by Eq. (2226). We see that the quantum correction leads to a significant improvement in the estimates for the cohesive energies when compared to experiment.

Let us turn next to the on-site vibrations as given in the Debye approximation. Inserting Eq. (2221) into Eq. (2210) and changing the integration variables to $x = \beta\epsilon$, we obtain

$$\langle \mathbf{u}^2(\mathbf{R}) \rangle = \frac{9\hbar^2}{4mk_B\Theta_D}\left(1 + 4t^2 \int_0^{1/t} \frac{x\,dx}{e^x - 1}\right). \tag{2231}$$

The length squared $9\hbar^2/4mk_B\Theta_D = R_{zp}^2$ is listed in Table 6.7 for the noble gases. The temperature dependence $\langle u^2(\mathbf{R}) \rangle / R_{zp}^2$ is shown in Fig. 6.10. In the classical limit, $t \gg 1$,

$$\langle \mathbf{u}^2(\mathbf{R}) \rangle = \frac{9k_B T}{mc^2 k_D^2} \quad \text{for} \quad T \gg \Theta_D. \tag{2232}$$

Note that this agrees with Eq. (2145) when it is evaluated using the Debye approximation. For low temperatures we need the integral, evaluated in Appendix L,

$$\int_0^{+\infty} \frac{dx\,x}{e^x - 1} = \zeta(2) = \frac{\pi^2}{6} \tag{2233}$$

TABLE 6.6 Zero-Point Energy Corrections to Ground-State Energies[a]

Element	u_0 (eV)	$|u_{zp}/u_0|$	$u_0 + u_{zp}$ (eV)	u_{expt} (eV)
Ne	−0.027	0.0269	−0.020	−0.020
Ar	−0.089	0.100	−0.080	−0.080
Kr	−0.120	0.0582	−0.113	−0.116
Ar	−0.089	0.0360	−0.166	−0.16

[a] The classical energy u_0 is taken from Table 6.4, u_z is given by Eq. (2226), the Debye temperatures in Table 6.5. The sum compares reasonably well with the experimental values taken from Ref. 27.

TABLE 6.7 Zero-Point Motion for Noble Gases

Element	Θ_D (K)	Atomic Number	R_{zp} (Å)
Ne	75	20.18	0.267
Ar	92	39.95	0.172
Kr	72	83.80	0.134
Ar	64	131.3	0.113

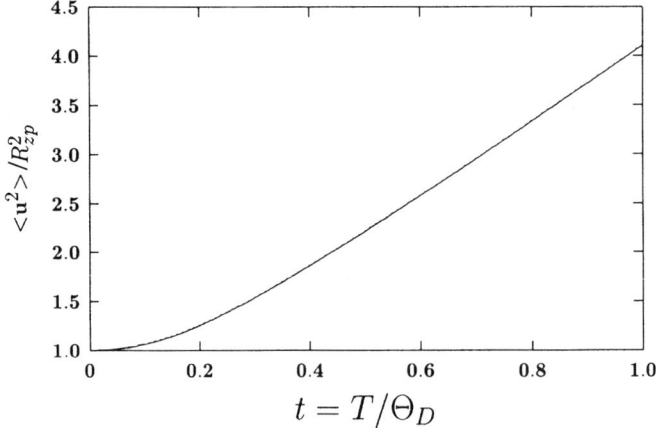

FIGURE 6.10 Displacement field fluctuations versus scaled temperature within the Debye approximation.

which leads to the result

$$\langle \mathbf{u}^2(\mathbf{R}) \rangle = R_{zp}^2 \left(1 + \frac{2\pi^2 t^2}{3} + \cdots\right) \quad (2234)$$

As expected, the quantum mechanical zero-point motion is the largest contribution at low temperatures.

Let us turn now to the correlation function displacement $G_{\alpha\mu}(\mathbf{R})$ for general $\mathbf{R}$. Inserting Eq. (2204) in Eq. (2201) and using Eqs. (2219) and (2214) gives

$$G_{\alpha\mu}(\mathbf{R}) = \frac{1}{n_D} \int \frac{d^3 k}{(2\pi)^3} \Theta(k_D - k) \frac{\hbar \delta_{\alpha\mu}}{2mck} \left(1 + \frac{2}{e^{\beta\hbar ck} - 1}\right) e^{+i\mathbf{k}\cdot\mathbf{R}}. \quad (2235)$$

The angular integration is easily carried out, and after introducing the dimensionless integration variable $x = \beta\hbar ck$, we obtain, after some algebra and an elementary integration,

$$G_{\alpha\mu}(\mathbf{R}) = \tfrac{2}{3} R_{zp}^2 F(k_D R, t) \delta_{\alpha\mu} \quad (2236)$$

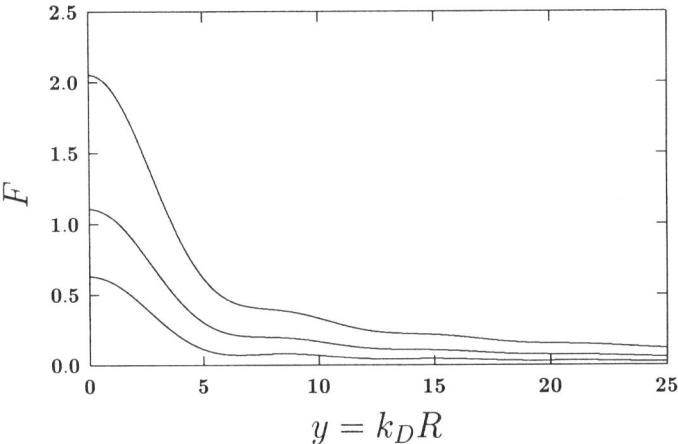

FIGURE 6.11 Scaled displacement field correlation function versus scaled distance between sites in Debye approximation. Curves from bottom to top correspond to temperatures $T/\Theta_D = 0.2, 0.6,$ and 1.0.

where, as above, $t = T/\Theta_D$ and

$$F(y,t) = \frac{1}{y}\left[\frac{1}{y}(1-\cos y) + 2t\int_0^{1/t}\frac{dx\sin ytx}{e^x - 1}\right]. \tag{2237}$$

We plot $F(y,t)$ as a function of $y = k_D R$ for various reduced temperatures t in Fig. 6.11. Several limiting cases can be worked out analytically. In the low-temperature limit with $t \ll 1$, we obtain

$$F(y,t) = \frac{1}{y}\left\{\frac{1}{y}(1-\cos y) + t\pi\left[\coth yt\pi - \frac{1}{yt\pi}\right]\right\}, \tag{2238}$$

while in the high-temperature limit

$$F(y,t) = \frac{2t}{y}\int_0^y \frac{dz}{z}\sin z. \tag{2239}$$

In the high-temperature limit we return to the classical result

$$G_{\alpha\mu}(R) = 3\delta_{\alpha\mu}\frac{k_B T}{mc^2 k_D^2}\frac{1}{k_D R}\int_0^{k_D R} dz\frac{\sin z}{z} \tag{2240}$$

where all of the factors of $\hbar$ have canceled. We see that correlations in the solid decay vary slowly with distance (R^{-1}) for large separations.

The large R limit for arbitrary t can be obtained by setting $z = xyt$ in the integral in Eq. (2237). We can then easily take the large y limit to obtain

$$\lim_{y \to \infty} yF(y,t) = 2t \int_0^{+\infty} dz \frac{\sin z}{z} = \frac{\pi}{2} 2t \qquad (2241)$$

and

$$G_{\alpha\mu}(\mathbf{R}) = \frac{2}{3}\delta_{\alpha\mu}R_{zp}^2 \frac{\pi t}{y} = \frac{3\pi}{2}\delta_{\alpha\mu}\frac{k_B T}{mc^2 k_D^2}\frac{1}{k_D R} \qquad k_D R \gg 1. \qquad (2242)$$

The key point here is that the displacements remain correlated over large distances, the $G_{\alpha\mu}(\mathbf{R})$ falling off only as R^{-1} for large distances. This simply reflects the long-wavelength behavior of $G_{\alpha\mu}(\mathbf{k})$, which goes as k^{-2} as $k \to 0$. Notice the strong analogy here between the lattice dynamics and phonons and electrostatics and photons. In electrostatics it is photons with energies $\approx \hbar ck$ that lead to the Coulomb potential $\approx 1/r$ between charged particles. Similarly, phonons with energies $\approx \hbar ck$ lead to a correlation between atoms on a lattice that falls off as $\approx 1/r$.

The average density follows from Eqs. (2195), (2198), and (2211):

$$\bar{n}(\mathbf{x}) = n_0 \sum_{\mathbf{Q}} e^{i\mathbf{Q}\cdot\mathbf{x}} e^{-W(\mathbf{Q})}. \qquad (2243)$$

In the Debye approximation $G_{\alpha\mu}(0)$ is diagonal in α and μ and

$$W(\mathbf{Q}) = \tfrac{1}{2}Q^2 \sum_{\alpha} G_{\alpha\alpha}(0). \qquad (2244)$$

Then $\bar{n}(\mathbf{x})$ clearly has the periodicity of the lattice. We have plotted $\bar{n}(\mathbf{x})$ in Fig. 6.12 along the x-axis for several temperatures with the on-site fluctuation given by Eq. (2231). Notice that the zero-point motion leads to broadening even at zero temperature.

The structure factor is given by Eqs. (2196), (2199), and (2211), in the form

$$S(\mathbf{k}) = e^{-2W} n_0 \sum_{\mathbf{R}} e^{+i\mathbf{k}\cdot\mathbf{R}} \exp\left[\sum_{\alpha\mu} k_\alpha k_\mu G_{\alpha\mu}(\mathbf{R})\right]. \qquad (2245)$$

This quantity is typically of interest because one is looking at the wavenumber dependence across the Brillouin zone. For these purposes one should not trust the Debye approximation. A realistic treatment depends on the details of the lattice structure and the normal modes. In practice, a more useful experimental probe is *inelastic scattering* [28], where typically, one uses neutron scattering to measure the *dynamic structure factor* $S(\mathbf{k}, \omega)$, where ω maps out the frequencies of collective modes, such as $\omega_i(\mathbf{k})$, in a system. $S(\mathbf{k})$ is simply the integral over all frequencies.

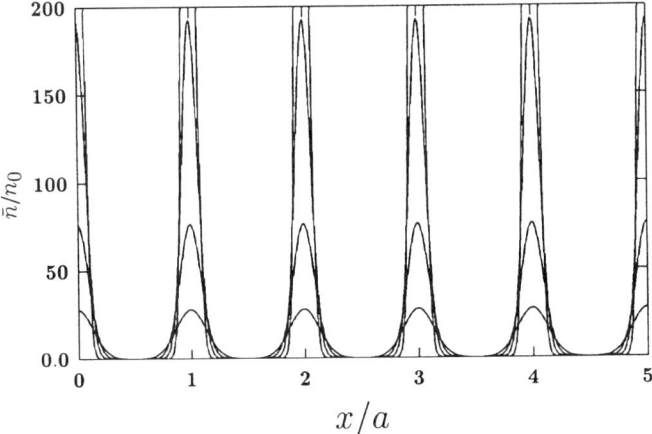

FIGURE 6.12 Average density, normalized to the uniform density, in a solid along the *x*-axis. The different curves, in the Debye approximation, correspond to different temperatures. As one goes from the broadest to the narrowest curves the temperatures are given by $T/\Theta_D = 2.0$, 1.0, 0.5, and 0.1.

6.6 ELASTIC THEORY OF SOLIDS

6.6.1 Single-Component Systems

Earlier in this chapter we discussed the long-wavelength limit for the vibrational part of the potential energy for a simple single-component system. This led naturally to the introduction of the elastic constants. In this section we continue with this simple model to show how one is led to the elastic theory of solids, which allows one to treat the macroscopic deformation of solids. Later in this section we indicate how these ideas, developed at the microscopic level in terms of a single- component system with a radial pair potential, can be extended to more general systems.

In our development of a theory of lattice vibrations, we assumed implicitly that the solid was unstressed and that there was no net displacement from equilibrium:

$$\langle \mathbf{u}(\mathbf{R}) \rangle = 0 \tag{2246}$$

and no net strain,

$$\langle U_{\alpha\beta} \rangle = 0. \tag{2247}$$

Elastic theory is concerned with the distortions of a solid generated by applied stress. Let us return to our microscopic analysis in Section 6.4.1 of a single-component system interacting via a radial pair potential. Let us look again at the Hamiltonian in the harmonic approximation given by Eq. (2078). We assume that the displacement at site $\mathbf{R}$ can be written in the form

$$u_\alpha(\mathbf{R}) = \langle u_\alpha(\mathbf{R}) \rangle + \psi_\alpha(\mathbf{R}) \tag{2248}$$

where, by definition, $\langle \psi_\alpha(\mathbf{R}) \rangle = 0$, and we can write [29]

$$\langle u_\alpha(\mathbf{R}) \rangle = \sum_\beta R_\beta U_{\alpha\beta} \tag{2249}$$

where $U_{\alpha\beta}$ is the associated net strain tensor. Inserting Eq. (2248), with $\langle u_\alpha(\mathbf{R}) \rangle$ given by Eq. (2249), into the Hamiltonian given by Eq. (2078) leads to

$$H = U_0 + K + \frac{1}{2} \sum_{\mathbf{R} \neq \mathbf{R}'} \sum_{\alpha\beta} \left[\sum_\mu R_\mu U_{\alpha\mu} + \psi_\alpha(\mathbf{R}) \right]$$
$$\times \left[\sum_\nu R'_\nu U_{\beta\nu} + \psi_\beta(\mathbf{R}') \right] D_{\alpha\beta}(\mathbf{R} - \mathbf{R}') \tag{2250}$$

where U_0 is the ground-state energy and K is the kinetic energy. Let us look at the various terms involving the dynamical matrix D. First we have the nonfluctuating contribution

$$F[U] \equiv \frac{1}{2} \sum_{\mathbf{R} \neq \mathbf{R}'} \sum_{\alpha\beta\mu\nu} R_\mu U_{\alpha\mu} R'_\nu U_{\beta\nu} D_{\alpha\beta}(\mathbf{R} - \mathbf{R}')$$
$$= \frac{V}{2} \sum_{\alpha\beta\mu\nu} U_{\alpha\mu} U_{\beta\nu} C_{\alpha\mu\beta\nu} \tag{2251}$$

and the elastic constants are given in this development by

$$C_{\alpha\mu\beta\nu} = \frac{1}{V} \sum_{\mathbf{R} \neq \mathbf{R}'} R_\mu R'_\nu D_{\alpha\beta}(\mathbf{R} - \mathbf{R}'). \tag{2252}$$

Since we know from Eq. (2113) that $\sum_\mathbf{R} D_{\alpha\beta}(\mathbf{R} - \mathbf{R}') = 0$, we can rewrite Eq. (2252) as

$$C_{\alpha\mu\beta\nu} = \frac{1}{V} \sum_{\mathbf{R} \neq \mathbf{R}'} (R_\mu - R'_\mu) R'_\nu D_{\alpha\beta}(\mathbf{R} - \mathbf{R}')$$
$$= \frac{1}{V} \sum_{\mathbf{R} \neq \mathbf{R}'} (R'_\mu - R_\mu) R_\nu D_{\alpha\beta}(\mathbf{R}' - \mathbf{R})$$

$$= \frac{1}{V} \sum_{\mathbf{R} \neq \mathbf{R}'} (R'_\mu - R_\mu) R_\nu D_{\alpha\beta}(\mathbf{R} - \mathbf{R}')$$

$$= -\frac{1}{2V} \sum_{\mathbf{R} \neq \mathbf{R}'} (R_\mu - R'_\mu)(R_\nu - R'_\nu) D_{\alpha\beta}(\mathbf{R} - \mathbf{R}')$$

$$= -\frac{N}{2V} \sum_{\mathbf{R}} R_\mu R_\nu D_{\alpha\beta}(\mathbf{R}). \tag{2253}$$

After inserting Eq. (2079) in this result, we see that it reduces to the result given by Eq. (2118) derived earlier. The cross terms in Eq. (2250) vanish because

$$\sum_{\mathbf{R} \neq \mathbf{R}'} \sum_{\mu\beta} R_\mu D_{\alpha\beta}(\mathbf{R} - \mathbf{R}') \psi_\beta(\mathbf{R}') = \sum_{\mathbf{R} \neq \mathbf{R}'} \sum_{\mu\beta} (R_\mu - R'_\mu) D_{\alpha\beta}(\mathbf{R} - \mathbf{R}') \psi_\beta(\mathbf{R}')$$
$$= \sum_{\mathbf{R}} \sum_{\mu\beta} R_\mu D_{\alpha\beta}(\mathbf{R}) \sum_{\mathbf{R}'} \psi_\beta(\mathbf{R}') = 0, \tag{2254}$$

where in the last step we used $D_{\alpha\beta}(-\mathbf{R}) = D_{\alpha\beta}(\mathbf{R})$. The Hamiltonian at the harmonic level then takes the form

$$H = U_0 + K + F[U] + H_p[\psi], \tag{2255}$$

where H_p is the phonon contribution to the Hamiltonian given by Eq. (2078), with $\mathbf{u}$ replaced by the fluctuating quantity $\psi = \mathbf{u} - \langle \mathbf{u} \rangle$. At low temperatures, where the fluctuations due to the kinetic energy and phonons can be neglected, the average energy of the system is given by

$$E = U_0 + F[U]. \tag{2256}$$

From our thermodynamic treatment of solids given earlier we know that the average stress tensor is given by

$$\bar{\sigma}_{\alpha\beta} = -\left(\frac{\partial E}{\partial (V U_{\alpha\beta})}\right)_{S,N}. \tag{2257}$$

After inserting our result for the energy given by Eqs. (2256) and (2251) into this equation, we obtain Hooke's law [30]:

$$\bar{\sigma}_{\alpha\beta} = -\sum_{\mu\nu} C_{\alpha\beta\mu\nu} U_{\mu\nu}. \tag{2258}$$

In Chapter 1 we discussed the role of the stress tensor in treating the mechanical properties of a many-particle system. In particular, we arrived, in Eq. (353), at a general expression for the stress tensor $\sigma_{\alpha\beta}$ for a single-component system interacting via a pairwise additive potential. It is instructive to see how Hooke's law follows from working directly with the average of Eq. (353).

The kinetic energy contribution to the stress tensor is given by its ideal gas form:

$$\bar{\sigma}^K_{\alpha\beta} = \frac{1}{V}\int d^3x\, \sigma^K_{\alpha\beta}(\mathbf{x})$$

$$= \frac{1}{V}\int d^3x \left\langle \sum_\mathbf{R} \frac{p_\alpha(\mathbf{R})p_\beta(\mathbf{R})}{m} \delta(\mathbf{x}-\mathbf{R}) \right\rangle = \delta_{\alpha\beta} k_B T n_0. \qquad (2259)$$

At low temperatures this contribution is negligible and we focus here on the potential contribution to the stress tensor. From our work in Chapter 1 and Appendix F, we know that the potential energy contribution to the stress tensor for a classical system with two-body interactions is given by

$$\sigma_{\alpha\beta}(\mathbf{x}) = -\frac{1}{2}\int_0^1 ds \int d^3y\, \frac{y_\alpha y_\beta}{y} V'(y)$$

$$\times \sum_{i\neq j} \delta\left(\mathbf{x}-\frac{\mathbf{y}}{2}(s+1)-\mathbf{r}^i\right)\delta\left(\mathbf{x}-\frac{\mathbf{y}}{2}(s-1)-\mathbf{r}^j\right). \qquad (2260)$$

Let us focus on the uniform part given by the spatial average

$$\bar{\sigma}_{\alpha\beta} = \frac{1}{V}\int d^3x\, \sigma_{\alpha\beta}(\mathbf{x}) = -\frac{1}{2}\frac{1}{V}\int_0^1 ds \int d^3y\, \frac{y_\alpha y_\beta}{y} V'(y)$$

$$\times \sum_{i\neq j} \delta\left[\frac{\mathbf{y}}{2}(s+1)+\mathbf{r}^i-\frac{\mathbf{y}}{2}(s-1)-\mathbf{r}^j\right]$$

$$= -\frac{1}{2V}\int d^3y\, \frac{y_\alpha y_\beta}{y} V'(y) \sum_{i\neq j} \delta(\mathbf{y}-\mathbf{r}^{ij})$$

$$= -\frac{1}{2V}\sum_{i\neq j} \frac{r^{ij}_\alpha r^{ij}_\beta}{r^{ij}} V'(r^{ij}) \qquad (2261)$$

where $\mathbf{r}^{ij} = \mathbf{r}^i - \mathbf{r}^j$. In a solid we can again write the particle coordinates in the form

$$\mathbf{r}^i \longrightarrow \mathbf{R} + \mathbf{u}(\mathbf{R}) \qquad (2262)$$

$$\mathbf{r}^j \longrightarrow \mathbf{R}' + \mathbf{u}(\mathbf{R}'), \qquad (2263)$$

ELASTIC THEORY OF SOLIDS 499

and expand in the *small* displacement $\mathbf{u}(\mathbf{R})$. The expanded stress tensor can then be written as

$$\bar{\sigma}_{ij} = -\frac{1}{2V} \sum_{\mathbf{R} \neq \mathbf{R}'} \left[\frac{r_\alpha r_\beta}{r} V'(r) \right]_{\mathbf{r}=\mathbf{R}-\mathbf{R}'}$$
$$- \frac{1}{2V} \sum_{\mathbf{R} \neq \mathbf{R}', \gamma} \left[\left(\frac{\partial}{\partial r_\gamma} \frac{r_\alpha r_\beta}{r} V'(r) \right) \right]_{\mathbf{r}=\mathbf{R}-\mathbf{R}'} [u_\gamma(\mathbf{R}') - u_\gamma(\mathbf{R})] + \mathcal{O}(u^2). \quad (2264)$$

Consider the leading term of $\mathcal{O}(1)$, which, using the translational invariance of the lattice sites, can be put in the form

$$\bar{\sigma}_{\alpha\beta}^{(0)} = -\frac{N}{2V} \sum_{\mathbf{R}} \frac{R_\alpha R_\beta}{R} V'(R). \quad (2265)$$

If we remember Eq. (2039) establishing the condition determining the lattice constant, we see, as expected, that the ground state is stressless in the absence of applied forces:

$$\bar{\sigma}_{\alpha\beta}^{(0)} = 0. \quad (2266)$$

We have, after letting $\mathbf{R}' = \mathbf{R} + \mathbf{r}$ in the next-order contribution to $\bar{\sigma}_{\alpha\beta}$, that

$$\bar{\sigma}_{\alpha\beta} = -\frac{1}{2V} \sum_{\mathbf{r}, \gamma} \left[\frac{\partial}{\partial r_\gamma} \left(r_\alpha r_\beta \frac{V'(r)}{r} \right) \right] \sum_{\mathbf{R}} [u_\gamma(\mathbf{R} + \mathbf{r}) - u_\gamma(\mathbf{R})]. \quad (2267)$$

The point here is to realize that at zero temperature, if there is no net strain, the system will have zero stress. If there is nonzero strain in the system, we can use Eq. (2248) in Eq. (2267). One must be careful in doing this sum. Depending on the displacement, the factor in Eq. (2267) can be written using Eq. (2249), as

$$\sum_{\mathbf{R}} [u_\gamma(\mathbf{R} + \mathbf{r}) - u_\gamma(\mathbf{R})] = \sum_{\mathbf{R}} \left[\sum_\mu r_\mu U_{\gamma\mu} + \psi_\gamma(\mathbf{R} + \mathbf{r}) - \psi_\gamma(\mathbf{R}) \right]. \quad (2268)$$

If the sum $\sum_{\mathbf{R}} \psi_\gamma(\mathbf{R})$ is well behaved, then

$$\sum_{\mathbf{R}} [\psi_\gamma(\mathbf{R} + \mathbf{r}) - \psi_\gamma(\mathbf{R})] = 0, \quad (2269)$$

and from a different starting point, we regain Hooke's law:

$$\bar{\sigma}_{\alpha\beta} = -\frac{N}{2V} \sum_{\mathbf{R}, \gamma\mu} \left[\frac{\partial}{\partial r_\gamma} \left(r_\alpha r_\beta \frac{V'(r)}{r} \right) \right]_{\mathbf{r}=\mathbf{R}} R_\mu U_{\gamma\mu}$$
$$= -\sum_{\gamma\mu} C_{\alpha\beta\gamma\mu} U_{\gamma\mu}, \quad (2270)$$

where the elastic tensor $C_{\alpha\beta\gamma\mu}$ is given in this derivation for a system with a two-body central potential, by

$$C_{\alpha\beta\gamma\mu} = \frac{n_0}{2} \sum_R R_\mu \left[\frac{\partial}{\partial r_\gamma}\left(r_\alpha r_\beta \frac{V'(r)}{r}\right)\right]_{r=R}. \tag{2271}$$

We obtain compatibility with our earlier results for the elastic tensor if we recognize that the two derivative terms giving Kronecker delta functions vanish due to conditions resulting from minimizing the energy:

$$\sum_R R_\mu \delta_{\alpha\gamma} R_\beta \frac{V'(R)}{R} = 0. \tag{2272}$$

Equation (2271) reduces to the symmetric form

$$C_{\alpha\beta\gamma\mu} = \frac{n_0}{2} \sum_R R_\mu R_\alpha \hat{R}_\beta \hat{R}_\gamma A(R) \tag{2273}$$

where, as in Eq. (2104),

$$A(r) = r\frac{\partial}{\partial r}\left(\frac{V'(r)}{r}\right). \tag{2274}$$

We see then that there is compatibility between the two ways of obtaining Hooke's law in our microscopic descriptions.

6.6.2 General Treatment of Elastic Theory

Our development of elastic theory in the preceding section was microscopic and direct, but limited to a simple system at low temperatures. Here we want to develop the theory more generally. The analysis in the preceding section was carried out in the microcanonical ensemble, where the strain tensor $VU_{\alpha\beta}$ was treated as one of the microvariables and the stress tensor derived, as usual, as the conjugate force as given by Eq. (2257). It is more physical to make a set of Legendre transformations which takes one over to the canonical ensemble, where temperature and *applied forces*, the stress tensor, are controlled. In this case the free energy is a function of temperature and the applied stress and is given by

$$F[T,\sigma] = -\beta^{-1} \ln Z \tag{2275}$$

where the partition function is defined by

$$Z = \text{Tr}_{\text{cl}}^N \, e^{-\beta H_\sigma} \tag{2276}$$

where

$$H_\sigma = H[\mathbf{u}] + \sum_{\alpha\beta} \int d^3x\, \sigma_{\alpha\beta}(\mathbf{x}) U_{\alpha\beta}(\mathbf{x}). \tag{2277}$$

In this expression $H[\mathbf{u}]$ is the original Hamiltonian expressed in terms of the displacements $\mathbf{u}(\mathbf{R})$, and the *strain-tensor field*, $U_{\alpha\beta}(\mathbf{x})$, is defined by

$$U_{\alpha\beta}(\mathbf{x}) = \tfrac{1}{2}\left[\nabla_\alpha u_\beta(\mathbf{x}) + \nabla_\beta u_\alpha(\mathbf{x})\right], \tag{2278}$$

where the displacement field is given by

$$u_\alpha(\mathbf{x}) = v_0 \sum_{\mathbf{R}} u_\alpha(\mathbf{R}) \delta(\mathbf{x}-\mathbf{R}) \tag{2279}$$

where the volume per unit cell is just the inverse particle density, $v_0 = 1/n_0$.
After computing $F[\sigma]$, we find that the average distortion is given by

$$V U_{\alpha\beta} = \frac{\partial F[\sigma]}{\partial \bar{\sigma}_{\alpha\beta}}. \tag{2280}$$

Thus we have a microscopic formulation of the problem more general than the low-temperature harmonic limit. Making the Legendre transformation $\bar{\sigma} \to U$ gives the free energy [31]

$$F[U] = F[\sigma] - \sum_{\alpha\beta} \int d^3x\, \bar{\sigma}_{\alpha\beta}(\mathbf{x}) U_{\alpha\beta}(\mathbf{x}) \tag{2281}$$

and assuming a uniform stress,

$$\bar{\sigma}_{\alpha\beta} = -\frac{\partial F[U]}{\partial(V U_{\alpha\beta})}. \tag{2282}$$

This general development allows for a more fundamental definition of the elastic constants:

$$C_{\alpha\beta\mu\nu} = -\frac{\partial \bar{\sigma}_{\alpha\beta}}{\partial U_{\mu\nu}} \tag{2283}$$

$$= \frac{1}{V_0}\frac{\partial^2 F[U]}{\partial U_{\alpha\beta}\, \partial U_{\mu\nu}}. \tag{2284}$$

This also suggests introducing the *compliance tensor*,

$$\kappa_{\alpha\beta\mu\nu} = -\frac{\partial U_{\alpha\beta}}{\partial \bar{\sigma}_{\mu\nu}} \tag{2285}$$

$$= -\frac{1}{V_0} \frac{\partial^2 F[\bar{\sigma}]}{\partial \bar{\sigma}_{\alpha\beta} \partial \bar{\sigma}_{\mu\nu}} \tag{2286}$$

which is essentially the matrix inverse of the elastic tensor. This is discussed further below.

Finally, we want to show that our result for $F[T, \bar{\sigma}]$ reduces to our previous results in the low-temperature harmonic limit. Let us assume a uniform average stress. We then want to evaluate the partition function given by Eq. (2276), where the Hamiltonian is given by Eq. (2277) in the harmonic expansion. After expanding the displacements about their average values and ignoring the fluctuating contributions to $F[T, \bar{\sigma}]$, which are small at low temperatures, we easily find that

$$F[T, \bar{\sigma}] = U_0 + F[U] + V \sum_{\alpha\beta} \bar{\sigma}_{\alpha\beta} U_{\alpha\beta} \tag{2287}$$

where $F[U]$ is determined by the zero-temperature elastic constants and corrections depend on fluctuations that vanish as $T \to 0$.

More generally [32], the elastic part of the free energy $F[U]$ is taken to be of the form

$$F[U] = F_0 + \tfrac{1}{2} \int d^3x \sum_{\alpha\beta\mu\gamma} C_{\alpha\beta\mu\gamma} U_{\alpha\beta}(\mathbf{x}) U_{\mu\gamma}(\mathbf{x}) \tag{2288}$$

where the elastic tensor $C_{\alpha\beta\mu\gamma}$ typically have lower symmetry than for the single-component system interacting by a radial pair potential. In Eq. (2288), F_0 gives the contribution to the free energy independent of the strain. In the low-temperature limit the free energy goes over to the energy and we recover the microscopic results given by Eq. (2255). Within elastic theory (which neglects higher-order terms in the strain) the stress tensor is given by

$$\bar{\sigma}_{\alpha\beta} = -\sum_{\mu\gamma} C_{\alpha\beta\mu\gamma} U_{\mu\gamma} \tag{2289}$$

where any spatial variation of the stress and strain tensors is assumed to be on the macroscopic scale.

The symmetry of $C_{\alpha\beta\gamma\mu}$ under exchange of $\beta \leftrightarrow \alpha$ or $\mu \leftrightarrow \gamma$ is very general. The energy of a crystal does not depend on its orientation and must be unaffected by a rigid rotation. As shown in Appendix B, the free energy will be independent of any rigid rotation of the system if we construct the free energy in terms of a symmetric strain tensor. Depending on the underlying lattice, the elastic tensor has certain other simplifying symmetries. These symmetries allow us to reorganize things in a slightly different form. The matrix $C_{\alpha\beta\gamma\mu}$ depends on the indices α and β only through the six pairs xx, yy, zz, yz, zx, xy with a similar dependence for the pair γ and μ. It is then convenient to introduce Voigt notation [33] and order the pairs $\alpha\beta$ and $\gamma\mu$ as six component vectors:

$$xx \to 1, \quad yy \to 2, \quad zz \to 3$$
$$yz \to 4, \quad xz \to 5, \quad xy \to 6$$

This means that we can rewrite $C_{\alpha\beta\gamma\mu}$ as a two-component matrix C_{ij}, the elastic stiffness matrix, where i and j range from 1 to 6. Thus we have that there are $6 \times 6 = 36$ independent components of C_{ij}. In this notation the elastic free energy can be rewritten in the form

$$F_E = \frac{1}{2} \int d^3x \sum_{ij} U_i U_j C_{ij}. \tag{2290}$$

It is clear, by symmetrizing with respect to the indices i and j, that one can enforce the symmetry

$$C_{ij} = C_{ji}. \tag{2291}$$

In Voigt notation we can rewrite Eq. (2283) in the form

$$\frac{\partial \bar{\sigma}_i}{\partial U_j} = -C_{ij}, \tag{2292}$$

and Hooke's law takes the form

$$\bar{\sigma}_i = -\sum_j C_{ij} U_j. \tag{2293}$$

Let us define the *elastic compliance matrix* κ_{ij} as the matrix inverse of the elastic stiffness coefficients C_{ij}:

$$\sum_k \kappa_{ik} C_{kj} = \delta_{ij}. \tag{2294}$$

Multiplying Hooke's law by κ_{ki}, summing over i, and using Eq. (2294), we easily obtain

$$U_k = -\sum_i \kappa_{ki} \bar{\sigma}_i \tag{2295}$$

or

$$\frac{\partial U_i}{\partial \sigma_j} = -\kappa_{ij}. \quad (2296)$$

This is clearly the expression of Eq. (2285) in Voigt notation. Equation (2295) is analogous to the paramagnetic case where a field induces a magnetization and κ plays the role of the magnetic susceptibility.

The number of independent components of a symmetric 6×6 matrix are simply the six diagonal components plus the $(36-6)/2 = 15$ independent off-diagonal components, for a total of 21 independent components. Consideration of the symmetries associated with the seven crystal groups allows us to reduce the number of independent elastic constants. The establishment of these general results is discussed by Landau and Lifshitz [34]. We discuss here the simple case of those systems with cubic symmetry (examples of such systems are given in Section 6.1.2). These systems are invariant under (1) cyclic permutations $x \to y \to z \to x$, (2) reflections, $x \to -x, y \to -y, z \to -z$, and (3) interchange of pairs $(x,y), (x,z), (y,z)$. Using (1) and (3), we can easily establish the following equalities among the 21 independent elastic constants:

$$C_{xxxx} = C_{yyyy} = C_{zzzz} \quad (2297)$$

$$C_{xxyy} = C_{xxzz} = C_{yyzz} \quad (2298)$$

$$C_{xyxy} = C_{xzxz} = C_{yzyz} \quad (2299)$$

$$C_{xxxy} = C_{zzzy} = C_{yyyx} = C_{yyyz} = C_{xxxz} = C_{zzzx} \quad (2300)$$

$$C_{xxyz} = C_{yyxz} = C_{zzxy} \quad (2301)$$

$$C_{xyxz} = C_{yxyz} = C_{zxzy}. \quad (2302)$$

This reduces the number of independent elastic constants to six. If we use condition (2), it is easy to see that all the elastic constants with a single index corresponding to a particular direction must vanish. Thus Eqs. (2300)–(2302) can be set equal to zero and we are then left with three independent elastic constants for systems with cubic symmetry: $C_{xxxx} = C_{11}$, $C_{xxyy} = C_{12}$, and $C_{xyxy} = C_{yzyz} = C_{44}$. In Table 6.8 we list values for these elastic constants for various systems possessing cubic symmetry.

One point of interest is to go back to the microscopic expression for the elastic constants given by Eq. (2273). Notice that this expression is symmetric under exchange of any two indices. This symmetry is in addition to the symmetries discussed above and leads (as discussed in Problem 6.16) to the *Cauchy relations* [42]:

$$\begin{array}{lll} C_{23} = C_{44} & C_{14} = C_{56} & C_{31} = C_{55} \\ C_{25} = C_{64} & C_{12} = C_{66} & C_{36} = C_{45} \end{array} \quad (2303)$$

TABLE 6.8 Elastic Constants for Some Cubic Crystals, Measured in Units of 10^{12} dyn·cm^{-2} at 300 K

Element	C_{11}	C_{12}	C_{44}	Reference
Li	0.148	0.125	0.108	35
Na	0.070	0.061	0.045	36
Cu	1.68	1.21	0.75	37
Ag	1.24	0.93	0.46	37
Au	1.86	1.57	0.42	37
Al	1.07	0.61	0.28	38
Pb	0.46	0.39	0.144	39
Ge	1.29	0.48	0.67	35
Si	1.66	0.64	0.80	37
Fe	2.34	1.36	1.18	40
Ni	2.45	1.40	1.25	41

In the case of cubic crystals, these all reduce to $C_{12} = C_{44}$. As discussed in some detail in Jones and March [43], the Cauchy relations are not satisfied in metals. It is reasonable that we must, at some level, include the conduction electrons in the treatment of the elastic properties of metals.

6.6.3 Bulk Modulus for Cubic Systems

Let us consider the variation of the pressure at low temperatures. Quite generally the average pressure for a system in equilibrium is given by

$$p = \tfrac{1}{3} \sum_{\alpha} \bar{\sigma}_{\alpha\alpha}. \qquad (2304)$$

We also know that the general expression for the volume change during a distortion is given by

$$\delta V = V \sum_{\alpha} U_{\alpha\alpha}. \qquad (2305)$$

Starting from Hooke's law,

$$\bar{\sigma}_{\alpha\beta} = -\sum_{\mu\nu} C_{\alpha\beta\mu\nu} U_{\mu\nu}, \qquad (2306)$$

the pressure at low temperatures is given by

$$p = -\tfrac{1}{3} \sum_{\alpha\mu\nu} C_{\alpha\alpha\mu\nu} U_{\mu\nu}. \qquad (2307)$$

It is clear from reflection symmetry for a cubic system that $C_{\alpha\alpha\mu\nu} = \delta_{\mu\nu}C_{\alpha\alpha\mu\mu}$ and

$$p = -\frac{1}{3}\sum_{\alpha\mu} C_{\alpha\alpha\mu\mu} U_{\mu\mu}. \tag{2308}$$

It is left as a Problem 6.17 to show that in the cubic case, $\sum_\alpha C_{\alpha\alpha\mu\mu}$ is independent of μ, and

$$p = -\frac{1}{3}\sum_{\alpha} C_{\alpha\alpha\mu\mu} \sum_{\mu} U_{\mu\mu}. \tag{2309}$$

Eliminating $\sum_\mu U_{\mu\mu}$ in terms of the volume change, using Eq. (2305), we find that

$$p = -\frac{1}{3}\sum_{\alpha} C_{\alpha\alpha\mu\mu} \frac{\delta V}{V}. \tag{2310}$$

The bulk modulus, which is defined by Eq. (2138), is given in this case by

$$B = \frac{1}{3}\sum_{\alpha} C_{\alpha\alpha\mu\mu} = \frac{1}{3}(C_{11} + 2C_{12}). \tag{2311}$$

We can obtain a microscopic expression for the bulk modulus in the case of a single-component system described by a radial pair potential by inserting our expression for the elastic constants given by Eq. (2273) into Eq. (2311). We obtain

$$B = \frac{1}{3}\frac{n_0}{2}\sum_R A(R)R_x^2\left(\hat{R}_x^2 + \hat{R}_y^2 + \hat{R}_z^2\right)$$
$$= \frac{n_0}{6}\sum_R A(R)R_x^2 = \frac{n_0}{18}\sum_R A(R)R^2.$$

Inserting the expression for $A(R)$ given by Eq. (2274) and using the condition for equilibrium given by Eq. (2053), we obtain

$$B = \frac{n_0}{18}\sum_R V''(R)R^2. \tag{2312}$$

It is shown in Problem 6.18, in the case of the Lennard-Jones potential, using the set of summations introduced in Section 6.1.3, that the bulk modulus can be written as

$$B = 4\epsilon n_0 \frac{A_6^2}{A_{12}}. \tag{2313}$$

ELASTIC THEORY OF SOLIDS

To complete the theoretical determination of B we must evaluate n_0. For an fcc lattice, it is easy to see that the volume per particle is

$$n_0 = \frac{\sqrt{2}}{a^3} \tag{2314}$$

where a^3 is the nearest-neighbor distance which, given the result from Eq. (2060), reduces to

$$n_0 = \frac{1}{\sigma^3} \left(\frac{A_6}{A_{12}} \right)^{1/2}, \tag{2315}$$

and the bulk modulus for an fcc system is given by

$$B = \frac{4\epsilon A_6^{5/2}}{\sigma^3 A_{12}^{3/2}}. \tag{2316}$$

The comparison of this result with experiment for the noble gases is given in Table 6.4.

At this stage we can develop one further test of the Debye theory. From Eqs. (2220) and (2217) we have a connection between the Debye temperature and the elastic constants:

$$k_B \Theta_D = \left(\frac{18\pi^2 n_0 \hbar^3}{c_L^{-3} + 2c_T^{-3}} \right)^{1/3}. \tag{2317}$$

From Problem 6.19 we have for cubic crystals that the longitudinal and transverse speeds of sound are given by $c_L = \sqrt{C_{11}/\rho_0}$ and $c_T = \sqrt{C_{44}/\rho_0}$. For a discussion of this comparison, the two determinations of Θ_D, see Ref. 44.

6.6.4 Isotropic Limit

The elastic part of the free energy for the case of cubic symmetry, using the Voigt matrix notation, is given by Eq. (2290):

$$F_E = F_0 + \frac{1}{2} \int d^3x \left(C_{11} \sum_\alpha U_{\alpha\alpha}^2 + C_{12} \sum_{\alpha \neq \beta} U_{\alpha\alpha} U_{\beta\beta} + 2C_{44} \sum_{\alpha \neq \beta} U_{\alpha\beta}^2 \right). \tag{2318}$$

If we remove the $\alpha \neq \beta$ constraint in the last two sums, this can be reexpressed as

$$F_E = F_0 + \frac{1}{2} \int d^3x \left[(C_{11} - C_{12} - 2C_{44}) \sum_\alpha U_{\alpha\alpha}^2 \right.$$
$$\left. + C_{12} \sum_{\alpha\beta} U_{\alpha\alpha} U_{\beta\beta} + 2C_{44} \sum_{\alpha\beta} U_{\alpha\beta}^2 \right]. \tag{2319}$$

It is useful to rewrite this in terms of Fourier transforms of the displacement field,

$$U_{\alpha\beta}(\mathbf{k}) = \frac{i}{2}[k_\alpha u_\beta(\mathbf{k}) + k_\beta u_\alpha(\mathbf{k})], \tag{2320}$$

and

$$F_E = F_0 + \frac{1}{2}\int d^3k \left\{ (C_{11} - C_{12} - 2C_{44}) \sum_\alpha k_\alpha^2 |u_\alpha(\mathbf{k})|^2 + C_{12}|\mathbf{k} \cdot \mathbf{u}(\mathbf{k})|^2 \right.$$
$$\left. + C_{44}\left[k^2|\mathbf{u}(\mathbf{k})|^2 + |\mathbf{k} \cdot \mathbf{u}(-\mathbf{k})|^2 \right] \right\}. \tag{2321}$$

The last three terms in Eq. (2321) are clearly invariant under rotations, since they can be written in terms of vector dot products. The first term is not invariant under rotations, and therefore, for an isotropic system, its coefficient must vanish:

$$C_{11} - C_{12} - 2C_{44} = 0. \tag{2322}$$

This condition can be used to establish which crystalline systems are approximately isotropic. There are situations where we have a polycrystalline arrangement or a glassy solid which are isotropic on the longest-length scales. The treatment of isotropic solids is a very established topic (pioneered by Cauchy and Poisson [45] in the 1820s). We have from Eqs. (2319) and (2322) that the elastic part of the free energy is given by

$$F_E = F_0 + \frac{1}{2}\int d^3x \left[C_{12}\left(\sum_\alpha U_{\alpha\alpha}\right)^2 + 2C_{44} \sum_{\alpha\beta} U_{\alpha\beta}^2 \right]. \tag{2323}$$

It is conventional, for isotropic systems, to introduce the Lamé coefficients

$$\lambda = C_{12} \tag{2324}$$

and

$$\mu = C_{44}. \tag{2325}$$

We showed above [see Eq. (2305)] that the change in volume associated with a deformation is proportional to $\sum_\alpha U_{\alpha\alpha}$. If this sum is zero, the volume of the body is unchanged by the deformation, only its shape is altered. Such a deformation is called a *pure shear* and μ is called the shear modulus. The opposite limit corresponds to volume changes in a shape-preserving fashion. Such deformations are called *hydrostatic compressions*. Any isotropic deformation can be written as the sum of a pure shear and a hydrostatic compression. If we define a traceless matrix,

$$W_{\alpha\beta} = U_{\alpha\beta} - \tfrac{1}{3}\delta_{\alpha\beta}\sum_{\gamma} U_{\gamma\gamma}, \tag{2326}$$

we can write

$$\sum_{\alpha\beta} U_{\alpha\beta}^2 = \sum_{\alpha\beta} W_{\alpha\beta}^2 + \tfrac{1}{3}\left(\sum_{\alpha} U_{\alpha\alpha}\right)^2, \tag{2327}$$

and the isotropic elastic free energy takes the form

$$F_E = F_0 + \tfrac{1}{2}\int d^3x \left[B\left(\sum_{\alpha} U_{\alpha\alpha}\right)^2 + 2\mu \sum_{\alpha\beta} W_{\alpha\beta}^2 \right] \tag{2328}$$

where

$$B = \lambda + \tfrac{2}{3}\mu. \tag{2329}$$

If we insert Eqs. (2322), (2324), and (2325) in the cubic expression for the bulk modulus, given by Eq. (2311), we find that

$$\begin{aligned} B &= \tfrac{1}{3}(C_{11} + 2C_{12}) = \tfrac{1}{3}(C_{12} + 2C_{44} + 2C_{12}) \\ &= \lambda + \tfrac{2}{3}\mu. \end{aligned} \tag{2330}$$

Thus we find agreement for the bulk modulus for an isotropic system.

6.6.5 Compliance Coefficients

Let us consider Hooke's law for the simple case of an isotropic system. In this case the elastic constants are given by

$$C_{ii} = C_{11} = \lambda + 2\mu \quad \text{for} \quad i = 1, 2, 3 \tag{2331}$$
$$C_{ij} = C_{12} = \lambda \quad \text{for} \quad i \neq j = 1, 2, 3 \tag{2332}$$
$$C_{ii} = C_{44} = \mu \quad \text{for} \quad i = 4, 5, 6 \tag{2333}$$
$$C_{ij} = 0 \quad \text{for} \quad i \neq j = 4, 5, 6. \tag{2334}$$

Hooke's law, Eq. (2293), then takes the form

$$\bar{\sigma}_i = \begin{cases} -2\mu U_i - \lambda \sum_{i=1}^{3} U_i & i = 1, 2, 3 \tag{2335} \\ -\mu U_i & i = 4, 5, 6. \tag{2336} \end{cases}$$

This can be rewritten in terms of the original tensor notation as

$$\bar{\sigma}_{\alpha\beta} = -\lambda \delta_{\alpha\beta} \sum_{\gamma} U_{\gamma\gamma} - 2\mu U_{\alpha\beta}. \tag{2337}$$

It is simple to invert Eqs. (2335) and (2336) to obtain the deformation associated with a given stress:

$$U_i = \begin{cases} -\dfrac{1}{2\mu}\bar{\sigma}_i + \dfrac{\lambda}{6\mu B}\sum_{i=1}^{3}\bar{\sigma}_i & i = 1, 2, 3 \tag{2338} \\ -\dfrac{1}{\mu}\bar{\sigma}_i & i = 4, 5, 6. \tag{2339} \end{cases}$$

If we return to the tensor notation, we have

$$U_{\alpha\beta} = -\frac{1}{9B}\delta_{\alpha\beta}\sum_{\gamma}\bar{\sigma}_{\gamma\gamma} - \frac{1}{2\mu}\left(\bar{\sigma}_{\alpha\beta} - \frac{1}{3}\delta_{\alpha\beta}\sum_{\gamma}\bar{\sigma}_{\alpha\alpha}\right). \tag{2340}$$

If we consider a uniform axial stress, which is a simple elongation or compression of a rod, then $\bar{\sigma}_{zz} = \sigma_0$ and all other $\bar{\sigma}_{\alpha\beta} = 0$. In this case Eq. (2340) reduces to

$$U_{\alpha\beta} = -\delta_{\alpha\beta}\sigma_0\left(\frac{2\mu - 3B}{18B\mu} + \delta_{\alpha z}\frac{1}{2\mu}\right). \tag{2341}$$

It is conventional to write

$$U_{zz} = -\frac{\sigma_0}{E} \tag{2342}$$

where E is the modulus of extension or Young's modulus, given by

$$E^{-1} = \frac{2\mu + 6B}{18B\mu}. \tag{2343}$$

The strain transverse to the stress is given by

$$U_{xx} = U_{yy} = -\frac{2\mu - 3B}{18B\mu}\sigma_0. \tag{2344}$$

Poisson's ratio is defined by

$$\sigma = -\frac{U_{xx}}{U_{zz}} \tag{2345}$$

$$= \frac{1}{2}\left(\frac{3B - 2\mu}{3B + \mu}\right). \tag{2346}$$

TABLE 6.9 Young's Modulus and Poisson's Ratio for Systems at Room Temperature and Atmospheric Pressure[a]

Substance	E	σ
Al	72.5	0.31
Ar	4.8	0.18
Si	130	0.28
NaCl	53.5	0.20
Granite	46	0.21
Fused silica	73.1	0.17

Source: Data from Ref. 46.
[a] E is measured in units of GPa, where $1\ \text{Pa} = 1\ \text{N} \cdot \text{m}^{-2}$.

For the system to be stable the coefficients of the *diagonal* terms in the elastic free energy, Eq. (2328), must be positive. Thus B and μ must be positive. We can then conclude that

$$-1 \leq \sigma \leq \tfrac{1}{2}. \tag{2347}$$

A table of values for σ for various substances is given in Table 6.9. In practice Poisson's ratio varies only between 0 and $\tfrac{1}{2}$. There are no known substances that expand transversely when stretched longitudinally ($\sigma < 0$).

6.6.6 Distortion by an Elastic Solid

Given an understanding of the relation between stress and strain in a solid, we can turn to the practical question of how to use elastic theory to determine the macroscopic distortion in a solid due to an imposed external force. As we know from the discussion in Chapter 1, the mechanical response of solids to applied forces is governed by the momentum density equation of motion in the steady state:

$$\frac{\partial \langle g_\alpha(\mathbf{x}) \rangle}{\partial t} = -\sum_\beta \nabla_\beta \langle \bar{\sigma}_{\alpha\beta}(\mathbf{x}) \rangle + F_\alpha(\mathbf{x}) = 0 \tag{2348}$$

where $F_\alpha(\mathbf{x})$ is the force density applied to the solid.

We restrict our discussion here to the case where the external forces are applied to the surface of the solid body. Therefore, in the body of the solid, we have the simple equation to be satisfied:

$$\sum_\beta \nabla_\beta \bar{\sigma}_{\alpha\beta}(\mathbf{x}) = 0. \tag{2349}$$

The applied force comes in as a boundary condition on the equation of equilibrium. If **F** is the external force acting on a surface element $d\mathbf{A} = \hat{n}\, dA$, it must be balanced by the force $\sum_\beta \bar{\sigma}_{\alpha\beta}\, dA_\beta$ due to the stresses in the solid:

$$\sum_\beta \bar{\sigma}_{\alpha\beta} \hat{n}_\beta\, dA = F_\alpha\, dA, \tag{2350}$$

or

$$\sum_\beta \bar{\sigma}_{\alpha\beta} \hat{n}_\beta = F_\alpha \tag{2351}$$

evaluated on the surface.

In the case of an isotropic solid the equation governing the displacement field in the bulk solid follows from inserting Hooke's law, Eq. (2337),

$$\sigma_{\alpha\beta} = -\frac{E}{1+\sigma}\left(U_{\alpha\beta} + \frac{\sigma}{1-2\sigma}\delta_{\alpha\beta}\sum_\gamma U_{\gamma\gamma}\right) \tag{2352}$$

into Eq. (2349). In Eq. (2352) we have followed the standard convention and written the elastic constants in terms of E and σ (see Problem 6.21). Expressing the strain tensor in terms of the displacement field and canceling overall multiplicative factors gives the basic equation governing the displacement field in the bulk:

$$(1 - 2\sigma)\nabla^2 \mathbf{u} + \nabla(\nabla \cdot \mathbf{u}) = 0. \tag{2353}$$

We now want to show how all of this fits together by taking a physically simple example [47]. This example shows that elastic theory can be rather challenging mathematically. Consider the case of a solid bounded by a half-plane. For example, there is solid for $z > 0$ and vacuum for $z < 0$. Suppose that we then impose a force at the surface which is localized at the origin:

$$F_\alpha(\mathbf{x}_\perp) = F_\alpha \delta(\mathbf{x}_\perp) \tag{2354}$$

where $\mathbf{x}_\perp = x\hat{x} + y\hat{y}$. Thus we must solve Eq. (2353) for $z \geq 0$ with the boundary condition Eq. (2351) at $z = 0$:

$$\sigma_{\alpha z}(\mathbf{x}_\perp) = -\frac{E}{1+\sigma}\left(U_{\alpha\beta} + \frac{\sigma}{1-2\sigma}\delta_{\alpha\beta}\sum_\gamma U_{\gamma\gamma}\right) = F_\alpha \delta(\mathbf{x}_\perp). \tag{2355}$$

We solve this problem by Fourier transformation in the x–y plane parallel to the surface:

$$u_\alpha(\mathbf{k}, z) = \int d^2 x_\perp\, e^{i\mathbf{k}\cdot\mathbf{x}_\perp} u_\alpha(\mathbf{x}_\perp, z). \tag{2356}$$

After Fourier transformation the bulk equation, Eq. (2353), becomes

$$\left(\delta_{\alpha z}\frac{d}{dz} + ik_\alpha\right)(u'_z + i\phi) + (1 - 2\sigma)(u''_\alpha - k^2 u_\alpha) = 0 \quad (2357)$$

where $u'_\alpha = \frac{d}{dz}u_\alpha(\mathbf{k}, z)$ and $\phi \equiv \mathbf{k}\cdot\mathbf{u}(\mathbf{k}, z)$. The Fourier transform of the boundary condition, Eq. (2355), reads

$$\frac{1}{2}\left(\delta_{\alpha z}\frac{d}{dz} + ik_\alpha\right)u_z + \frac{1}{2}u'_\alpha + \frac{\sigma}{1 - 2\sigma}\delta_{\alpha z}(u'_z + i\phi) = -\frac{1+\sigma}{E}F_\alpha \quad (2358)$$

where $z = 0$.

Let us solve the bulk equation, Eq. (2357), first. We break this up into the three orthogonal directions $\hat{z}, \hat{k}$, and $\hat{c} = \hat{z} \times \hat{k}$. The $\hat{c}$ direction decouples from the other two directions and $u_c = \hat{c}\cdot\mathbf{u}$ satisfies

$$u''_c - k^2 u_c = 0. \quad (2359)$$

This has the solution, which vanishes as $z \to \infty$, given by

$$u_c = Ce^{-kz}. \quad (2360)$$

The constant C is left to be determined by the boundary conditions.

The two remaining equations that result from projecting Eq. (2357) onto the $\hat{z}$ and $\hat{k}$ directions are

$$i\phi' + 2(1-\sigma)u''_z - k^2(1-2\sigma)u_z = 0 \quad (2361)$$

and

$$u'_z - \frac{i}{k^2}(1-2\sigma)\phi'' + 2i(1-\sigma)\phi. \quad (2362)$$

If we take another z derivative of Eq. (2361), it then depends on u'_z and u'''_z. These quantities can be expressed in terms of ϕ and its derivatives through Eq. (2362) and derivatives with respect to z. In this way we obtain a equation strictly in terms of ϕ. u_z is obtained, once ϕ is known, by solving Eq. (2362), which is a simple first-order equation. After a significant amount of algebra (see Problem 6.23) the equation satisfied by ϕ is the fourth-order differential equation

$$\phi'''' - 2k^2\phi'' + k^4\phi = 0. \quad (2363)$$

One can see by direct insertion back into Eq. (2363) that its general solution, with the appropriated behavior at $z \to \infty$, is

$$\phi = (A + Bzk)e^{-kz} \tag{2364}$$

where A and B are undetermined. We can then use this expression for ϕ to integrate Eq. (2362) and obtain

$$u_z = -i(A + B(1 + zk + 2(1 - 2\sigma)))e^{-kz}. \tag{2365}$$

Thus we have determined u_c, $\mathbf{k} \cdot \mathbf{u}$, and u_z in the bulk in terms of the constants A, B, and C, which must be determined by the boundary conditions. Going back to the boundary condition, Eq. (2358), we must evaluate $u(\mathbf{k}, z)$ and its z derivative at $z = 0$. It is then a straightforward problem to invert the set of linear equations to obtain the unknown set of coefficients

$$A = i(1 - 2\sigma)\tilde{F}_z + 2(1 - \sigma)\hat{\mathbf{k}} \cdot \tilde{\mathbf{F}} \tag{2366}$$

$$B = -i\tilde{F}_z - \hat{\mathbf{k}} \cdot \tilde{\mathbf{F}} \tag{2367}$$

$$C = \frac{2}{k}\hat{\mathbf{c}} \cdot \tilde{\mathbf{F}} \tag{2368}$$

where

$$\tilde{\mathbf{F}} \equiv \frac{1+\sigma}{E}\mathbf{F}. \tag{2369}$$

Thus we have the solution for the Fourier transform of the displacement field $\mathbf{u}(\mathbf{k}, z)$. Finally, we must carry out the inverse Fourier transform. We have, for example, for the z-component of the displacement:

$$u_z(\mathbf{x}) = \int \frac{d^2k}{(2\pi)^2} e^{i\mathbf{k}\cdot\mathbf{x}_\perp} e^{-kz}(-i)[A + B(1 + zk + 2(1 - 2\sigma))]$$

$$= G_{zz}(\mathbf{x})\tilde{F}_z + \mathbf{G}_{z\perp}(\mathbf{x}) \cdot \tilde{\mathbf{F}}, \tag{2370}$$

where the coefficients are given by

$$G_{zz}(\mathbf{x}) = \int \frac{d^2k}{(2\pi)^2} e^{i\mathbf{k}\cdot\mathbf{x}_\perp} e^{-kz} \frac{i}{k}[-i(2\sigma - 1) - i(1 + zk + 2(1 - 2\sigma))] \tag{2371}$$

$$G_{z\perp}^\alpha(\mathbf{x}) = \int \frac{d^2k}{(2\pi)^2} e^{i\mathbf{k}\cdot\mathbf{x}_\perp} e^{-kz} \frac{i}{k}[2\hat{k}_\alpha(1 - \sigma) - \hat{k}_\alpha(1 + zk + 2(1 - 2\sigma))]. \tag{2372}$$

Notice that each of these integrals can be expressed in terms of the set of integrals:

$$I_n(\mathbf{x}) = \int \frac{d^2k}{(2\pi)^2} e^{i\mathbf{k}\cdot\mathbf{x}_\perp} \frac{e^{-kz}}{k^n} \tag{2373}$$

where n is an integer, since

$$G_{zz}(\mathbf{x}) = 2(1-\sigma)I_1(\mathbf{x}) + zI_0(\mathbf{x}) \tag{2374}$$

and

$$G_{z\perp}^\alpha(\mathbf{x}) = -\nabla_\perp^\alpha[(1-2\sigma)I_2(\mathbf{x}) + zI_1(\mathbf{x})]. \tag{2375}$$

In a quite similar fashion the transverse displacement can be written

$$u_\perp^\alpha(\mathbf{x}) = G_{\perp z}^\alpha(\mathbf{x})\tilde{F}_z + \sum_\beta G_{\perp\perp}^{\alpha\beta}(\mathbf{x})\tilde{F}_\beta \tag{2376}$$

where

$$G_{\perp z}^\alpha(\mathbf{x}) = (1-2\sigma)\nabla_\perp^\alpha I_2(\mathbf{x}) - z\nabla_\perp^\alpha I_1(\mathbf{x}) \tag{2377}$$

$$G_{\perp\perp}^{\alpha\beta}(\mathbf{x}) = -2(1-\sigma)\nabla_\perp^\alpha \nabla_\perp^\beta I_3(\mathbf{x}) + z\nabla_\perp^\alpha \nabla_\perp^\beta I_2(\mathbf{x})$$
$$- 2\sum_{\gamma\mu} \epsilon_{\alpha z\gamma}\epsilon_{\beta z\mu}\nabla_\perp^\gamma \nabla_\perp^\mu I_3(\mathbf{x}) \tag{2378}$$

and $\epsilon_{\alpha z\gamma}$ is the usual antisymmetric tensor.

The rest of the calculation involves carrying out the Fourier transforms for the I_n and their derivatives. It should be noted that some care is required since the I_n are not well behaved for $n > 1$. However, it is the gradients of these quantities that are needed, and these are well behaved (see Problem 6.26). The key result, from which all of the other integrals can be worked out with a bit of algebra, is

$$I_1(\mathbf{x}) = \frac{1}{2\pi r} \tag{2379}$$

where $r = \sqrt{z^2 + x_\perp^2}$. After putting all of this together, we obtain the final results:

$$u_z(\mathbf{x}) = \frac{1+\sigma}{2\pi E}\left[\left(\frac{2(1-\sigma)}{r} + \frac{z^2}{r^3}\right)F_z + \left(\frac{1-2\sigma}{r(r+z)} + \frac{z}{r^3}\right)\mathbf{x}_\perp \cdot \mathbf{F}\right] \tag{2380}$$

$$\mathbf{u}_\perp(\mathbf{x}) = \frac{1+\sigma}{2\pi E}\left[\left(\frac{z}{r^3} - \frac{1-2\sigma}{r(r+z)}\right)\mathbf{x}_\perp F_z + \frac{2(1-\sigma)r+z}{r(r+z)}\mathbf{F}_\perp\right.$$
$$\left. + \frac{2r(\sigma r+z)+z^2}{r^3(r+z)^2}\mathbf{x}_\perp \mathbf{x}_\perp \cdot \mathbf{F}\right]. \tag{2381}$$

If we restrict the force to be perpendicular to the surface and look at the compression in the z-direction, we find that

$$u_z(\mathbf{x}) = \frac{1+\sigma}{2\pi E}\left(\frac{2(1-\sigma)}{r} + \frac{z^2}{r^3}\right)F_z. \tag{2382}$$

We see the characteristic $1/r$ decay of the perturbation at long distances. Thus we have the connection between a macroscopic deformation and the elastic constants. These elastic constants, as first pointed out by Einstein [48], are directly connected with the *eigenfrequencies of the atoms*—the lattice vibrations.

REFERENCES

[1] Exceptional cases are discussed in D. Frenkel, "Entropy driven phase transitions," *Physica A* **263**, 26 (1999).

[2] M. J. Buerger, *Introduction to Crystal Geometry*, McGraw-Hill, New York, 1971.

[3] A. Bravais (1845) was the first to properly enumerate the 14 different kinds of Bravais lattices.

[4] There is a good discussion in N. Ashcroft and N. D. Mermin, *Solid State Physics*, Holt, Rinehart, and Winston, New York, 1976, p. 115.

[5] The space groups and symmetry are discussed in N. Ashcroft and N. D. Mermin, *Solid State Physics*, Holt, Rinehart and Winston, New York, 1976; and L. D. Landau and E. M. Lifshitz, *Statistical Physics*, 3rd ed., Pergamon Press, Oxford, 1980, Chap. XIII.

[6] J. E. Lennard-Jones and A. E. Ingham, *Proc. R. Soc. A* **107**, 636 (1925).

[7] N. Ashcroft and N. D. Mermin, *Solid State Physics*, Holt, Rinehart and Winston, New York, 1976, p. 401.

[8] See, for example, the discussion in Appendix II in M. Born and K. Huang, *Dynamical Theory of Crystal Lattices*, Oxford University Press, London, 1954, where the Ewald method is used to analyze the Madalung energy. See also P. P. Ewald, *Ann. Phys.* **54**, 519 (1917); **64**, 253 (1921); and E. Madalung, *Phys. Z.* **18**, 524 (1918).

[9] J. M. Ziman, *Principles of the Theory of Solids*, Cambridge Univesity Press, Cambridge, 1965, p. 37.

[10] M. Born and K. Huang, *Dynamical Theory of Crystal Lattices*, Oxford University Press, London, 1954, Chap. 1.

[11] We assume that we have reflection symmetry about certain axes. For an fcc lattice, for example, $R_\alpha \rightarrow -R_\alpha$, with the other two orthogonal components R_β, and R_μ unchanged, is effected by letting $n_\beta \rightarrow -n_\mu, n_\mu \rightarrow -n_\beta$ and $n_\alpha \rightarrow n_\alpha + n_\beta + n_\mu$.

[12] M. Born and T. von Kármán, *Phys. Z.* **13**, 297 (1912).

[13] P. L. Dulong and A. T. Petit, *Ann. Chim. Phys.* **10**, 395 (1819) determined their result empirically.

[14] The origins of the problems are the same as in treating the photon gas.

[15] J. Goldstone, *Nuovo Cimento* **19**, 154 (1961); Y. Nambu, *Phys. Rev.* **117**, 648 (1960), P. W. Anderson, *Phys. Rev.* **130**, 62 (1963). See Chapter 7 in D. Forster, *Hydrodynamic*

Fluctuations, Broken Symmetry and Correlation Functions, W.A. Benjamin, Reading, Mass., 1975.
[16] F. A. Lindemann, *Phys. Z.* **11**, 609 (1910).
[17] One should check that the general structure of the harmonic theory can be developed for general dimensionality d.
[18] B. I. Halperin and D. R. Nelson, *Phys. Rev. Lett.* **41**, 121 (1978); **41**, 519 (1978); D. R. Nelson and B. I. Halperin, *Phys. Rev. B* **19**, 2457 (1979).
[19] A. P. Young, *J. Phys. C* **11**, L453 (1978).
[20] J. M. Kosterlitz and D. J. Thouless, *J. Phys. C* **6**, 1181 (1973); J. M. Kosterlitz, *J. Phys. C* **7**, 10461 (1974), J. M. Kosterlitz and D. J. Thouless, *Progress Low Temperature Physics*, Vol. VII-B, ed. D. F. Brewer, North-Holland, Amsterdam, 1978; V. L. Berezinskii, *Zh. Eksp. Teor. Fiz.* **61**, 114 (1971) [*Sov. Phys. JETP* **34**, 610 (1972)].
[21] P. Debye, *Ann. Phys.* **43**, 49 (1914); I. Waller, Uppsala University, Uppsala, Sweden, 1925.
[22] J. Fenkel, *Wave Mechanics, Elementary Theory*, Oxford University Press, Oxford, 1936, p. 267.
[23] For a discussion of lattice dynamics, see, for example, J. Callaway, *Quantum Theory of the Solid State*, Academic Press, New York, 1976.
[24] P. Debye, *Ann. Phys.* **39**, 789 (1912); translation in *The Collected Papers of Peter J. W. Debye*, Interscience, New York, 1954, p. 650.
[25] A. Münster, *Statistical Thermodynamics*, Vol. II, Springer-Verlag, Berlin, 1974, p. 10.
[26] C. Kittel, *Introduction to Solid State Physics*, 7th ed., Wiley, New York, 1996, p. 126.
[27] C. Kittel, *Introduction to Solid State Physics*, 7th ed., Wiley, New York, 1996, p. 157.
[28] The 1994 Nobel Prize in Physics was awarded to Bertrain Brockhouse and Clifford Shull for pioneering contributions to the development of neutron scattering techniques for studies of condensed matter.
[29] In the continuum and the case of uniform strain we need to invert

$$U_{\alpha\beta} = \tfrac{1}{2}\bigl[\nabla_\alpha u_\beta(\mathbf{x}) + \nabla_\beta u_\alpha(\mathbf{x})\bigr]$$

to obtain, up to a constant,

$$u_\alpha(\mathbf{x}) = \sum_\beta x_\beta U_{\alpha\beta}.$$

Thus when we look at the average displacement on the lattice, we obtain Eq. (2249). Note also that this expression is equivalent to our initial introduction of the strain in Chapter 1 as given by Eq. (302).
[30] The initial discussion of the response of solids to applied forces was due to Galileo Galilei, *Discorsi e Dimostrazioni Matematiche*, Leiden, 1638. This work was followed by the discovery of Hooke's law in 1660 by Robert Hooke, *De Potential Restitutiva*, London, 1678.
[31] In writing this expression we have allowed for the possibility that the stress can vary slowly with position $\mathbf{x}$. As discussed in Section 6.6, this allows for the application of macroscopic forces to a solid that will lead to variation of the stress and strain with position. For our purposes in this section, this variation complicates the analysis mathematically and we will, for the sake of simplicity, assume that the applied stress is uniform.

[32] L. D. Landau and E. M. Lifshitz, *Theory of Elasticity*, Pergamon Press, Oxford, 1986, p. 23.
[33] W. Voigt, *Lehrbuch der Kristall Physik*, Teubner, Berlin, 1910.
[34] L. D. Landau and E. M. Lifshitz, *Theory of Elasticity*, Pergamon Press, Oxford, 1986.
[35] H. B. Huntington, *Solid State Phys.* **7**, 214 (1958).
[36] P. Ho and A. L. Ruoff, *J. Phys. Chem. Solids* **29**, 2101 (1968).
[37] J. deLaunay, *Solid State Phys.* **2**, 220 (1956).
[38] P. Ho and A. L. Ruoff, *J. Appl. Phys.* **40**, 3 (1969).
[39] P. Ho and A. L. Ruoff, *J. Appl. Phys.* **40**, 51 (1969).
[40] J. A. Rayne and B. S. Chandrasekhar, *Phys. Rev.* **122**, 1714 (1961).
[41] G. A. Alers et al., *J. Phys. Chem. Solids* **13**, 40 (1960).
[42] M. Born and K. Huang, *Dynamical Theory of Crystal Lattices*, Oxford University Press, London, 1954, p. 136.
[43] W. Jones and N. March, *Theoretical Solid State Physics*, Vol. 1, Wiley, London, 1973.
[44] A. Münster, *Statistical Thermodynamics*, Vol. 2, Springer-Verlag, Berlin, 1969, p. 26.
[45] The framework of governing differential equations for elastic theory was developed by Navier in 1821 [Paris, *Mem. Acad. Sci.* **7** (1827)]. Cauchy discovered most of the elements of the pure theory of elasticity and presented this work to the Paris Academy in September 1822. Poisson came to a similar development along a different path, and his results were read before the Paris Academy in April 1828. The theory of elasticity established by Poisson and Cauchy was then applied to numerous problems well in advance of experimental tests. For a more complete discussion of the history of theory of elasticity, see A. E. H. Love, *A Treatise on the Mathematical Theory of Elasticity*, Cambridge University Press, Cambridge, 1920.
[46] G. Simmons and H. Wang, *Single Crystal Elastic Constants and Calculated Aggregate Properties: A Handbook*, 2nd ed., MIT Press, Cambridge, Mass., 1971.
[47] This problem is treated in L. D. Landau and E. M. Lifshitz, *Theory of Elasticity*, Pergamon Press, Oxford 1986, using a very different method. This problem was solved originally by J. Boussinesq, *Application des potentials*, Paris, 1885.
[48] A. Einstein, *Ann. Phys.*, **35**, 679 (1911).

PROBLEMS

6.1. Show that the only lattice structures in two dimensions that are invariant under a set of rotations are those where $a_1 = a_2$ and

$$\theta = \frac{2\pi}{n}$$

and $n = 3, 4,$ and 6.

6.2. Assume that **k** is along one of the cubic axes in a cubic crystal. Show that in the long-wavelength limit, one can solve the eigenvalue problem defined by Eq. (2092) explicitly. In particular, determine the longitudinal (along **k**) and transverse (perpendicular to **k**) speeds of sound.

6.3. Using the general properties of the dynamical matrix $D_{\alpha\beta}(\mathbf{R})$, show that its Fourier transform, given by Eq. (2091), can be written in the form given by Eq. (2114).

6.4. Using the nearest-neighbor model introduced in Section 6.4.3, show that the bulk modulus is given by

$$B = \frac{n_0}{3} V''(a) a^2.$$

6.5. With the choice $J_\beta(\mathbf{R}') = -ik_\beta \delta_{\mathbf{R},\mathbf{R}'}$, show that Eq. (2896) in Appendix O leads to Eq. (2153).

6.6. Using the techniques developed in Problem 6.5 and Appendix O, show that the average for classical harmonic solids needed to evaluate the structure factor is given by Eq. (2161).

6.7. Show that the operators defined by Eqs. (2166) and (2167) satisfy the commutation relations given by Eqs. (2169) and (2170). Go on to show that the creation and annihilation operators defined by Eqs. (2171) and (2172) satisfy Eqs. (2173) and (2174).

6.8. Show that in the case of nonrelativistic massive particles, where $\hbar\omega(\mathbf{k}) = \hbar^2 k^2/2m$, the more general definition of the density of states given by Eq. (2191) reduces to the result found in Chapter 3.

6.9. The identity given by Eq. (2197) can be proven using the techniques developed in Chapter 3. Define the quantity

$$Z[\beta, \lambda] = \mathrm{Tr}\, e^{-\beta H_0} e^{-\lambda A}$$

where A is a Hermitian operator that is linear in the momentum and position operators. Since each of these operators is linear in the boson creation, a_i^+, and annihilation, a_i, operators, we can write

$$A = \sum_i (\eta_i^* a_i + a_i^+ \eta_i)$$

where i labels all single-particle states and will not need to be specified further.

If we define

$$\langle a_i \rangle_\lambda = \frac{\mathrm{Tr}\, e^{-\beta H_0} e^{-\lambda A} a_i}{Z[\beta, \lambda]}$$

and similarly for $\langle a_i^+ \rangle_\lambda$, show that

$$-\frac{\partial}{\partial \lambda} \ln Z[\beta, \lambda] = \sum_i (\eta_i^* \langle a_i \rangle_\lambda + \langle a_i^+ \rangle_\lambda \eta_i).$$

Evaluate $\langle a_i \rangle_\lambda$ and $\langle a_i^+ \rangle_\lambda$ as functions of λ using the methods of section 3.5. Use this result in the derivative of the ln Z to derive the identity Eq. (2197).

6.10. Suppose that we use as the defining average phonon speed in the Debye theory the expression

$$c^2 = \frac{1}{3} \sum_i \int \frac{d\Omega_k}{4\pi} c_i^2(\hat{k}).$$

For a cubic system, express c^2 in terms of the elastic constants.

6.11. The Hamiltonian governing a solid consisting of N_s particles can be written in the *Einstein model* as

$$H = -N_s \epsilon_0 + \hbar \omega_0 \sum_i a_i^+ a_i$$

where ϵ_0 is the binding energy per particle and all the phonons have the same frequency ω_0. The normalization is such that $\sum_i = N_s$. Compute the free energy F for this system. Suppose that this solid is coexisting with its gas phase and the gas can be described by the ideal gas equations of state. Find the equation governing coexistence between the two phases.

6.12. Consider a two-dimensional solid, with total area A, where the atoms have a Debye density of states

$$D(\epsilon) = \frac{1}{\pi \hbar^2} \frac{\epsilon}{c^2} \theta(\epsilon_D - \epsilon)$$

where c is the speed of sound (which we assume is the same in all directions) and ϵ_D is the Debye energy. We assume that the dispersion relation for the phonons is simply

$$\epsilon = c\hbar k$$

and the eigenfunctions $e_\nu^l(k)$ are diagonal: $e_\nu^l(k) = \delta_{\nu,l}$. Compute the average squared displacements

$$L^2 = \langle [\mathbf{u}(\mathbf{R})]^2 \rangle$$

where $\mathbf{u}(\mathbf{R})$ is quantum displacement of an atom at site $\mathbf{R}$. You should be able to obtain an explicit expression for L^2 for temperatures $T > 0$. What can you conclude from this result?

6.13. Calculate the heat capacity of a one-dimensional system of phonons at low temperatures. We know that $\lim_{t \to 0} C = 0$. Find the first finite temperature correction to this result.

6.14. The mean-square amplitude of fluctuations in the high-temperature classical limit is given by Eq. (2232). Using the Lindemann criterion—that one has melting at a temperature T_M where the root-mean-square amplitude approximately equal to 0.25 the lattice-spacing—estimate the melting temperatures for the noble gases. Compare with the experimental values: Ne (24.56 K), Ar (83.81 K), Kr (115.8 K), and Xe (161.4 K).

6.15. Suppose that the atomic displacement field is given by

$$u_\alpha(\mathbf{x}) = \sum_\beta \sum_\mathbf{R} R_\beta \bar{U}_{\alpha\beta} \delta(\mathbf{x} - \mathbf{R})$$

where $\bar{U}_{\alpha\beta}$ is a constant. Since the strain field is given by

$$U_{\alpha\beta}(\mathbf{x}) = \tfrac{1}{2}\bigl[\nabla_\alpha u_\beta(\mathbf{x}) + \nabla_\beta u_\alpha(\mathbf{x})\bigr],$$

show that

$$\int d^3 x\, U_{\alpha\beta}(\mathbf{x}) = V \bar{U}_{\alpha\beta}.$$

6.16. Show that the permutation symmetry satisfied by the elastic constants for systems with radial pair potential leads to the Cauchy relations given by Eqs. (2303). Show that for cubic systems these all reduce to $C_{12} = C_{44}$.

6.17. Show that in the case of a system with cubic symmetry, $\sum_\alpha C_{\alpha\alpha\mu\mu}$ is independent of μ.

6.18. Show that the expression, Eq. (2312), for the bulk modulus for cubic systems can be reduced to Eq. (2313) if one has a single-component system interacting via a radial pair potential.

6.19. Work out the long-wavelength form for the dynamical matrix in the isotropic case in terms of λ, μ, and the mass density ρ_0. Find the longitudinal and transverse speeds of sound in this case.

6.20. Establish the relationship between the elastic constants for an isotropic system given by

$$2\mu = \frac{E}{1+\sigma}$$

$$\lambda = \frac{E}{1+\sigma} \frac{\sigma}{1-2\sigma}.$$

6.21. Show that Hooke's law, given by Eq. (2337), can be written in the form of Eq. (2352) in terms of E and σ.

6.22. Determine the deformation of an infinite isotropic elastic medium when a force **F** is applied to a small region in it.

6.23. Show that the solutions of Eqs. (2361) and (2362) can be reduced to the single equation given by Eq. (2363).

6.24. Show that Eq. (2364) is indeed a solution of Eq. (2363).

6.25. Show that the boundary condition given by Eq. (2358) determines the constants A, B, and C given by Eqs. (2366)–(2368).

6.26. Show that $I_1(\mathbf{x})$, defined by Eq. (2373), is given by Eq. (2379). Evaluate all of the other I_n as needed in Eqs. (2373)–(2378).

6.27. Given the expressions derived in Problems 6.25 and 6.26 work out explicitly G_{zz}, $G_{z\perp}^{\alpha}$, and $G_{\perp\perp}^{\alpha\beta}$ as defined by Eqs. (2373)–(2378).

APPENDICES

A INTRODUCTION TO PROBABILITY THEORY

The treatment of a large number of degrees of freedom requires introduction of statistical methods. In this appendix we develop the elements of probability theory [1] useful in our statistical mechanical analysis.

Permutations and Combinations

Consider the case of a set of N distinct objects. A *permutation* is any specific ordering of this set. We can obtain the number of different permutations of N distinct objects in the following way. Suppose that we have N ordered boxes in addition to the N objects. There are N different objects that can fill the first box. Since one object has been used, the second box can be filled in $(N-1)$ distinct ways. Clearly, one can continue and the N boxes can be filled in $N(N-1)(N-2)\ldots 2 \times 1 = N!$ ways. Thus the number of different permutations of N distinct objects is $N!$.

Suppose that one selects n objects from the larger set N. What is the number of permutations P_n^N of this process? This case corresponds to having n boxes and filling these boxes sequentially. The first box can be filled in N ways the second box in $(N-1)$ ways, and so on until the nth in $(N-n+1)$ ways. Clearly, P_n^N is given by

$$P_n^N = N(N-1)\cdots(N-n+1) = \frac{N!}{(N-n)!}. \tag{2383}$$

Suppose that we do not distinguish among the n objects selected in determining P_n^N. Then we have $n!$ equivalent permutations. The number of *combinations* C_n^N of N objects taken n at a time is given by

$$C_n^N = \frac{P_n^N}{N!} = \frac{N!}{(N-n)!\,n!}. \tag{2384}$$

If the set of N objects is divided into l subsets of identical objects where $n_i(i=1,2,\ldots,l)$ is the number of identical elements in subset i, the number of permutations is

$$P_{n_1,n_2,\ldots n_l}^N = \frac{N!}{n_1!n_2!\cdots n_l!} \tag{2385}$$

where

$$\sum_{i=1}^{l} n_i = N. \tag{2386}$$

Probability Distributions

Definitions. As discussed in Chapter 1, probability is the quantification of the likeliness of an outcome in an experiment. If A is one possible outcome in an experiment, we do the experiment N times and obtain the result A n_A times, then the probability of obtaining the result A in a single experiment is defined by

$$P(A) = \lim_{N \to \infty} \frac{n_A(N)}{N}. \tag{2387}$$

The configuration space of an experiment is a set, S, of elements such that any result of the experiment corresponds to one element of the set. We can also define a set of results B, and $P(B)$ is the probability of obtaining a result in a single experiment that is a member of the set B. For any set B in S, we have

$$P(B) \leq P(S) = 1. \tag{2388}$$

The *union* of two sets of results B and C is denoted $B \cup C$. $B \cup C$ is the set of all points belonging to B or C or both. The *intersection* of B and C, denoted $B \cap C$, is the set of all points belonging to both B and C. If events B and C are *mutually exclusive*, then $B \cap C = 0$, where 0 is the empty set (contains no points).

We then have that $P(B \cup C)$ is the probability that members of set B or C or both occur as the outcome of an experiment. Similarly, $P(B \cap C)$ is the probability that both B and C occur as a result of an experiment.

The sets B and C are said to be *independent* if

$$P(B \cap C) = P(B)P(C). \tag{2389}$$

Note that since $P(B \cap C) \neq 0$, B and C have some points in common. Therefore, independent events are not mutually exclusive events.

The *conditional probability* $P(B|C)$, giving the probability that a result in set B occurs given that a result in set C also occurs, is defined as

$$P(B|C) = \frac{P(B \cap C)}{P(C)}. \tag{2390}$$

Discrete Stochastic Variables. A *stochastic variable* X, defined on a configuration space S, is a function that maps elements of the set of results from experiments A of S onto the real numbers X_A.

INTRODUCTION TO PROBABILITY THEORY 525

Let X be a stochastic variable on S which can take on a countable set of values $X = (x_1, x_2, \ldots, x_n)$. Thus any result A in S will produce one of the values of x_i for the stochastic variable X. The probabilities P_i of obtaining the value x_i in an experiment must satisfy the conditions

$$P_i \geq 0 \tag{2391}$$

for all i, and

$$\sum_{i=1}^{n} P_i = 1. \tag{2392}$$

The mean of *average* value of X is defined as

$$\langle X \rangle = \sum_{i=1}^{n} x_i P_i. \tag{2393}$$

The *autocorrelation* of X is defined as

$$\langle (\delta X)^2 \rangle = \sum_{i=1}^{n} (x_i - \langle X \rangle)^2 P_i = \langle X^2 \rangle - \langle X \rangle^2. \tag{2394}$$

Continuous Stochastic Variables. Let X be a stochastic variable that can take on a continuous set of real values in the range $x_0 \leq X \leq x_f$. Although it does not make sense to talk of the probability of obtaining a particular value for X in a measurement (this is clearly zero), it does make sense to discuss the probability $P(a \leq X \leq b)$ that we find a value of X in the range $a \leq X \leq b$. This quantity is related to the probability density of X, $P(x)$, by

$$P(a \leq X \leq b) = \int_a^b P(x)\, dx. \tag{2395}$$

The probability density must satisfy the conditions

$$P(x) \geq 0 \tag{2396}$$

and

$$\int_{x_0}^{x_f} P(x)\, dx = 1. \tag{2397}$$

The mean or average value of X is defined by

$$\langle X \rangle = \int_{x_0}^{x_f} x P(x)\, dx. \tag{2398}$$

The autocorrelation of X is defined by

$$\langle (\delta X)^2 \rangle = \int_{x_0}^{x_f} (x - \langle X \rangle)^2 P(x)\, dx = \langle X^2 \rangle - \langle X \rangle^2. \tag{2399}$$

Joint Probability Distributions. Let X and Y be stochastic variables in a configuration space S with discrete values $X(S) = (x_1, x_2, \ldots, x_{n_x})$ and $Y(S) = (y_1, y_2, \ldots, y_{n_y})$, respectively. We can then define the joint probability distribution $P(x_i, y_j)$, which is the probability of obtaining both x_i and y_j in an experiment.

As straightforward generalizations of Eqs. (2391), and (2392), we require that

$$P(x_i, y_j) \geq 0 \tag{2400}$$

and

$$\sum_{i=1}^{n_x} \sum_{j=1}^{n_y} P(x_i, y_i) = 1. \tag{2401}$$

The averages of X and Y are defined by

$$\langle X \rangle = \sum_{i=1}^{n_x} \sum_{j=1}^{n_y} x_i P(x_i, y_i) \tag{2402}$$

$$\langle Y \rangle = \sum_{i=1}^{n_x} \sum_{j=1}^{n_y} y_i P(x_i, y_i). \tag{2403}$$

The *correlation* of X and Y is defined by

$$\langle \delta X \delta Y \rangle = \sum_{ij} (x_i - \langle X \rangle)(y_i - \langle Y \rangle) P(x_i, y_i)$$
$$= \langle XY \rangle - \langle X \rangle \langle Y \rangle. \tag{2404}$$

The probability distribution for X, independent of any determination of Y, is simply

$$P(x_i) = \sum_{j=1}^{n_y} P(x_i, y_j). \tag{2405}$$

The generalization to continuous joint probability distributions is straightforward.

Examples of Distributions

There are several important examples of probability distributions that are worthy of review.

Binomial Distributions. Let us carry out a sequence of N statistically independent trials and assume that each trial can have only one of the outcomes, $+1$ or -1. Let us denote the probability of outcome $+1$ by p and the probability of outcome -1 by q. Then $p + q = 1$. This is just the coin-flipping example generalized to the case $p \neq q$.

In a given sequence of N trials, the outcome $+1$ can occur n_1 times and the outcome -1 can occur n_2 times, where $N = n_1 + n_2$. For a given permutation of n_1 outcomes $+1$ and n_2 outcomes -1, the probability is $p^{n_1} q^{n_2}$ since the N trials are statistically independent. The number of combinations for obtaining these results in $N!/(n_1! n_2!)$. Consequently, the probability of any combination of n_1 outcomes $+1$ and n_2 outcomes -1 is the *binomial distribution*,

$$P_N(n_1) \equiv \frac{N!}{n_1! n_2!} p^{n_1} q^{n_2}. \tag{2406}$$

Using the binomial theorem, we can show that the distribution is normalized:

$$\sum_{n_1=0}^{N} P_N(n_1) = \sum_{n_1=0}^{N} \frac{N!}{n_1!(N-n_1)!} p^{n_1} q^{N-n_1} = (p+q)^N = 1. \tag{2407}$$

The mean value of obtaining outcome $+1$ is

$$\langle n_1 \rangle = \sum_{n_1=0}^{\infty} n_1 P_N(n_1) = \sum_{n_1=0}^{\infty} \frac{n_1 N!}{n_1!(N-n_1)!} p^{n_1} q^{N-n_1}$$

$$= p \frac{\partial}{\partial p} \left[\sum_{n_1=0}^{N} P_N(n_1) \right]_q = p \frac{\partial}{\partial p} (p+q)^N = pN. \tag{2408}$$

In a similar manner, we obtain

$$\langle n_1^2 \rangle = \sum_{n_1=0}^{N} n_1^2 P_N(n_1) = (Np)^2 + Npq. \tag{2409}$$

The autocorrelation function (variance) is given by

$$\langle (\delta n_1)^2 \rangle = \langle n_1^2 \rangle - \langle n_1 \rangle^2 = Npq. \tag{2410}$$

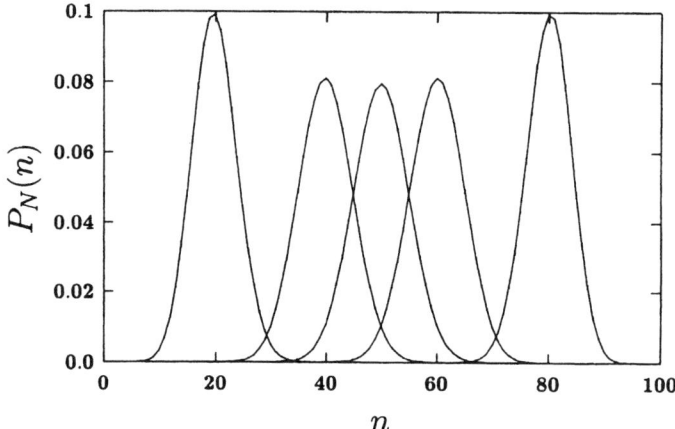

FIGURE A.1 Binomial probability distribution versus n for $N = 100$; from left to right, $p = 0.2, 0.4, 0.5, 0.6$, and 0.8.

The standard deviation is

$$\sigma_N = \sqrt{Npq}. \tag{24.11}$$

Gaussian (or Normal) Distribution. The limit of the binomial distribution for large N is very interesting. As shown in Fig. A.1, if we plot the binomial distribution for fixed, large N as a function of n_1, we find that it is strongly peaked at the average value $n_{\max} = Np$. We also see that as N increase, we can treat n_1 as a continuous variable. We can then expand $P_N(n_1)$ in a power series about its maximum value. We first note from Problem 1.24 that for large N, where we can use Stirling's approximation, the value of the distribution at its maximum is given by

$$P_N(n_{\max}) = \frac{1}{\sqrt{2\pi Npq}}. \tag{24.12}$$

Since $P_N(n_1)$ must fall off rapidly for large n_1 we expand, instead of $P_N(n_1)$, $\ln P_N(n_1)$. We can take the $1/\sqrt{N}$ behavior of the maximum into account by expanding

$$\ln[\sqrt{2\pi N} P_N(n_1)] = \sum_{s=0}^{\infty} \frac{A_s}{s!}(n_1 - Np)^s \tag{24.13}$$

where the Taylor series coefficients are given as usual by

$$A_s = \frac{d^s}{dn_1^s} \ln[\sqrt{2\pi N} P_N(n_1)]\Big|_{n_1 = Np}. \tag{24.14}$$

It is shown in Problem 1.24 that

$$A_1 = 0 \tag{2415}$$

$$A_2 = -\frac{1}{Npq} \tag{2416}$$

and in general,

$$|A_s| = \frac{1}{(Npq)^{s-1}}. \tag{2417}$$

Then, to leading nontrivial order in powers of $1/N$,

$$\ln[\sqrt{2\pi N}\, P_N(n_1)] = A_0 - \frac{1}{Npq}(n_1 - Np)^2 \tag{2418}$$

and

$$P_N(n_1) = \frac{e^{A_0}}{\sqrt{2\pi N}} e^{-(1/Npq)(n_1 - Np)^2} \tag{2419}$$

and the constant A_0 can be determined by requiring that the probability distribution be normalized. The final form for the distribution function is the Gaussian or normal distribution function

$$P_N(n_1) = \frac{1}{\sqrt{2\pi pqN}} e^{-(1/Npq)(n_1 - Np)^2}. \tag{2420}$$

Notice that the value at the maximum given by Eq. (2412) is preserved.

Poisson Distribution. The Poisson distribution can be obtained from the binomial distribution in the limit $N \to \infty, p \to 0$ such that $Np \equiv a \approx n_1 \ll N$, where a is a finite constant. Using Stirling's formula and the result $e^z = \lim_{n \to \infty}(1 + z/n)^n$, we can write

$$\frac{N!}{(N-n_1)!} \approx \frac{\sqrt{N}}{\sqrt{N-n_1}} \frac{(N/e)^N}{[(N-n_1)/e]^{N-n_1}} = \left(\frac{N}{e}\right)^{n_1}\left(1 - \frac{n_1}{N}\right)^{n_1 - N}$$

$$= \left(\frac{N}{e}\right)^{n_1} e^{n_1} = N^{n_1}, \tag{2421}$$

and as $p \to 0$,

$$(1-p)^{N-n_1} \approx \left(1 - \frac{a}{N}\right)^N = e^{-a}. \tag{2422}$$

using these results in the binomial distribution, we obtain the Poisson distribution

$$P_N(n_1) = \frac{a^{n_1} e^{-a}}{n_1!} \tag{2423}$$

valid in the limit $p \to 0$ and $N \to \infty$ such that the product $Np \to a$. The Poisson distribution applies when many experiments are carried out, but the result $+1$ has only a small probability of occurring. The expected number of outcomes $+1$ will be a.

The Poisson distribution is normalized to 1:

$$\sum_{n_1=0}^{\infty} \frac{a^{n_1} e^{-a}}{n_1!} = e^a e^{-a} = 1. \tag{2424}$$

It is easy to see that $\langle n_1 \rangle = a$. We note that the Poisson distribution for n_1 is determined entirely in terms of the mean value of n_1.

B TRANSFORMATION THEORY, CONSERVATION LAWS, AND INVARIANCE PRINCIPLES

Phase-Space Transformations

In this appendix we set up the formal machinery needed to investigate the invariance and symmetry principles in classical mechanics. To begin this discussion of transformation theory, we look at the motion of a system through phase space as a function of time. Thus we look at the trajectory of a phase-space coordinate $q(0)$ as a function of time: $q(0) \to q(t)$. Here we simplify the notation further and drop the subscript N on $q_N(t)$. It is useful to think of this time evolution as a sequence of coordinate transformations. As we show below, this transformation is an example of a special set of coordinate transformations known as *canonical transformations* [2]. Then we show that there are various other canonical transformations that can be identified with physical symmetry operations in Γ-space (translations, rotations, etc.). These symmetry operations are used to establish invariance or symmetry principles that connect each invariance with a conservation law. These symmetry principles and the associated conservation laws are the most important ingredients in building a theory of statistical physics.

Time Translations

Let us consider the time evolution of an isolated system of N particles whose equations of motion are given by Hamilton's equations [3],

$$\dot{r}_i^\alpha = \frac{\partial H}{\partial p_i^\alpha} \tag{2425}$$

$$\dot{p}_i^\alpha = -\frac{\partial H}{\partial r_i^\alpha}, \tag{2426}$$

where $H = H[q]$ is the time-independent Hamiltonian, i labels the particles, and α is the vector component label. If we consider the time evolution of any dynamical variable $A(q(t))$, we obtain, using the chain rule for differentiation,

$$\frac{dA(q(t))}{dt} = \sum_{i,\alpha} \left[\frac{\partial A}{\partial r_i^\alpha(t)} \frac{d}{dt} r_i^\alpha(t) + \frac{\partial A}{\partial p_i^\alpha(t)} \frac{d}{dt} p_i^\alpha(t) \right]. \quad (2427)$$

Combining this with Hamilton's equations, Eqs. (2425) and (2426), we obtain

$$\frac{dA}{dt} = \{A, H\} \quad (2428)$$

where the Poisson bracket $\{\cdot, \cdot\}$ for any dynamical variables A and B is defined [4] by

$$\{A, B\} = \sum_{i,\alpha} \left[\frac{\partial A}{\partial r_i^\alpha(t)} \frac{\partial B}{\partial p_i^\alpha(t)} - \frac{\partial A}{\partial p_i^\alpha(t)} \frac{\partial B}{\partial r_i^\alpha(t)} \right]. \quad (2429)$$

The fundamental Poisson brackets among the basic phase-space coordinates are given by

$$\{r_i^\alpha(t), p_j^\beta(t)\} = \delta_{ij}\delta_{\alpha\beta} \quad (2430)$$

$$\{r_i^\alpha(t), r_j^\beta(t)\} = \{p_i^\alpha(t), p_j^\beta(t)\} = 0. \quad (2431)$$

We can view the time evolution of a system as a series of coordinate transformations $q(t = 0) \to q(t = t_1) \to q(t = t_2)$. If this sequence of jumps is small ($t' = t + \Delta$, where Δ is small), we can write Hamilton's equations in the form

$$p_i^\alpha(t') = p_i^\alpha(t) - \Delta \frac{\partial H}{\partial r_i^\alpha(t)}$$

$$r_i^\alpha(t') = r_i^\alpha(t) + \Delta \frac{\partial H}{\partial p_i^\alpha(t)}$$

up to terms of $\mathcal{O}(\Delta^2)$. It is then easy to show that the fundamental Poisson brackets, evaluated at time t', have the same form, given by Eqs. (2430) and (2431), as at time t except that t is replaced by t'. Coordinate transformations that preserve the fundamental Poisson bracket relations are known as *canonical transformations*. One can show that the Poisson bracket operation is itself invariant under canonical transformations: The Poisson bracket is the same whether the derivatives are performed using the original or the transformed phase-space coordinates.

Canonical Transformations

There are many ways of constructing canonical transformations [5]. We shall be interested in generating such transformations through application of infinitesimal canonical transformations. Let s be some real continuous parameter, and let G be a function of the phase-space coordinates q. Then, for sufficiently small s, the transformation

$$p_i^\alpha(s) = p_i^\alpha - s \frac{\partial G(q)}{\partial r_i^\alpha}$$

$$r_i^\alpha(s) = r_i^\alpha + s \frac{\partial G(q)}{\partial p_i^\alpha}$$

is a canonical transformation. The proof is direct. We compute the Poisson bracket

$$\begin{aligned}
\{r_i^\alpha(s), p_j^\beta(s)\} &= \{r_i^\alpha, p_j^\beta\} - s\left\{r_i^\alpha, \frac{\partial G}{\partial r_j^\beta}\right\} \\
&\quad + s\left\{\frac{\partial G}{\partial p_i^\alpha}, p_j^\beta\right\} + \mathcal{O}(s^2) \\
&= \delta_{ij}\delta_{\alpha\beta} - s\frac{\partial^2 G}{\partial p_i^\alpha \partial r_j^\beta} + s\frac{\partial^2 G}{\partial r_j^\beta \partial p_i^\alpha} + \mathcal{O}(s^2) \\
&= \delta_{ij}\delta_{\alpha\beta} + \mathcal{O}(s^2)
\end{aligned} \tag{2432}$$

and see that $q(s) \equiv (\mathbf{r}_i(s), \mathbf{p}_i(s))$ satisfy canonical Poisson bracket relations up to $\mathcal{O}(s^2)$. G is called the *generator* of the transformation. We can then use a series of such transformations to build up the continuous variation of $q(s)$ with s. Therefore, in going from s to $s + \Delta s$, we have

$$p_i^\alpha(s + \Delta s) = p_i^\alpha(s) - \Delta s \frac{\partial G(q(s))}{\partial r_i^\alpha(s)} + \mathcal{O}((\Delta s)^2)$$

$$r_i^\alpha(s + \Delta s) = r_i^\alpha(s) + \Delta s \frac{\partial G(q(s))}{\partial p_i^\alpha(s)} + \mathcal{O}((\Delta s)^2)$$

and therefore,

$$\lim_{\Delta s \to 0} \frac{[p_i^\alpha(s + \Delta s) - p_i^\alpha(s)]}{\Delta s} = \frac{dp_i^\alpha(s)}{ds} = -\frac{\partial G(q(s))}{\partial r_i^\alpha(s)} \tag{2433}$$

$$\lim_{\Delta s \to 0} \frac{[r_i^\alpha(s + \Delta s) - r_i^\alpha(s)]}{\Delta s} = \frac{dr_i^\alpha(s)}{ds} = \frac{\partial G(q(s))}{\partial p_i^\alpha(s)}. \tag{2434}$$

Thus we can determine $q(s)$ as a function of s by solving Eqs. (2433) and (2434). It is then easy to go on and determined the variation of some arbitrary function $A(q(s))$ with the parameter s. We have immediately that

$$\begin{aligned}
\frac{dA}{ds} &= \sum_{i,\alpha} \left[\frac{\partial A}{\partial r_i^\alpha(s)} \frac{dr_i^\alpha(s)}{ds} + \frac{\partial A}{\partial p_i^\alpha(s)} \frac{dp_i^\alpha(s)}{ds}\right] \\
&= \sum_{i,\alpha} \left[\frac{\partial A}{\partial r_i^\alpha(s)} \frac{\partial G}{\partial p_i^\alpha(s)} - \frac{\partial A}{\partial p_i^\alpha(s)} \frac{\partial G}{\partial r_i^\alpha(s)}\right] = \{A, G\},
\end{aligned} \tag{2435}$$

and the variation of A with s is driven by the Poisson bracket of A with the generator of the transformation G.

Comparing Eqs. (2428) and (2435), it is obvious that the Hamiltonian generates time translations. There are other interesting choices for the generator G. If

$$G = p_j^\beta, \qquad (2436)$$

where $\mathbf{p}_j$ is the momentum of the jth particle, then Eqs. (2433) and (2434) become

$$\frac{dr_i^\alpha(s)}{ds} = \frac{\partial G}{\partial p_i^\alpha} = \delta_{ij}\delta_{\alpha\beta}$$

and

$$\frac{dp_i^\alpha(s)}{ds} = -\frac{\partial G}{\partial r_i^\alpha} = 0.$$

We can easily integrate these equations to obtain

$$r_i^\alpha(s) = r_i^\alpha(0) + s\delta_{ij}\delta_{\alpha\beta} \qquad (2437)$$
$$p_i^\alpha(s) = p_i^\alpha(0). \qquad (2438)$$

s then has the interpretation as the length of a translation of the ith particle along the β direction. Similarly, the choice $G = -r_j^\beta$, generates translations of the momenta of the jth particle.

Symmetry Principles and Constants of the Motion

General Case. Note that if the generator of the transformation does not depend explicitly on the associated parameter s, then it is independent of the parameter s. This follows directly from Eq. (2435) since

$$\frac{dG}{ds} = \{G, G\} = 0 \qquad (2439)$$

and, for example, reflects the time independence of the Hamiltonian in the case where $G = H$. If the generator $G(q)$ is a constant of the motion,

$$\frac{dG}{dt} = 0, \qquad (2440)$$

then, using Eq. (2435), it has zero Poisson bracket with the Hamiltonian

$$\{G, H\} = 0. \qquad (2441)$$

This also means that the Hamiltonian is invariant under the transformation generated by G:

$$\frac{dH}{ds} = \{H, G\}$$
$$= -\{G, H\} = -\frac{dG}{dt} = 0. \qquad (2442)$$

Therefore, for each constant of the motion there is an invariance principle [22]. An invariance principle requires that the physical trajectories generated by Newton's law or Hamilton's equation, Eqs. (2425) and (2426), are unaffected by the associated transformation. Typically, this invariance results from the invariance of the Hamiltonian under the transformation. This is not always the case, as we discuss below.

We can then go through the various conserved quantities and identify the associated invariance principle:

Momentum Conservation. In the absence of external forces, the total momentum,

$$\mathbf{P}_T = \sum_{i=1}^{N} \mathbf{p}_i, \qquad (2443)$$

is conserved:

$$\frac{d\mathbf{P}_T}{dt} = 0 \qquad (2444)$$

and

$$\{\mathbf{P}_T, H\} = 0. \qquad (2445)$$

As indicated by Eqs. (2437) and (2438), the momenta generate translations in space. If $G = P_T^\beta$, then

$$\frac{dr_i^\alpha(s)}{ds} = \frac{\partial P_T^\beta}{\partial p_i^\alpha} = \delta_{\alpha,\beta} \qquad (2446)$$

$$\frac{dP_i^\alpha(s)}{ds} = -\frac{\partial}{\partial r_i^\alpha} P_T^\beta = 0 \qquad (2447)$$

and the total momentum in the β-direction generates translations of all coordinates by an amount s in the β-direction:

$$r_i^\alpha(s) = r_i^\alpha(0) + s\delta_{\alpha,\beta} \qquad (2448)$$
$$p_i^\alpha(s) = p_i^\alpha(0). \qquad (2449)$$

Thus $\{\mathbf{P}_T, H\} = 0$ indicates that H is invariant under translations in space:

$$\frac{\partial H}{\partial s} = 0. \tag{2450}$$

Clearly, the trajectories resulting from Hamilton's equations in the translated coordinate system are the same as those in the untransformed coordinate system. This should be contrasted with the case where just one particle is translated by a distance s. In this case, subsequent trajectrories in phase space will be modified due to the modified collision sequence in the fluid.

Angular Momentum Conservation. The total angular momentum,

$$\mathbf{L}_T = \sum_{i=1}^{N} \mathbf{r}_i \times \mathbf{p}_i, \tag{2451}$$

is conserved for systems interacting with radial forces:

$$\frac{d\mathbf{L}_T}{dt} = \{\mathbf{L}_T, H\} = 0. \tag{2452}$$

If we choose $G = L_T^\beta$, then Eqs. (2433) and (2434) give

$$\frac{dr_i^\alpha(s)}{ds} = \sum_\gamma \epsilon_{\alpha\beta\gamma} r_i^\gamma(s) \tag{2453}$$

$$\frac{dp_i^\alpha(s)}{ds} = \sum_\gamma \epsilon_{\alpha\beta\gamma} p_i^\gamma(s) \tag{2454}$$

where $\epsilon_{\beta\gamma\alpha}$ is the usual antisymmetric Levi-Civita tensor [6]. If, for example, $\beta = z$, then

$$\frac{dr_i^x(s)}{ds} = -r_i^y(s) \tag{2455}$$

$$\frac{dr_i^y(s)}{ds} = r_i^x(s) \tag{2456}$$

$$\frac{dr_i^z(s)}{ds} = 0. \tag{2457}$$

The solution to these differential equations is

$$r_i^x(s) = r_i^x \cos s + r_i^y \sin s \tag{2458}$$
$$r_i^y(s) = -r_i^y \cos s + r_i^x \sin s \tag{2459}$$

and s is the angle of rotation relative to the x-axis about the z-axis. The momentum is rotated in the same fashion. The conservation of total angular momentum implies that $\partial H/\partial s = 0$ and that the Hamiltonian is invariant under rotations.

Energy Conservation. This case is straightforward since energy conservation, $dH/dt = 0$, implies that H is invariant under time translations.

Galilean Invariance. In classical mechanics we know that the equations of motion are invariant under transformations where the velocities are translated by a constant vector. This serves to define the notion of an inertial reference frame. More formally, such a Galilean transformation [7] is generated by

$$G^\beta = \sum_i (m_i r_i^\beta - t p_i^\beta). \tag{2460}$$

This quantity is conserved when the total momentum is conserved,

$$\frac{dG^\beta}{dt} = \sum_i \left(p_i^\beta - p_i^\beta - t \frac{dp_i^\beta}{dt} \right)$$

$$\frac{dG^\beta}{dt} = -t \frac{dP_T^\beta}{dt} = 0. \tag{2461}$$

G^β generates the transformations,

$$\frac{dr_i^\alpha(s)}{ds} = \frac{\partial G^\beta}{\partial p_i^\alpha(s)} = -t \delta_{\alpha\beta} \tag{2462}$$

$$\frac{dp_i^\alpha(s)}{ds} = -\frac{\partial G^\beta}{\partial r_i^\alpha(s)} = -m_i \delta_{\alpha\beta} \tag{2463}$$

so

$$r_i^\alpha(s) = r_i^\alpha(0) - st \delta_{\alpha\beta} \tag{2464}$$

$$p_i^\alpha(s) = p_i^\alpha(0) - m_i s \delta_{\alpha\beta} \tag{2465}$$

and the transformation is such that velocities are translated by an amount $-s$ and positions by an amount $-st$ in the β-direction. Thus we transform into a coordinate system moving with the constant velocity $-s\hat{x}^\beta$. This symmetry is slightly different from the previous cases because the generator depends explicitly on time and Eq. (2435) must be replaced by

$$\frac{dH}{ds} = \{H, G^\beta\}$$

$$= -\{G^\beta, H\}$$

$$= -\frac{dG^\beta}{dt} + \frac{\partial G^\beta}{\partial t}. \tag{2466}$$

While Eq. (2461) shows that G^β is time independent and conserved, we have from Eq. (2460) that

$$\frac{\partial G^\beta}{\partial t} = -P_T^\beta(s)$$
$$= -\sum_i [P_i^\beta(t) - m_i s] \qquad (2467)$$

and

$$\frac{dH}{ds} = -P_T^\beta + Ms \qquad (2468)$$

where M is the total mass. In this case the Hamiltonian is not invariant under the transformation. If we let $s = V_\beta$, the β-component of the translation velocity, then Eq. (2468) can be rewritten as

$$\frac{dH}{dV_\beta} = -P_T^\beta + MV_\beta \qquad (2469)$$

and we can integrate to obtain

$$H(V) = H(0) - \mathbf{V} \cdot \mathbf{P}_T + \tfrac{1}{2} M \mathbf{V}^2. \qquad (2470)$$

This result also follows directly by inserting Eqs. (2464) and (2465) directly into the Hamiltonian and using the fact that H is invariant under translations in space. The last two terms in Eq. (2470) result from the kinetic energy term in the Hamiltonian. Even though the Hamiltonian is not invariant under Galilean transformations, we know that the equations of motion are invariant.

The importance of these and other symmetries and invariances is central to the development of statistical mechanics. Our results here are summarized in Table B.1.

Distortions

In this section we discuss how transformation theory can lead to a clean analysis of mechanical forces for systems in equilibrium. We build here on the discussion in Section 1.12.2, where we introduce the strain tensor. We are particularly interested in

TABLE B.1 Symmertries, Conservation Laws, and Generators

Generator	Symmetry	Conservation law
Total momentum	Translation	Momentum
Total angular momentum	Rotation	Angular momentum
Hamiltonian	Time translation	Energy
G	Galilean invariance	None

systems, such as solids, in which a distortion at the walls can be communicated to the bulk of the system. Such distortions can be treated [8] using the transformation theory we developed above.

Let us consider an infinitesimal transformation of the particle coordinates of the form

$$r_i^\alpha \to r_i'^\alpha = r_i^\alpha + \sum_\beta U_{\alpha\beta} r_i^\beta, \qquad (2471)$$

where $U_{\alpha\beta}$ is the strain tensor. As indicated in our general discussions of transformations, we look for transformations that are canonical. This means that we must transform the momentum variables $\mathbf{p}_i \to \mathbf{p}_i'$ such that the new set of phase-space coordinates $\mathbf{p}_i'$ and $\mathbf{r}_i'$ satisfies the canonical Poisson bracket relations given by Eq. (2430). It is easily seen that the new variables are canonical (for infinitesimal transformations) if the momenta transform as

$$p_i^\alpha \to p_i'^\alpha = p_i^\alpha - \sum_\beta U_{\alpha\beta} p_i^\beta. \qquad (2472)$$

It is also easy to see that the generator of the transformation is given by

$$S = \sum_i \sum_{\alpha\beta} p_i^\alpha U_{\alpha\beta} r_i^\beta. \qquad (2473)$$

The nature of the transformation generated by S can be appreciated by looking at the averages of Eqs. (2471) and (2472) for a solid. In this case we have $\langle p_i^\alpha \rangle = 0$, whereas due to the localization of particles near a lattice site, $\langle r_i^\alpha \rangle = R_\alpha$, where $\mathbf{R}$ is a lattice point in the solid. Then Eq. (2472) reduces to an identity and Eq. (2471) reads

$$R_\alpha' = R_\alpha + \sum_\beta U_{\alpha\beta} R_\beta. \qquad (2474)$$

This is the result studied in Chapter 1.

How does the Hamiltonian transform under such a distortive transformation? We obtain the first-order change in the Hamiltonian under this transformation by using the chain rule for differentiation for the change in the phase-space coordinates, Eqs. (2471) and (2472), and the explicit form for S given by Eq. (2473), to obtain

$$H[q'] - H[q] \equiv \delta H = \{H, S\}. \qquad (2475)$$

Note, however, that we can rewrite the generator (2473) in terms of the momentum density $\mathbf{g}(\mathbf{x})$ using Eq. (129), to obtain

$$S = \int d^3x \sum_{\alpha\beta} x^\beta U_{\alpha\beta} g^\alpha(\mathbf{x}). \qquad (2476)$$

Notice that if we choose the strain tensor to be antisymmetric,

$$U_{\alpha\beta} = \sum_{\gamma} \epsilon_{\alpha\beta\gamma}\theta_{\gamma}, \qquad (2477)$$

the generator takes the form

$$\begin{aligned} S_R &= \int d^3x \sum_{\alpha\beta} x^{\beta} \sum_{\gamma} \epsilon_{\alpha\beta\gamma}\theta_{\gamma} g^{\alpha}(\mathbf{x}) \\ &= -\int d^3x \, \mathbf{L}(\mathbf{x}) \cdot \boldsymbol{\theta} = -\mathbf{L}_T \cdot \boldsymbol{\theta} \end{aligned} \qquad (2478)$$

where $\mathbf{L}(\mathbf{x})$ is the angular momentum density and $\mathbf{L}_T$ the total angular momentum. Thus S_R generates uniform rotations about $\hat{\theta}$ through an angle θ. If we work with distortions that are symmetric, antisymmetric rotations such as that in Eq. (2477) are not included.

Returning to out general discussion, we have that the change in the Hamiltonian due to a distortion is given by

$$\delta H = \int d^3x \sum_{\alpha\beta} x^{\beta} U_{\alpha\beta} \{H, g^{\alpha}(\mathbf{x})\}. \qquad (2479)$$

We have already evaluated the Poisson bracket of the momentum density with the Hamiltonian when working out the equation of motion for $\mathbf{g}(\mathbf{x})$, as given in Eq. (352), in terms of the stress tensor:

$$\{g^{\alpha}(\mathbf{x}), H\} = -\sum_{\beta} \nabla^{\beta} \sigma_{\alpha\beta}(\mathbf{x}). \qquad (2480)$$

Inserting this result into the equation for δH above and integrating by parts, assuming that we obtain no contribution at the boundaries, we obtain the key result:

$$\delta H = -\int d^3x \sum_{\alpha,\beta} U_{\alpha\beta} \sigma_{\alpha\beta}(\mathbf{x}) \qquad (2481)$$

which is dicussed in some detain in Chapter 1.

The work balance result given by Eq. (352) can be obtained in another complementary manner using transformation theory. If we remember that the system is stationary in equilibrium, then, for any observable $A(q)$, we have

$$\frac{d\langle A \rangle}{dt} = 0. \qquad (2482)$$

If we choose $A = S$, the generator of the distortive transformation, in the equation of motion, we obtain

$$\frac{d\langle S \rangle}{dt} = \langle \{S, H_T\} \rangle = 0, \tag{2483}$$

where H_T is the total Hamiltonian, including the walls containing the fluid. The external contribution, using Eq. (341), can be written as

$$\langle \delta H_E \rangle = \langle \{H_E, S\} \rangle = -\int d^3x \, U_{\alpha\beta} x_\alpha \langle F_E^\beta(\mathbf{x}) \rangle, \tag{2484}$$

and the balance between work done on the system and the work done by the system reduces to

$$\langle \delta H \rangle = -\int d^3x \, U_{\alpha\beta} \langle \sigma_{\alpha\beta} \rangle = -\langle \delta H_E \rangle = \int d^3x \, U_{\alpha\beta} x_\alpha \langle F_E^\beta(\mathbf{x}) \rangle \tag{2485}$$

which is the same result found in Chapter 1 [see Eq. (351)].

Evolution Equations and Averages

In this section we want to work out some useful formal results needed in Chapter 1 and Appendix C. We showed above that changes in a function $A(q)$ of the phase-space coordinates generated by the function G and associated with the parameter s are governed by the equation

$$\frac{dA}{ds} = \{A, G\}. \tag{2486}$$

It will be convenient to introduce the *operator* $\hat{L}_G$ defined by

$$\hat{L}_G A \equiv -i\{A, G\} \tag{2487}$$

such that Eq. (2486) can be rewritten as

$$\frac{dA}{ds} = i\hat{L}_G A. \tag{2488}$$

Since $\hat{L}_G$ is independent [9] of the parameter s, we can formally integrate this equation to obtain the evolution equation:

$$A(s) = \hat{U}_G(s) A(s=0), \tag{2489}$$

where the evolution operator is given formally by

$$\hat{U}_G(s) = e^{i\hat{L}_G s}. \tag{2490}$$

For the case of time evolution, where the generator is the Hamiltonian, and $s = t$ (time), $\hat{L}_G \equiv \hat{L}$ is the Liouville operator, and

$$\hat{U}_H(t) \equiv \hat{U}(t) = e^{i\hat{L}t} \tag{2491}$$

is known as Koopman's operator [10]. Clearly, $\hat{U}(t)$ propagates points in Γ-space forward in time. If G is the total momentum P_T^z, then $U_{P_T^z}(z_o)$ translates the coordinates of the system a distance z_o along the z-axis (see Problem 1.45).

This formal development allows us to view our averages of observables in a different manner. The average value of a variable A at time t is given by Eq. (29). Using Eq. (2489), we can rewrite Eq. (29) in the form

$$\langle A(t) \rangle = \int dq(0) P[q(0)] \hat{U}(t) A(q(0)) \tag{2492}$$

where, for simplicity, we use $q(0)$ rather than $q_N(0)$. Note however, since $\hat{L}$ is a first-order differential operator, we easily find, after integrating by parts and canceling some terms, that for general functions A and B,

$$\int dq(0) A(q(0)) \hat{L} B(q(0)) = - \int dq(0) B(q(0)) \hat{L} A(q(0)). \tag{2493}$$

Inserting the series representation for $\hat{U}(t)$,

$$\hat{U}(t) = \sum_{n=0}^{\infty} \frac{(i\hat{L}t)^n}{n!}, \tag{2494}$$

in Eq. (2492), followed by reated use of Eq. (2493), leads to the result

$$\langle A(t) \rangle = \int dq(0) \sum_{n=0}^{\infty} \frac{(it)^n}{n!} P[q(0)] (\hat{L})^n A(q(0))$$

$$= \int dq(0) \sum_{n=0}^{\infty} \frac{(it)^n}{n!} \left(-\hat{L} P[q(0)] \right) (\hat{L})^{n-1} A(q(0))$$

$$= \int dq(0) \sum_{n=0}^{\infty} \frac{(it)^n}{n!} ((-\hat{L})^n P[q(0)]) A(q(0))$$

$$= \int dq(0) A(q(0)) e^{-i\hat{L}t} P[q(0)]$$

$$= \int dq(0) A(q(0)) \hat{U}(-t) P[q(0)]$$

$$\langle A(t) \rangle = \int dq(0) A(q(0)) P[q(-t)]. \tag{2495}$$

Comparing Eqs. (2492) and (2495), we see that we can look at this situation in two ways. In the first, variables change in time and the initial probability distribution is fixed. This is known [11] in quantum mechanics as the Heisenberg representation. In the second view, the *Schrödinger representation*, the probability distribution evolves in time and the observables are time independent. Notice that the probability distribution in this case evolves backward in time and obeys the Liouville equation [12]

$$\frac{d}{dt} P[q(-t)] = -i\hat{L} P[q(-t)]. \tag{2496}$$

C MOTION IN PHASE SPACE

Deterministic Mechanics

The basic program in classical mechanics is to map out flows in Γ-space resulting from a solution of Newton's laws. Given the initial point in phase space, $q(0) = q_0$, we can unambiguously determine the position in phase space at a later time t by propagating the system forward in time. As discussed in Chapter 1, a particular trajectory can be characterized by $6N - 1$ constants of the motion. Let us look at this here in more detail. Suppose, as suggested in Chapter 1, that we construct the $6N - 1$ constants or *integrals* of the motion,

$$\phi_i = \phi_i(q(t)) \tag{2497}$$

for $i = 1$ to $6N - 1$ and an additional equation giving $t = t(q(t))$. Of course, the energy,

$$E = H(q(t)), \tag{2498}$$

can be chosen as one of these constants of the motion in an isolated system.

Knowledge of these $6N - 1$ constants of the motion is equivalent to a determination of the trajectory of the system in phase space. This follows since each constant of the motion restricts the motion of the system to a $(6N - 1)$-dimensional hypersurface. For example, conservation of energy requires that for a given and fixed energy, there is a surface in phase space on which the system must remain. The intersection of the $6N - 1$ hypersurfaces defines a line in phase space corresponding to the actual trajectory of the system. The time when the system is at a given point is given by $t = t(q(t))$.

Example: Noninteracting Particles. A very simple example may help. Suppose that we have a single noninteracting particle in two dimensions. The solution to the equations of motion is, of course, given in terms of the *initial* positions $\mathbf{r}_0$ and momentum $\mathbf{p}_0$ by

$$x(t) = x_0 + \frac{p_0^x}{m} t \tag{2499}$$

$$y(t) = y_0 + \frac{p_0^y}{m} t \tag{2500}$$

$$p^x(t) = p_0^x \tag{2501}$$

$$p^y(t) = p_0^y. \tag{2502}$$

We can invert these equations to obtain

$$x_0 = x(t) - \frac{p_0^x(t)}{m} t \tag{2503}$$

$$y_0 = y(t) - \frac{p_0^y(t)}{m} t \tag{2504}$$

$$p_0^x = p^x(t) \tag{2505}$$

$$p_0^y = p^y(t). \tag{2506}$$

We can then solve Eq. (2503) for t, to obtain

$$t = \frac{m}{p_0^x(t)} [x(t) - x_0] \tag{2507}$$

and we can then insert this result into Eqs. (2504)–(2506) to obtain the constants of motion,

$$\Phi_1 = y(t) - y_0 - \frac{p_0^y(t)}{p_0^x(t)} [x(t) - x_0] \tag{2508}$$

$$\Phi_2 = p^x(t) - p_0^x \tag{2509}$$

$$\Phi_3 = p^y(t) - p_0^y. \tag{2510}$$

Φ_2 and Φ_3 are trivial, but Φ_1 tells us that the trajectory in the x–y plane is a straight line with slope p_0^y/p_0^x.

In principle, we can reduce the problem of deterministic classical mechanics to the determination of the constants of the motion. Although this program seems attractive, we shall see that in practice, the situation is extremely complicated even in very simple situations. In particular, we will see that all of these constants of the motion should *not* be treated on the same footing.

Types of Flow in Phase Space

The study of the time evolution or flows in Γ-space was pioneered [13] by von Neumann and Birkhoff. They were motivated by the need to understand the works of

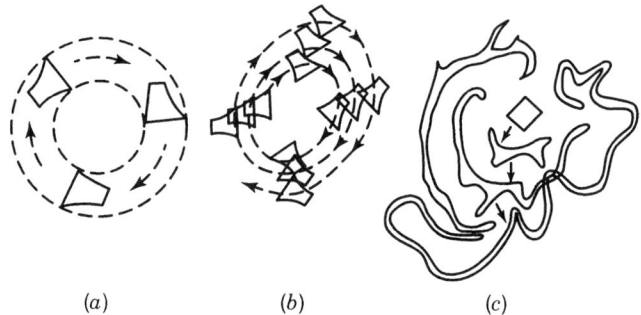

FIGURE C.1 Types of flow in phase space: (a) nonergodic; (b) ergodic but not mixing; (c) mixing. (From R. Balescu, *Equilibrium and Nonequilibrium Statistical Mechanics*, Wiley, New York, 1975.)

Boltzmann [14] and Gibbs [15], which were considered basic to statistical mechanics but which did not make sense in a rigorous mathematical formulation. The basic approach is to study domains in phase space and analyze the motion of these sets of points as the system evolves in time.

Consider a domain γ_0 of phase space (see Fig. C.1) at a time t_0. The *volume* of phase space occupied by γ_0 is simply

$$V(\gamma_0) = \int_{\gamma_0} dq_0. \tag{2511}$$

It will be useful to introduce a function D_γ, defined in phase space q, that projects out the volume γ in phase space:

$$D_\gamma = \begin{cases} 1 & q \in \gamma \\ 0 & \text{otherwise}. \end{cases} \tag{2512}$$

Then we can rewrite Eq. (2511) as

$$V(\gamma_0) = \int_\Gamma D_{\gamma_0} dq_0. \tag{2513}$$

As time evolves, the domain γ_0 is mapped (essentially on a point-by-point basis) onto the region γ_t at time t. The volume of the region γ_t is given by

$$V(\gamma_t) = \int_\Gamma D_{\gamma_t} dq_0. \tag{2514}$$

Whether we use $dq_0 = dq(t=0)$ or $dq(t)$ is irrelevant since [16] $dq_0 = dq(t)$. This implies, of course, that the magnitude of the Jacobian of the transformation

$q_0 \to q(t)$ is 1. Since we have in general, using Eq. (2489) from Appendix B,

$$D_{\gamma t} = \hat{U}(t) D_{\gamma 0}, \qquad (2515)$$

where $\hat{U}(t)$ is the time evolution operator, we can rewrite Eq. (2514) as

$$V(\gamma_t) = \int_\Gamma (\hat{U}(t) D_{\gamma 0}) dq_0. \qquad (2516)$$

Since the derivatives in the Liouville operator $\hat{U}(t)$ can be taken with respect to q_0, we can integrate repeatedly by parts and obtain

$$\begin{aligned} V(\gamma_t) &= \int_\Gamma dq_0 D_{\gamma 0} \hat{U}(t) \\ &= \int_\Gamma dq_0 D_{\gamma 0} = V(\gamma_0). \end{aligned} \qquad (2517)$$

Therefore, the volume of any domain of phase space is preserved by the motion. This is a statement of Liouville's theorem [17].

As one looks at the point-by-point mapping with time of a smooth volume γ_0, one finds that the bulk of γ_0 is mapped onto a smooth portion γ_t. There may be erratic points or *areas* of points that do not map onto the bulk volume of γ_t. Such points form a subset of zero volume (or more precisely, zero measure) and are ignored in the following developments, which are based on the concept of measure. We should keep in mind that we are throwing out various *pathological* initial conditions, but these may, in certain circumstances, be the interesting physical situations. If we ignore these points of measure zero, we can make a number of statements about the general, global properties of the motion in phase space.

As we proceed, there is one question to keep in mind. Consider two nearby points in Γ-space at time t_0; how do they move relative to one another as time evolves? Does the trajectory of one remain close to the other, or do they deviate from each other systematically?

There are three basic types of flow in phase space: nonergodic, ergodic, and mixing. Let us first discuss them from a qualitative point of view. The type of flow shown in Fig. C.1*a* is called *nonergodic*. In this case the domain γ_0 moves through Γ without distortion and is restricted to a particular portion of Γ. Typically, the motion will be quasiperiodic, and as time evolves, the motion is restricted to a finite fraction of Γ-space. In this case there will be a part of phase space that will not be visited by γ_0. Examples are given later.

In the motion depicted in Fig. C.1*b* the domain γ_0 is only slightly distorted as the system moves with time, but the distortion is significant enough that the volume does not return to its initial position in Γ-space. As time evolves, the volume element sweeps out essentially the entire phase space (with the exception of regions of measure zero). This type of flow is called *ergodic* [18].

Finally, we have the type of flow shown in Fig. C.1c. In this case the shape of the volume element is violently distorted as it evolves in time. Individual points that start out close together in phase space have widely diverging trajectories. This type of flow is called *mixing* [19], and the mixing property is associated with systems in which the distance between initially neighboring points diverges exponentially [20] in time. Eventually, in a mixing system, a volume $V(\gamma_0)$ will evolve into extremely fine filaments that spread uniformly over the entire Γ-space.

Ergodic Theorems

Let us now discuss the mathematical ramifications of these different types of motion. Physically, nonergodic motion is distinct from ergodic and mixing motions. In the nonergodic case things are periodic and there is no seperation of nearby points with time. The system just keeps repeating itself. However, in the ergodic and mixing cases the system appears to become more distorted and chaotic as time evolves. Thus, the position of the system in Γ-space at long times for these cases will be essentially randomly distributed.

As discussed in Chapter 1, Boltzmann originally introduced the ergodic hypothesis by suggesting that every system will, if left to itself, sooner or later, pass through every point in Γ-space consistent with conservation of energy. Rosenthal and Plancherel [21] showed in 1913 that a one-dimensional trajectory (moving point), even if it fills densely a higher-dimensional domain, cannot go through every point in that domain. Although Boltzmann's ergodic hypothesis failed in detail, it inspired a *quasi* ergodic theorem [22] which can be proven rigorously.

Theorems concerning ergodic flow can be proven only under certain restrictive conditions. The most important condition is that the probability distribution be stationary. We must be able to prepare our system such that

$$P[q(-t)] = P[q_0]. \tag{2518}$$

We describe this situation as being controlled by an *invariant ensemble*. If we have an invariant ensemble, averages are time independent:

$$\langle A(t) \rangle = \int dq_0 \, A(q_0) P[q(-t)]$$
$$= \int dq_0 \, A(q_0) P[q_0]. \tag{2519}$$

We can think of constructing $P[q_0]$ as a function of the constants of the motion Φ. Later we must come back and discuss the physical reasonableness of preparing such an invariant ensemble.

Next we must develop the concept of measure a bit further. The measure of an infinitesimal domain of phase space is defined as

$$d\mu = P[q] \, dq \tag{2520}$$

where we require that

$$P[q] \geq 0. \tag{2521}$$

The measure of a finite volume γ is defined as

$$\mu(\gamma) = \int_\gamma d\mu = \int_\Gamma dq\, P[q] D_\gamma. \tag{2522}$$

It is convenient to choose $P[q]$ such that the measure of the entire Γ-space is 1:

$$\mu(\Gamma) = \int_\Gamma dq\, P[q] = 1. \tag{2523}$$

With these definitions, the measure of any part γ of Γ-space is a number satisfying

$$0 \leq \mu(\gamma) \leq 1. \tag{2524}$$

For the remainder of this section we restrict our discussion to the case of an invariant ensemble [see Eq. (2518)]. With this restriction, we find directly that

$$\begin{aligned}\mu(\gamma_t) &\equiv \int_\Gamma dq_0 P[q_0] D_{\gamma t} = \int_\Gamma dq_0 P[q_0] \hat{U}(t) D_{\gamma 0} \\ &= \int_\Gamma dq_0 D_{\gamma 0} \hat{U}(-t) P[q_0] = \int_\Gamma dq_0 D_{\gamma 0} P[q_0, -t] \\ &= \int_\Gamma dq_0 D_{\gamma 0} P[q_0] = \mu(\gamma_0),\end{aligned} \tag{2525}$$

and any *invariant* measure of a phase element is conserved under time evolution.

In this case Birkhoff [23] was able to prove a revised form of the ergodic theorem: If $A(q)$ is a dynamical function that is integrable over all phase space,

$$\int_\Gamma d\mu |A(q)| < \infty, \tag{2526}$$

the time integral

$$\bar{A}(q) = \lim_{T \to \infty} \frac{1}{T - t_0} \int_{t_0}^T dt\, A(q(t)) \tag{2527}$$

has a finite limit independent of the lower limit of integration for *almost all* $q(t_0)$ (except for a set of measure zero). Furthermore, he showed that if the phase space of the system cannot be partitioned into invariant regions, then (1) $\bar{A}(q)$ is constant almost everywhere in Γ (independent of q) and (2)

$$\bar{A}(q) = \langle A(q) \rangle = \int_\Gamma d\mu A(q) \qquad (2528)$$

and the time average equals the phase-space average.

The relation, given by Eq. (2528), between time and space averages essentially *defines* what we mean by ergodic flow and reflects the fact that for such systems almost every trajectory of the system spends equal times in equal regions everywhere in Γ-space. The ergodic theorem is important, when it holds, because it tells us that experimentally, the process of time averaging (over a time-dependent system) is equivalent to ensemble averaging. Clearly, ensemble averaging requires much more work since we must prepare the system many times. In the time average we prepare it once and let it evolve in time. An important consequence [23] of the ergodic theorem is that there is only *one* invariant probability distribution for a system that is ergodic, and that is

$$P_0[q] = P_0 \qquad (2529)$$

where P_0 is a constant in both phase space and time. This means simply that if the system is ergodic and stationary, we are equally likely to find the system anywhere in the available space (up to points of zero measure).

The ergodic theorem proven by Birkhoff holds if and only if the phase space cannot be separated into two invariant regions, say γ_1 and γ_2, whose measures are different from zero or 1. Thus it is required that no trajectory can be confined to a finite portion of phase space but rather, must wander through the entire Γ-space. This brings us back to our discussion of the constants of the motion for a system. Apparently, the constants of the motion can be divided into two kinds, isolating and nonisolating. *Isolating constants* define a hypersurface that divides Γ-space into invariant regions, and a system with such isolating constants is not ergodic. *Nonisolating constants* do not partition Γ-space; apparently, their associated hypersurfaces over *essentially* all of the phase space.

A discrete set of isolating constants such as the energy is not a serious problem, since we can limit our analysis to the motion of systems on a surface of constant energy and ask: Is the motion on *this* surface ergodic? If there are other isolating constants, the process of reducing the region of phase space sampled is continued. One of the main problems in ergodic theory is to determine the number of isolating constants in a given system. One encouraging result (which we discuss a bit more later) is that for N hard spheres in a box, the only isolating constant is the energy. In general, however, it is extremely difficult to prove rigorously that the flow for a given system is isolating or nonisolating.

The revised ergodic theorem due to Birkhoff requires not only that the associated probability distribution be invariant, but that it be a constant independent of the phase-space coordinates. Can we prepare a system in an ensemble where $P[q, -t_0]$ is a constant independent of the phase-space coordinates, q? This is impossible for the simplest of reasons—at the microscopic level the particles move and the probability distribution must reflect this motion. Thus we cannot prepare our system in an invariant ensemble! It appears that quasiergodicity itself is not a very useful property of a system since we cannot prepare such systems in an invariant ensemble where the theorems of interest hold!

It is at this stage that we must remember those physical situations where the average value of a quantity (e.g., the pressure of a gas) is independent of time. Such situations exist for mixing systems. The mixing property can be stated mathematically for two functions, $A(q)$ and $B(q)$, that are square integrable on Γ, in the form

$$\lim_{t \to \pm\infty} \int_\Gamma A(q_0) B(q,t) \, dq_0 = \frac{\int_\Gamma B(q_0) \, dq_0 \int_\Gamma A(q_0) \, dq_0}{\int_\Gamma dq_0} \tag{2530}$$

where $q(0) = q_0$. To understand the meaning of this result, let us choose

$$B[q(-t)] = P[q(-t)], \tag{2531}$$

so that (2530) becomes

$$\lim_{t \to \pm\infty} \int_\Gamma A(q_0) P[q(-t)] dq_0 \equiv \lim_{t \to \pm\infty} \langle A(t) \rangle$$

$$= \frac{\int_\Gamma P[q_0] dq_0 \int_\Gamma A(q_0) dq_0}{\int_\Gamma dq_0}. \tag{2532}$$

However, since $P[q_0]$ is normalized,

$$\int_\Gamma P[q_0] dq_0 = 1, \tag{2533}$$

we have

$$\lim_{t \pm \infty} \langle A(t) \rangle = \frac{\int_\Gamma A(q_0) dq_0}{\int_\Gamma dq_0} = \int_\Gamma A(q_0) d\mu_0. \tag{2534}$$

Therefore, the long-time average of an observable in a mixing system is equal to the invariant ensemble average where the ensemble is governed by the constant probability distribution

$$P[q_0] = \frac{1}{\int_\Gamma dq_0} \equiv P_0. \tag{2535}$$

If we prepare a mixing system in a state governed by the probability distribution $P[q, -t_0]$ at time $t = -t_0$ and wait a long time t_0 to the time $t = 0$, we can assume that the system has mixed to such a degree that averages can be computed using the constant invariant ensemble. It is important to remember that mixing gives a *coarse-grained* approach to a stationary state. This follows from our earlier discussion that $P[q(-t)]$ must, on the most detailed level, depend on time, since the particles continue to move about. However, if we average over small volumes of Γ-space, coarse-grain, the motion looks stationary.

A standard example of this phenomenon is to visualize a beaker containing oil and water. We prepare the oil and water so they are initially separated and assume that they cannot diffuse into one another. We then stir the liquids and, while the total volumes of the oil and water remain fixed, to some degree, they *mix*. On average, the density of oil is nearly uniform throughout the beaker. In principle, if we are careful enough, we can stir the oil back into its original shape. In this sense mixing behavior remains microscopically time reversible.

Let us now connect the mathematical statement of mixing with a more geometrical picture. Suppose that a function D_B projects out of Γ-space a specified volume labeled B:

$$\mu_0(B) = \int_\Gamma D_B d\mu_0. \tag{2536}$$

Then we can look at the quantity $\int_\Gamma D_B D_{A_t} d\mu_0$, which gives the overlap between a volume in phase space B and a volume A that has evolved in time to a volume A_t. We can write

$$\mu_0(A_t \cap B) = \int_\Gamma D_B D_{A_t} d\mu_0, \tag{2537}$$

which is the measure of the intersection of the two volumes at time t. Our mixing property (2530) tells us that

$$\lim_{t \to \pm\infty} \int_\Gamma D_B D_{A_t} dq_0 = \frac{\int_\Gamma D_B dq_0 \int_\Gamma D_{A_t} dq_0}{\int_\Gamma dq_0}. \tag{2538}$$

If we multiply both sides of Eq. (2538) by the invariant probability distribution P_0, we obtain for the left-hand side,

$$\lim_{t \to \pm\infty} \int_\Gamma D_B D_{A_t} P_0 \, dq_0 = \lim_{t \to \pm\infty} \mu_0(A_t \cap B), \tag{2539}$$

while for the right-hand side, using Eq. (2522), we obtain

$$\frac{\int_\Gamma D_B P_0 \, dq_0 \int_\Gamma D_A P_0 \, dq_0}{\int_\Gamma P_0 \, dq_0} = \mu_0(B)\mu_0(A). \tag{2540}$$

Our mixing property can then be written in the form

$$\lim_{t \to \pm\infty} \mu_0(A_t \cap B) = \mu_0(B)\mu_0(A). \tag{2541}$$

We can imagine that as A_t evolves in time it produces a progressively finer nonintersecting web. Eventually, as time evolves and A spreads throughout Γ and is distributed more uniformly, the measure of the intersection $\mu_0(A_t \cap B)$ will be in the same ratio to the measure $\mu_0(B)$ as the total measure $\mu_0(A_t)$ is to the measure of the complete space:

$$\frac{\mu_0(A_t \cap B)}{\mu_0(B)} = \frac{\mu_0(A_t)}{\mu_0(\Gamma)}. \tag{2542}$$

Since the flow is measure preserving,

$$\mu_0(A_t) = \mu_0(A) \tag{2543}$$

and from Eq. (2523), $\mu_0(\Gamma) = 1$, so Eq. (2542) reduces to Eq. (2541).

It is easy to show that mixing implies ergodicity. Let us suppose that there exists an invariant subset of phase space A:

$$A_t = A. \tag{2544}$$

Then

$$\mu_o(A_t \cap A) = \mu_0(A \cap A) = \mu_0(A). \tag{2545}$$

Let $B = A$ in our statement of mixing, Eq. (2541); then as $t \to \pm\infty$,

$$\mu_0(A) = [\mu_0(A)]^2, \tag{2546}$$

and $\mu_0(A)$ equals 0 or 1. Therefore, there is *no* invariant subset with nonzero measure, and the system is therefore ergodic. The converse is not true. Ergodicity does not imply mixing.

We have now classified certain types of flows in phase space and discussed the long-time *random* properties of ergodic and mixing systems. However, we have not addressed the important question: Given a dynamical system, whose Hamiltonian, $H(q)$, is known, is it ergodic? Is it mixing? The answer to these questions in general is very difficult. Let us look at the simplest case of a set of oscillators.

Physical Systems

Single Harmonic Oscillator. We start with the simplest example of an interacting dynamical system, a single harmonic oscillator described by the Hamiltonian

$$H = \frac{p^2}{2m} + \frac{1}{2}kx^2. \tag{2547}$$

In this case the energy E, a particular value of H, is clearly an isolating constant. The other variable governing the motion in phase space is an angle ϕ that is related to the momentum p and position x by

$$p = (2mE)^{1/2} \cos \phi \tag{2548}$$

and

$$x = -\left(\frac{2E}{k}\right)^{1/2} \sin \phi. \tag{2549}$$

It is easy to see that this angle has the simple time dependence

$$\phi = \omega t + \phi_0 \tag{2550}$$

where the oscillator frequency is given by

$$\omega^2 = \frac{k}{m} \tag{2551}$$

and ϕ_0 is the value of ϕ at $t = 0$. We see that as ϕ increases, the system periodically maps out all of the available phase space. Clearly, this system is ergodic on the energy surface. However, the system is not mixing, since a fixed volume A will move without distortion and its intersection with a fixed domain B is alternately empty and nonempty, and its measure does not have a limit as $t \to \infty$.

Two Uncoupled Harmonic Oscillators. Consider next the case of two uncoupled harmonic oscillators described by the Hamiltonian

$$H = \tfrac{1}{2}(p_1^2 + \omega_1^2 x_1^2) + \tfrac{1}{2}(p_2^2 + \omega_2^2 x_2^2) \tag{2552}$$

where we set the masses equal to 1 and introduce the frequencies ω_1 and ω_2. Let us immediately make the change or variables

$$p_i = (2\omega_i J_i)^{1/2} \cos \phi_i \tag{2553}$$

$$x_i = -\left(\frac{2J_i}{\omega_i}\right)^{1/2} \sin \phi_i \tag{2554}$$

for $i = 1$ and 2, so that the Hamiltonian given by Eq. (2552) reduces to

$$H = \omega_1 J_1 + \omega_2 J_2. \tag{2555}$$

In this case J_i and ϕ_i are action-angle variables [24] and it is easy to verify that we have the equations of motion

$$\dot{J}_i = -\frac{\partial H}{\partial \phi_i} = 0 \tag{2556}$$

$$\dot{\phi}_i = \frac{\partial H}{\partial J_i} = \omega_i. \tag{2557}$$

Clearly, J_1 and J_2 are constants of the motion (in this case the individual energies $E_i = \omega_i J_i$ are conserved). The angles are given simply by

$$\phi_i = \omega_i t + \phi_i^0. \tag{2558}$$

It is very convenient to describe the motion in terms of a torus. We measure J_1 along the interior axis and J_2 along the radius of the transverse cross section. Since J_1 and J_2 are isolating constants, the motion on the total energy surface <u>is *not* ergodic</u>. What about the motion on the torus? There are two possibilities.

1. ω_1/ω_2 *is rational.* Suppose that $\omega_2 = 2\omega_1$. Clearly, the ϕ_2 variable goes through two cycles to each 2π cycle for ϕ_1. A more detailed analysis, discussed in Problem 1.15, shows that there is a third and final constant of the motion, Φ_3, which is a two-branched function that is isolating. The motion is not ergodic. Similarly, any rational ratio ω_2/ω_1 leads to periodic motion, and Φ_3 is an isolating constant corresponding to multivalued function with a finite number of branches.
2. ω_1/ω_2 *is irrational.* If the frequencies ω_1 and ω_2 are *incommensurate*, it is clear that after starting at some points (ϕ_1^0, ϕ_2^0), the system will *never* come back to the initial position. The trajectory fills the surface of the torus densely and never closes on itself. The constant of the motion Φ_3 exists, but it is a very pathological function. It is a multivalued function with an infinite number of branches. Clearly, it is a nonisolating constant of the motion, and the system is ergodic on the surface of the torus.

We have seen that the direct program of classical mechanics is somewhat illusory. It is not true in general that finding $2N - 1$ constants of the motion (for N oscillators) will confine the motion of the system to a nice regular curve in phase space. For a system of N uncoupled oscillators, only N constants are isolating; the remaining $N - 1$ constants are, in general, nonisolating (up to a subset of zero measure). Hence the trajectory of the system densely fills an N-dimensional region of phase space.

Simple Pendulum. It is, of course, true that harmonic oscillators are very simple systems and one might expect more complex behavior from more complex systems. The simplest generalization of a harmonic oscillator is the simple pendulum [25] described by the Hamiltonian

$$H = \frac{p^2}{2} - \omega_0^2 \cos \phi \tag{2559}$$

which leads to the equations of motion

$$\dot{\phi} = \frac{\partial H}{\partial p} = p \tag{2560}$$

and

$$\dot{p} = -\frac{\partial H}{\partial \phi} = -\omega_0^2 \sin\phi. \tag{2561}$$

Clearly, for small displacements, these reduce to those for the harmonic oscillator. The solution of Eqs. (2560) and (2561) can be written in the closed form [25]

$$\sin\frac{\phi}{2} = k\,\mathrm{sn}(\omega_0 t, k) \tag{2562}$$

where sn is the Jacobi elliptic function [26] and

$$k^2 = \frac{1}{2}\left(\frac{E}{\omega_0^2} + 1\right). \tag{2563}$$

As the energy $H = E$ ranges from $-\omega_0^2$ to ω_0^2, k goes from 0 to 1 and we have bounded *pendulum motion* with the period given by

$$T(k) = \frac{4K}{\omega_0} \tag{2564}$$

where K is the elliptic integral

$$K = \int_0^{\pi/2} d\theta (1 - k^2 \sin^2\theta)^{1/2}. \tag{2565}$$

However, for $E > \omega_0^2$ one has free rotation and unbounded motion for ϕ. The important point here is that as we put more energy into the system, we can obtain a qualitative change in the behavior from bounded to unbounded motion. The ergodic properties of the simple pendulum are discussed in Problem 1.14.

Coupled Anharmonic Oscillators. Clearly, a set of N harmonic oscillators is *not* ergodic on the hypersurface corresponding to conserved energy. Before the middle 1950s, it was widely believed that this lack of ergodicity would be eliminated for coupled oscillators by any small amount of anharmonicity in the system [27].
The idea is that the anharmonicity will cause the various N invariants (corresponding to the energy of each oscillator) to be coupled and will create paths through the Γ-space compatible only with conservation of total energy. Soon after the development of modern computers, Fermi, Ulam and Pasta, [28] set out to demonstrate this *equipartition of energy* phenomena by studying the interaction of 64 oscillators with anharmonic couplings. As demonstrated by the following example, the situation is more complicated than they had thought.

Consider the improbable *Toda* Hamiltonian [29]

$$H_T = \tfrac{1}{2}(p_x^2 + p_y^2) + \tfrac{1}{24}\left(e^{2x - 2y\sqrt{3}} + e^{2x + 2y\sqrt{3}} + e^{-4x}\right) - \tfrac{1}{8} \tag{2566}$$

governing momenta p_x and p_y and positions x ad y. A study of the Poincaré section [30] for this Hamiltonian shows that the curves do not fill out a three-dimensional volume. This implies that there is another constant of the motion that has been found to be given by

$$I = 8\dot{y}(\dot{y}^2 - 3\dot{x}^2) + (\dot{y} + \dot{x}\sqrt{3})e^{2x-2y\sqrt{3}} + (\dot{y} - \dot{x}\sqrt{3})e^{2x+2y\sqrt{3}} - 2\dot{y}e^{-4x}. \quad (2567)$$

In this case one has nested tori with H_T and I serving as constants associated with J_1 and J_2. This is clearly a nontrivial integrable system and one does not have equipartition of energy.

Suppose now that we look at the related system obtained by expanding the Toda Hamiltonian, H_T, in powers of x and y:

$$H = \tfrac{1}{2}(p_x^2 + p_y^2) + \tfrac{1}{2}(x^2 + y^2) + xy^2 - \tfrac{1}{3}x^3. \quad (2568)$$

This Hamiltonian, studied by Henon and Heiles [31], models the motion of a star in the average field of a galaxy! This is reduced to the motion of a single particle under the influence of a cylindrically symmetric potential. Henon and Heiles studied trajectories for a variety of energies. For the energy $E = \tfrac{1}{12}$ they found *only* smooth curves indicating that to computer accuracy there is an additional isolating integral and the motion is not ergodic. This is in agreement with the findings of Fermi, Ulam, and Pasta and those for the Toda Hamiltonian. A small amount of anharmonicity does *not* eliminate all of the isolating integrals of the motion beyond energy! As one increases the energy (and the effects of anharmonicity) to $E = \tfrac{1}{8}$, the simple picture found at lower energies begins to break down. One finds that some trajectories are nonergodic and some are ergodic. For the higher energy $E = \tfrac{1}{6}$, almost no stable motion remains. A single trajectory wanders over almost the entire surface. In a very small energy range, the system has undergone a transition from stable to chaotic behavior.

It is important to note, comparing the Toda and Henon and Heiles Hamiltonians, that small changes in a completely integrable Hamiltonian can make significant qualitative changes in the dynamics. Even in the case where we have only four degrees of freedom and energy conservation, we see that the question of ergodicity is complicated and depends on the value of the energy. For low energies the system is nonergodic, but at high energies it is *probably* ergodic.

KAM Theorem

In the low-energy regions of our previous examples the system is nonergodic. It turns out that this behavior is rather general, and at just about the same times as the Fermi, Ulam, and Pasta work, Kolmogorov conjectured, and Arnold and Moser independently proved, the *KAM theorem* [32]. The KAM theorem applies to systems governed by a Hamiltonian of the form

$$H(J, \phi) = H_0(J) + \lambda H_1(J, \phi) \quad (2569)$$

where we have a set of N constant J's $(J_1, J_2, \ldots, J_N)$, and N phase variables $(\phi_1, \phi_2, \ldots, \phi_N)$. $H_0(J)$ is the Hamiltonian describing a completely integrable system (say, a set of uncoupled oscillators), and $H_1(J, \phi)$ is anharmonic in the variables J and ϕ that couple the invariant subspaces. We assume that we can treat H_1 using perturbation theory for small λ. Such perturbation theory calculations are familiar from celestial mechanics, where $H_0(J)$ describes the independent motion of the planets about the sun, and H_1 describes the perturbation due to the interaction among the planets. This perturbation theory shares with quantum mechanical perturbation theory the feature that it breaks down near a region of degeneracy associated with the fundamental frequencies $\omega_i = \partial H_0(J)/\partial J_i$ characteristic of the N-uncoupled periodic systems. This degeneracy occurs when we have the *resonance*

$$\sum_{i=1}^{N} n_i \omega_i = 0 \tag{2570}$$

where the n_i are any set of integers. This resonance or degeneracy is important in perturbation theory because the first-order correction to the *constant* action variables J_i can be shown [33] to be given by (for $N = 2$)

$$J_1 = J_1^0 + \sum_{n_1, n_2} \lambda V_{n_1, n_2} \frac{\cos(n_1 \phi_1 + n_2 \phi_2)}{n_1 \omega_1 + n_2 \omega_2} \tag{2571}$$

where we assume that the perturbation has the Fourier representation

$$H_1(J_1, J_2, \phi_1, \phi_2) = \sum_{n_1, n_2} V_{n_1, n_2} \cos(n_1 \phi_1 + n_2 \phi_2) \tag{2572}$$

and J_1^0 is the value of the action variable to zeroth order in λ.

When $n_1 \omega_1 + n_2 \omega_2 = 0$, the perturbation theory *correction* is not small. Generally, ω_1 and ω_2 are functions of J_1 and J_2, and there will be limited regions of phase space corresponding to particular values of J_1 and J_2 where the resonance condition is satisfied. These regions occupy small but finite measure regions of phase space where perturbation theory is not applicable. In the rest of phase space, perturbation theory should apply. The KAM theorem says roughly that if λ is sufficiently small and $H_1(J, \phi)$ is analytic in the J's and ϕ's in a given domain, phase space can be divided into the two types of volumes of nonzero measure mentioned above. As $\lambda \to 0$, the volume associated with the resonances shrinks to zero. In this case the system is clearly *not* ergodic for small enough λ. As λ increases, the volume associated with the resonances increases. When two resonance volumes overlap, the system apparently becomes chaotic. There is evidence that the sharp transition from nonergodic to ergodic behavior we mentioned earlier can be associated with the critical energy associated with the overlap of resonances.

One can understand how resonant regions expand and overlap by focusing on the region of phase space near a resonance. In this domain one can make [34] a change of variables and take advantage of the resonance condition $\sum_i \omega_i m_i = 0$ to find that the new variables separate into a set of rapidly varying phases and a pair of slow variables. After averaging over the rapidly varying phases, one finds that the slow variables, an action variable I, measuring the distance to the resonant value of the action and a phase variable θ, are described by an effective Hamiltonian

$$\bar{H} = \frac{1}{2}I^2 - \frac{\lambda}{2}u\cos\theta \qquad (2573)$$

which is just the Hamiltonian for a simple pendulum. λ is again the strength of the perturbation in Eq. (2569), and u is proportional to the average of H_1 over the fast phases.

The key point here is that for low energies these resonances will have low energy and one will have stable pendulumlike motion. As one increases the energy, the energy in the resonant motion will increase, and for $\lambda u > 1$, the motion delocalizes and moves away from the resonance and one can expect overlap to begin occurring.

From the point of view of celestial mechanics, the KAM theorem reassures us that we should be able to prove the stability of planetary motion (an unsolved problem). However, from the point of view of statistical mechanics, the situation is much less clear. At present very little is known about the magnitude of the critical energy above which a system is ergodic for systems with many degrees of freedom. We can guess, however, that as we increase the number of degrees of freedom, the number of possible resonances increases. Consequently, as we increase the number of degrees of freedom, we expect the critical energy to decrease. Although we cannot prove it, we expect anharmonic oscillator systems to be ergodic in the large-N limit for any nonzero λ.

System of Hard Spheres

Given the extreme complexity of a set of anharmonically coupled oscillators and our inability to establish whether or not they are ergodic, we might be very discouraged about the entire field. Fortunately, in 1962, Sinai [35] announced a proof that a system of $N(\geq 2)$ hard spheres enclosed in a box with hard walls is a mixing system. This property holds even in the limit of very small (two spheres!) systems. Consequently, the mixing property is satisfied in at least one semirealistic system. Note that since the interaction between hard spheres is infinitely repulsive, the KAM theory is *not* applicable in this case.

D INEQUALITY USED TO MAXIMIZE ENTROPY

Consider the function

$$f(x) = 1 + x(\ln x - 1) \qquad (2574)$$

in the range $x \geq 0$. As $x \to 0$, $f(x) \to 1$. As x increases in the range $0 < x < 1$, $\ln x - 1$ is monotonically decreasing and $f(x)$ is monotonically decreasing. Since

$f(1) = 0$, we have that in the range $0 \leq x \leq 1$, $f(x) \geq 0$. For $x = 1 + \epsilon$ we can expand in a power series in ϵ and obtain

$$f(1 + \epsilon) = \frac{\epsilon^2}{2} + \cdots \tag{2575}$$

and $f(x)$ increases away from the minimum at $x = 1$ and increases monotonically as $|x - 1|$ increases. We then have that $f(x) \geq 0$ for $x \geq 0$. Having established

$$1 + x(\ln x - 1) \geq 0 \tag{2576}$$

we can divide by x for $x > 0$, to obtain

$$\frac{1}{x} - 1 + \ln x \geq 0. \tag{2577}$$

Subtracting $\ln x$ from both sides of this equation gives the final result

$$\frac{1}{x} - 1 \geq -\ln x, \tag{2578}$$

which we use to advantage in Chapter 1.

E VOLUME AND SURFACE AREA OF A d-DIMENSIONAL SPHERE OF UNIT RADIUS

In this appendix we compute the volume of a d-dimensional sphere with unit radius

$$C_d = \int d^d x \, \theta \left(1 - \sum_{i=1}^{d} x_i^2 \right) \tag{2579}$$

where θ is the usual step function, which is 1 for positive arguments and zero for negative arguments. The key point in this analysis is to realize that in any dimension we can write

$$d^d x = r^{d-1} dr \, d\Omega_d \tag{2580}$$

where r is the radial coordinate, defined as

$$r^2 = \sum_{i=1}^{d} x_i^2, \tag{2581}$$

VOLUME AND SURFACE AREA OF A d-DIMENSIONAL SPHERE

and $d\Omega$ is the angular volume element. Equation (2579) can then be written as

$$C_d = \int_0^\infty r^{d-1} dr \int d\Omega_d \theta(1 - r^2)$$
$$= K_d \int_0^1 r^{d-1} dr = \frac{K_d}{d}. \tag{2582}$$

The surface area in d dimensions,

$$K_d \equiv \int d\Omega_d, \tag{2583}$$

can be evaluated by considering the integral

$$J \equiv \int d^d x \, e^{-\sum_{i=1}^d x_i^2}. \tag{2584}$$

We can evaluate J in two ways. First we go over to radial coordinates, as above, to obtain

$$J = \int_0^\infty r^{d-1} dr \int d\Omega_d \, e^{-r^2}$$
$$= K_d \int_0^{+\infty} r^{d-1} dr \, e^{-r^2}. \tag{2585}$$

If we change integration variables to $y = r^2$ we obtain

$$J = K_d \frac{1}{2} \int_0^{+\infty} y^{(d/2-1)} e^{-y} dy$$
$$= K_d \frac{1}{2} \Gamma\left(\frac{d}{2}\right) \tag{2586}$$

where $\Gamma(x)$ is the usual Γ-function. We can evaluate J a second way by noting that it is a product of d identical integrals,

$$J = \int dx_1 e^{-x_1^2} \int dx_2 \, e^{-x_2^2} \cdots \int dx_d e^{-x_d^2}$$
$$= \left(\int_{-\infty}^\infty dx \, e^{-x^2}\right)^d = \pi^{d/2}. \tag{2587}$$

Equating the two expressions for J leads to

$$K_d = \frac{2\pi^{d/2}}{\Gamma(d/2)}. \tag{2588}$$

Inserting this result in Eq. (2582), we obtain

$$C_d = \frac{2\pi^{d/2}}{d\Gamma(d/2)}. \tag{2589}$$

If we remember the property of the Γ-function,

$$\Gamma(n+1) = n\Gamma(n), \tag{2590}$$

or

$$\Gamma\left(\frac{d}{2}+1\right) = \frac{d}{2}\Gamma\left(\frac{d}{2}\right), \tag{2591}$$

C_d can be reduced to

$$C_d = \frac{\pi^{d/2}}{\Gamma(d/2+1)}. \tag{2592}$$

If we note that $\Gamma(3/2) = \sqrt{\pi}/2, \Gamma(2) = 1, \Gamma(\frac{5}{2}) = 3\sqrt{\pi}/4$, and $\Gamma(3) = 2$, we easily obtain that $C_1 = 2, C_2 = \pi, C_3 = 4\pi/3$, and $C_4 = \pi^2/2$. Certainly, C_1, C_2, and C_3 are what we expect.

F LOCAL CONSERVATION LAWS

We discussed in some detail in Chapter 1 the crucial role of the conservation laws. In particular, we discussed the importance of the local nature of the conservation laws. In this appendix we show how we can explicitly construct the continuity equations governing the mass, momentum, angular momentum, and energy densities for a collection of classical particles interacting via a two-body potential.

These densities constitute a set of *local observables*. It is convenient to express them in terms of a *fundamental phase-space density*

$$f(\mathbf{x}, \mathbf{p}) = \sum_{i=1}^{N} \delta(\mathbf{x} - \mathbf{r}_i)\delta(\mathbf{p} - \mathbf{p}_i). \tag{2593}$$

The particle, momentum, and angular momentum densities can be expressed as momentum integrals of $f(\mathbf{x}, \mathbf{p})$:

$$n(\mathbf{x}) = \int d^3p\, f(\mathbf{x}, \mathbf{p}) \tag{2594}$$

$$\mathbf{g}(\mathbf{x}) = \int d^3p\, \mathbf{p} f(\mathbf{x}, \mathbf{p}) \tag{2595}$$

$$\mathbf{l}(\mathbf{x}) = \int d^3p\, (\mathbf{x} \times \mathbf{p}) f(\mathbf{x}, \mathbf{p}). \tag{2596}$$

LOCAL CONSERVATION LAWS

The evaluation of the energy density, as discussed in Problem 1.57, can be written in the form

$$\epsilon(\mathbf{x}) = \int d^3p \, \frac{\mathbf{p}^2}{2m} f(\mathbf{x},\mathbf{p}) + \frac{1}{2}\int d^3p \, d^3p_1 \int d^3x_1 V(\mathbf{x}-\mathbf{x}_1) f_2(\mathbf{x},\mathbf{p},\mathbf{x}_1,\mathbf{p}_1) \quad (2597)$$

where we have introduced the two-particle phase-space density

$$f_2(\mathbf{x},\mathbf{p},\mathbf{x}_1,\mathbf{p}_1) = \sum_{i\neq j=1}^{N} \delta(\mathbf{x}-\mathbf{r}_i)\delta(\mathbf{p}-\mathbf{p}_i)\delta(\mathbf{x}_1-\mathbf{r}_j)\delta(\mathbf{p}_1-\mathbf{p}_j). \quad (2598)$$

The two-particle phase-space density can be expressed in terms of the one-particle phase-space density,

$$f_2(\mathbf{x},\mathbf{p},\mathbf{x}_1,\mathbf{p}_1) = f(\mathbf{x},\mathbf{p})f(\mathbf{x}_1\mathbf{p}_1) - \delta(\mathbf{x}-\mathbf{x}_1)\delta(\mathbf{p}-\mathbf{p}_1)f(\mathbf{x},\mathbf{p}). \quad (2599)$$

Since all of the densities can be expressed in terms of the one-particle phase-space density, we first look at its equations of motion given by the Poisson bracket:

$$\frac{\partial}{\partial t} f(\mathbf{x},\mathbf{p}) = \{f(\mathbf{x},\mathbf{p}), H\} \quad (2600)$$

where the Hamiltonian is just the space integral of the energy density,

$$H = \int d^3x \, \epsilon(\mathbf{x}). \quad (2601)$$

Once we have evaluated the fundamental Poisson brackets

$$\{f(\mathbf{x},\mathbf{p}), f(\mathbf{x}',\mathbf{p}')\} = -\nabla_x f(\mathbf{x},\mathbf{p}) \cdot \nabla_p[\delta(\mathbf{p}-\mathbf{p}')\delta(\mathbf{x}-\mathbf{x}')] \\ + \nabla_p f(\mathbf{x},\mathbf{p}) \cdot \nabla_x[\delta(\mathbf{p}-\mathbf{p}')\delta(\mathbf{x}-\mathbf{x}')], \quad (2602)$$

all other Poisson brackets follow easily and we obtain

$$\frac{\partial f(\mathbf{x},\mathbf{p})}{\partial t} = -\frac{\mathbf{p}\cdot\nabla_x}{m} f(\mathbf{x},\mathbf{p}) \\ - \int d^3x' d^3p' \nabla_x V(\mathbf{x}-\mathbf{x}') \cdot \nabla_p f_2(\mathbf{x},\mathbf{p},\mathbf{x}',\mathbf{p}'). \quad (2603)$$

This equation can be used as the basis for the development of kinetic theory. For our purposes here, we use it as a convenient way of deriving the equations satisfied by the conserved densities. In the case of conservation of particle number, the analysis is simple. If we integrate Eq. (2603) over momentum, we note that the term involving the two-body potential vanishes, and using Eqs. (2594) and (2595), we obtain the local statement of conservation of mass [36]:

$$\frac{\partial n}{\partial t} = -\nabla \cdot \frac{\mathbf{g}}{m}. \tag{2604}$$

The continuity equation for the momentum density $\mathbf{g}(\mathbf{x})$ is obtained by multiplying Eq. (2603) by p_α and integrating over $\mathbf{p}$ to obtain

$$\frac{\partial g_\alpha(\mathbf{x})}{\partial t} = -\sum_\beta \nabla_x^\beta \int d^3p \frac{p_\alpha p_\beta}{m} f(\mathbf{x}, \mathbf{p})$$

$$+ \int d^3x_1 \, d^3p \, d^3p_1 [\nabla_x^\alpha V(\mathbf{x} - \mathbf{x}_1)] f_2(\mathbf{x}, \mathbf{p}, \mathbf{x}, \mathbf{p}_1). \tag{2605}$$

The last term, which depends on the two-particle potential, can be rewritten, as shown in Problem 1.58, in the form

$$\int d^3x_1 \nabla_x^\alpha V(\mathbf{x} - \mathbf{x}_1) n_2(\mathbf{x}, \mathbf{x}_1) = \frac{1}{4} \sum_\beta \nabla_x^\beta \int d^3r \int_{-1}^{+1} ds \frac{r_\beta r_\alpha}{r} \frac{\partial V(r)}{\partial r}$$

$$\times n_2 \left[\mathbf{x} + \tfrac{1}{2}(s+1)\mathbf{r}, \mathbf{x} + \tfrac{1}{2}(s-1)\mathbf{r} \right] \tag{2606}$$

where

$$n_2(\mathbf{x}, \mathbf{x}_1) = \int d^3p \, d^3p_1 f_2(\mathbf{x}, \mathbf{p}, \mathbf{x}_1, \mathbf{p}_1) = \sum_{i \neq j=1}^{N} \delta(\mathbf{x} - \mathbf{r}_i) \delta(\mathbf{x}_1 - \mathbf{r}_j). \tag{2607}$$

Using this result in Eq. (2605) leads to the continuity equation for the momentum density, g_α, written in the conventional form

$$\frac{\partial g_\alpha(\mathbf{x})}{\partial t} = -\sum_\beta \nabla_x^\beta \sigma_{\alpha\beta}(\mathbf{x}). \tag{2608}$$

The advantage of the manipulation given by Eq. (2606) is that we can explicitly extract the microscopic *stress tensor* given by

$$\sigma_{\alpha\beta}(\mathbf{x}) = \int d^3p \frac{p_\alpha p_\beta}{m} f(\mathbf{x}, \mathbf{p}) - \frac{1}{4} \int_{-1}^{+1} ds \int d^3r \frac{r_\alpha r_\beta}{r} \frac{\partial V(r)}{\partial r}$$

$$\times n_2 \left[\mathbf{x} + \tfrac{1}{2}(s+1)\mathbf{r}, \mathbf{x} + \tfrac{1}{2}(s-1)\mathbf{r} \right]. \tag{2609}$$

The angular momentum density satisfies the equation of motion

$$\frac{\partial l_\alpha(x)}{\partial t} = \sum_{\beta,\gamma} \epsilon_{\alpha\beta\gamma} x_\beta \frac{\partial g_\gamma(\mathbf{x})}{\partial t}$$

$$= \sum_{\beta\gamma} \epsilon_{\alpha\beta\gamma} [\nabla_\mu(x_\beta \sigma_{\gamma\mu}(\mathbf{x})) - \sigma_{\gamma\mu}(\mathbf{x})\delta_{\beta\mu}]. \tag{2610}$$

LOCAL CONSERVATION LAWS

If, as shown above for a simple fluid, the stress tensor is symmetric,

$$\sigma_{\gamma\mu} = \sigma_{\mu\gamma}, \tag{2611}$$

the angular momentum density satisfies the continuity equation,

$$\frac{\partial}{\partial t} l_\alpha(\mathbf{x}) = -\sum_\mu \nabla_\mu M_{\alpha\mu}(\mathbf{x}) \tag{2612}$$

where the angular momentum current is given by

$$M_{\alpha\mu}(\mathbf{x}) = \sum_{\beta,\gamma} \epsilon_{\alpha\beta\gamma} x_\beta \sigma_{\gamma\mu}(\mathbf{x}). \tag{2613}$$

Finally, we can work out the equation of motion for the energy density. Using the techniques described in Problem 1.58, you are asked to show in Problem 1.59 that

$$\frac{\partial \epsilon(\mathbf{x})}{\partial t} = -\nabla \cdot \mathbf{J}_\epsilon(x) \tag{2614}$$

where the energy current is given by

$$J_\epsilon^\alpha(\mathbf{x}) = \int d^3p \left[\frac{p^2}{2m} + \int d^3r \frac{1}{2} V(\mathbf{x}-\mathbf{r}) n(\mathbf{r}) \right] p_\alpha f(\mathbf{x},\mathbf{p})$$
$$- \frac{1}{4} \int_{-1}^{+1} ds \int d^3r \int d^3p\, d^3p' \sum_\beta \frac{r_\alpha r_\beta}{r} \frac{\partial V(r)}{\partial r} (p_\beta + p'_\beta)$$
$$\times f_2\left(\mathbf{x} - \tfrac{1}{2}(s+1)\mathbf{r}, \mathbf{p}, \mathbf{x} - \tfrac{1}{2}(s-1)\mathbf{r}, \mathbf{p}'\right). \tag{2615}$$

Note that $\sigma_{\alpha\beta}$, $M_{\alpha\beta}$, and J_ϵ^α are nonlocal quantities with a range determined by the range of the potential.

The microscopic expression for the stress tensor is true generally, independent of the thermodynamic state of the system. Let us consider the average of $\sigma_{\alpha\beta}(\mathbf{x})$ in the case where the system is in an equilibrium fluid state. The only properties of fluids we need in this analysis is that they are translationally and rotationally invariant. We then have that

$$\langle \sigma_{\alpha\beta}(\mathbf{x}) \rangle = \int d^3p \frac{p_\alpha p_\beta}{m} \langle f(\mathbf{x},\mathbf{p}) \rangle \tag{2616}$$
$$- \frac{1}{4} \int_{-1}^{+1} ds \int d^3r \frac{r_\alpha r_\beta}{r} \frac{\partial V(r)}{\partial r} \left\langle n_2\left[\mathbf{x} - \tfrac{1}{2}(s+1)\mathbf{r}, \mathbf{x} - \tfrac{1}{2}(s-1)\mathbf{r}\right] \right\rangle.$$

In a translationally invariant system,

$$\langle f(\mathbf{x}, \mathbf{p}) \rangle = f(\mathbf{p}), \qquad (2617)$$

and in a rotationally invariant system,

$$f(\mathbf{p}) = f(p). \qquad (2618)$$

Then

$$\int d^3 p \frac{p_\alpha p_\beta}{m} f(p) = \delta_{\alpha,\beta} \frac{1}{3} \int d^3 p \frac{p^2}{m} f(p) = \frac{2}{3} \delta_{\alpha,\beta} n \epsilon_K \qquad (2619)$$

where ϵ_K is the average kinetic energy per particle. Next consider the average over the part of the stress tensor containing the potential. Using translational invariance, we have

$$\langle n_2(\mathbf{x}, \mathbf{y}) \rangle = \langle n_2(\mathbf{x} + \mathbf{z}, \mathbf{y} + \mathbf{z}) \rangle. \qquad (2620)$$

Choosing $\mathbf{z}$ appropriately, we have

$$\begin{aligned}
&\langle n_2[\mathbf{x} - \tfrac{1}{2}(s+1)\mathbf{r}, \mathbf{x} - \tfrac{1}{2}(s-1)]\mathbf{r} \rangle \\
&= \langle n_2[\mathbf{0}, \mathbf{x} - \tfrac{1}{2}(s-1)\mathbf{r} - \mathbf{x} + \tfrac{1}{2}(s+1)\mathbf{r}] \rangle \\
&= \langle n_2(\mathbf{0}, \mathbf{r}) \rangle \equiv g(r) n^2
\end{aligned} \qquad (2621)$$

where in the last step we recognize that the average of n_2 is proportional to the radial distribution function $g(r)$, which is discussed in detail in Chapter 4. Inserting this result back into the second term in Eq. (2616) gives

$$-\frac{1}{4} \int_{-1}^{+1} ds \int d^3 r \frac{r_\alpha r_\beta}{r} \frac{\partial V(r)}{\partial r} g(r) n^2 = -\frac{1}{2} \delta_{\alpha\beta} \frac{1}{3} \int d^3 r \, r \frac{\partial V(r)}{\partial r} g(r) n^2 \qquad (2622)$$

and finally, the average stress tensor is given by

$$\langle \sigma_{\alpha\beta}(\mathbf{x}) \rangle = \delta_{\alpha\beta} p \qquad (2623)$$

where

$$p = \frac{2}{3} \epsilon_K - \frac{1}{6} \int d^3 r \, r \frac{\partial V(r)}{\partial r} g(r) n^2. \qquad (2624)$$

We identify p as the pressure in Chapter 1. It is left as an exercise (Problem 1.50) to show that Eq. (2624) can be rewritten as

$$pV = \left\langle \sum_{i=1}^{N} \frac{\mathbf{p}_i^2}{3m} \right\rangle - \frac{1}{6} \left\langle \sum_{i \neq j=1}^{N} |\mathbf{r}_i - \mathbf{r}_j| \frac{\partial V(|\mathbf{r}_i - \mathbf{r}_j|)}{\partial |\mathbf{r}_i - \mathbf{r}_j|} \right\rangle. \qquad (2625)$$

G METHOD OF LAGRANGE MULTIPLIERS

Suppose that we want to find the extremum (maximum or minimum) of the function $f(x_1, x_2, \ldots, x_n)$, where the n variables $x_1, x_2, \ldots, x_n$ satisfy the constraining equation

$$h(x_1, x_2, \ldots, x_n) = 0. \tag{2626}$$

The condition that f is at an extremum point is, as usual, given by

$$df = \sum_{i=1}^{n} \frac{\partial f}{\partial x_i} dx_i = 0. \tag{2627}$$

Equation (2626) gives the additional condition

$$dh = \sum_{i=1}^{n} \frac{\partial h}{\partial x_i} dx_i = 0. \tag{2628}$$

If the x_i were all independent, we could vary them independently and conclude that the condition for an extremum is

$$\frac{\partial f}{\partial x_i} = 0 \tag{2629}$$

for all i. The x_i are *not* independent, due to the constraining condition given by Eq. (2626). One way of dealing with this constraint is to solve Eq. (2626) to find, for example,

$$x_n = g(x_1, x_2, \ldots, x_n - 1), \tag{2630}$$

and then insert x_n into f and then vary the $n - 1$ independent variables $x_1, x_2, \ldots, x_n - 1$. This method is impractical in situations where f and h are complicated. An alternative method is due to Lagrange. Go back to Eq. (2628), multiply it by some unknown parameter λ, and add it to Eq. (2627), to obtain

$$\sum_{i=1}^{n} \left(\frac{\partial f}{\partial x_i} + \lambda \frac{\partial h}{\partial x_i} \right) dx_i = 0. \tag{2631}$$

Although only $(n - 1)$ of the differentials dx_i are independent (e.g., $dx_1, \ldots, dx_n - 1$), the parameter λ is at one's disposal. If we choose λ such that

$$\frac{\partial f}{\partial x_n} + \lambda \frac{\partial h}{\partial x_n} = 0 \tag{2632}$$

the dx_n differential is eliminated from Eq. (2631) and we have

$$\sum_{i=1}^{n-1} \left(\frac{\partial f}{\partial x_i} + \lambda \frac{\partial h}{\partial x_i} \right) dx_i = 0. \tag{2633}$$

Since the differentials dx_i for $i = 1, 2, \ldots, n-1$ can be treated as independent, we have that the coefficient of each must be zero, and for $i = 1, 2, \ldots, n-1$,

$$\frac{\partial f}{\partial x_i} + \lambda \frac{\partial h}{\partial x_i} = 0. \tag{2634}$$

Summarizing: The condition for stationarity is

$$\frac{\partial f}{\partial x_i} + \lambda \frac{\partial h}{\partial x_i} = 0 \tag{2635}$$

for *all* x_i. The extremum condition is to vary

$$W = f + \lambda h \tag{2636}$$

with respect to the x_i and choose

$$\lambda = -\frac{\partial f}{\partial x_n} \bigg/ \frac{\partial h}{\partial x_n} \tag{2637}$$

using the most convenient choice for x_n.

H STATIONARY-PHASE TREATMENT OF THE PARTITION FUNCTION IN THE MICROCANONICAL ENSEMBLE

Consider a large system governed by the Hamiltonian H. We are concerned here with the associated microanonical ensemble partition function

$$Z_M(E) = \text{Tr}\, \delta(E - H) E_T \tag{2638}$$

where Tr is the dimensionless sum over all degrees of freedom in the system. E_T is an uninteresting constant normalization energy making $Z_M(E)$ dimensionless. The entropy in the microcanonical ensemble is given by

$$S(E, N, V) = k_B \ln Z_M(E). \tag{2639}$$

We want to evaluate $Z_M(E)$ using the method of stationary phase. The first step is

STATIONARY-PHASE TREATMENT OF THE PARTITION FUNCTION

to use the integral representation for the δ-function, to obtain

$$Z_M(E) = \text{Tr} \int_{-\infty}^{+\infty} \frac{dk}{2\pi} e^{+ik(E-H)} E_T$$

$$= \int_{-\infty}^{+\infty} \frac{dk}{2\pi} e^{+ikE} \text{Tr}\, e^{-ikH} E_T. \quad (2640)$$

If we define

$$e^{W[k]} = \text{Tr}\, e^{-ikH} \quad (2641)$$

then

$$Z_M(E) = \int_{-\infty}^{+\infty} \frac{dk}{2\pi} e^{+ikE+W[k]} E_T$$

$$= \int_{-\infty}^{+\infty} \frac{dk}{2\pi} e^{G(k)} E_T \quad (2642)$$

where

$$G(k) = ikE + W[k]. \quad (2643)$$

The important point is that we are interested in those cases where E, and therefore $G(k)$, is very large and one has rapid oscillations in the integral except for those values of k in the complex plane, where G is more slowly varying. This is just the situation where the method of steepest decent [37] is applicable. The main contribution to the integral comes from those values of k near the value k_0 where $G(k)$ is a maximum:

$$\frac{dG(k)}{dk} = 0 \quad \text{for} \quad k = k_0. \quad (2644)$$

We then Taylor-series expand G about $k = k_0$:

$$G(k) = G(k_0) + \left(\frac{dG}{dk}\right)_{k=k_0} (k - k_0) + \frac{1}{2}\left(\frac{d^2 G}{dk^2}\right)_{k=k_0} (k - k_0)^2 + \ldots, \quad (2645)$$

define

$$M = -\left(\frac{\partial^2 G}{\partial k^2}\right)_{k=k_0} \quad (2646)$$

and use Eq. (2644) to obtain

$$G(k) = G(k_0) - \frac{M}{2}(k - k_0)^2 + \cdots \quad (2647)$$

If $G(k_0)$ is to be a maximum, we must have, as we show below, that $M > 0$. Inserting Eq. (2647) in Eq. (2642), we obtain

$$\begin{aligned} Z_M(E) &= e^{G(k_0)} \int_{-\infty}^{+\infty} \frac{dk}{2\pi} e^{-(1/2)M(k-k_0)^2} E_T \\ &= e^{G(k_0)} \int_{-\infty}^{+\infty} \frac{dx}{2\pi} e^{-(1/2)Mx^2} E_T \\ &= \frac{e^{G(k_0)} E_T}{(2\pi M)^{1/2}} \left[1 + \mathcal{O}\left(\frac{1}{\sqrt{N}}\right) \right]. \end{aligned} \qquad (2648)$$

We must then explicitly determine k_0, $G(k_0)$, and M. The quantity $G(k_0)$ appearing in Eq. (2645) is given by Eq. (2643) with k set equal to k_0:

$$G(k_0) = ik_0 E + W(k_0). \qquad (2649)$$

k_0 is then determined by Eq. (2644), which we can rewrite as

$$\left. \frac{dW(k)}{dk} \right|_{k=k_0} = -iE. \qquad (2650)$$

Physically, we expect the entropy S to grow linearly with the size of the system, and since from Eq. (2639),

$$Z_M \approx e^{S/k_B} \qquad (2651)$$

we also expect $\ln Z_M$ to grow linearly with the size of the system. Therefore, in the stationary-phase method we expect the main contribution to the integral in Eq. (2642) to be from along the imaginary axis in the upper plane. Thus we define

$$\beta \equiv ik_0 \qquad (2652)$$

and write Eq. (2650), using Eq. (2641), in the form

$$E \operatorname{Tr} e^{-\beta H} = \operatorname{Tr} e^{-\beta H} H. \qquad (2653)$$

If we write $\beta = \beta' + i\beta''$ and the imaginary β'' is nonzero, we have rapid oscillations since H is very large and there is no obvious robust solution. If we assume that β'' is small and we can expand Eq. (2653) in powers of β'', the only self-consistent solution we obtain is that β'' is zero. Thus our stationary-phase point is $k_0 = -i\beta$, where β is real and positive. With this result we can then write

$$G(k_0) = \beta E + W(\beta) \qquad (2654)$$

where $W(\beta)$ is defined as a function of β by

$$W(\beta) = \ln \operatorname{Tr} e^{-\beta H} \qquad (2655)$$

STATIONARY-PHASE TREATMENT OF THE PARTITION FUNCTION 569

and β is determined by inverting, for fixed E,

$$\frac{\partial W(\beta)}{\partial \beta} = -E. \tag{2656}$$

We can also evaluate the coefficient of the second-order term in the Taylor series expansion, M, defined by Eq. (2646), using Eqs. (2641) and (2643),

$$M = -\left(\frac{\partial^2 G}{\partial k^2}\right)_{k=k_0} = -\left(\frac{\partial^2 W}{\partial k^2}\right)_{k=k_0=-i\beta}$$
$$= \frac{\partial^2 W}{\partial \beta^2}. \tag{2657}$$

This can be made more explicit by taking the derivatives inside the average in Eq. (2655):

$$M = -\frac{\partial}{\partial \beta}\left(\frac{\mathrm{Tr}\, e^{-\beta H} H}{\mathrm{Tr}\, e^{-\beta H}}\right)$$
$$= \frac{\mathrm{Tr}\, e^{-\beta H} H^2}{\mathrm{Tr}\, e^{-\beta H}} - \left(\frac{\mathrm{Tr}\, e^{-\beta H} H}{\mathrm{Tr}\, e^{-\beta H}}\right)^2$$
$$= \langle (\delta H)^2 \rangle > 0, \tag{2658}$$

as promised. Combining Eqs. (2648), (2654), and (2657) gives

$$Z_M(E) = \frac{e^{\beta E + W(\beta)} E_T}{[2\pi \partial^2 W/\partial \beta^2]^{1/2}}. \tag{2659}$$

We can write Eq. (2659) in a more transparent form. Note that the Helmholtz free energy, the potential in the canonical ensemble, can be written as

$$\beta F = -\ln \mathrm{Tr}\, e^{-\beta H} = -W(\beta). \tag{2660}$$

If we use Eq. (2660) in Eq. (2659) and take the logarithm, we have, according to Eq. (2639), an expression for the entropy to leading order in N, given by

$$S = k_B \ln Z_M = \frac{1}{T}(E - F). \tag{2661}$$

Using this expression back in Eq. (2659) gives our final result:

$$Z_M(E) = \frac{e^{S(E)/k_B} E_T}{[2\pi \partial^2 W/\partial \beta^2]^{1/2}}. \tag{2662}$$

We have expressed the partition function in the microcanonical ensemble in terms of quantities (F, β) appropriate to the canonical ensemble.

I PROPERTIES OF JACOBIANS

In this appendix we discuss some simple properties of Jacobians of use in thermodynamics. The discussion is restricted to the case of two variables, but the results generalize, as indicated in Chapter 2, to n variables. Let us assume that u and v are functions of x and y. The Jacobian in this case is defined by

$$\frac{\partial(u,v)}{\partial(x,y)} = \det \begin{bmatrix} \dfrac{\partial u}{\partial x} & \dfrac{\partial u}{\partial y} \\ \dfrac{\partial v}{\partial x} & \dfrac{\partial v}{\partial y} \end{bmatrix}$$

$$= \frac{\partial u}{\partial x}\frac{\partial v}{\partial y} - \frac{\partial u}{\partial y}\frac{\partial v}{\partial x}. \tag{2663}$$

It is trivial to see from Eq. (2663) that one has the antisymmetry:

$$\frac{\partial(u,v)}{\partial(x,y)} = -\frac{\partial(v,u)}{\partial(x,y)}. \tag{2664}$$

If we assume that $u = u(r,s)$ and $v = v(r,s)$, where $r = r(x,y)$ and $s = s(x,y)$ are another set of variables, we can use the chain rule for differentiation, to obtain

$$\begin{aligned}
\frac{\partial(u,v)}{\partial(x,y)} &= \frac{\partial u}{\partial x}\frac{\partial v}{\partial y} - \frac{\partial v}{\partial x}\frac{\partial u}{\partial y} \\
&= \left(\frac{\partial u}{\partial r}\frac{\partial r}{\partial x} + \frac{\partial u}{\partial s}\frac{\partial s}{\partial x}\right)\left(\frac{\partial v}{\partial r}\frac{\partial r}{\partial y} + \frac{\partial v}{\partial s}\frac{\partial s}{\partial y}\right) - \left(\frac{\partial v}{\partial r}\frac{\partial r}{\partial x} + \frac{\partial v}{\partial s}\frac{\partial s}{\partial x}\right) \\
&\quad \times \left(\frac{\partial u}{\partial r}\frac{\partial r}{\partial y} + \frac{\partial u}{\partial s}\frac{\partial s}{\partial y}\right) \\
&= \frac{\partial u}{\partial r}\left(\frac{\partial r}{\partial x}\left(\frac{\partial v}{\partial r}\frac{\partial r}{\partial y} + \frac{\partial v}{\partial s}\frac{\partial s}{\partial y}\right) - \frac{\partial r}{\partial y}\left(\frac{\partial v}{\partial r}\frac{\partial r}{\partial x} + \frac{\partial v}{\partial s}\frac{\partial s}{\partial x}\right)\right) \\
&\quad + \frac{\partial u}{\partial s}\left(\frac{\partial s}{\partial x}\left(\frac{\partial v}{\partial r}\frac{\partial r}{\partial y} + \frac{\partial v}{\partial s}\frac{\partial s}{\partial y}\right) - \frac{\partial s}{\partial y}\left(\frac{\partial v}{\partial r}\frac{\partial r}{\partial x} - \frac{\partial v}{\partial s}\frac{\partial s}{\partial x}\right)\right) \\
&= \frac{\partial u}{\partial r}\left(\frac{\partial r}{\partial x}\frac{\partial v}{\partial s}\frac{\partial s}{\partial y} - \frac{\partial r}{\partial y}\frac{\partial v}{\partial s}\frac{\partial s}{\partial x}\right) + \frac{\partial u}{\partial s}\left(\frac{\partial s}{\partial x}\frac{\partial v}{\partial r}\frac{\partial r}{\partial y} - \frac{\partial s}{\partial y}\frac{\partial v}{\partial r}\frac{\partial r}{\partial x}\right) \\
&= \frac{\partial u}{\partial r}\frac{\partial v}{\partial s}\frac{\partial(r,s)}{\partial(x,y)} + \frac{\partial u}{\partial s}\frac{\partial v}{\partial r}\frac{\partial(s,r)}{\partial(x,y)} \\
&= \frac{\partial(u,v)}{\partial(r,s)}\frac{\partial(r,s)}{\partial(x,y)} \tag{2665}
\end{aligned}$$

as desired.

J ABSTRACT VECTOR SPACES

There are various physical situations where we are interested in quantities defined on an abstract vector space. The treatment of Hilbert space in quantum mechanics is the most prominent example, but there are many other examples in matrix algebra. Thus it is useful to assume that a matrix B_{ij} can be treated as the matrix elements of an operator $\hat{B}$ acting on an abstract vector space spanned by a set of basis vectors $|j\rangle$:

$$B_{ij} = \langle i|\hat{B}|j\rangle. \tag{2666}$$

If the operator is Hermitian or real and symmetric, the basis vectors $|j\rangle$ can be constructed to be a complete and orthonormal set. In the Dirac notation we use here, the quantity $|j\rangle$ is called a *ket*, and its adjoint, $\langle j|$, is called a *bra*. These basis vectors are orthonormal in the sense that their inner product is defined such that

$$\langle i|j\rangle = \delta_{ij}. \tag{2667}$$

In the general case of a complex Hilbert space, we require that

$$\langle i|j\rangle = \langle j|i\rangle^*. \tag{2668}$$

The completeness of the basis set $|i\rangle$ means that an arbitrary vector defined on the space can be specified by the complex numbers giving the projection onto the basis set, that is,

$$|A\rangle = \sum_i |i\rangle\langle i|A\rangle, \tag{2669}$$

or, completeness can be written as

$$1 = \sum_i |i\rangle\langle i|. \tag{2670}$$

The dual of $|A\rangle$ is defined by

$$\langle A| = \sum_i \langle A|i\rangle\langle i|$$
$$= \sum_i \langle i|A\rangle^* \langle i|. \tag{2671}$$

The inner product between two arbitrary vectors is given by

$$\langle A|B\rangle = \sum_{ij} \langle A|i\rangle\langle i|j\rangle\langle j|B\rangle$$
$$= \sum_i \langle A|i\rangle\langle i|B\rangle = \sum_i \langle i|A\rangle * \langle i|B\rangle. \tag{2672}$$

A simple but very useful example of a set of basis vectors are the unit vectors $\hat{x}_i$ associated with an orthogonal three-dimensional coordinate system. Clearly, the inner product (the dot product in this case) is orthonormal,

$$\hat{x}_i \cdot \hat{x}_j = \delta_{ij}, \tag{2673}$$

and the set is complete since an arbitrary vector **A** can be expressed as

$$\mathbf{A} = \sum_i \hat{x}_i (\hat{x}_i \cdot \mathbf{A}). \tag{2674}$$

Clearly, the inner product for two arbitrary vectors is given in this case by

$$\mathbf{A} \cdot \mathbf{B} = \sum_i (\hat{x}_i \cdot \mathbf{A})(\hat{x}_i \cdot \mathbf{B}). \tag{2675}$$

If we know the components of a vector $|A\rangle$ in one basis, $|i\rangle$, it can be expressed in terms of another complete and orthonormal basis, $|\theta_i\rangle$, if we know the matrix elements, $\langle i|\theta_j\rangle$, connecting the two basis sets for all i and j. This follows by inserting the completeness relation given by Eq. (2670) into the definition of the projection of $|A\rangle$ onto the set $|\theta_i\rangle$:

$$\langle \theta_i | A \rangle = \sum_i \langle \theta_i | i \rangle \langle i | A \rangle. \tag{2676}$$

Let us return to Eq. (2666), where we interpreted the matrix B_{ij} as a matrix element of an abstract operator: $\hat{B}$. In this approach we treat matrix multiplication by writing

$$\sum_k A_{ik} B_{kj} = \sum_k \langle i|\hat{A}|k\rangle \langle k|\hat{B}|j\rangle$$
$$= \langle i|\hat{A}\hat{B}|j\rangle, \tag{2677}$$

where again the set $|k\rangle$ is complete. The trace of a matrix is defined by

$$\sum_i A_{ii} = \sum_i \langle i|\hat{A}|i\rangle \equiv \operatorname{Tr}\hat{A}. \tag{2678}$$

In our development we will need to use two simple properties of matrices. First we need the theorem that the trace of a matrix is independent of the representation used to evaluate the trace. Assume that we have two independent complete and orthonormal basis sets $|i\rangle$ and $|\theta_i\rangle$. Using the completeness of the set $|\theta_i\rangle$,

$$1 = \sum_i |\theta_i\rangle \langle \theta_i|, \tag{2679}$$

we can rewrite Eq. (2678) as

$$\begin{aligned}\operatorname{Tr}\hat{A} &= \sum_i \langle i|\hat{A}|i\rangle = \sum_{ijk} \langle i|\theta_j\rangle\langle\theta_j|\hat{A}|\theta_k\rangle\langle\theta_k|i\rangle \\ &= \sum_{j,k} \langle\theta_j|\hat{A}|\theta_k\rangle\langle\theta_k| \sum_i |i\rangle\langle i|\theta_j\rangle \\ &= \sum_{j,k} \langle\theta_j|\hat{A}|\theta_k\rangle\langle\theta_k|\theta_j\rangle \\ &= \sum_{j,k} \langle\theta_j|\hat{A}|\theta_k\rangle\delta_{k,j} = \sum_j \langle\theta_j|\hat{A}|\theta_j\rangle. \end{aligned} \qquad (2680)$$

This proves the proposition.

It is very useful to be able to diagonalize matrices. From the point of view taken here, the diagnonalization of a matrix $B_{ij} = \langle i|\hat{B}|j\rangle$ corresponds to a transformation from the basis set $|i\rangle$ to the set $|b_i\rangle$, where

$$\langle b_i|\hat{B}|b_j\rangle = b_i \delta_{ij}. \qquad (2681)$$

Construction of the basis set $|b_i\rangle$ corresponds to solving the eigenvalue problem

$$\hat{B}|b_i\rangle = b_i|b_i\rangle \qquad (2682)$$

where b_i is the ith eigenvalue and $|b_i\rangle$ the ith eigenvector. The construction of the eigenvalues and eigenfunctions associated with a given matrix B_{ij} comes from taking the inner product of Eq. (2682), with b_i set equal to some unknown value b, with the bra $< j|$, and introducing a complete set of states to obtain

$$\sum_j \langle i|\hat{B}|j\rangle\langle j|b\rangle = b\langle i|b\rangle. \qquad (2683)$$

Since the matrix elements $B_{ij} = \langle i|\hat{B}|j\rangle$ are assumed to be known, this is a linear homogeneous equation for the $\langle j|b\rangle$. It follows from the *fundamental theorem* for solving homogeneous linear equations [38] that solutions exist only for values of b satisfying

$$\det(B - b) = 0. \qquad (2684)$$

This discrete set of solutions gives the eigenvalues b_i, and the eigenvectors

$$|b_k\rangle = \sum_i |i\rangle\langle i|b_k\rangle \qquad (2685)$$

are constructed from Eq. (2683) by choosing in order the solutions of Eq. (2684) for each k, and $b = b_k$, and solving the associated linear set of equations for the $\langle i|b_k\rangle$.

K SECOND QUANTIZATION

Spin and Statistics

Suppose that $|\psi\rangle$ is an abstract quantum state describing a system of n identical particles. Then in the coordinate representation the wavefunction corresponding to that state is given by

$$\psi(1, 2, \ldots, i, \ldots, n) = \langle 1, 2, \ldots, i, \ldots, n | \psi \rangle, \qquad (2686)$$

where the index i refers to both space and spin coordinates of particle i. If these particles have half-integral spin, they are called *fermions*. If they have integral spin, they are called *bosons*. It is a fact of nature that the wavefunction for fermions is antisymmetric under the interchange of two particles,

$$\psi(1, \ldots, j, \ldots, i, \ldots, n) = -\psi(1, \ldots, i, \ldots, j, \ldots, n) \qquad \text{fermions,} \qquad (2687)$$

whereas the wavefunction is symmetric under the interchange of two particles for bosons,

$$\psi(1, \ldots, j, \ldots, i, \ldots, n) = \psi(1, \ldots, i, \ldots, j, \ldots, n) \qquad \text{bosons.} \qquad (2688)$$

This symmetrization requirement is technically difficult to handle when constructing states using the usual Schrödinger description. Use of the method of second quantization [39] allows us to incorporate the correct symmetrization at each step.

Creation and Annihilation Operators

We now want to develop a formalism where the central objects are not the wavefunctions, but operators that have the effect of creating or destroying particles in certain single-particle quantum states.

Bosons. We consider first the case of bosons. Suppose that in the absence of interactions, we can introduce well-defined single-particle states and label these states with the single-particle quantum numbers i. i may include a momentum index $\mathbf{p}$ as well as a spin index σ. Thus a single-particle state is determined by $i = (\mathbf{p}, \sigma)$. We then introduce the creation operator a_i^+, which will have the property of creating a particle in the quantum state i, and the annihilation operator a_i, which destroys or removes a particle from the quantum state i. By definition, for bosons, these operators obey the commutation relations

$$[a_i, a_j^+] = a_i a_j^+ - a_j^+ a_i = \delta_{ij} \qquad (2689)$$

and

$$[a_i, a_j] = [a_i^+, a_j^+] = 0 \qquad (2690)$$

and a_i^+ is the Hermitian conjugate of a_i. The essence of the mathematics of second quantization are contained in these commutation relations. To see this, we first investigate the eigenvalue spectrum for the *number operator* for state i:

$$n_i = a_i^+ a_i. \quad (2691)$$

The effect of this operator acting on a quantum state is first to take a particle out of state i and then put one back. This operation, it turns out, is sensitive to the number of particles present in state i—hence the name *number operator*.

We assume that there exists a normalized eigenvector $|\alpha_i\rangle$ of n_i such that

$$n_i|\alpha_i\rangle = a_i^+ a_i |\alpha_i\rangle = \alpha_i|\alpha_i\rangle \quad (2692)$$

and $\langle \alpha_i | \alpha_i \rangle = 1$. To save writing, let us for the time being drop the index i and focus on a particular single-particle state. If we define a state $|\alpha'\rangle$ by applying the annihilation operator to the state $|\alpha\rangle$,

$$|\alpha'\rangle = a|\alpha\rangle, \quad (2693)$$

then taking the inner product with its dual $\langle \alpha'|$, we obtain

$$\langle \alpha' | \alpha' \rangle = \langle \alpha | a^+ a | \alpha \rangle = \langle \alpha | n | \alpha \rangle = \alpha \langle \alpha | \alpha \rangle$$
$$= \alpha \geq 0. \quad (2694)$$

The last inequality follows since the *norm* of a nonzero vector is positive.

To go further, we need some operator identities. Any operators A, B, and C satisfy the operator identities

$$[AB, C] = A[B, C] + [A, C]B. \quad (2695)$$

Using this result with the commutation relations given by Eqs. (2689) and (2690), we obtain

$$[a^+ a, a] = -a \quad (2696)$$

and

$$[a^+ a, a^+] = a^+. \quad (2697)$$

These are equivalent to

$$(a^+ a) a = a(a^+ a - 1) \quad (2698)$$
$$(a^+ a) a^+ = a^+(a^+ a + 1). \quad (2699)$$

We then use Eq. (2698), after allowing the number operator n to act on the state $|\alpha'\rangle$,

$$\begin{aligned} n|\alpha'\rangle &= (a^+a)a|\alpha\rangle = a(a^+a - 1)|\alpha\rangle \\ &= a(\alpha - 1)|\alpha\rangle \\ &= (\alpha - 1)a|\alpha\rangle = (\alpha - 1)|\alpha'\rangle. \end{aligned} \qquad (2700)$$

This shows that $|\alpha'\rangle$ is an eigenvector for $n = a^+a$ with eigenvalue $\alpha - 1$ unless $|\alpha'\rangle = a|\alpha\rangle = 0$. Similarly, we can show that

$$(a^+a)a^+|\alpha\rangle = (\alpha + 1)a^+|\alpha\rangle \qquad (2701)$$

unless

$$a^+|\alpha\rangle = 0. \qquad (2702)$$

Note that $|\alpha'\rangle$ is not normalized to 1 since

$$\langle \alpha'|\alpha'\rangle = \langle \alpha|a^+a|\alpha\rangle = \alpha. \qquad (2703)$$

Now suppose that $a^n|\alpha\rangle \neq 0$ for all $n \geq 0$. Then, inspired by Eq. (2698), we investigate the quantity

$$\begin{aligned} a^+aa^n &= (aa^+ - 1)a^n \\ &= a(a^+a - 1)a^{n-1} \\ &= -a^n + a(a^+a)a^{n-1}. \end{aligned} \qquad (2704)$$

On the right-hand side of Eq. (2704), we have $(a^+a)a^{n-1}$, so we have a *recursion relation* we can iterate:

$$\begin{aligned} a^+aa^n &= -a^n + a(-a^{n-1} + a(a^+a)a^{n-2}) \\ &= -2a^n + a^2a^+aa^{n-2}. \end{aligned} \qquad (2705)$$

Continuing this process, one eventually finds that

$$a^+aa^n = -na^n + a^na^+a. \qquad (2706)$$

This operator equation immediately translates to the eigenvalue equation result,

$$\begin{aligned} (a^+a)a^n|\alpha\rangle &= (-na^n + a^na^+a)|\alpha\rangle \\ &= (\alpha - n)a^n|\alpha\rangle. \end{aligned} \qquad (2707)$$

Thus $a^n|\alpha\rangle$ is an eigenstate of the number operator with eigenvalue $\alpha - n$. However, for sufficiently large n, the result

$$\langle\alpha|(a^+)^{n+1}a^{n+1}|\alpha\rangle = \langle\alpha|(a^+)^n a^+ a a^n|\alpha\rangle$$
$$= (\alpha - n)\langle\alpha|(a^+)^n a^n|\alpha\rangle \geq 0 \qquad (2708)$$

presents the possibility of contradiction. The way out of this contradiction is to assume that there must exist some nonnegative integer $\bar{n}$ such that

$$a^{\bar{n}}|\alpha\rangle \neq 0 \qquad (2709)$$

but

$$a^{\bar{n}+1}|\alpha\rangle = 0. \qquad (2710)$$

Then setting $n = \bar{n}$ in Eq. (2707), we obtain

$$a^+ a^{\bar{n}+1}|\alpha\rangle = (\alpha - \bar{n})a^{\bar{n}}|\alpha\rangle = 0, \qquad (2711)$$

which is satisfied by

$$\alpha = \bar{n}. \qquad (2712)$$

The state $\alpha = \bar{n}$ is the *ground state*, $|0\rangle$, which satisfies

$$a|0\rangle = 0, \qquad (2713)$$

and

$$a^+ a|0\rangle = 0. \qquad (2714)$$

The ground state therefore contains no particles, so we cannot destroy a particle that is not there.

We can build up multiparticle states by applying a^+ repeatedly to the ground state. The assertion that

$$(a^+)^n|0\rangle \qquad (2715)$$

is an *n*-particle state is equivalent to the statement that this state is an eigenstate of the number operator $a^+ a$ with eigenvalue n. We can define the normalized eigenstates

$$|n\rangle = \frac{1}{\sqrt{n!}}(a^+)^n|0\rangle. \qquad (2716)$$

The proof that these states are indeed orthonormal,

$$\langle m|n\rangle = \delta_{mn}, \tag{2717}$$

is discussed in Problem 3.37.

We can easily show that these basis states satisfy

$$a^+|n\rangle = \sqrt{n+1}|n+1\rangle \tag{2718}$$
$$a|n\rangle = \sqrt{n}|n-1\rangle \tag{2719}$$
$$a^+a|n\rangle = n|n\rangle. \tag{2720}$$

These states are determined up to a phase factor. We may choose the phase to be zero. We can see from these equations why we call a^+ and a creation and annihilation operators. a^+ increases the number of particles in our basis set by one, while a reduces it by one. The number operator samples the number of particles by first taking out a particle and then putting it back in. This basis, $|n\rangle$, is known as the *occupation number basis*. This is because we keep track of the number of particles with a particular set of quantum numbers i. Note that this basis followed from the commutation relations for the operators.

The completeness of our basis set

$$1 = \sum_{n=0}^{\infty} |n\rangle\langle n| \tag{2721}$$

follows from the fact that the state of a system with one energy state is specified completely if we give the number of particles in that state.

If we have many, M, possible single-particles states, we can write the occupation basis set

$$|n_1 n_2 \cdots n_i \cdots n_M\rangle = \frac{(a_1^+)^{n_1}}{\sqrt{n_1!}} \cdots \frac{(a_i^+)^{n_i}}{\sqrt{n_i!}} \cdots \frac{(a_M^+)^{n_M}}{\sqrt{n_M!}} |0\rangle \tag{2722}$$

and a_i^+ creates a particle in the ith single-particle quantum state. It is clear that this basis spans all possible numbers of particles in all possible single-particle states.

Fermions. The case of fermions can be treated in a similar manner except for one major difference. The creation and annihilation operators for fermions satisfy the *anticommutation* relations $[a_i, a_j^+]_+ = \delta_{ij}$, $[a_i, a_j]_+ = [a_i^+, a_j^+]_+ = 0$, where

$$[A, B]_+ = AB + BA. \tag{2723}$$

Again, we will be interested in the eigenvalue spectrum of the number operator

$$n_i = a_i^+ a_i. \tag{2724}$$

SECOND QUANTIZATION

The discussion in this case is considerably simplified by the result

$$[a_i, a_i]_+ = 2a_i a_i = 0 = 2a_i^+ a_i^+. \tag{2725}$$

This result, which is essentially the Pauli exclusion principle, tells us that we cannot create two fermions in the same state.

We can now look at the eigenvalue spectrum asociated with the number operator. As in the boson case we begin by looking at a single energy state and suppressing the index i. Let us assume that $|\alpha\rangle$ is an eigenstate of $n = a^+a$ with a nonzero eigenvalue α:

$$n|\alpha\rangle = \alpha|\alpha\rangle. \tag{2726}$$

Let us apply the operator a to this equation:

$$an|\alpha\rangle = \alpha a|\alpha\rangle. \tag{2727}$$

On the left-hand side we have

$$an|\alpha\rangle = aa^+a|\alpha\rangle = (1 - a^+a)a|\alpha\rangle = a|\alpha\rangle = \alpha a|\alpha\rangle. \tag{2728}$$

This tells us that the only nonzero eigenvalue is $\alpha = 1$, corresponding to the state $|1\rangle$. If we construct $|\alpha'\rangle = a|1\rangle$, this state has eigenvalue zero,

$$n|\alpha'\rangle = a^+aa|1\rangle = 0, \tag{2729}$$

so $|\alpha'\rangle$ must be the vacuum state. So for the ith quantum state, we have only two possibilities, with either 0 or 1 particles present.

For a system with many (M) quantum states, we have the many-particle basis states

$$|n_1 n_2 \cdots n_M\rangle = (a_1^+)^{n_1} (a_2^+)^{n_2} \cdots (a_M^+)^{n_M} |0\rangle. \tag{2730}$$

This basis set can be seen to be complete and orthonormal within the appropriate space.

Field Operators

There are situations where it is more convenient to create a particle at a particular space point as opposed to creating it in a particular single-particle eigenstate. We define

$$\psi^+(\mathbf{r}) = \sum_i \phi_i^*(\mathbf{r}) a_i^+ \tag{2731}$$

$$\psi(\mathbf{r}) = \sum_i \phi_i(\mathbf{r}) a_i \tag{2732}$$

where the ϕ_i's are a complete and orthonormal set of single-particle wave functions obtained from a solution of the single-particle Schrödinger equation. $\psi^+(\mathbf{r})$ adds a particle to the system at the point $\mathbf{r}$ and $\psi(\mathbf{r})$ removes a particle from the system at point **r**. ψ^+ and ψ are called *field operators* and, for bosons, satisfy the commutation relations

$$[\psi(\mathbf{r}), \psi^+(\mathbf{r}')] = \sum_{ij} \phi_i^*(\mathbf{r})\phi_j(\mathbf{r}')[a_i, a_j^+]$$
$$= \sum_i \phi_i^*(\mathbf{r})\phi_i(\mathbf{r}') = \delta^3(\mathbf{r}-\mathbf{r}') \qquad \text{bosons.} \qquad (2733)$$

The field operators for fermions satisfy the anticommutation relations

$$[\psi(\mathbf{r}), \psi^+(\mathbf{r}')]_+ = \delta^3(\mathbf{r}-\mathbf{r}') \qquad \text{fermions.} \qquad (2734)$$

We can now form the *N*-particle states:

$$|\mathbf{r}_1\mathbf{r}_2\cdots\mathbf{r}_N\rangle = \frac{1}{\sqrt{N!}}\psi^+(\mathbf{r}_N)\psi^+(\mathbf{r}_{N-1})\cdots\psi^+(\mathbf{r}_1)|0\rangle. \qquad (2735)$$

This basis set can be seen to be complete and orthonormal within the appropriate space.

If we have a single particle in state *i*, the quantum state can be given as, $|1_i\rangle = a_i^+|0\rangle$. The probability of finding this particle at position $\mathbf{r}_1$ is given by

$$P_i(\mathbf{r}_1) = |\langle \mathbf{r}_1|1_i\rangle|^2 \qquad (2736)$$

where

$$\langle \mathbf{r}_1|1_i\rangle = \langle 0|\psi(\mathbf{r}_1)a_i^+|0\rangle = \sum_j \langle 0|\phi_j(\mathbf{r}_1)a_ja_i^+|0\rangle$$
$$= \sum_j \phi_j(\mathbf{r}_1)\langle 0|(\delta_{ij} + a_i^+a_j)|0\rangle = \phi_i(\mathbf{r}_1). \qquad (2737)$$

Similarly, the associated two-particle wavefunction is given by

$$P_{i\ell}(\mathbf{r}_1,\mathbf{r}_2) = \langle \mathbf{r}_1\mathbf{r}_2|1_i1_\ell\rangle = \langle 0|\psi(\mathbf{r}_1)\psi(\mathbf{r}_2)a_i^+a_\ell^+|0\rangle$$
$$= \sum_{j,k} \phi_j(\mathbf{r}_1)\phi_k(\mathbf{r}_2)\langle 0|a_ja_ka_i^+a_\ell^+|0\rangle. \qquad (2738)$$

It is left to Problem 3.38 to show that

$$\langle \mathbf{r}_1\mathbf{r}_2|1_i1_\ell\rangle = \phi_\ell(\mathbf{r}_1)\phi_i(\mathbf{r}_2) + \eta\phi_i(\mathbf{r}_1)\phi_\ell(\mathbf{r}_2) \qquad (2739)$$

which is symmetric under exchange of two bosons ($\eta = 1$) and antisymmetric under exchange of two fermions ($\eta = -1$). Thus we are able to create multiparticle states with the correct permutation symmetries.

L EVALUATION OF BOSE INTEGRALS

We need to evaluate *Bose integrals* of the form

$$I_B(n) = \int_0^\infty \frac{dx\, x^n}{e^x - 1} = \int_0^\infty dx\, \frac{x^n e^{-x}}{1 - e^{-x}}$$

$$= \int_0^\infty dx\, x^n \sum_{k=0}^\infty e^{-(k+1)x}$$

$$= \sum_{k=0}^\infty \int_0^\infty dx\, x^n e^{-(k+1)x}$$

$$= \sum_{k=0}^\infty \frac{1}{(k+1)^{n+1}} \int_0^\infty dy\, y^n e^{-y}. \tag{2740}$$

If we introduce the Riemann zeta function

$$\zeta(x) = \sum_{k=0}^\infty \frac{1}{(k+1)^x}, \tag{2741}$$

and the Γ-function

$$\Gamma(x) = \int_0^\infty dy\, y^{x-1} e^{-y}, \tag{2742}$$

we have

$$I_B(n) = \zeta(n+1)\Gamma(n+1). \tag{2743}$$

If n is an integer, this reduces to

$$I_B(n) = n!\,\zeta(n+1). \tag{2744}$$

Some useful results [40]:

$$\zeta(\tfrac{3}{2}) \approx 2.612 \tag{2745}$$

$$\zeta(2) = \frac{\pi^2}{6} \approx 1.645 \tag{2746}$$

$$\zeta(\tfrac{5}{2}) \approx 1.341 \tag{2747}$$

$$\zeta(3) \approx 1.202 \tag{2748}$$

$$\zeta(4) = \frac{\pi^4}{90} \approx 1.082 \tag{2749}$$

$$\zeta(6) = \frac{\pi^6}{945} \tag{2750}$$

$$\Gamma(\tfrac{1}{2}) = \sqrt{\pi} \tag{2751}$$

M FERMI INTEGRALS AT LOW TEMPERATURES

Consider the integral

$$I = \int_0^{+\infty} d\epsilon\, n(\epsilon) \frac{d\phi(\epsilon)}{d\epsilon} \tag{2752}$$

where

$$n(\epsilon) = (e^{\beta(\epsilon-\mu)} + 1)^{-1} \tag{2753}$$

is the Fermi–Dirac distribution and $\phi(\epsilon)$ is some smooth function of ϵ such that $\phi(0) = 0$. We follow Sommerfeld and develop I as a power series expansion in $k_B T$ which will be valid for low temperatures. We first perform an integration by parts, to obtain

$$I = n(\epsilon)\phi(\epsilon)\Big|_0^\infty - \int_0^\infty \phi(\epsilon) \frac{dn}{d\epsilon} d\epsilon. \tag{2754}$$

Since $\phi(0) = 0$ and we assume that $\lim_{\epsilon\to\infty} e^{-\beta\epsilon}\phi(\epsilon) = 0$, we have

$$I = -\int_0^\infty \phi(\epsilon) \frac{dn(\epsilon)}{d\epsilon}. \tag{2755}$$

We note, however, that

$$\frac{dn(\epsilon)}{d\epsilon} = \frac{-\beta}{(e^{\beta(\epsilon-\mu)} + 1)(e^{-\beta(\epsilon-\mu)} + 1)} \tag{2756}$$

is very sharply peaked near $\epsilon = \mu$ for large β. More precisely,

$$\lim_{\beta\to\infty} \frac{dn(\epsilon)}{d\epsilon} = -\delta(\epsilon - \mu). \tag{2757}$$

It therefore makes sense to Taylor-series expand $\phi(\epsilon)$ about the point $\epsilon = \mu$ and write

$$\phi(\epsilon) = \sum_{n=0}^{\infty} \frac{1}{n!} (\epsilon - \mu)^n \phi^{(n)}(\mu) \tag{2758}$$

where

$$\phi^{(n)}(\mu) = \frac{d^n}{d\mu^n} \phi(\mu). \tag{2759}$$

Our integral can then be written in the form

$$I = -\sum_{n=0}^{\infty} \frac{\phi^{(n)}(\mu)}{n!} \int_0^{\infty} d\epsilon (\epsilon - \mu)^n \frac{dn(\epsilon)}{d\epsilon}. \tag{2760}$$

If we change integration variables to $x = \beta(\epsilon - \mu)$, we find that

$$I = -\sum_{n=0}^{\infty} \frac{\phi^{(n)}(\mu)}{n!} (\beta^{-1})^{n+1} \int_{-\beta\mu}^{\infty} dx\, x^n \frac{-\beta}{(e^x + 1)(e^{-x} + 1)}$$

$$= \sum_{n=0}^{\infty} \frac{\phi^{(n)}(\mu)}{n!} (k_B T)^n I_n(\beta\mu). \tag{2761}$$

The integral

$$I_n(\beta\mu) = \int_{-\beta\mu}^{\infty} \frac{dx\, x^n e^x}{(e^x + 1)^2} \tag{2762}$$

can be written as the sum of two parts:

$$I_n(\beta\mu) = I_n(\infty) - \bar{I}_n(\beta\mu) \tag{2763}$$

where

$$I_n(\infty) = \int_{-\infty}^{\infty} dx \frac{x^n e^x}{(e^x + 1)^2} \tag{2764}$$

and

$$\bar{I}_n(\beta\mu) = \int_{-\infty}^{-\beta\mu} dx \frac{x^n e^x}{(e^x + 1)^2}. \tag{2765}$$

Letting $x \to -x$ in $\bar{I}_n$ gives

$$\bar{I}_n(\beta\mu) = (-1)^n \int_{\beta\mu}^{\infty} \frac{dx\, x^n e^{-x}}{(e^{-x} + 1)^2}. \tag{2766}$$

Then since

$$\frac{1}{(e^{-x} + 1)^2} \leq 1 \tag{2767}$$

we have

$$\begin{aligned}|\bar{I}_n(\beta\mu)| &\leq \int_{\beta\mu}^{\infty} dx\, x^n e^{-x} \\ &= e^{-x} \sum_{r=0}^{n} \frac{(-1)^r n! x^{n-r}}{(n-r)!(-1)^{r+1}}\bigg|_{\beta\mu}^{\infty} \\ &= e^{-\beta\mu} n! \sum_{r=0}^{n} \frac{(\beta\mu)^{n-r}}{(n-r)!} \\ &= e^{-\beta\mu} n! \sum_{r=0}^{n} \frac{(\beta\mu)^r}{r!}.\end{aligned} \qquad (2768)$$

Consequently, since we obtain self-consistently for fermions at low temperatures that $\mu > 0$, the contributions of $\bar{I}_n$ are exponentially small for large β, small T. We then have that

$$I = \sum_{n=0}^{\infty} \frac{\phi^{(n)}(\mu)}{n!} (k_B T)^n I_n(\infty) + \mathcal{O}(e^{-\beta\mu}). \qquad (2769)$$

We can evaluate $I_n(\infty)$ as follows. We first note that I_n vanishes if n is odd because the rest of the integrand multiplying x^n is even:

$$I_{2n+1}(\infty) = 0. \qquad (2770)$$

We can then write for even n,

$$\begin{aligned}I_n(\infty) &= 2 \int_0^{\infty} \frac{x^n e^x dx}{(e^x+1)^2} \\ &= -2 \int_0^{\infty} x^n dx \frac{d}{dx} \frac{1}{e^x+1}\end{aligned}$$

Then for $n = 0$,

$$I_0(\infty) = 1, \qquad (2772)$$

while for $n > 0$, we have

$$\begin{aligned}I_n(\infty) &= 2n \int_0^{\infty} \frac{x^{n-1} dx}{e^x+1} = 2n \int_0^{\infty} dx \frac{x^{n-1} e^{-x}}{1+e^{-x}} \\ &= 2n \int_0^{\infty} dx\, x^{n-1} \sum_{k=0}^{\infty} (-1)^k e^{-(k+1)x} \\ &= 2n \sum_{k=0}^{\infty} (-1)^k \int_0^{\infty} dx\, x^{n-1} e^{-(k+1)x}.\end{aligned} \qquad (2773)$$

Let $y = (k+1)x$ in the integral; then

$$I_n(\infty) = 2n \sum_{k=0}^{\infty} \frac{(-1)^k}{(k+1)^n} \int_0^{\infty} dy\, y^{n-1} e^{-y}$$

$$= 2n \sum_{k=0}^{\infty} \frac{(-1)^k}{(k+1)^n} (n-1)! = 2n! \sum_{k=0}^{\infty} \frac{(-1)^k}{(k+1)^n}. \tag{2774}$$

It is shown in Problem 3.12 that the sum

$$\sum_{k=0}^{\infty} \frac{(-1)^k}{(k+1)^n} = (1 - 2^{1-n})\zeta(n) \tag{2775}$$

where $\zeta(n)$ is the Riemann zeta function defined in Appendix L. Putting all of this together yields

$$I_n(\infty) = 2n!(1 - 2^{1-n})\zeta(n), \tag{2776}$$

so, returning to Eq. (2761), we find that

$$I = \phi^{(0)}(\mu) I_0 + \sum_{n=1}^{\infty} \frac{\phi^{(2n)}(\mu)}{(2n)!} (k_B T)^{2n} I_{2n}(\beta\mu) + \mathcal{O}(e^{-\beta\mu})$$

$$= \phi^{(0)}(\mu) + 2\sum_{n=1}^{\infty} \phi^{(2n)}(\mu)(k_B T)^{2n}(1 - 2^{1-2n})\zeta(2n) + \mathcal{O}(e^{-\beta\mu}). \tag{2777}$$

If we now define

$$g(\epsilon) = \frac{d\phi(\epsilon)}{d\epsilon}, \tag{2778}$$

we have the very useful result

$$\int_0^{\infty} d\epsilon\, n(\epsilon) g(\epsilon) = \int_0^{\mu} g(\epsilon) d\epsilon$$
$$+ \sum_{n=1}^{\infty} 2(1 - 2^{1-2n})\zeta(2n)(k_B T)^{2n} g^{(2n+1)}(\mu) + \mathcal{O}(e^{-\beta\mu}),$$

which gives us a power series in the temperature. Some values for the Riemann zeta function are listed in Appendix L. Using the analytic values for the low-order terms leads to the expansion

$$\int_0^{\infty} d\epsilon\, n(\epsilon) g(\epsilon) = \int_0^{\mu} d\epsilon\, g(\epsilon) + \frac{\pi^2}{6}(k_B T)^2 g^{(1)}(\mu) + \frac{7\pi^4}{360}(k_B T)^4 g^{(3)}(\mu) + \cdots$$

$$\tag{2779}$$

N THERMAL PERTURBATION THEORY

Introduction

Suppose that we have a Hamiltonian of the form

$$H = H_0 + \lambda H_I, \tag{2780}$$

where H_0 is a Hamiltonian we can treat, while H_I is considered a small perturbation. We want to compute the free energy

$$F(\lambda) = -\beta^{-1} \ln \operatorname{Tr} e^{-\beta(H_0 + \lambda H_I)}. \tag{2781}$$

Let us define the unnormalized density matrix

$$\rho(\beta) = e^{-\beta(H_0 + \lambda H_I)} \tag{2782}$$

which is a function of β. Our goal is to determine $\rho(\beta)$ and $F(\lambda)$ in a power series expansion in the *small* quantity λ. To this end we first take the derivative of ρ with respect to β:

$$\frac{\partial \rho(\beta)}{\partial \beta} = -(H_0 + \lambda H_I)\rho(\beta). \tag{2783}$$

It is useful to define the auxiliary quantity $\rho_I(\beta)$ via

$$\rho(\beta) = \rho_0(\beta)\rho_I(\beta) \tag{2784}$$

where

$$\rho_0(\beta) = e^{-\beta H_0} \tag{2785}$$

is the unperturbed density matrix. We then insert Eq. (2784) into Eq. (2783), to obtain

$$-H_0 \rho(\beta) + \rho_0(\beta) \frac{\partial \rho_I(\beta)}{\partial \beta} = -(H_0 + \lambda H_I)\rho(\beta). \tag{2786}$$

After canceling the terms proportional to H_0, we obtain an equation for $\rho_I(\beta)$:

$$\frac{\partial \rho_I(\beta)}{\partial \beta} = -e^{\beta H_0} \lambda H_I e^{-\beta H_0} \rho_I(\beta). \tag{2787}$$

Using the boundary condition $\rho(0) = 1$, we can integrate Eq. (2787) to obtain

$$\rho_I(\beta) = 1 - \int_0^\beta d\beta' \lambda H_I(\beta') \rho_I(\beta') \tag{2788}$$

where

$$H_I(\beta) \equiv e^{\beta H_0} H_I e^{-\beta H_0}. \tag{2789}$$

If we *iterate* Eq. (2788), by substituting for $\rho_I(\beta')$ in terms of the right-hand side of Eq. (2788), we obtain

$$\rho_I(\beta) = 1 - \int_0^\beta d\beta' \lambda H_I(\beta') + \int_0^\beta d\beta' \lambda H_I(\beta') \int_0^{\beta'} d\beta'' \lambda H_I(\beta'') \rho_1(\beta''). \tag{2790}$$

Clearly, this iteration process can be continued and we obtain the desired perturbation theory expansion in powers of λ,

$$\rho(\beta) = \rho_0(\beta)[1 - \int_0^\beta d\beta' \lambda H_I(\beta') + \int_0^\beta d\beta' \lambda H_i(\beta') \int_0^{\beta'} d\beta'' \lambda H_I(\beta'') + \mathcal{O}(\lambda^3)]. \tag{2791}$$

We now want to use this expression to obtain an expansion for the free energy $F(\lambda)$ in powers of λ. We have immediately that

$$F(\lambda) = -\beta^{-1}\ln \text{Tr}\rho(\beta) = -\beta^{-1}\ln \text{Tr}\rho_0(\beta)\rho_I(\beta) \tag{2792}$$

$$= -\beta^{-1}\ln(Z_0 \langle \rho_I(\beta) \rangle_0), \tag{2793}$$

where $\langle \cdot \rangle_0$ indicates an average over the unperturbed density matrix

$$\langle A \rangle_0 = \text{Tr} \frac{\rho_0(\beta) A}{Z_0}, \tag{2794}$$

and

$$Z_0 = \text{Tr}\rho_0(\beta) = \text{Tr}\, e^{-\beta H_0} \tag{2795}$$

is the unperturbed partition function. Equation (2793) can be written in the form

$$F(\lambda) = F_0 - \beta^{-1}\ln\langle \rho_I(\beta) \rangle_0 \tag{2796}$$

where

$$F_0 = -\beta^{-1}\ln Z_0 \tag{2797}$$

is the unperturbed free energy. A key point, which was stressed in Chapter 4, is that we should expand quantities, such as the free energy, with a simple (linear) volume dependence. If we Taylor-series expand the free energy,

$$F(\lambda) = \sum_{n=0}^{\infty} \frac{\lambda^n}{n!} F^{(n)}(\beta), \qquad (2798)$$

then

$$F^{(n)}(\beta) = \left(\frac{\partial^n}{\partial \lambda^n} F(\lambda)\right)_{\lambda=0}. \qquad (2799)$$

We then easily find the first two derivatives:

$$\frac{\partial F(\lambda)}{\partial \lambda} = -\beta^{-1} \frac{1}{\langle \rho_I \rangle_0} \frac{\partial}{\partial \lambda} \langle \rho_I \rangle_0, \qquad (2800)$$

$$\frac{\partial^2 F(\lambda)}{\partial \lambda^2} = \beta^{-1} \frac{1}{\langle \rho_I \rangle_0^2} \left(\frac{\partial}{\partial \lambda} \langle \rho_I \rangle_0\right)^2 - \beta^{-1} \frac{1}{\langle \rho_I \rangle_0} \frac{\partial^2}{\partial \lambda^2} \langle \rho_I \rangle_0. \qquad (2801)$$

Using the explicit expression for ρ_I given by Eq. (2788), we easily obtain

$$F^{(1)}(\beta) = \beta^{-1} \int_0^\beta d\beta' \langle H_I(\beta') \rangle_0 \qquad (2802)$$

$$F^{(2)}(\beta) = \beta^{-1} \left\{ \left[\int_0^\beta d\beta' \langle H_I(\beta') \rangle_0 \right]^2 - 2 \int_0^\beta d\beta' \int_0^{\beta'} d\beta'' \langle H_I(\beta') H_I(\beta'') \rangle_0 \right\}. \qquad (2803)$$

$F^{(1)}(\beta)$ can be simplified using the cyclic invariance of the trace to obtain

$$F^{(1)}(\beta) = \beta^{-1} \int_0^\beta \langle H_I(\beta') \rangle_0 d\beta'$$
$$= \beta^{-1} \int_0^\beta d\beta' \frac{1}{Z_0(\beta)} \text{Tr} e^{-\beta H_0} e^{\beta' H_0} H_I e^{-\beta' H_0}$$
$$= \langle H_I \rangle_0. \qquad (2804)$$

$F^{(2)}(\beta)$, given by Eq. (2803), can be rewritten as

$$F^{(2)}(\beta) = -2\beta^{-1} \int_0^\beta d\beta' \int_0^{\beta'} d\beta'' \langle \delta H_I(\beta') \delta H_I(\beta'') \rangle_0 \qquad (2805)$$

where

$$\delta H_I(\beta) = H_I(\beta) - \langle H_I(\beta) \rangle_0. \qquad (2806)$$

This follows since

$$2\beta^{-1} \int_0^\beta d\beta' \int_0^{\beta'} d\beta'' \langle H_I(\beta') \rangle_0 \langle H_I(\beta'') \rangle_0 = 2\beta^{-1} \langle H_I \rangle_0^2 \int_0^\beta d\beta' \beta'$$
$$= \beta \langle H_I \rangle_0^2. \qquad (2807)$$

THERMAL PERTURBATION THEORY

$F^{(2)}(\beta)$ can be rewritten in a more explicit form by introducing the complete set of eigenstates $|n\rangle$ associated with H_0:

$$H_0|n\rangle = E_n|n\rangle. \tag{2808}$$

We can then write, using the cyclic invariance of the trace,

$$\begin{aligned}
\langle \delta H_I(\beta')\delta H_I(\beta'')\rangle &= \frac{1}{Z_0}\operatorname{Tr} e^{-\beta H_0} e^{\beta' H_0} \delta H_I e^{(\beta''-\beta')H_0} \delta H_I e^{-\beta'' H_0} \\
&= \frac{1}{Z_0}\operatorname{Tr} e^{-\beta H_0} e^{(\beta'-\beta'')H_0} \delta H_I e^{(\beta''-\beta')H_0} \delta H_I \\
&= \frac{1}{Z_0}\sum_{nm}\langle n|e^{-\beta H_0} e^{(\beta'-\beta'')H_0} \delta H_I|m\rangle\langle m|e^{(\beta''-\beta')H_0} \delta H_I|n\rangle \\
&= \frac{1}{Z_0}\sum_{nm} e^{-\beta E_n} e^{(\beta'-\beta'')E_n} \langle n|\delta H_I|m\rangle e^{(\beta''-\beta')E_m} \langle m|\delta H_I|n\rangle \\
&= \frac{1}{Z_0}\sum_{nm} e^{-\beta E_n} |\langle n|\delta H_I|m\rangle|^2 e^{\beta'(E_n-E_m)} e^{\beta''(E_m-E_n)}. \tag{2809}
\end{aligned}$$

Inserting Eq. (2809) into Eq. (2805) one has, with $\Delta E_{nm} = E_n - E_m$, the integrations

$$\begin{aligned}
\int_0^\beta d\beta' \int_0^{\beta'} d\beta'' e^{(\beta'-\beta'')\Delta E_{nm}} &= \int_0^\beta d\beta' e^{\beta'\Delta E_{nm}} \frac{1-e^{-\beta'\Delta E_{nm}}}{\Delta E_{nm}} \\
&= \frac{1}{\Delta E_{nm}}\left(\frac{e^{\beta\Delta E_{nm}}-1}{\Delta E_{nm}}-\beta\right) \tag{2810}
\end{aligned}$$

for $n \neq m$, while for $n = m$ the integrations are trivial and Eq. (2805) reduces to

$$\begin{aligned}
F^{(2)}(\beta) = {}& 2\beta^{-1}\frac{1}{Z_0}\sum_{n\neq m} e^{-\beta E_n}|\langle n|H_I|m\rangle|^2 \frac{1}{\Delta E_{nm}}\left(\beta+\frac{1-e^{\beta\Delta E_{nm}}}{\Delta E_{nm}}\right) \\
& - \beta\frac{1}{Z_0}\sum_n e^{-\beta E_n}|\langle n|\delta H_I|n\rangle|^2. \tag{2811}
\end{aligned}$$

The first term in Eq. (2811) reduces significantly if we note that

$$\begin{aligned}
\sum_{n\neq m} e^{-\beta E_n}\frac{|\langle n|H_I|m\rangle|^2}{(\Delta E_{nm})^2}(1-e^{\beta\Delta E_{nm}}) \\
= \sum_{n\neq m}\frac{|\langle n|H_I|m\rangle|^2}{(\Delta E_{nm})^2}(e^{-\beta E_n}-e^{-\beta E_m}) = 0 \tag{2812}
\end{aligned}$$

after exchanging $n \leftrightarrow m$ in the argument of the sums. Then

$$F^{(2)}(\beta) = \frac{2}{Z_0} \sum_{n \neq m} e^{-\beta E_n} \frac{|\langle n|H_I|m\rangle|^2}{\Delta E_{nm}} - \beta \sum_n \frac{e^{-\beta E_n}}{Z_0} |\langle n|\delta H_I|n\rangle|^2. \qquad (2813)$$

Putting this all together, we have for the free energy, up to terms of second order in λ,

$$F(\lambda) = F_0 + \lambda \langle H_I \rangle_0 + \lambda^2 \left(\frac{1}{Z_0} \sum_{n \neq m} \frac{e^{-\beta E_n}}{\Delta E_{nm}} |\langle n|H_I|m\rangle|^2 \right.$$
$$\left. - \frac{\beta}{2} \frac{1}{Z_0} \sum_n e^{-\beta E_n} |\langle n|\delta H_I|n\rangle|^2 \right). \qquad (2814)$$

This can be put in a more compact form if we introduce some notation. Define, for general quantum operators A and B, the quantities

$$\langle \delta A \delta B \rangle_D = \sum_n P_n^0 (\delta A)_{nn} (\delta B)_{nn}, \qquad (2815)$$

and

$$\langle AB \rangle_s = \sum_{n \neq m} A_{nm} B_{mn} \frac{P_n^0 - P_m^0}{E_m^0 - E_n^0}, \qquad (2816)$$

where

$$P_n^0 = \frac{e^{-\beta E_n^0}}{Z_0} \qquad (2817)$$

$$(\delta A)_{nn} = A_{nn} - \sum_m P_m^0 A_{mm} \qquad (2818)$$

and

$$A_{mn} = \langle \psi_m^0 | A | \psi_n^0 \rangle. \qquad (2819)$$

The expansion for the free energy can then be written in the compact form

$$F(\lambda) = F_0 + \lambda \langle H_I \rangle_0 - \frac{\lambda^2}{2} \left[\beta \langle (\delta H_I)^2 \rangle_D + \langle H_I^2 \rangle_S \right]. \qquad (2820)$$

In the low-temperature regime the physics is governed by the ground state $|0\rangle$ with energy eigenvalue E_G^0 separated from the first excited state $|1\rangle$ with eigenvalue

E_1^0. In this case one can reduce Eq. (2820) further. By assumption,

$$E_1^0 - E_G^0 \equiv \Delta > 0 \tag{2821}$$

and for example, the partition function becomes, as $T \leftrightarrow 0$,

$$\begin{aligned} Z_0 &= g_0 e^{-\beta E_G^0} + g_1 e^{-\beta E_1^0} + \cdots \\ &= g_0 e^{-\beta E_G^0}\left(1 + \frac{g_1}{g_0} e^{-\beta \Delta} + \cdots\right) \end{aligned} \tag{2822}$$

where g_0 and g_1 are the degeneracies of the associated states. Then

$$\lim_{\beta \to \infty} F_0 = \lim_{\beta \to \infty} -\beta^{-1} \ln Z_0 = E_G^0. \tag{2823}$$

Similarly, for the first correction term,

$$\lim_{\beta \to \infty} F^{(1)} = \langle 0|H_I|0\rangle. \tag{2824}$$

Going further requires consideration of the ground state. Suppose that the ground state is degenerate with g states $|g\rangle$ but the operator H_I is also degenerate with respect to the states $|g\rangle$; then we can use the results from Problem 5.1 to show that $\lim_{T \to 0} \langle(\delta H_I)^2\rangle_D \to 0$ exponentially. In the remaining term in Eq. (2820), $\langle H_I^2\rangle_S$, the $\beta \to \infty$ limit picks out one of the ground-state terms, and one finally obtains

$$\lim_{T \to 0} F(\lambda) = E_G = E_G^0 + \lambda\langle 0|H_I|0\rangle + \lambda^2 \sum_{m>0} \frac{|\langle 0|H_I|m\rangle|^2}{E_0 - E_m} + \cdots \tag{2825}$$

Effective Hamiltonians

Another useful application of perturbation theory is the determination of an effective Hamiltonian in the case where there is a separation of energy scales between different degrees of freedom. Here we develop the basic formal machinery. Specific examples are given in the text.

Let us assume that we have a system governed by a Hamiltonian of the form

$$H = H_0^F + H_0^s + \lambda H_I \tag{2826}$$

where the Hamiltonian H_0^F is associated with very fast or high-frequency degrees of freedom, while H_0^s is associated with the very slow or low-frequency degrees of freedom. We assume that these sets of degrees of freedom are such that

$$[H_0^F, H_0^s] = 0 \tag{2827}$$

and the Hilbert space associated with H can be written as a product space

$$H_0^F |n_F, n_s\rangle = E_n^F |n_F, n_s\rangle \tag{2828}$$

$$H_0^s |n_F, n_s\rangle = E_n^s |n_F, n_s\rangle \tag{2829}$$

$$|n_F, n_s\rangle = |n_F\rangle \times |n_s\rangle. \tag{2830}$$

The part of the Hamiltonian H_I couples the fast and slow degrees of freedom.

The basic idea is that we want to average over the fast degrees of freedom to obtain an effective Hamiltonian governing the slow degrees of freedom. Let us define the effective Hamiltonian for the slow degrees of freedom, H_s, via

$$e^{-\beta H_s} = \text{Tr}_F \, e^{-\beta H} \tag{2831}$$

where

$$\text{Tr}_F A = \sum_{n_F} \langle n_F | A | n_F \rangle. \tag{2832}$$

More explicitly, we can rewrite Eq. (2831) as

$$H_s(\lambda) = -\beta^{-1} \ln \text{Tr}_F \, e^{-\beta(H_0^F + H_0^s + \lambda H_I)} \tag{2833}$$

and again we want to construct $H_s(\lambda)$ as a power series in λ. For $\lambda = 0$, since H_0^F and H_0^s commute,

$$e^{-\beta(H_0^F + H_0^s)} = e^{-\beta H_0^F} e^{-\beta H_0^s} \tag{2834}$$

and

$$H_s(0) = -\beta^{-1} \ln e^{-\beta H_0^s} Z_0^F = H_0^s + F_0^F \tag{2835}$$

where

$$F_0^F = -\beta \ln Z_0^F \tag{2836}$$

is the free energy associated with H_0^F and is a constant. We set up the perturbation theory expansion in much the same way as in the first half of this appendix. First we define

$$\rho(\beta) = e^{-\beta H} = \rho_0(\beta) \rho_I(\beta) \tag{2837}$$

where

$$\rho_0(\beta) = e^{-\beta H_0^F} e^{-\beta H_0^s}. \tag{2838}$$

Clearly ρ_I is given by Eq. (2791) where $H_0 = H_0^F + H_0^S$ and

$$H_s(\lambda) = H_s^0 - \beta^{-1} \ln \mathrm{Tr}_F \, e^{-\beta H_F^0} \rho_I(\lambda, \beta) = \sum_{n=0}^{\infty} \frac{\lambda^n H_s^{(n)}(\beta)}{n!}. \tag{2839}$$

The first-order contribution is given by

$$H_s^{(1)}(\beta) = \left[\frac{\partial}{\partial \lambda} H_s(\lambda)\right]_{\lambda=0} = \beta^{-1} \int_0^\beta d\beta' \langle H_I(\beta') \rangle_0^F. \tag{2840}$$

Going to second order, we have

$$H_s^{(2)}(\beta) = \left[\frac{\partial^2 H_s(\lambda)}{\partial \lambda^2}\right]_{\lambda=0} = \beta (H_s^{(1)}(\beta))^2$$

$$- 2\beta^{-1} \int_0^\beta d\beta' \langle H_I(\beta') \int_0^{\beta'} d\beta'' H_I(\beta'') \rangle_0^F. \tag{2841}$$

If we let $H_I(\beta) = \delta H_I(\beta) + \langle H_I(\beta) \rangle_0^F$ in the second term in Eq. (2841), it is straightforward to show that

$$H_s^{(2)}(\beta) = \beta (H_s^{(1)}(\beta))^2 - 2\beta^{-1} \int_0^\beta d\beta' \beta' \langle H_I(\beta') \rangle_0^F H_s^{(1)}(\beta')$$

$$- 2\beta^{-1} \int_0^\beta d\beta' \langle \delta H_I(\beta') \int_0^{\beta'} d\beta'' \delta H_I(\beta'') \rangle_0^F. \tag{2842}$$

It is in this form where it is easiest to make a comparison with the classical result.

At this stage we have not yet used the separation of energy scales. Consider the quantity

$$H_I(\beta) = e^{\beta(H_F^0 + H_S^0)} H_I e^{-\beta(H_F^0 + H_S^0)} \tag{2843}$$

which occurs prominently in the effective Hamiltonian. The dependence on H_S^0 complicates the β dependence of the $H_s^{(2)}$ considerably. However, if we assume that matrix elements of the set of slow degrees of freedom between $|m\rangle_0^S$ and $|n\rangle_0^S$ satisfy $\beta|E_m - E_n| \ll 1$, we can safely write

$$H_I(\beta) = e^{\beta H_F^0} H_I e^{-\beta H_F^0}. \tag{2844}$$

This means immediately that we can show that

$$\langle H_I(\beta') \rangle_0^F = \langle H_I \rangle_0^F \tag{2845}$$

$$H_s^{(1)}(\beta) = \langle H_I \rangle_0^F \tag{2846}$$

and

$$H_s^{(2)}(\beta) = -2\beta^{-1} \int_0^\beta d\beta' \int_0^{\beta'} d\beta'' \langle \delta H_I(\beta') \delta H_I(\beta'') \rangle_0^F \tag{2847}$$

and the β-dependence of $\delta H_I(\beta)$ is given by Eq. (2844). $H_s^{(2)}(\beta)$ is now of the same general form as $F^{(2)}(\beta)$ given by Eq. (2805) and we can follow that same analysis in the first part of this appendix, to obtain

$$H_S(\lambda) = H_S^0 + F_F^0 + \lambda \langle H_I \rangle_0^F - \frac{\lambda^2}{2}[\langle (\delta H_I)^2 \rangle_{s,0}^F + \beta \langle (\delta H_I)^2 \rangle_{D,0}^F]. \tag{2848}$$

It is then easy to show, following the earlier development, that in the low-temperature limit,

$$H_s = H_s^0 + F_0^F + \lambda_0^F \langle 0|H_I|0\rangle_0^F + \lambda^2 \sum_{m>0} \frac{|_0^F\langle 0|H_I|m\rangle_0^F|^2}{E_0 - E_m} \tag{2849}$$

up to terms of $\mathcal{O}(\lambda^2)$.

O STATISTICAL MECHANICS OF SYSTEMS WITH A CLASSICAL QUADRATIC HAMILTONIAN

There are several cases in this book where we consider classical statistical mechanical systems described by a Hamiltonian (or effective Hamiltonian) of the general form

$$H = \tfrac{1}{2} \sum_{ij} K_{ij} \phi_i \phi_j - \sum_i h_i \phi_i \tag{2850}$$

where ϕ_i is a real variable with values ranging from $-\infty$ to $+\infty$. The index i can simply label a particular particle (as for the momentum in a one-dimensional ideal gas $\phi_i \to P_i$), or it may include a vector index (P_i^α). For solids we consider local displacements on a lattice and $\phi_i \to \phi_\alpha(\mathbf{R})$, where $\mathbf{R}$ is a lattice site and α a vector index. In writing Eq. (2850) we assume that K_{ij} is symmetric, since it can always be constructed to be so by interchanging $\phi_i \leftrightarrow \phi_j$ in the summation.

We are interested in computing averages over the probability distribution proportional to $e^{-\beta H}$. In particular, we have the partition function

$$Z(K,h) = \int \left(\prod_i d\phi_i\right) e^{-\beta H}, \tag{2851}$$

the associated free energy,

$$F(K,h) = -\beta^{-1} \ln Z(K,h) \tag{2852}$$

the correlation functions

$$C_i = \langle \phi_i \rangle \tag{2853}$$
$$C_{ij} = \langle \delta\phi_i \delta\phi_j \rangle \tag{2854}$$

and

$$W(J) = \langle e^{\sum_i J_i \phi_i} \rangle \tag{2855}$$

where the averages are given by

$$\langle A \rangle = \frac{1}{Z} \int \left(\prod_i d\phi_i \right) e^{-\beta H} A. \tag{2856}$$

It will turn out that F, Z, and W are slightly more difficult to evaluate than C_i and C_{ij}, so we discuss C_i and C_{ij} first.

Let us consider the two related identities

$$\int \left(\prod_k d\phi_k \right) \frac{\partial}{\partial \phi_i} e^{-\beta H} = 0 \tag{2857}$$

$$\int \left(\prod_k d\phi_k \right) \frac{\partial}{\partial \phi_i} (\phi_j e^{-\beta H}) = 0 \tag{2858}$$

which follow from evaluating the integral as $|\phi_i| \to \infty$, where $e^{-\beta H}$ vanishes rapidly (as long as K_{ij} is defined properly, as discussed below). If we simply differentiate in each of these expression, we obtain

$$\sum_j K_{ij} \langle \phi_j \rangle = h_i \tag{2859}$$

and

$$\sum_k K_{ik} \langle \phi_k \phi_j \rangle - h_i \langle \phi_j \rangle = \delta_{ij} k_B T. \tag{2860}$$

Remembering the definitions given by Eqs. (2853) and (2854), Eq. (2859) reads

$$\sum_j K_{ij} C_j = h_i, \tag{2861}$$

while Eq. (2860) can be rewritten as

$$\sum_k K_{ik} C_{kj} + \sum_k K_{ik} C_k C_j - h_i C_j = \delta_{ij} k_B T. \tag{2862}$$

The second two terms on the left of Eq. (2862) cancel due to Eq. (2861), and one has the key result:

$$\sum_k K_{ik} C_{kj} = \delta_{ij} k_B T. \qquad (2863)$$

To go further, one must be able to diagonalize the matrix K_{ij}. Let us formally introduce the set of eigenvectors ϵ_i^ν such that

$$\sum_k K_{ik} \epsilon_k^\nu = \lambda_\nu \epsilon_i^\nu \qquad (2864)$$

where λ_ν are the real eigenvalues. The explicit construction of the ϵ_k^ν and λ_ν depends on the specific problem. We assume here that the ϵ_k^ν are complete and orthonormal.

$$\sum_k \epsilon_k^\mu \epsilon_k^\nu = \delta_{\mu\nu} \qquad (2865)$$

$$\sum_\nu \epsilon_i^\nu \epsilon_j^\nu = \delta_{ij}. \qquad (2866)$$

We can then expand

$$C_i = \sum_\nu \epsilon_i^\nu C_\nu, \qquad (2867)$$

and

$$C_{ij} = \sum_{\mu,\nu} \epsilon_i^\nu C_{\nu\mu} \epsilon_j^\mu \qquad (2868)$$

and insert these expressions into Eqs. (2861) and (2863), to obtain

$$\sum_j K_{ij} \sum_\nu \epsilon_j^\nu C_\nu = \sum_\nu \lambda_\nu \epsilon_i^\nu C_\nu = h_i \qquad (2869)$$

and

$$\sum_{\nu,\mu} \lambda_\nu \epsilon_i^\nu C_{\nu\mu} \epsilon_j^\mu = \delta_{ij} k_B T. \qquad (2870)$$

Then multiply Eq. (2869) by ϵ_i^α and sum over i, multiply Eq. (2870) by $\epsilon_i^\alpha \epsilon_j^\beta$ and sum over i and j. Using Eq. (2865), we obtain

$$C_\alpha = \frac{h_\alpha}{\lambda_\alpha} \qquad (2871)$$

and
$$C_{\alpha\beta} = \frac{\delta_{\alpha\beta}k_B T}{\lambda_\alpha} \qquad (2872)$$

where
$$h_\alpha = \sum_i \epsilon_i^\alpha h_i. \qquad (2873)$$

Finally, we have the results for the averages of the original fields:
$$C_i = \sum_j h_j \sum_\alpha \frac{\epsilon_i^\alpha \epsilon_j^\alpha}{\lambda_\alpha} \qquad (2874)$$

while
$$C_{ij} = k_B T \sum_\nu \frac{\epsilon_i^\nu \epsilon_j^\nu}{\lambda_\nu}. \qquad (2875)$$

As a simple example, suppose that the Hamiltonian consists of only the kinetic energy in one dimension,
$$K_{ij} = \frac{1}{m}\delta_{ij}. \qquad (2876)$$

Then
$$\epsilon_i^\nu = \delta_{i\nu} \qquad (2877)$$
$$\lambda_\nu = \frac{1}{m} \qquad (2878)$$

and
$$C_{ij} = \langle P_i P_j \rangle = k_B T \sum_\nu \delta_{i\nu}\delta_{j\nu} m = m(k_B T)\delta_{ij} \qquad (2879)$$

as we expect.

Turning to the determination of the free energy, we can use the result, which follows from Eq. (2852), that
$$\frac{\partial}{\partial h_i} F(K, h) = -C_i \qquad (2880)$$

together with Eq. (2874), to obtain

$$\frac{\partial F}{\partial h_i} = -\sum_j h_j \sum_\alpha \frac{\epsilon_i^\alpha \epsilon_j^\alpha}{\lambda_\alpha}. \qquad (2881)$$

We can then integrate with respect to h, with the result

$$F(K,h) = F(K,0) - \frac{1}{2}\sum_{ij} h_i h_j \sum_\alpha \frac{\epsilon_i^\alpha \epsilon_j^\alpha}{\lambda_\alpha}. \qquad (2882)$$

We must then evaluate F for $h = 0$. We do this through the use of a formal device. Let us introduce the auxiliary Hamiltonian,

$$H_g = \tfrac{1}{2}\sum_{ij} \phi_i[gK_{ij} + (1-g)\delta_{ij}]\phi_j. \qquad (2883)$$

For $g = 1$ we regain our original Hamiltonian with $h = 0$, while for $g = 0$ it corresponds to a set of uncoupled harmonic oscillators:

$$H_0 = \tfrac{1}{2}\sum_i \phi_i^2. \qquad (2884)$$

Consider then

$$F(K,g) = -\beta^{-1}\ln \int \left(\prod_i d\phi_i\right) e^{-\beta H_g}. \qquad (2885)$$

We have that

$$\frac{\partial F}{\partial g} = \left\langle \frac{\partial H_g}{\partial g} \right\rangle_g \qquad (2886)$$

where $\langle \ \rangle_g$ indicates an average weighted by H_g. We have, however, that

$$\frac{\partial F}{\partial g} = \frac{1}{2}\sum_{ij} G_{ij}^g (K_{ij} - \delta_{ij}) \qquad (2887)$$

and $G_{ij}^g = \langle \phi_i \phi_j \rangle_g$ is given by Eq. (2875), where we replace $\lambda_\nu \to g\lambda_\nu + (1-g)$ since

$$\sum_j [gK_{ij} + (1-g)\delta_{ij}]\epsilon_j^\nu = [g\lambda_\nu + (1-g)]\epsilon_i^\nu. \qquad (2888)$$

Inserting this result back into Eq. (2887), we obtain

$$\begin{aligned}\frac{\partial F}{\partial g} &= \frac{1}{2}k_BT\sum_{ij}\sum_\nu \frac{\epsilon_i^\nu \epsilon_j^\nu}{g\lambda_\nu + (1-g)}(K_{ij} - \delta_{ij}) \\ &= \frac{1}{2}k_BT\sum_\nu \frac{1}{g\lambda_\nu + 1 - g}(\lambda_\nu - 1). \\ &= \frac{1}{2}k_BT\frac{\partial}{\partial g}\sum_\nu \ln(g\lambda_\nu + 1 - g). \end{aligned} \qquad (2889)$$

Integrating this last equation with respect to g from 0 to 1, we obtain

$$F(K, h=0) = F_{g=0} + \tfrac{1}{2}k_BT\sum_\nu \ln \lambda_\nu. \qquad (2890)$$

The Gaussian contribution for $g=0$ is easily obtained as

$$F_{g=0} = -\beta^{-1}\ln\prod_i\left(\int d\phi_i\, e^{-\beta\phi_i^2/2}\right) = \frac{N}{2}\beta^{-1}\ln 2\pi\beta \qquad (2891)$$

where $\sum_i = N$ is the total number of fields ϕ_i. Thus, combining Eqs. (2882), (2890), and (2891), we obtain

$$\begin{aligned}F(K, h) &= \frac{N}{2}\ln 2\pi\beta + \frac{1}{2}k_BT\sum_\nu \ln \lambda_\nu \\ &\quad - \frac{1}{2}\sum_{ij}h_ih_j\sum_\alpha \frac{\epsilon_i^\alpha \epsilon_j^\alpha}{\lambda_\alpha}.\end{aligned} \qquad (2892)$$

In the example of lattice vibrations in Chapter 6, the eigenvectors $\epsilon_i(\mathbf{k})$ are labeled by a wavenumber $\mathbf{k}$ with N values, and a vector label i with three values. Therefore, there are $3N$ degrees of freedom and there is no applied force, $h_i = 0$, so the free energy is given by

$$F = \tfrac{3}{2}N\ln 2\pi\beta + \tfrac{1}{2}k_BT\sum_\mathbf{k}\sum_i \ln \lambda_i(\mathbf{k}). \qquad (2893)$$

Finally, we turn to the quantity

$$W(J) = \langle e^{\sum_i J_i\phi_i}\rangle \qquad (2894)$$

which arises in Chapter 6 in our discussion of density correlations in a solid. $W(J)$

can be written in terms of a ratio of partition functions,

$$W(J) = \frac{Z(K, -J/\beta)}{Z(K, 0)}$$
$$= e^{-\beta(F(K,-J/\beta) - F(K,J=0))} \quad (2895)$$
$$= \exp\left(\frac{k_B T}{2} \sum_{ij} J_i J_j \sum_\alpha \frac{\epsilon_i^\alpha \epsilon_j^\alpha}{\lambda_\alpha}\right).$$
$$= \exp\left(\frac{1}{2} \sum_{ij} J_i J_j C_{ij}\right). \quad (2896)$$

In the analysis of the average density in a harmonic solid we need the quantity $\langle e^{-i\mathbf{k}\cdot\mathbf{u}(\mathbf{R})}\rangle$. If we write the argument of the exponential in Eq. (2894) as

$$\sum_i J_i \phi_i \rightarrow \sum_\beta \sum_{\mathbf{R}'} J_\beta(\mathbf{R}') u_\beta(\mathbf{R}') = -i\mathbf{k} \cdot \mathbf{u}(\mathbf{R}), \quad (2897)$$

we obtain the desired average from Eq. (2896) by choosing

$$J_\beta(\mathbf{R}') = -ik_\beta \delta_{\mathbf{R},\mathbf{R}'}. \quad (2898)$$

It is left as a problem to show that this leads to Eq. (2153).

REFERENCES

[1] For a more complete discussion, see L. Reichl, *A Modern Course in Statistical Physics*, University of Texas Press, Austin, Texas, 1980, Chap. 5.

[2] Canonical transformations are the classical versions of the unitary transformations in quantum mechanics. This is discussed by H. Goldstein, *Classical Mechanics*, 2nd ed., Addison-Wesley, Reading, Mass., 1980; and by E. C. G. Sudarshan and N. Mukunda, *Classical Dynamics*: a modern perspective. Wiley New York: (1974).

[3] Newtonian mechanics can be formulated using a variety of formal structures, as shown, for example, in L. D. Landau and E. M. Lifshitz, *Mechanics*, 2nd ed., Pergamon Press, Oxford, 1969; or H. Goldstein, *Classical Mechanics*, 2nd ed., Addison-Wesley, Reading, Mass., 1980. Newton's equations of motion can be reexpressed in terms of a Lagrangian or Hamiltonian formulation. For the purposes of statistical mechanics, it is the Hamiltonian formulation that is particularly useful.

[4] As we shall see as we proceed, it is very useful to introduce the Poisson bracket formalism. A good introduction is given in Chapter 2 in R. J. Finkelstein, *Nonrelativistic Mechanics*, W.A. Benjamin, Reading, Mass., 1973. The utility of this approach is that it will allow us to develop transformation theory, in close analogy to the more familiar analysis in quantum mechanics. Indeed, one of the deepest connections between quantum and classical mechanics is through the mapping of Lie algebras satisfied by commutators in quantum theory onto that satisfied by Poisson brackets in classical theory.

[5] A general discussion is given by H. Goldstein, *Classical Mechanics*, 2nd ed., Addison-Wesley, Reading, Mass., 1980.

[6] ϵ_{ijk} is the completely antisymmetric Levi-Civita tensor. ϵ_{ijk} is zero if any two subscripts are equal. ϵ_{ijk} changes sign if one interchanges any two indices $\epsilon_{ijk} = -\epsilon_{jik}$. Finally, one has the convention $\epsilon_{123} = 1$.

[7] The complete group of invariances associated with nonrelativistic systems, including Galilean invariance, is discussed by R. J. Finkelstein, *Nonrelativistic Mechanics*, W.A. Benjamin, Reading, Mass., 1973.

[8] Our analysis here follows that of Professor Paul Martin, lecture notes, Harvard University, 1968–1969.

[9] The operator $\hat{L}_G$ is independent of the parameter s because the Poisson bracket is invariant under canonical transformations. Thus the Poisson bracket $\{A, B\}$ is independent of whether one uses the phase-space coordinates $[r_i, p_j]$ or the set $[r_i(s), p_j(s)]$, related to the set $[r_i, p_j]$ by a canonical transformation parameterized by s, to take the derivatives.

[10] In the particular case where $G = H$, $L_H \equiv L$, called the *Liouville operator*, which was introduced by the B. O. Koopman, *Proc. Natl. Acad. Sci. U.S.A.* **17**, 315 (1931). In the same paper the evolution operator $U_H = U(t) = e^{iLt}$, known as *Koopman's operator*, was introduced. For a discussion of the Liouville operator, see R. Balescu, *Equilibrium and Nonequilibrium Statistical Mechanics*, Wiley, New York, 1975. Additional references: J. G. Kirkwood, *J. Chem. Phys.* **14**, 180 (1946); **15**, 72 (1947); J. O. Hirschfelder, J. O. Curtiss, and R. B. Bird, *The Molecular Theory of Gases and Liquids*, Wiley, New York, 1954; D. G. Curve, T. F. Jordan, and E. C. G. Sudarshan, *Rev. Mod. Phys.* **35**, 350 (1963); R. Balescu et al., *Physics* **33**, 558, 581 (1967).

[11] For a rather complete discussion of the quantum mechanical case, see A. Messiah, *Quantum Mechanics*, Vol. I, Wiley, New York, 1966.

[12] The first use of the Liouville equation in statistical mechanics was by Boltzmann (1968): L. Boltzmann, *Wien. Ber.* **58**, 517 (1868). See also M. Born and H.S. Green, *Proc. R. Soc. A* **188**, 10 (1946); **191**, 168 (1947); J. G. Kirkwood, *J. Chem. Phys.* **15**, 72 (1947); I. Prigogine, *Non-equilibrium Statistical Mechanics*, Wiley, New York, 1962; R. Zwanzig, *Phys. Rev.* **144**, 170 (1966); B. O. Koopman, *Proc. Natl. Acad. Sci. U.S.A.* **17**. 315 (1931); and J. von Neumann, *Ann. Math.* **33**, 587, 789 (1932).

[13] This general field is known as *ergodic theory*, and general overviews are given by J. Lebowitz and O. Penrose, *Phys. Today*, Feb. 1973; R. Balescu, *Equilibrium and Nonequilibrium Statistical Mechanics*, Wiley, New York, 1975, App.; and L. E. Reichl, *A Modern Course in Statistical Physics*, University of Texas Press, Austin, Texas, (1980). Much of all this resulted from the desire to *derive* thermodynamics from mechanics. In the course of the development here it should become clear that this is a difficult task for each particular physical system and not possible in general. Part of the problem is the practical consideration that detailed calculations are typically possible only for systems with a small number of degrees of freedom and with idealized boundary conditions. The definitive foundations of a workable ergodic theorem are associated with J. von Neumann, *Proc. Natl. Acad. Sci. U.S.A.* **18**, 70, 263 (1932); and G. D. Birkoff, *Proc. Natl. Acad. Sci. U.S.A.* **17**, 650, 656 (1931).

[14] See the discussion in Ref. 20 in Chapter 1.

[15] J. W. Gibbs, *Elementary Principles in Statistical Mechanics*, Ox Bow Press, Woodbridge, Conn., 1981, p. 144, introduced the notion of a mixing system. Clearly, the

relevance of ensemble theory to physical systems was not on a firm footing in 1902, and Gibbs gave a qualitative discussion of how the ensemble averages reflect a coarse-grained time average.

[16] The invariance of the Jacobian for the transformation $q(0) \to q(t)$ is clearly shown in L. Reichl, *A Modern Course in Statistical Physics*, University of Texas Press, Austin, Texas, 1980, p. 190. There is also a clear proof in L. D. Landau and E. M. Lifshitz, *Mechanics*, 2nd ed., Pergamon Press, Oxford, 1969, p. 146.

[17] According to A. Münster, *Statistical Thermodynamics*, Vol. II, Springer-Verlag, Berlin, 1974, p. 10, the original reference to Liouville's theorem is J. Liouville, *J. Math. Phys. Appl.* **3**, 348 (1838). L. D. Landau and E. M. Lifshitz, *Mechanics*, 2nd ed., Pergamon Press Oxford, 1969, gives a very clear proof of this result.

[18] The word *ergodic* was first used by Boltzmann but in a different sense; the term was given its present meaning by P. T. Ehrenfest; see S. G. Brush, *Statistical Physics and the Atomic Theory of Matter: From Boyle and Newton to Landau and Onsager*, Princeton University Press, Princeton, N.J., 1983.

[19] The mathematical definition of mixing was introduced by J. von Neumann, *Ann. Math.* **33**, 587 (1932), and developed by E. Hopf, *J. Math. Phys.* **13**, 51 (1934); *Ergodic Theorie*, Springer, Berlin, 1937. The roots of mixing go back to J. W. Gibbs, *Elementary Principles in Statistical Mechanics*, Ox Bow Press, Woodbridge, Conn., 1981, p. 144. He understood the need for a *coarse-grained* view of equilibration.

[20] If a system is allowed to evolve from two slightly differing initial states, x and $x + \epsilon$, then after time t their divergence may be characterized approximately as $\epsilon(t) \approx \epsilon e^{\lambda t}$, where λ is the Lyapunov exponent and gives the average rate of divergence. If $\lambda < 0$, one has convergence of the trajectories. If $\lambda > 0$, one has rapid divergence and sensitivity to initial conditions. A. M. Lyapunov (1857–1918) was a Russian mathematician. For a more technical discussion, see S. N. Rasband, *Chaotic Dynamics of Nonlinear Systems*, Wiley, New York, 1990.

[21] A. Rosenthal, *Ann. Phys.* **42**, 796 (1913), and H. Plancharel, *Ann. Phys.* **42**, 1016 (1913), proved that strict ergodicity cannot exist for a completely isolated system with time-independent forces. For a discussion, see J. E. Mayer and M. G. Mayer, *Statistical Mechanics*, 2nd ed., Wiley, New York, 1977

[22] P. Ehrenfest and T. Ehrenfest, *Enzyklopädie Mathematischen Wissenschaften*, Bd. IV, Teil 32, Leipzig, 1911, introduced the quasiergodic hypothesis, which states that a trajectory starting from almost any point P must come arbitrarily near any point Q on the energy surface, but need not pass through it.

[23] G. D. Birkoff, *Proc. Natl. Acad. Sci. U.S.A.* **17**, 650, 656 (1931); J. von Neumann, *Proc. Natl. Acad. Sci. U.S.A.* **18**, 70, 263 (1932).

[24] Action variables are treated in H. Goldstein, *Classical Mechanics*, 2nd ed., Addison-Wesley, Reading Mass., 1980, and L. D. Landau and E. M. Lifshitz, *Mechanics*, 2nd ed., Pergamon Press, Oxford, 1969.

[25] For a rather detailed discussion of the pendulum within this context, see B. V. Chirikov, *Phys. Rep.* **52**, 263 (1979). It is also discussed by S. N. Rasband, *Chaotic Dynamics of Nonlinear Systems*, Wiley, New York, 1990, Chap. 8, Sec. 2.

[26] M. Abramowitz and I. Stegun, *Handbook of Mathematical Functions*, National Bureau of Standards, Washington, D.C., 1964. Jacobi elliptic functions are discussed on page 567.

[27] It is not sufficient simply to couple N oscillators in a quadratic fashion, since, in principle, we can carry out a linear transformation to a new set of *normal* coordinates that do correspond to N uncoupled oscillators.

[28] See Enrico Fermi, *Collected Papers*, Vol. II, University of Chicago Press, Chicago, 1965, p. 978.

[29] The Toda Hamiltonian is discussed in S.N. Rasband, *Chaotic Dynamics of Nonlinear Systems*, Wiley, New York, 1990, Sec. 8.3; and J. Ford, in *Fundamental Problems in Statistical Mechanics*, ed. E. G. D. Cohen, North-Holland, Amsterdam, 1975.

[30] The Poincaré section is a device invented by Henri Poincaré (1854–1912) as means of simplifying phase-space diagrams of complicated systems. The basic idea is to view the phase-space diagram stroboscopically, so that the motion is observed periodically. See the discussion in G. L. Baker and J. P. Gollub, *Chaotic Dynamics: An Introduction*, Cambridge University Press, Cambridge, 1990.

[31] M. Henon and C. Heiles, *Astron. J.* **69**, 73 (1964). See the Discussion on page 171 of S. N. Rasband, *Chaotic Dynamics of Nonlinear Systems*, Wiley, New York, 1990.

[32] V. I. Arnold, *Russ. Math. Surv.* **18**, 9 (1963); 85 (1963); J. Moser, *Nachr. Acad. Weiss. Goettingen II, Math Phys. I*, **1** (1962). See G. H. Walker and J. Ford, *Phys. Rev.* **188**, 416 (1969); A. N. Kolmogorov, "Address to the 1954 International Congress of Mathematicians," translated in R. Abrahams, *Foundations of Mechanics*, W.A. Benjamin, New York, 1967, App. D.

[33] L. Reichl (*A Modern Course in Statistical Physics*, University of Texas Press, Austin, Texas, 1980, p. 228) derives this result and gives detailed background.

[34] This mapping is discussed in some detail by S. N. Rasband, *Chaotic Dynamics of Nonlinear Systems*, Wiley, New York, 1990, Sec. 8.4.

[35] Y. G. Sinai, in *The Boltzman Equation*, ed. E. G. D. Cohen and W. Thirring, Springer-Verlag, Vienna, 1973; Y. G. Sinai, *Sov. Math. Dokl.* **4**, 1818 (1963); Y. G. Sinai, "Ergodicity of Boltzmann's equations," in *Statistical Mechanics: Foundations and Applications*, ed. T. A. Bak, W.A. Benjamin, New York, 1967.

[36] There are, of course, much simpler ways of obtaining this result, as shown in Chapter 1. If we define the mass density $\rho = mn$, local conservation of mass takes the form

$$\frac{\partial \rho}{\partial t} = -\nabla \cdot \mathbf{g}.$$

[37] G. Arfken, *Mathematical Methods for Physicists*, Academic Press, New York, 1985, p. 428.

[38] For a full discussion, see P. Dennery and A. Krzywicki, *Mathematics for Physicists*, Harper & Row, New York, 1967.

[39] For a more detailed discussion of second quantization, see G. Baym, *Lectures in Quantum Mechanics*, W.A. Benjamin, New York, 1969, p. 411.

[40] M. Abramowitz and I. Stegun, *Handbook of Mathematical Functions*, Washington, National Bureau of Standards, D.C. 1964, p. 807.

INDEX

absolute temperature scale, 99, 100
additivity, 40, 118
adiabatic compressibility, 143
adiabatic processes, 85, 86, 89
adiabatic speed of sound, 192
Alder, B. J., 94
Alers, G. A., 518
Anderson, M. H., 250
Anderson, P. W., i, 1, 94, 100, 101, 516
anharmonic oscillators, 554
anticommutation relations, 206, 579
antiferromagnets, 158, 406, 409
applied stress, 495, 518
Arfken, G., 603
Arnold, V. I., 555, 603
Ashcroft, N., 249, 516
Ashcroft, N. N., 339
atomic polarizability, 327
autocorrelation function, 5, 525, 527
Avogadro's number, 177

Baker, G. L., 36, 603
Balescu, R., 601
Bardeen, J., 191
Baym, G., 248, 249, 603
BCC lattice, 457
Becker, R., 99, 102, 189
Berezinskii, V. L., 517
Bernoulli, D., 101
Berry, R. S., 102
big bang, 63, 100, 224
Binder, K., 443
binomial distribution, 527
Bird, R. B., 601
Birkoff, G. D., 543, 547, 548, 601, 602
blackbody radiation, 222
Bloch, F., 189
Boltzmann, L., 15, 95, 249, 294, 339, 544, 546, 601, 602
Boltzmann paradigm, 13, 36, 90
Boltzmann's constant, 45, 178
Boltzmann's principle, 33, 97
Bonhoeffer, K. F., 268, 338
Born, M., 340, 466, 516, 601
Born–Oppenheimer approximation, 328

Born–von Karman boundary conditions, 468
Bose–Einstein condensation, 236, 244, 250, 256
Bose–Einstein distribution, 219
Bose, S. N., 249
bosons, 206, 209, 217, 240, 244, 253, 266, 574
boundary conditions, 16, 17, 18, 29, 36, 37, 79, 84, 95, 215, 379, 381, 391, 416, 437, 468, 478, 512, 601
Boussinesq, J., 518
Boyle, R., 94, 99, 101, 249
Boyle's law, 99
Boyle temperature, 310
Bradley, C.C., 250
Bravais, A., 516
Bravais lattice, 454
Brillouin function, 371, 384
Brillouin, L., 96, 441
Brockhouse, B., 517
broken continuous symmetry, i, ii, 85, 102, 477
broken rotational symmetry, 60, 68, 409
broken symmetry, i, 18, 64, 67, 94, 101, 119, 156, 189
broken translational symmetry, i, ii, 68, 100, 102
Brownian motion, 14
Brueckner, K., 249
Brush, S.G., 94, 96, 97, 101, 249, 268, 328, 338, 339, 340, 602
Buerger, M. J., 516
bulk modulus, 476, 506, 509

Callaway, J., 517
Callen, H. B., i, 189
canonical ensemble, 40, 54, 98, 131, 416, 474, 569
canonical transformations, 530, 531, 601
Carnot, S., 97, 189
Cauchy, A. L., 508, 518
Cauchy relations, 504
Caugin, T., 97
Chandrasekhar, B. S., 518
chemical potential, 43, 50, 53, 89, 99, 118, 120, 142, 147, 155, 158, 220, 223, 227, 235, 237, 245
chemical reactions, 118, 154, 232, 332
Cheung. P. S. Y., 94
Chirikov, B. V., 602
Christie, D. E., 190

605

606 INDEX

Clapeyron, E., 190
classical limit, 218
Clausius–Clapeyron Equation, 147
Clausius–Mossotti relation, 395
Clausius, R., 97, 101, 189, 190, 442
clock model, 103, 197
clockwork universe, 95
coarse graining, i, 14, 16, 262, 321, 327, 347, 359, 361, 362, 375, 396, 440, 441, 550, 602
coexistence curve, 149, 175, 183, 310, 452
cohesion energy, 466
combinations, 523
commutation relations, 206, 207, 214, 250, 252, 480, 574, 580
commutators, 101, 600
completeness, 199, 418, 571, 578
compliance tensor, 502
compression, 69, 71, 74, 78, 84, 509, 510, 516
condensation, 3, 163
condensation energy, 163
conditional probability, 524
configuration, 4, 6, 8, 10, 11, 21, 25, 30, 45, 69, 95, 152, 199, 200, 202, 363, 376, 431, 524
conjugate forces, 120, 121
connected graphs, 281, 282
conservation laws, i, 1, 14, 59, 170, 530, 560
conservation of angular momentum, 17, 18, 30, 535, 536, 537, 560
conservation of energy, 1, 15, 37, 39, 64, 96, 98, 122, 365, 536, 542
conservation of momentum, 1, 39, 64, 534
conservation of particle number, 18, 30, 37, 39, 118, 250
continuity equation, 39, 77, 82, 98, 560, 562
Cook, G. A., 190
Cooper, L. N., 191
Cornell, E. A., 250
correlation function, 5, 66, 190, 297, 366, 375, 476, 485, 595
correlation length, 448
Courant, R., 190
creation-annihilation operators, 206, 208, 210, 481, 574, 578
critical index, 160, 187, 199, 402, 403, 404, 425, 439
critical phenomena, i, 150, 157, 158, 160, 161, 163, 173, 185, 186, 188, 189, 191, 304, 310, 321, 404, 424, 430, 438
critical point, 150, 157, 160, 185, 189, 303, 438
crystal lattice, 448
cumulants, 285
Curie constant, 370
Curie law, 370, 388, 404, 421

Curie, P., 441
Curie temperature, 157, 405
Curie–Weiss law, 404, 438
Curtiss, J. O., 601
Curve, D. G., 601
Cuthbertson, C., 441
Cuthbertson, M., 441
cyclotron frequency, 379

Darwin, C. S., 100
Davis, K. B., 250
de Broglie wavelength, 51, 53, 100, 152, 155, 217, 219, 221
de Gennes, P. G., 95
De Haas, W. J., 442
De Haas–Van Alphen effect, 389
de Launay, J., 518
Debye–Hückel Theory, 25, 319, 339
Debye–Langevin Theory, 347, 375, 441
Debye, P. D., ii, 441, 517
Debye screening, 318, 337
Debye temperature, 488, 491
Debye theory, 487, 507, 517
Debye–Waller factor, 479, 487, 517
defects, 454
degenerate phases, 163, 167
DeLeener, M., 94
Dennery, P., 603
density correlation function, 297
density matrix, 204
density of states, 214, 217, 220, 222, 249, 379, 484, 487
detailed balance, 433
diamagnetic susceptibility, 354
diamagnetism, 357, 388
diatomic molecule, 60, 262, 320, 327, 328, 338, 341, 354
dielectric constant, 394
dielectric materials, 117, 321, 347, 373, 374, 389, 394
dipolar interactions, 348, 392, 396
Dirac notation, 199, 571
Dirac, P. A. M., 248, 249, 442
direct correlation function, 305, 307, 319
director field, 85
disconnected graphs, 281
dispersion, 5, 53, 110
dispersion force, 326
displacement correlation function, 476
displacement field, 394
dissociation, 10, 260, 262, 332, 465
distortions, 71, 85, 495, 537
Domb, C., 443

INDEX 607

Duhem, P., 190
Dulong–Petit law, 475
Dulong, P. L., 516
dynamical matrix, 468
Dyson, F. J., 314, 339

early universe, 63, 100, 235, 252
effective free energy, 158, 161, 162, 170
effective hamiltonians, 320
Egelstaff, P., 339
Ehrenfest, P., 95, 102, 191, 602
Ehrenfest, T., 95, 602
Eigenschitz, R., 339
eigenvalue problem, 139, 204, 329, 418, 469, 573, 575, 596
Einstein, A., 97, 249, 516, 518
elastic constants, 495, 501, 504, 516
elastic tensor, 472, 473, 496, 501, 503, 504
elastic theory, i, 495, 502, 511, 518
electric charge, 10, 29, 209, 313, 318, 324, 337, 347, 348, 363, 376, 389, 466
electric dipole moment, 11, 321, 325, 347, 349, 352, 355, 359, 363, 364, 389, 442
electric susceptibility, 352
electromagnetic radiation, 210, 222, 347
electronic specific heat, 231
energy, 1, 7, 15, 17, 18, 35, 37, 39, 40, 45, 51, 52, 53, 64, 78, 84, 86, 96, 98, 118, 121, 122, 201
energy current, 39, 111, 563
energy representation, 121, 124
ensemble, 4, 15, 16, 25, 54, 90, 93, 94, 117, 122, 124, 133, 190, 546, 548, 602
Ensher, J. R., 250
enthalpy, 143, 194
entropy, 29, 31, 33, 35, 40, 44, 47, 51, 58, 88, 89, 93, 97, 98, 99, 117, 118, 119, 120, 121, 122, 124, 130, 135, 151, 154, 189, 199, 200, 226, 252, 365, 516, 568
entropy of mixing, 151
entropy principle, 29, 40, 117, 135, 147, 154, 189
entropy representation, 122
Epstein, P. S., 94
equations of state, 44, 118, 123, 131, 156, 189
equilibration, 18, 29, 96, 602
equilibrium state, 20, 21, 29, 30, 33, 46, 71, 84, 93, 118, 203, 205
equipartition theorem, 261, 265, 475, 554
ergodic behavior, 14, 16, 17, 105, 545, 551
ergodic theorem, 548
ergodic theory, ii, 14, 16, 17, 18, 36, 40, 95, 545, 546, 547, 548, 554, 601, 602
Erpenbeck, J. J., 94
Euler equation, 100, 122

Euler, L., 98
Ewald method, 516
Ewald, P. P., 516
exchange interaction, 348, 395, 398, 405
extensive variables, 35, 50, 54, 118, 120, 124, 135, 244, 274
external constraints, 18, 93, 117, 118, 199
external forces, 1, 2, 79, 511, 534
Eyring, E. M., 441, 442
Eyring, H., 363, 441, 442

Fano, U., 248
FCC lattice, 456
Feller, W., 94
Fenkel, J., 517
Fermi–Dirac distribution, 219
Fermi energy, 227, 235
Fermi, E., 340, 554, 555, 603
Fermi, F., 249
fermions, 206, 209, 226, 253, 266, 574, 578
ferromagnets, 1, 68, 156, 158, 159, 253, 352, 396, 397, 405, 421, 442
Fetter, A., 248
Fetter, A. L., 250
field operators, 207, 210, 579
finite size corrections, 419
Finkelstein, R. J., 600
first law of thermodynamics, 122
Fisher, M. E., 443
flipping probability, 433
fluctuating dipole model, 373
fluctuation, 5, 53, 135, 190, 297, 478
fluid structure factor, 296, 302, 303, 319
Ford, J., 603
Forster, D., 516
Fourier, J., 98
Fowler, R. H., 100, 338, 340
fractional quantum Hall effect, 389
Frank, A., 441
Frenkel, D., 516
Frölich, H., 442
fugacity, 272, 299

Galilean invariance, 536, 601
Galilei, G., 517
Gamma space, 8, 14, 16, 17, 95, 530, 541, 542, 543, 548
Gaussian distribution, 528
Gay–Lussac's law, 99
Gell–Mann, M., 249
generator, 17, 64, 65, 76, 531, 533, 536, 537, 541
generic variables, 19, 20, 96, 190
Gibbs–Duhem relation, 123, 138, 182

Gibbs ensemble, 145, 155, 289
Gibbs free energy, 110
Gibbs, J. W., 94, 96, 97, 98, 99, 189, 190, 190, 544, 601, 602
Gibb's paradox, 152
Gibb's phase rule, 195
Gibbs potential, 131, 147
Girvin, S. M., 442
Glauber, R. J., 436, 443
Goldstein, H., 95, 96, 190, 600, 601, 602
Goldstone, J., 102, 516
Gollub, J. P., 603
Gottfried, K., 248, 340, 440
Goudsmit, S., 248
Gould, H., 94, 443
Graben, H. W., 97
grand canonical ensemble, 40, 45, 50, 51, 67, 98, 129, 201, 208, 221, 269, 364
grand potential, 51, 202, 211, 272
Green, H. S., 601
Green, M. S., 443
ground state lattice structure, 458
Gubbins, K. E., 327, 339
Guggenheim, E. A., 191, 309, 338, 339
Guldberg, C. M., 190

Halperin, B. I., 478, 517
Hamiltonian, 7, 16, 17, 30, 33, 65, 66, 68, 76, 79, 89, 126, 208, 211, 271, 314, 320, 328, 348, 359, 373, 396, 414, 415, 466, 536, 537, 538, 541, 551, 555, 591, 594, 600
Hamilton's equations, 126, 530
harmonic oscillator, 105, 210, 252, 340, 379, 441, 453, 474, 552, 554, 598
Harrison, S. F., 339
Harteck, P., 268, 338
HCP lattice, 458
heat bath, 57, 90, 99, 135, 144, 189, 194, 432
heat capacity, 2, 192, 243, 264
Heiles, C., 555, 603
Heisenberg representation, 542
Heisenberg, W., 442
helium, 209, 250, 357, 452, 458
Helmholtz free energy, 58, 100, 114, 125, 132, 142, 159
Helmholtz, H. von, 100, 189
Henderson, D., 363, 441, 442
Henon, M., 555, 603
Herring, C., 396, 442
hexagonal lattice, 457
high temperature expansions, 425
Hilbert, D., 190
Hilbert space, 199, 571
Hirschfelder, J. O., 601

Ho, P., 518
Holton, G., 99
honeycomb lattice, 456
Hooke, R., 517
Hooke's law, 77, 113, 497, 499, 509, 512, 517
Hoover, W. G., 102, 339
Hopf, E., 602
Huang, K., 466, 516
Huang, Kerson, 100
Hückel, E., 339
Hulet, R. G., 250
Hund's rules, 361
Huntington, H. B., 518
hydrodynamics, i

ideal gas, 33, 35, 36, 44, 51, 53, 98, 99, 108, 109, 122, 211, 220
identical particles, 19, 152, 190, 200
information theory, 20, 21, 23, 25, 26, 27, 96, 109, 112
Ingham, A. E., 516
initial conditions, 3, 9, 14, 16, 17, 18, 20, 542, 545, 602
integer quantum Hall effect, 389
intensive variables, 36, 44, 45, 89, 93, 120, 124, 126, 137, 140, 151, 459
intersection, 524
intrinsic spin, 347, 348, 377, 384
invariance principles, 1, 17, 58, 65, 69, 96, 100, 530, 534, 601
invariant ensemble, 546
ionization, 332, 337, 338
Ising, E., 414, 442
Ising model, 11, 13, 106, 159, 348, 414, 415, 421, 425, 432, 436, 439, 440, 442
isolating constants, 15, 16, 548
isothermal compressibility, 2, 110, 131, 143, 143, 151, 187, 289, 304, 344, 404
isotropic fluid, 83

Jacobians, 90, 133
Jaynes, E. T., 96
Jeans, J., 101
Jellinek, J., 102
joint probability, 526
Jones, W., 505, 518
Jordan, T. F., 601
Joule–Thomson Effect, 194

Kadanoff, L. P., 249, 443
KAM Theorem, 555, 603
Katz, A., 96
Kaufmann, B., 442
Keesom, W. H., 339

Ketterle, W., 250
kinetic energy, 6, 17, 18, 19, 60, 82, 90, 208, 211, 217, 259, 266, 270, 328, 466, 475, 498, 537
kinetic theory, 83, 101, 338, 561
Kingston, A. E., 339
Kirkwood, J. G., 601
Kittel, C., 442, 517
Kolmogorov, A. N., 555, 603
Koopman, B. O., 601
Koopman's operator, 541, 601
Kosterlitz, J. M., 517
Kosterlitz–Thouless phase, 478
Kramers, H. A., 422, 442
Krzywicki, A., 603
Kubo, R., 285, 339, 441
Kurn, M. R., 250

Lagrange multipliers, 46, 50, 201, 365, 565
Lagrangian mechanics, 126
Lamé coefficients, 508
Landau and Lifshitz, 40, 98, 102, 151, 190, 338, 440, 442, 504, 516, 518, 600, 602
Landau diamagnetism, 388
Landau gauge, 378
Landau, L. D., i, 94, 98, 191, 197, 249, 321, 338, 442
Landau levels, 379, 389
Landau theory, 156, 158, 164, 173, 185, 188, 312, 402, 403, 404, 425, 438
Landberg, P. T., 98
Landé g-facto, 360
Landé, L., 441
Langevin equation, 102
Langevin function, 368, 384
Langevin, P., 441
Langevin paradigm, 18, 36, 90
Langmuir isotherm, 196
Laplace, P, 95, 95
Larmor diamagnetic susceptibility, 354
latent heat, 147, 185
lattice coordination, 462, 465
lattice gas, 343, 415
lattice structure, 3, 59, 67, 117, 427, 439, 454, 458, 460, 464, 465
lattice vibrations, 466
law of mass action, 156
law of rectilinear diameters, 312
Lebowitz, J., 601
Lee, D. L., 249
Legendre transforms, 124, 126, 127, 189, 190
Lekner, J., 339
Lenard, A., 314, 339
Lennard–Jones, J. E., 339, 516
Lennard–Jones potential, 209, 271, 272, 327, 330

Lenz, W., 414
Levi–Civita tensor, 601
Lieb, E. H., 314, 339
Lindemann criterion, 477
Lindemann, F. A., 517
linear response theory, 347
linked cluster expansion, 274, 282
Liouville equation, 542, 601
Liouville, J., 602
Liouville operator, 601
Liouville's theorem, 545, 602
liquid crystals, 3, 71, 85, 95, 158, 320, 321
local conservation laws, 14, 36, 64, 560
London, F., 326, 327, 339, 340
long range interactions, 98, 117, 326, 348
Lorentz approximation, 391, 392, 442
Lorentz, H., 442
Lorentz, L., 442
Loschmidt, J., 96
Love, A. E. H., 518
Luttinger, J. M., 442
Lyapunov, A. M., 602

Ma, S., 443
macrostate, 18, 21, 93, 119
macrovariables, i, 13, 17, 18, 20, 21, 38, 39, 85, 93, 117, 118, 151, 159, 189
Madalung, E., 516
Madelung energy, 466
magnetic flux quanta, 381
magnetic lattice models, 395
magnetic materials, 347
magnetic moments, 10, 13, 111, 112, 156, 251, 320, 348, 359, 361, 364, 366, 370, 395, 414
magnetic structure factor, 366, 375, 376
magnetic susceptibility, 352, 366
magnetization, 1, 68, 113, 118, 121, 157, 158, 251, 352, 364, 365, 383, 389, 397, 409, 420, 424
many-body theory, 209, 214, 305
March, N., 505, 518
Margeneau, H., 327, 339
Martin, P. C., 101, 249, 601
Massieu, M., 189
master equation, 435
Matthews, M. R., 250
Mattis, D. C., 396, 442
Maxwell–Boltzmann distribution, 219, 221
Maxwell construction, 181, 183
Maxwell relations, 132
Maxwell velocity distribution, 270
Maxwell's equations, 210
Maxwell, J. C., 94, 190, 191, 249, 338, 339
Mayer cluster expansion, 270, 288

Mayer, J. E., 101, 282, 339, 602
Mayer, M. G., 101, 602
McCoy, B. M., 422, 442
McQuarrie, D. A., 262, 338
mean-field theory, 312, 391, 398, 405, 406, 425, 438, 442
measurements, 3, 19, 20
mechanical equilibrium, 69
mechanical force, ii, 59, 85, 88, 537
Mermin, N. D., 249, 516
mesoscopic physics, 321
mesoscopic scale, i, 320
Messiah, A., 601
method of stationary phase, 56, 100, 566
Metropolis, N., 436, 443
Mewes, M.-O., 250
microcanonical ensemble, 30, 33, 35, 36, 53, 88, 97, 432, 566
microcanonical trace, 31
microvariables, 17, 18, 20, 21, 30, 36, 38, 39
Milne–Thomson, L. M., 98
missing information, 21, 26, 27, 29, 119, 151, 199
mixing behavior, 14, 16, 17, 33, 86, 545, 549, 550, 601, 602
mixtures, 10, 11, 118, 151, 153
molecular polarizabilities, 355
momentum density, 38, 77, 79, 82, 110, 511, 538, 539, 562
Monte Carlo simulations, 431
Monte Carlo steps, 437
Moser, J., 555, 603
Mossotti, O., 442
Mukunda, N., 600
Münster, A., 97, 102, 191, 517, 518, 602
mutually exclusive, 524

n-vector model, 439
Nambu–Goldstone modes, 85, 101, 189, 477, 516
Nambu, Y., 102, 516
Narnhofer, H., 314, 339
Navier, C. L. M. H., 98, 518
Néel, L., 442
Néel temperature, 409
Nelson, D. R., 478, 517
Nernst's theorem, 130
neutron scattering, 366, 406, 494, 517
Newmann, J. R., 100
Newtonian mechanics, 1, 9, 90, 96, 600
noble gases, 150, 271, 324, 327, 357, 454, 458, 491
Noether, E., 96

nonergodic behavior, 14, 545
nonmagnetic materials, 354
nonpolar materials, 354
Nosé dynamics, 90, 102, 432

observable, 3, 9, 19, 20, 30, 199, 201, 539, 542
occupation basis state, 205, 208, 578
one-component plasma, 314
Onnes, K., 190, 339
Onsager, L., 94, 422, 424, 439, 442
open systems, 14, 18, 40, 54, 64, 90
Oppenheimer, J. R., 340
order parameter, 156, 157, 166, 173, 409, 438
order parameter susceptibility, 159, 160
Ornstein–Zernike equation, 306
orthohydrogen, 268
Osheroff, D. D., 249
osmotic pressure, 153

parahydrogen, 268
paramagnet, 1, 68, 157, 163, 347, 354, 357, 361, 369, 371, 377
partition function, 33, 46, 50, 51, 57, 93, 100, 123, 202, 203, 208, 211, 261, 270, 299, 333, 356, 566, 594
Pasta, 554, 555, 603
Pathria, R. K., 339
Pauli exclusion principle, 227, 579
Pauling, L., 441
Pauli paramagnetism, 388
Pauli, W., 249, 441
Peierls, R., 422, 442
Penrose, O., 601
Percus, J. K., 339
Percus–Yevick approximation, 306
permutation, 523
Perron, J., 14, 95
perturbation theory, 211, 270, 282, 304, 348, 361, 556, 586
Petit, A. T., 516
phase coexistence, 43, 144, 149, 167, 175, 181
phase separation, 11, 151, 169, 181
phase space, 3, 6, 8, 10, 14, 65, 199, 364, 376, 431, 530, 542
phase-space density, 560
phonons, 216, 217, 453, 483
photons, 210, 216, 217, 222, 232
Plancherel, M., 546
Planck distribution, 223
Planck, M., 30, 33, 97, 189, 223, 225, 249
plasma, 10, 313, 316, 347, 377
Poincaré cycles, 29, 96
Poincaré section, 555, 603

INDEX **611**

Poincaré, H., 96, 603
Poisson, S. D., 508, 518
Poisson bracket, 65, 68, 101, 112, 531, 532, 533, 561, 600, 601
Poisson distribution, 529
Poisson's ratio, 510
Poisson–Sum formula, 445
polarization, 351, 352, 374, 391, 395
polymers, 11, 320
potential energy, 17, 18, 87, 88, 92, 98, 115, 170, 208, 209, 266, 271, 309, 314, 316, 340, 390, 458, 459, 466, 470, 472, 473, 495
Power, H., 101
pressure, 2, 13, 16, 44, 50, 59, 69, 81, 82, 84, 89, 101, 121, 287, 299
Prigogine, I., 601
primitive unit cell, 455
primordial microwave background, 224
principle of corresponding states, 188, 308
probability distributions, 524
probability operator, 203
probability theory, 3, 8, 523
projection operator, 205
Proust, J., 95
pure state, 205

quantum statistics, 205, 219, 262, 338
quasi ergodic theorem, 546

radial distribution function, 82, 298, 299, 305, 316, 317, 339, 564
Radzig, A., 441
Rahman, A., 94
random variable, 20
Rasband, S. N., 602, 603
Ray, J., 97
Rayleigh–Jeans formula, 223
Rayleigh, Lord, 249
Rayne, J. A., 518
reciprocal lattice vectors, 468, 479
Ree, F. H., 339
Reed, T. M., 327, 339
Reichl, L., 600, 603
Reichl, L. E., 94, 601, 602
renormalization group, i, 321, 440
reproducibility, 3, 20
Resibois, P., 94
restricted ensemble, 67, 68
resummations, 305
Richardson, R. C., 249
rigidity, 1, 2, 94, 100, 101, 101
Ritchie, D. S., 443
Robertson, H. S., 96

Roller, D., 100
Rosenbluth, A., 436, 443
Rosenbluth, M., 436, 443
Rosenthal, A., 546, 602
rotation matrices, 67
rotational invariance, 64, 66, 67, 68
rotational motion, 11, 17, 65, 68, 262, 263, 320, 328, 341, 539
rotor–vibration model, 262, 268, 328
Ruoff, A. L., 518

Sacker, O., 98
Sackett, C. A., 250
Sackur–Tetrode equation, 36, 44, 98
scaling, 191, 308, 440
Schiff, L., 101
Schrieffer, J. R., 191
Schrödinger equation, 206, 330, 378, 579
Schrödinger representation, 542
Schwinger, J., 249
second law of thermodynamics, i, 29, 96, 97, 189, 189
second quantization, 13, 205, 248, 574, 603
Shannon, C. E., 96
Shaw, N., 190
shear, 2, 69, 71, 78, 81, 83, 85, 508
shear modulus, 508
Shoenberg, D., 389, 442
Shull, C., 517
Simmons, G., 518
simple pendulum, 553
Sinai, Y. G., 603, 603
single-particle states, 206, 211, 214, 244, 262, 378, 381, 574
Smirnov, B., 441
Sommerfeld, A., 97, 99, 229, 250, 582
Sommerfeld expansion, 229, 582
specific heats, 94, 109, 115, 131, 143, 231, 243, 261, 265, 424, 488
speed of sound, 483
spin-orbit coupling, 359
spin-spin correlation function, 448
spin-statistics theorem, 206, 266, 574
spinodal line, 167, 174
spontaneous symmetry breaking, 156, 236, 409
staggered magnetization, 158, 409
Stanley, H. E., 443, 443
Stefan–Boltzmann constant, 226, 249
Stefan, J., 249
Stickler, S. J., 340
Stirling's formula, 26, 35, 105
stochastic variables, 5, 6, 8, 9, 103, 104, 524, 525

Stokes, G. G., 98
Stover, B. J., 363, 441, 442
strain tensor, 73, 78, 102, 113, 118, 474, 474, 537
stress tensor, 39, 77, 79, 82, 83, 84, 121, 497, 497, 502, 539, 562, 563, 564
structural information, 294, 296
sublattice ordering, 12, 407
subsystems, 21, 40, 119, 138, 144, 195
Sudarshan, E. C. G., 600, 601
superconductors, 1, 3, 158
superfluids, 1, 3, 13, 18, 158, 189, 209, 249
superoperators, 101
symmetry principle, i, 17, 18, 59, 66, 207, 530

Teller, A., 436, 443
Teller, E., 436, 443
temperature, 1, 42, 45, 58, 60, 96, 98, 100, 124
temperature scale, 45, 98, 100
Tetrode, H., 98
thermal bath, 18
thermal expansion coefficient, 2, 131, 192, 289
thermodynamic limit, 35
thermodynamic potential, 48, 50, 52, 100
thermodynamic stability, 135, 139, 142, 147, 160, 314
thermodynamics, i, ii, 1, 117, 118, 120, 156, 177, 189, 321, 601
thermometer, 1, 45, 124
third law of thermodynamics, 130
Thomson, W., 96, 97, 100, 194
Thouless, D. J., 517
time average, 15, 20, 92, 548
time correlation functions, 66, 190
time translational invariance, 17, 66, 536, 537
Tobochnik, J., 94, 443
Toda Hamiltonian, 16, 96, 554, 555, 603
Tollett, J. J., 250
total angular momentum, 7, 17, 18, 30, 65, 66, 108, 113, 359, 535, 537, 539
total correlation function, 305
total momentum, 7, 17, 30, 64, 66, 108, 111, 534, 537, 541
Towneley, R., 101
trace, 31, 45, 110, 203, 205, 213, 367, 427, 572
transfer matrix, 418
transformation theory, 76, 101, 118, 530, 537, 539, 600
translational invariance, 18, 64, 66, 67, 100, 396
trapping potential, 244

triangular lattice, 456
triple point, 149, 150, 304

Uhlenbeck, G. E., 248
Ulam, S., 554, 555, 603
ultraviolet catastrophe, 224
uncertainty principle, 13, 199
universality, 440
Urey, H. C., 340

Van Alphen, P. M., 442
van der Waal's equation, 177, 191, 292, 308
van der Waal's interaction, 324, 326, 327, 339
Van der Waals, J. D., 191
van Leeuwen's theorem, 377
van t'Hoff's formula, 154
van Vleck, J. H., 359, 361, 441
van't Hoff, J. H., 190
vibrational motion, 11, 216, 253, 262, 320, 328, 452, 476, 483, 516
virial coefficients, 287, 291, 294, 310
virial expansion, 287
virial theorem, 84, 345
Voigt notation, 503
Voigt, W., 518
von Karman, Th., 516
von Neumann, J., 543, 601, 602

Waage, P., 190
Wainwright, T. E., 94
Walecka, J. D., 248
Walker, G. H., 603
Waller, I., 517
Wang, H., 518
Wang, S. C., 339
Wannier, G. H., 422, 442
Waterston, J. J., 338
Weinberg, S., 101, 249
Weiss, P., 442
Wertheim, M. S., 307, 339
Weyl, H., 100
Widom, B., 443
Wieman, C. E., 250
Wien law, 224
Wien, W., 249
Wiener, N., 96
Wigner, E. P., 100
Wigner–Eckart theorem, 360
Wilson, K. G., 440
Wood, W. W., 94
work, 1, 81, 122, 189, 539
Wu, T. T., 422, 442

x-ray scattering, 117, 294, 297, 304, 366
XY model, 10, 415, 439

Yang, C. N., 425, 442
Yevick, G. J., 339
Young's modulus, 510
Young, A. P., 478, 517

Zeeman term, 68, 254, 360, 365, 370, 387, 395, 409
Zeeman, P., 442
Zermelo, E., 96
zero-point energy, 237, 484, 490
Ziman, J. M., 516
Zwanzig, R., 601